S0-CMU-348

The Synapse: Function, Plasticity, and Neurotrophism

The Synapse: Function, Plasticity, and Neurotrophism

Motoy Kuno

Department of Physiology
Kyoto University Faculty of Medicine
Kyoto, Japan

Oxford New York Tokyo

OXFORD UNIVERSITY PRESS

1995

Oxford University Press, Walton Street, Oxford OX2 6DP
Oxford New York Toronto
Delhi Bombay Calcutta Madras Karachi
Kuala Lumpur Singapore Hong Kong Tokyo
Nairobi Dar es Salaam Cape Town
Melbourne Auckland Madrid

and associated companies in
Berlin Ibadan

Oxford is a trade mark of Oxford University Press

Published in the United States
by Oxford University Press Inc., New York

© *M. Kuno, 1995*

All rights reserved. No part of this publication may be reproduced, stored in a retrieval system, or transmitted, in any form or by any means, without the prior permission in writing of Oxford University Press. Within the UK, exceptions are allowed in respect of any fair dealing for the purpose of research or private study, or criticism or review, as permitted under the Copyright, Designs and Patents Act, 1988, or in the case of reprographic reproduction in accordance with the terms of licences issued by the Copyright Licensing Agency. Enquiries concerning reproduction outside those terms and in other countries should be sent to the Rights Department, Oxford University Press, at the address above.

This book is sold subject to the condition that it shall not, by way of trade or otherwise, be lent, re-sold, hired out, or otherwise circulated without the publisher's prior consent in any form of binding or cover other than that in which it is published and without a similar condition including this condition being imposed on the subsequent purchaser.

A catalogue record for this book is available from the British Library

Library of Congress Cataloging in Publication Data
Kuno, Motoy.
The synapse: function, plasticity, and neurotrophism / Motoy Kuno.
Includes bibliographical references.
1. Synapses. 2. Neuroplasticity. 3. Neurotrophic functions.
I. Title.
{DNLM: 1. Synapses—physiology. 2. Neuronal Plasticity—physiology. 3. Neuroregulators—physiology. 4. Nerve Regeneration. 5. Nerve Growth Factors. 6. Neurophysiology—trends. ®}
QP364.K86 1995 591.1'88—dc20 94—3340

ISBN 0 19 854687 4

Typeset by Footnote Graphics, Warminster, Wiltshire
Printed in Great Britain by
The Alden Press, Oxford

PREFACE

This book is a review of current concepts in neurobiology regarding the synapse, plasticity, and neurotrophism intended for graduate students and research workers. Throughout, I have kept two purposes in mind: first, to elucidate how these current ideas came to be accepted; second, to link the latest research in neurophysiology with that in molecular neurobiology. Each purpose stems from my concern for the future of research in neuroscience. Let me explain.

To achieve the first purpose, each part begins with a brief historical perspective, and in every new topic I have tried to include how this question was posed, how it was examined, and how its outcome relates to our current concepts. Familiarizing ourselves with the roads so far travelled is important in rapidly moving fields like neurobiology, where landmark interpretations of experimental results often shift repeatedly as new, unforeseen data emerge. For simplicity, most research workers are content to keep the 'big picture' in mind rather than its contributory experimental underlay. This is unfortunate. Not knowing the actual development of our current thought or the critical issue originally addressed makes it harder for us to grasp the rationale and the logic driving our current understanding. Moreover, awareness of past progress (and detours!) is sometimes precisely what we need to approach successfully the neurobiological problems encountered today.

My second purpose is to show where molecular neurobiology sheds light on our understanding of neuronal function. Neuroscience has benefited greatly by the remarkable recent progress in molecular biology, which in disclosing unexpected clues has provided novel insights into mechanisms of neuronal function. The identification of many new molecules associated with synaptic function, neuronal plasticity, and neurotrophism suggests novel schemes for molecular mechanisms of neuronal function. Yet many of these proposals remain speculative, lacking quantitative tests of a causal relationship between the molecules and the cell function. Indeed, Maddox (*Nature*, 355, 201, 1992) recently asked 'Is molecular biology yet a science?'—contending that as a largely qualitative inquiry it neglects the strong tradition of quantitative measurement on which arguments in neurophysiology traditionally rest. I agree that more quantitative analyses are required for identifying molecular mechanisms of neuronal function, and wherever possible the text points out specific gaps awaiting some willing researcher's attention. Throughout, because my emphasis is on neuronal function, not on molecular events *per se*, many interesting molecular events had to be omitted, unless demonstrably relevant to neuronal function.

The need for such a bridge as this book became clear to me when I had the privilege of collaborating with several distinguished molecular neurobiologists, Shosaku Numa, Shigetada Nakanishi, and their colleagues. This experience helped me to identify which concepts, phenomena, terminology, and techniques are a barrier to communication between molecular biologists and neurophysiologists, particularly electrophysiologists. The text defines these snags as they arise, or explores them in a separate *Box* in greater detail. No attempt was made to explain very basic principles in neurophysiology, such as the resting membrane potential and the action potential, however. For this the reader may wish to consult *From neuron to brain*, an excellent textbook by John G. Nicholls, A. Robert Martin, and Bruce G. Wallace (Sinauer Associates, Inc., Sunderland, 1992).

Kyoto, Japan M. K.
May, 1994

ACKNOWLEDGEMENTS

I have received the help of many people in preparing this book. Bob Martin, University of Colorado, and Hiromu Yawo, Kyoto University, read Part I of the manuscript; Tomoo Hirano, Kyoto University, and Josh Sanes, Washington University, read Part II; and Bill Snider, Washington University, and Tetsuo Yamamori, Frontier Research Program at Riken, reviewed Part III. They all gave me numerous valuable criticisms and suggestions. I am indebted greatly to Roberto Gallego, Universidad de Alicante, and Lee McIlwain, University of North Carolina, Chapel Hill, for locating historical literature by Magendie and by Foster. I would like to express my special appreciation to Ellen Curtin for her usual, excellent editorial assistance. I am also grateful to Miki Tatsumi for her patient help in printing the manuscript. I thank the editorial staff of Oxford University Press for their helpful advice and suggestions. Finally, I thank my wife, Aki, for her constant encouragement and for lending me her computer to prepare the manuscript at home.

CONTENTS

Part I The Synapse

Part II Plasticity

Part III Neurotrophism

Part I The Synapse

1

Historical perspective of the synapse

1.1 CONTIGUITY VERSUS CONTINUITY AT SYNAPTIC SITES

A neuronal signal travels from one neurone to another across the synapse between the two excitable cells. The term *synapse* was first introduced by Sherrington in *A text book of physiology*, Foster's 1897 edition. Figure 1.1 reproduces the illustration of synapses from Foster's book. Sherrington postulated that the connection of one nerve cell with another at the synapse is merely a contact without actual continuity of substance: 'So far as our present knowledge goes we are led to think that the tip of an axon is not continuous with the nerve cell on which it impinges . . . The lack of continuity offers an opportunity for some change in the nature of the nervous influence as it passes from the one cell to the other.' Again, in Sherrington's contribution to *Text book of physiology*, Schäfer's 1900 edition, we find: 'The nerve centre exhibits a valve-like function, allowing conduction to occur through it in one direction only . . . That the direction of nerve impulses is not reversible along the neural chains, may be a function of the synapse.' Evidently, Sherrington considered the synapse to have a specialized role in neuronal communication. The lack of protoplasmic continuity that he posited at synapses was probably inferred from the functional implication of synapses, rather than from morphological evidence.

Each presynaptic branch drawn by Sherrington terminates at the cell body or dendrite with a smooth tip (Fig. 1.1). Elsewhere, however, quite a different profile was being described. By the staining method with reduced silver nitrate, Held (1897) had observed that the axon endings of a distant neurone form bulbous end-feet (*Endfüsse*) on the cell body of the postsynaptic neurone. In sharp contrast with Sherrington's notion, Held saw a

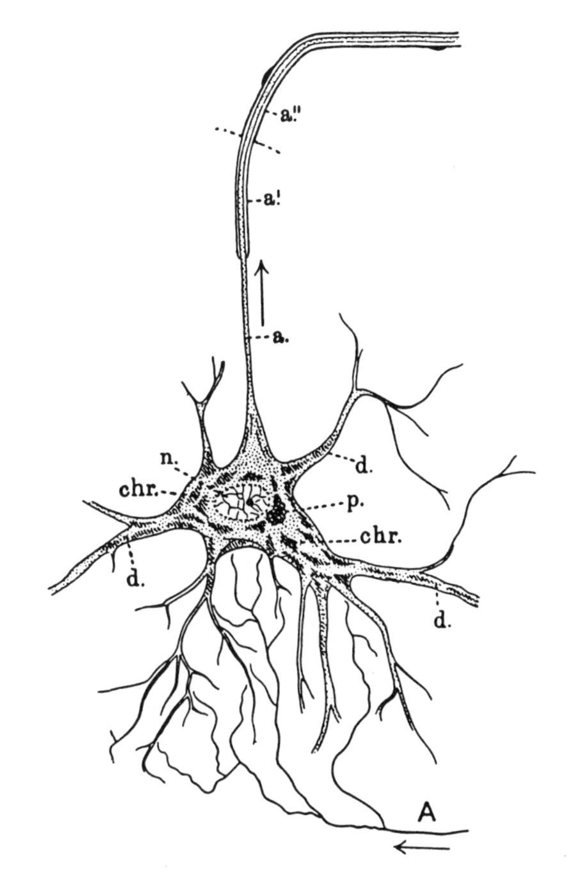

Fig. 1.1. A schematic diagram of synapses formed by a presynaptic fibre (A) on a spinal motor neurone, based on the morphology stained with silver or methylene blue. The neurone cell body contains nucleus (n), pigment (p) and chromatic substance (chr); d, dendrites. The motor axon (a) acquires a myelin sheath (a′), then, outside the cord, a primitive sheath (a″). (From Foster 1897.)

definite continuity or 'concrescence' between the two neurones. From its outset the search for the structure of the synapse was marked by controversy.

Granular material within the end-feet, which Held (1897) described as neurosomes, was eventually identified as mitochondria (Bodian 1937). Yet, the crucial question of whether two abutting neurones hold their protoplasmic relationship by continuity or contiguity remained unresolved far longer. Despite many meticulous studies, the morphology of synapses could not be observed adequately at the light microscopic level. So aware was Bodian of the risk of error in the silver impregnation method that he did not even review the general arguments on the synaptic structure in his 1937 article. In his 1942 review, Bodian stressed that only ultrastructural studies would resolve the detailed morphology of synapses. Ultimately, the introduction of electron microscopy proved Sherrington correct concerning the absence of protoplasmic continuity across the contact surface of the synaptic apparatus, when a gap of a few hundred Ångströms wide was directly sighted between the presynaptic and subsynaptic membranes (De Robertis and Bennett 1955; Palay 1956).

In retrospect, it is quite puzzling why the concept of cell connections in the autonomic ganglion had been ignored completely. Before the synapse was defined by Sherrington, Langley (1896) had described 'that each sympathetic ganglion is a *cell station*'. The sympathetic ganglion then drawn by Langley (Fig. 1.2) clearly illustrates our current view of the synapse. Both Langley and Sherrington contributed to Schäfer's 1900 textbook. Although Schäfer noticed similarity between central synapses and ganglionic cell connections, Sherrington did not refer to Langley's chapter; nor did Langley refer to Sherrington's.

1.2 CHEMICAL VERSUS ELECTRICAL SYNAPTIC TRANSMISSION

In Schäfer's textbook of 1900, Sherrington pointed out that the valve-like action at central synapses resembles the valve-like connection between motor nerve fibres and skeletal muscle, suggesting that similar mechanisms work in neuronal synapses and neuromuscular junctions. When acetylcholine was proposed by Dale in 1935 as the chemical transmitter substance operating at ganglionic synapses and neuromuscular junctions, it sparked fresh controversy concerning synaptic transmission. Is synaptic transmission electrical or chemical in nature? The principal question was whether a local current generated by the action potential at a presynaptic terminal is sufficient to elicit excitation in the postsynaptic cell, or whether the release of a chemical transmitter substance is a prerequisite for synaptic transmission. The vigour of the controversy at the 1939 symposium on the synapse leaps from Forbes' report (1939): 'Dale remarked that it was unreasonable to suppose that nature would provide for the liberation in the ganglion of acetylcholine . . . for the sole purpose of fooling physiologists. To this Monnier replied that it was likewise unreasonable to suppose action potentials would be delivered at the synapse . . . to fool physiologists.'

The swing of the pendulum between the electrical and chemical hypotheses and how this dispute was resolved experimentally is well documented (Eccles 1959, 1982; Feldberg 1977; Kandel *et al.* 1987). It was the development of techniques for intracellular recording that settled the debate in favour of the chemical hypothesis for neuromuscular junctions (Fatt and Katz 1951), for ganglionic transmission (Blackman *et al.* 1963), and for central excitatory and inhibitory synapses (Brock *et al.* 1952; Eccles *et al.* 1954). Ironically, the conclusion of this persistent debate was soon followed by the discovery of transmission through novel electrical synapses (Furshpan and Potter 1959; Bennett *et al.* 1963). Electrically mediated inhibition was also found (Furukawa and Furshpan 1963; Korn and Faber 1975). In some synapses, moreover, transmission proved to employ a dual mode, electrical as well as chemical (Martin and Pilar 1963; Rovainen 1967; Shapovalov and Shiriaev 1980).

It is now well established that the presence of the gap junction (Revel and Karnovsky 1967) is the morphological substrate for electrical synapses (Brightman and Reese 1969; Payton *et al.* 1969). Although electrical junctions indisputably comprise a 'respected' class of synapse, such electrically mediated synapses are so outnumbered by chemical synapses as to be considered to be 'unconventional' (Bennett 1974).

1.3 SYNAPTIC VESICLES AND TRANSMITTER QUANTA

Soon after the introduction of electron microscopy, the presence of a large number of synaptic vesicles in presynaptic terminals was reported independently by three groups (De Robertis and Bennett 1954; Palade 1954;

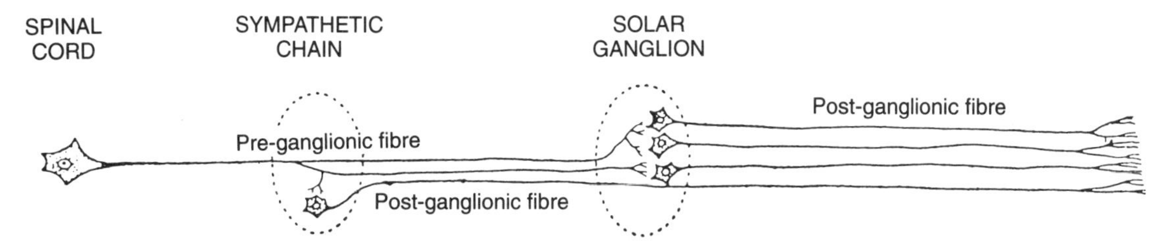

Fig. 1.2. A schematic diagram postulated for synaptic connection in the sympathetic ganglion (solar ganglion). Preganglionic sympathetic fibres arising from the spinal cord are assumed to be connected with the postganglionic neurones in the ganglion. (From Langley 1896.)

Palay 1954). Even prior to these scientific reports, a picture showing synaptic vesicles had in fact been casually circulating: The publicity brochure of the LKB ultramicrotome, with a micrograph of synapses containing synaptic vesicles in retinal neurones taken by Fritiof Sjöstrand, was distributed at the 1953 electron microscope meeting at Pocono (H. Stanley Bennett, personal communication). Sjöstrand (1953) described the vesicles as 'minute granules' but did not comment on their biological significance. De Robertis and Bennett (1955) suggested that synaptic vesicles perforate the presynaptic membrane and discharge their content (transmitter substance) into the synaptic cleft. Individual synaptic vesicles would thus act as a release unit as well as a storage unit for the transmitter substance. Fortuitously, in the same year, Del Castillo and Katz (1954a) proposed the quantum hypothesis of transmitter release for the frog neuromuscular junction, based on the statistical fluctuations in the amplitude of end-plate potentials evoked by consecutive nerve impulses. According to this hypothesis, the transmitter is released in a certain amount as a unit (quantum), and a nerve impulse normally releases some large integral number of the units.

Might the synaptic vesicles then be morphological counterparts of the transmitter quanta? This vesicular hypothesis (Del Castillo and Katz 1956; Katz 1962) predicts that transmitter release would be accompanied by two morphological alterations. First, presynaptic terminals would contain fewer synaptic vesicles after excessive transmitter release (depletion). Second, the presynaptic membrane facing the postsynaptic cell would enlarge its surface area as vesicular membranes fuse with the terminal membrane (exocytosis). Indeed, both a decrease in the number of synaptic vesicles and an increase in the surface area of the presynaptic membrane have since been observed in association with transmitter release (Ceccarelli *et al.* 1972; Heuser and Reese 1973). Furthermore, the number of exocytotic vesicles estimated by rapid-freezing seems to correlate with the number of transmitter quanta released (Heuser *et al.* 1979; Heuser 1989). Although compelling, the vesicular hypothesis is not universally accepted (Israel *et al.* 1979; Tauc 1982).

1.4 RECEPTORS FOR THE SYNAPTIC TRANSMITTER

The transmitter substance released from a presynaptic terminal diffuses across the synaptic cleft and acts on the postsynaptic cell. A receptor specialized for the transmitter on the postsynaptic cell membrane is thus the site for interaction between the two excitable cells.

Langley (1909) pioneered the concept of receptors for chemical mediators. He found that the contraction of skeletal muscle induced by applying nicotine can be blocked by curare; moreover, the sensitivity to nicotine proved higher at the nerve entry zone than in other regions of the muscle. He also noted that this nicotine sensitivity is preserved even after denervation of the muscle. Langley postulated that the innervation site of muscle fibres must have a 'receptive substance' that specifically binds both nicotine and curare. Where is the receptive substance located? Langley expected it to be in the cytoplasm, rather than on the membrane, of skeletal muscle.

Later, the receptor for a transmitter substance was considered to be a membrane protein (Nachmansohn 1959), and attempts were made to isolate the acetylcholine (ACh) receptor protein from the electroplax of the electric eel (Ehrenpreis 1959). The progress of this research was accelerated by the discovery of snake venom alpha toxins (Lee 1973), which bind the ACh receptor specifically and virtually irreversibly. In the early 1970s, several laboratories began to isolate and purify the nicotinic ACh receptor by use of labelled snake toxins (Karlin 1974; Rang 1975). The ACh receptor was found to contain four distinct subunits (Schmidt and Raftery 1973; Weill *et al.* 1974), termed α, β, γ, and δ (Hucho *et al.* 1976). Raftery *et al.* (1980) determined the sequence of about 50 amino acid residues for each of the four subunits, which permitted the construction of specific probes (synthetic oligonucleotides) for the complementary DNAs (cDNAs) that encode the receptor subunits. A library of cDNA clones derived from messenger RNAs (mRNAs) of the *Torpedo* electric organ was screened in this way. First, the entire amino acid residues were sequenced for the α-subunit (Noda *et al.* 1982), then the three other subunits were sequenced (Claudio *et al.* 1983; Numa *et al.* 1983). A new avenue leading to molecular neurobiology had now opened as the pursuit of the synapse neared its centenary.

2

The role of calcium in transmitter release

The Nobel lecture Katz (1971) delivered in Stockholm on 12 December, 1970 summarized the concept of neuromuscular transmission as having two stages, the release of acetylcholine (ACh) from the nerve terminal (stage I), and the action of ACh on the muscle (stage II):

$$\mathrm{N} \xrightarrow{\ \mathrm{I}\ } \mathrm{ACh} \xrightarrow{\ \mathrm{II}\ } \mathrm{M}$$

This scheme can be generalized to include any chemical synapse. To do so, we simply substitute a proper transmitter substance for ACh and replace muscle (M) with a postsynaptic cell responsive to a presynaptic nerve impulse (N). This and the subsequent chapters examine the first step of the transmission process, stage I. The principal mechanism discussed is how the arrival of a presynaptic impulse at the nerve terminal causes release of the transmitter. How the transmitter acts on the postsynaptic cell (stage II) will be discussed in Chapters 4 and 5.

2.1 CALCIUM IS AN ESSENTIAL CO-FACTOR FOR TRANSMITTER RELEASE

As early as 1936 Feng observed that the depression of neuromuscular transmission induced by curare can be reversed by increasing the external concentration of calcium (Ca^{2+}). The underlying mechanism remained unclear for some time, however. Del Castillo and Stark (1952) then found that the size of the postsynaptic response at neuromuscular junctions depends on the Ca^{2+} concentration in the bathing solution. The experiments illustrated in Fig. 2.1 show the behaviour of neuromuscular transmission of a frog muscle immersed in a Ca^{2+}-free solution. When an extracellular recording

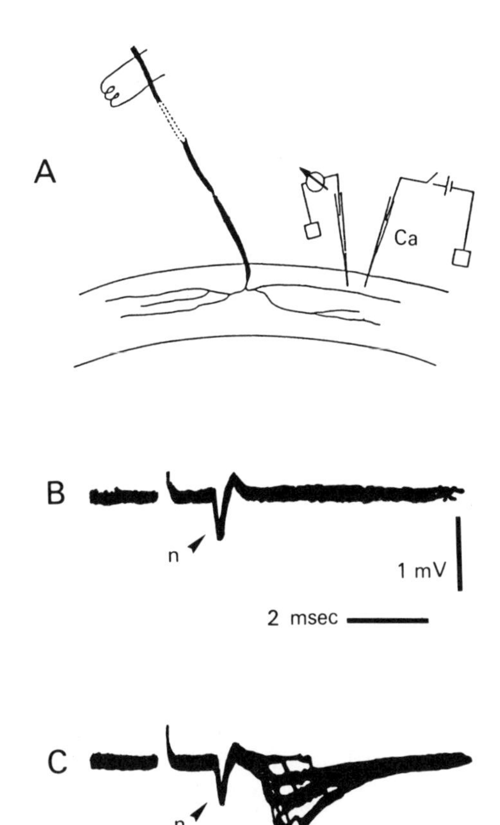

Fig. 2.1. Focal recordings of end-plate currents (e.p.c.s) with an extracellular electrode in a Ca^{2+}-free solution. A, schematic diagram of the recording procedure. B, without Ca^{2+} efflux from the Ca^{2+} pipette, nerve stimulation produces only a nerve terminal action current (n) not followed by the generation of e.p.c.s. C, when Ca^{2+} is applied from the pipette, nerve stimulation evokes e.p.c.s (e) as well as the nerve terminal action current (n). (Adapted from Katz and Miledi 1965c.)

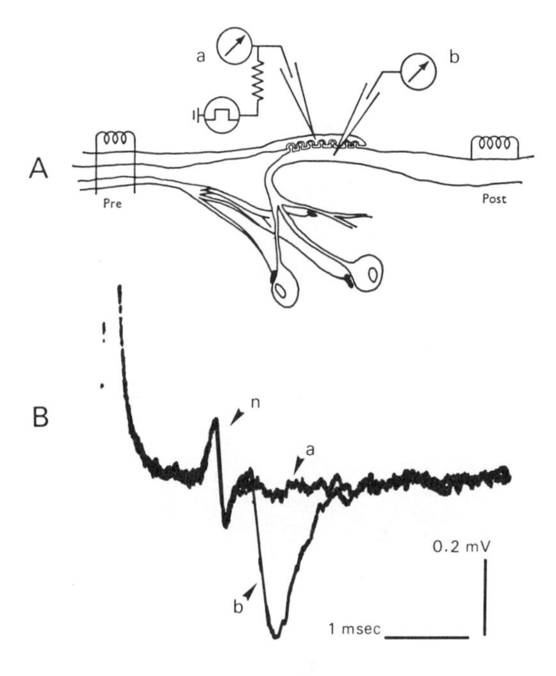

Fig. 2.2. Effects of Ca^{2+} on transmission in the giant synapse of the squid stellate ganglion. A, a pipette filled with a Ca^{2+}-rich solution (a) delivered Ca^{2+} focally while extracellular responses were recorded with another electrode (b). The response was elicited by stimulating the presynaptic axon (Pre). B, without Ca^{2+} efflux from the pipette presynaptic stimulation produced only a nerve action current (a) in a Ca^{2+}-free solution (a). When Ca^{2+} was applied from the pipette extracellularly, a synaptic response (b) followed the terminal action current (n). (Adapted from Miledi and Slater 1966.)

electrode was placed near the end-plate region, only the action current arriving in the motor nerve terminal was observed following nerve stimulation (Fig. 2.1B, n). But, when Ca^{2+} was applied to the same junction from a nearby pipette, arrival of the same nerve action current (n) now caused the release of ACh, thereby evoking end-plate currents (Fig. 2.1C, e). External Ca^{2+} thus appears to be an essential co-factor in the process of nerve-evoked release of the transmitter.

Where might the site of action of Ca^{2+} be in this process? For instance, might Ca^{2+} act on the external surface of the nerve terminal membrane or must it actually enter the terminal to trigger the transmitter release?

2.1.1 Calcium influx in nerve terminals

The approach to the issues raised above was found by way of the giant synapse in the stellate ganglion of the squid, whose size permits insertion of microelectrodes into both pre- and postsynaptic axons at or near the synaptic site (Hagiwara and Tasaki 1958). Miledi and Slater (1966) employed a micropipette filled with $CaCl_2$-rich solution to apply Ca^{2+} focally, while recording electrical activity with an extracellular electrode placed in the giant synapse (Fig. 2.2A). When the preparation was superfused with a Ca^{2+}-free solution, presynaptic stimulation evoked only the action current of the presynaptic axon (n), with no ensuing postsynaptic response (Fig. 2.2Ba). During ionophoretic release of Ca^{2+} to the focal extracellular space from the pipette, however, the presynaptic action current (n) was followed by a synaptic response elicited in the postsynaptic axon (Fig. 2.2Bb). These results mirror those observed at the frog neuromuscular junction (Fig. 2.1B,C). Next, Miledi and Slater (1966) inserted the Ca^{2+}-filled pipette into the presynaptic axon and injected Ca^{2+} ionophoretically. Now, presynaptic stimulation in a Ca^{2+}-free solution evoked no synaptic response. These results seemed to suggest that the trigger is accessible only from outside the presynaptic membrane.

However, in 1973 Miledi again pursued the effect of intracellular injections of Ca^{2+} into presynaptic axons in the squid giant synapse and showed that intraterminal injection of Ca^{2+} can, by itself, cause transmitter release. While monitoring intracellular potentials of the postsynaptic axon, he inserted a Ca^{2+}-filled pipette into the presynaptic axon; a small postsynaptic depolarization then ensued, and this response diminished when a negative bias current was applied through the pipette to halt the Ca^{2+} efflux. A slow depolarization reappeared in the postsynaptic axon whenever the negative bias current was switched off (Fig. 2.3, between the arrows).

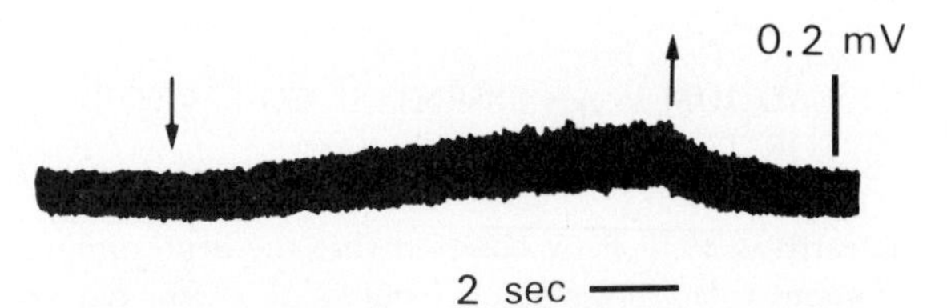

Fig. 2.3. A depolarizing response produced in the postsynaptic axon by injecting Ca^{2+} into the presynaptic axon in a squid giant synapse. The experimental procedure was similar to that illustrated in Fig. 2.2A, but the Ca^{2+}-filled pipette was inserted into the presynaptic axon, while recording the postsynaptic intracellular potentials with another electrode. The preparation was superfused with sea water containing tetrodotoxin. The bias current preventing Ca^{2+} efflux from the pipette was switched off during the period indicated by two arrows. (Adapted from Miledi 1973.)

Interestingly, the postsynaptic depolarization was associated with an increase in voltage fluctuations or 'noise' in the baseline (Miledi 1973). The amplitude and time course of the underlying discrete potentials producing this noise were evaluated by the analysis previously employed for 'elementary' voltages produced by the action of individual ACh molecules at neuromuscular junctions (Katz and Miledi 1972; see section 4.2). The amplitude (10–25 μV) and decay time constant (2 ms) estimated for the discrete potentials proved comparable to those of spontaneous miniature synaptic potentials recorded from the postsynaptic axon of the squid giant synapse (Miledi 1967). In short, leaking Ca^{2+} into the presynaptic axon seemed to increase the release probability of individual transmitter quanta (Miledi 1973).

Might enhancement of transmitter release induced by intraterminal injections of Ca^{2+} reduce the number of synaptic vesicles? Martin and Miledi (1978) examined structural details of the giant synapse following injections of Ca^{2+} into the presynaptic axon. Depletion of synaptic vesicles and the formation of clusters of vesicles were noted, but these changes were confined to the site of injection. Presumably, Ca^{2+} so injected has a highly localized action. This localized action would account for the lack of effect on transmitter release from the entire terminal, seen previously by Miledi and Slater (1966). If Ca^{2+} injection causes an asynchronous release of individual quanta of the transmitter from the restricted region, it still may not significantly affect the amount of transmitter released from the entire contact area of the giant synapse following a presynaptic impulse (Charlton *et al.* 1982). Taken together, these results indicate that increasing the intracellular Ca^{2+} concentration in a presynaptic terminal increases the probability of releasing the transmitter. Under normal conditions, then, transmitter release is presumably triggered by the entry of Ca^{2+} into the presynaptic terminal in response to the terminal depolarization (action potential).

The minuteness of a motor nerve terminal at the neuromuscular junction precludes insertion of a Ca^{2+}-filled pipette, so other means were sought to increase the intracellular Ca^{2+} concentration there. One approach was to fuse liposomes loaded with Ca^{2+} to the terminal membrane (Rahamimoff *et al.* 1978). Liposomes with an estimated diameter of 25 nm were prepared from phosphatidylcholine in a Ca^{2+}-rich (25 mM) solution. When Ca^{2+}-loaded liposomes were added to the bath immersing the frog neuromuscular preparation, both the end-plate potential amplitude and the frequency of spontaneous miniature end-plate potentials increased significantly within 1 to 2 min (Rahamimoff *et al.* 1978). Because the transmitter release was not affected when KCl-loaded liposomes were added to the bath, the factor responsible for increased transmitter release must have been the content of the liposomes, not the liposomes themselves. These intriguing observations, however, did not prove directly that the Ca^{2+} concentration had in fact risen in the motor nerve terminal.

The entry of Ca^{2+} into presynaptic terminals on arrival of a nerve impulse has been demonstrated directly by chemical agents sensitive to calcium. Llinás *et al.* (1972) injected a bioluminescent protein, aequorin, into the presynaptic terminal of the squid giant synapse. This substance emits light when exposed to low concentrations of ionized calcium, and the light emission can be measured with a photomultiplier. The intracellular free Ca^{2+} concentration of the presynaptic axon clearly increased during repetitive stimulation of the presynaptic axon.

The Ca^{2+}-sensitive dye, arsenazo III, has likewise mapped the Ca^{2+} transient in the squid presynaptic axon following a single nerve impulse (Charlton *et al.* 1982; Augustine *et al.* 1985). As shown in Fig. 2.4, when an arsenazo III-loaded presynaptic terminal of the squid giant synapse was depolarized, the presynaptic terminal displayed dye transients proportional in magnitude to the amount of terminal depolarization from a holding potential of −70 mV up to 0 mV. As depolarization exceeded 0 mV, however, the response progressively diminished (Fig. 2.4, at +20 mV and +40 mV); eventually, it disappeared at about +60 mV. This behaviour is expected, since the entry of positively charged calcium ions would be gradually obstructed by the increased electric field as the internal potential becomes more positive (Katz and Miledi 1967; see section 2.1.2). Hence, these results were consistent with the idea that Ca^{2+} enters nerve terminals in association with activation of the presynaptic terminal.

The sensitivity and spatial resolution of these Ca^{2+} transients have since been improved greatly by the introduction of fluorescent Ca^{2+} indicators (Tsien 1988). Regrettably, their temporal resolution is still inadequate for quantitative analysis of the relationship between Ca^{2+} influx and transmitter release at nerve terminals (Smith and Augustine 1988). How can we resolve this difficulty? This is the subject of the next section.

2.1.2 Calcium currents at nerve terminals

The *synaptic delay* is the time required for initiating the postsynaptic response once a nerve impulse arrives at the

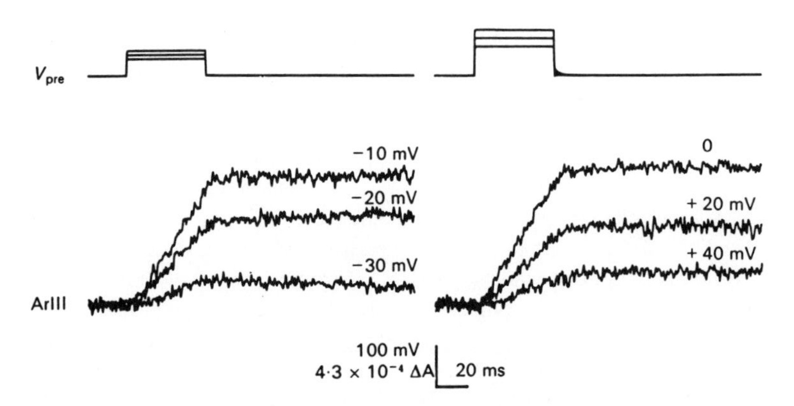

Fig. 2.4. Signals of arsenazo III (ArIII) detected in a presynaptic terminal of the squid giant synapse on depolarization of the presynaptic terminal. The presynaptic terminal was depolarized from a holding potential of −70 mV to varying levels indicated on each record under voltage clamp (V_{pre}). Note progressive increases in arsenazo III transients with stepwise depolarization to 0 mV, whereas the signals were reduced by further depolarization beyond 0 mV. (From Augustine *et al.* 1985.)

presynaptic terminal. This can be measured only by simultaneous recordings of the presynaptic impulse and the onset of the postsynaptic response (Figs 2.1 and 2.2). The synaptic delay measured 0.4–0.8 ms in both the squid giant synapse (Hagiwara and Tasaki 1958) and the frog neuromuscular junction (Katz and Miledi 1965a). That is, Ca^{2+} enters and launches the transmitter release process within a fraction of 1 ms after an impulse reaches at the presynaptic terminal. Any event whose time-scale is measured in milliseconds can be analysed only in terms of electrical responses. It is therefore appropriate to examine Ca^{2+} entry into a presynaptic terminal as a Ca^{2+} current.

We have seen that the amount of Ca^{2+} entering presynaptic terminals increases with the amount of depolarization applied to the terminal, at least up to a certain level (Fig. 2.4). The Ca^{2+} hypothesis for transmitter release would then predict that the amplitude of the postsynaptic response should also increase with the amount of presynaptic depolarization. This prediction was examined by recording synaptic potentials from the postsynaptic axon in response to depolarizing pulses applied to the presynaptic axon in the squid giant synapse (Bloedel *et al.* 1966; Katz and Miledi 1966). The preparations were exposed to tetrodotoxin (TTX) to block the initiation of action potentials by voltage-gated sodium (Na^{+})channels. As shown in Fig. 2.5, the postsynaptic response began to appear when the presynaptic terminal was depolarized by 20–25 mV, and the synaptic response (ordinate) enhanced with further presynaptic depolarization (abscissa). Because TTX blocks the Na^{+} conductance, the Na^{+} influx into presynaptic terminals must not be required for the depolarization-induced transmitter release. This result then suggests that depolarizing the presynaptic terminal by 20–25

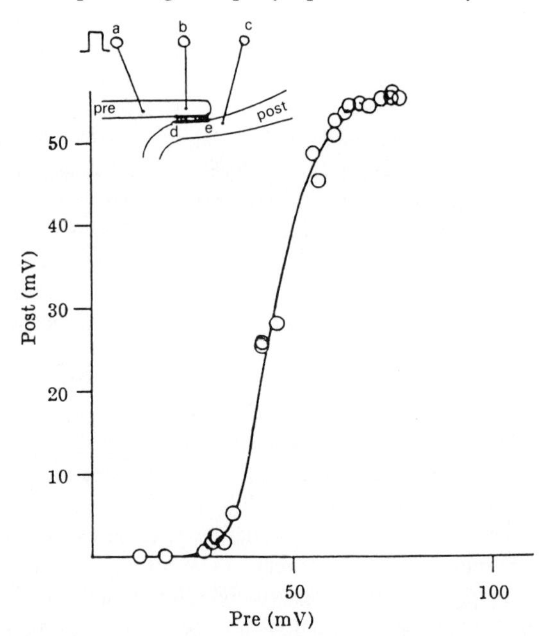

Fig. 2.5. The relation between the amplitude of postsynaptic response and the amount of depolarization in the presynaptic terminal at the squid giant synapse. Inset, the placement of intracellular electrodes; a, currents were applied; b, presynaptic depolarization was recorded; c, the postsynaptic response was monitored. d-e, length of synaptic contact (0.8 mm). a–d, 0.6 mm. d–b, 0.35 mm. e–c, 0.3 mm. Extracellular Ca^{2+} concentration, 58 mM. Na^{+} channels were blocked by tetrodotoxin. (From Katz and Miledi 1966.)

mV lets Ca^{2+} enter the terminal and trigger transmitter release.

Might the electrical changes induced by this Ca^{2+} influx be recorded from presynaptic terminals in response to terminal depolarization? To address this possibility, two microelectrodes were inserted into a presynaptic terminal of the squid giant synapse in the presence of TTX (Katz and Miledi 1969). Under normal conditions the steady depolarization recorded by one electrode failed to increase proportionately with an increase in the intensity of depolarizing current pulses applied through the second electrode (Fig. 2.6A). This inadequate voltage response was attributed to the depolarization-induced activation of potassium (K^+) channels. When the voltage-gated K^+ channels were blocked by injection of tetraethylammonium (TEA) ions into the presynaptic axon, the 'input–output relationship' sloped much more steeply (Fig. 2.6B). Moreover, when the presynaptic terminal was depolarized by 20–25 mV, the membrane potential now jumped at a critical intensity of the current (Fig. 2.6B). This suggests that a regenerative ionic current arises at a critical membrane potential of the presynaptic terminal. With the Na^+ and K^+ conductances blocked by TTX and TEA, respectively, the regenerative current responsible for the voltage jump appears to be Ca^{2+}.

We have seen that the entry of Ca^{2+} into a presynaptic terminal on depolarization is blocked as the terminal depolarization reaches a certain level (Fig. 2.4). What would happen to transmitter release under this condition? As shown in Fig. 2.7A, after an injection of TEA into a presynaptic terminal, a current pulse can maintain depolarization in the terminal (Pre), thereby eliciting a synaptic response in the postsynaptic axon (Post). This postsynaptic response was blocked, however, when the presynaptic depolarization was sufficiently large (Fig. 2.7B; Katz and Miledi 1969). Presynaptic depolarization *per se* is not, then, the primary factor for transmitter release. On the contrary, the postsynaptic response was generated when the large depolarization ended (Fig. 2.7B). How can we interpret this behaviour? It is due to a combination of two changes produced by depolarization; one is an increase in the Ca^{2+} conductance, and the other, changes in the driving force for Ca^{2+} currents. The Ca^{2+} conductance increases progressively when the terminal is depolarized by more than 20–25 mV. On the other hand, the driving force for Ca^{2+} currents (the difference between the terminal membrane potential and the Ca^{2+} equilibirum potential) is reduced as the level of terminal depolarization approaches the Ca^{2+} equilibrium potential (Box A). Although the Ca^{2+} influx is thus reduced or blocked during large terminal depolarization, the Ca^{2+} conductance continues to increase; on termination of the

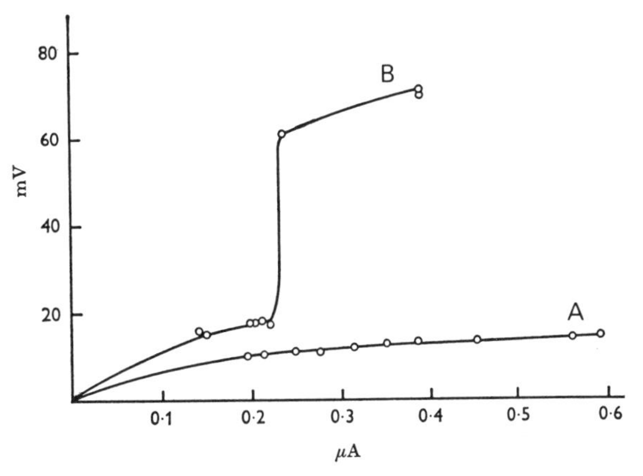

Fig. 2.6. The amount of depolarization as a function of the intensity of intracellularly applied current pulses in a presynaptic terminal of the squid giant synapse superfused with a Na^+-free isotonic Ca^{2+} solution. Before (A) and after (B) intracellular injection of tetraethylammonium ions. (Adapted from Katz and Miledi 1969.)

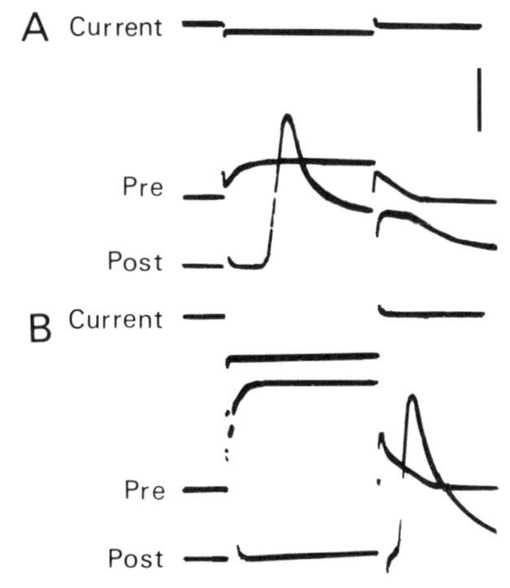

Fig. 2.7. Effects of relatively small (A) and large (B) presynaptic depolarizations on the postsynaptic response in the squid giant synapse. Current, depolarizing pulse (duration, 9.5 ms) applied to the presynaptic terminal. Pre, depolarization recorded from the presynaptic terminal. Post, synaptic response recorded from the postsynaptic axon. Note in B the postsynaptic response suppressed during strong presynaptic depolarization is generated immediately after termination of the depolarization. External Ca^{2+} concentration, 5.5 mM. Vertical calibration, 3.35 μÅ for current, 100 mV for Pre and 20 mV for Post. (Adapted from Katz and Miledi 1969.)

Box A The driving force for ionic currents

Ionic currents can be described by using Ohm's law. According to Ohm's law, the amount of current (I) flowing through a pathway is proprotional to the potential difference (ΔV) and inversely related to the resistance in the pathway (R). The ionic conductance (g) is the reciprocal of the resistance through which the ions move. Therefore, Ca^{2+} currents (I_{Ca}), for example, can be given by:

$$I_{Ca} = \Delta V_{Ca}/R_{Ca} = g_{Ca} \cdot \Delta V_{Ca}$$

where g_{Ca} is the Ca^{2+} conductance which is the reciprocal of the resistance of the Ca^{2+} channels through which Ca^{2+} flows (R_{Ca}). In this relationship, ΔV_{Ca} acts as a driving force for Ca^{2+}. How can we estimate this driving force?

The external Ca^{2+} concentration is more than 1000 times higher than the intracellular Ca^{2+} concentration. Therefore, when the Ca^{2+} conductance (g_{Ca}) increases by depolarization, Ca^{2+} flows into the cell by its concentration (or chemical) gradient. Ca^{2+} entry is also influenced by the electrical potential across the membrane, being accelerated when the inside is negative with respect to the outer surface, and retarded when the inside is positive. The algebraic sum of these two forces (chemical and electrical gradients) determines the net Ca^{2+} flow across the membrane. When the two forces are equal and opposite, there will be no net Ca^{2+} flux, calcium ions being in equilibrium. Under this condition, the Ca^{2+} equilibirum potential (E_{Ca}) exactly balances the diffusion force produced by the Ca^{2+} concentration gradient. E_{Ca} can be given by the following equation:

$$E_{Ca} = \frac{RT}{zF} \ln \frac{[Ca]_o}{[Ca]_i}$$

Here $[Ca]_o$ and $[Ca]_i$ are the external and internal Ca^{2+} concentrations, respectively; R, the gas constant; T, the absolute temperature; z, the valence of the ion, and F, the faraday. If the membrane potential (E_m) is held at E_{Ca}, there will be no net Ca^{2+} flux. At any other membrane potential, the driving force for Ca^{2+} currents is the difference between E_m and E_{Ca}. Hence:

$$I_{Ca} = g_{Ca} \cdot (E_m - E_{Ca})$$

Even though g_{Ca} is increased by depolarization, Ca^{2+} currents will not flow if the cell is depolarized up to E_{Ca}. If the cell is further depolarized beyond E_{Ca}, the Ca^{2+} currents flow in the outward direction. Since the direction of the ionic current is reversed at its equilibrium potential, the equilibrium potential is sometimes called the *reversal potential*.

From the above relationship it is clear that the intensity of I_{Ca} depends on g_{Ca} and its driving force ($E_m - E_{Ca}$). As the cell is depolarized beyond a certain level, g_{Ca} increases, but the driving force decreases. According to the balance of these two facors, I_{Ca} first increases and then is progressively diminished when the cell is further depolarized, eventually approaching zero at the equilibrium potential, as shown in Figs 2.9 and 2.12.

---

depolarization Ca^{2+} then rushes into the presynaptic terminal with the sudden increase of the driving force before the Ca^{2+} conductance diminishes, thereby precipitating the 'off response' in the postsynaptic axon (Fig. 2.7B).

These results plainly indicate that the terminal depolarization normally induced by an action potential reaching the presynaptic ending causes a regenerative Ca^{2+} influx in the terminal and that this Ca^{2+} influx is responsible for triggering the transmitter release. Because a similar TTX-resistant Ca^{2+} influx accompanies depolarization of the squid giant axon (Baker *et al.* 1971), voltage-gated Ca^{2+} channels must exist all along the giant axon. However, their density is clearly higher in the terminal membrane than in the axonal membrane.

We have seen that depolarization of a presynaptic terminal induces a surge in the Ca^{2+} conductance. This raises two questions: How quickly is the resulting Ca^{2+} current initiated, and how long does it last? The test of these questions requires recording the Ca^{2+} current directly. This was done by applying the voltage clamp technique to presynaptic terminals of the squid stellate ganglion (Llinás *et al.* 1981). Figure 2.8Aa illustrates a depolarizing pulse that produced an initial inward Na^+ current followed by an outward K^+ current in the presynaptic terminal, a well-documented ionic basis for excitation (Hodgkin and Huxley 1952). Blocking these Na^+ and K^+ currents by TTX and TEA, respectively, unmasked a separate, slow inward current (Fig. 2.8Ab). This slow current (Fig. 2.8Ba) was in turn eliminated

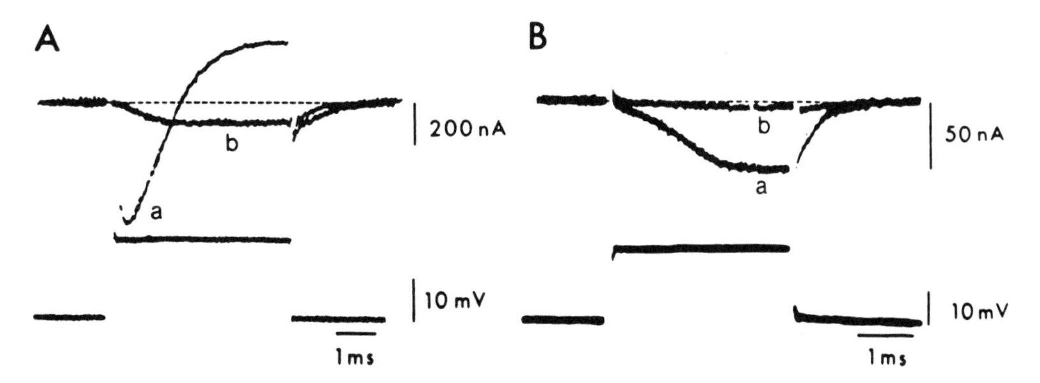

Fig. 2.8. Ca^{2+} currents recorded from a presynaptic terminal of the squid giant synapse under voltage clamp. The terminal was depolarized from a holding potential of −70 mV (lower traces). A: a, with low doses of TTX and 3–aminopyridine (3–AP), an initial inward Na^+ current followed by an outward K^+; b, a slow inward Ca^{2+} current was unmasked by additional TTX and 3–AP plus intracellularly injected TEA. B: a, with blockage of the Na^+ and K^+ conductances a depolarizing pulse elicited an inward Ca^{2+} current; b, the Ca^{2+} current was blocked by 1 mM Cd^{2+}. External Ca^{2+} concentration, 10 mM. (From Llinás *et al.* 1981.)

by adding cadmium (Cd^{2+})ions (Fig. 2.8Bb), a known blocker of Ca^{2+}channels. The Ca^{2+} currents induced in presynaptic terminals by depolarization were thus isolated (Llinás *et al.* 1981).

Figure 2.9 shows the magnitude of the Ca^{2+} currents as a function of depolarization in a presynaptic terminal of the squid. The Ca^{2+} current appeared when the terminal was depolarized to −50 to −45 mV from a holding potential of −70 mV, and it peaked at membrane potentials of −15 to −5 mV.

Llinás *et al.* (1982) went on to investigate the time course of Ca^{2+} currents elicited by a normal action potential reaching the presynaptic terminal. Their approach was essentially the method used by Bastian and Nakajima (1974) for demonstrating a propagating action potential in the transverse tubules of skeletal muscle. An action potential recorded from a presynaptic terminal with an intracellular electrode was first stored on a digital tape then imposed on the presynaptic terminal through the voltage clamp circuit after TTX and TEA blockade of voltage-gated Na^+ and K^+ conductances. Figure 2.10 shows the Ca^{2+} current (I_{Ca}) elicited in the presynaptic terminal under voltage clamp and the waveform of the replicated action potential (AP, truncated), together with the synaptic response (Post) recorded simultaneously from the postsynaptic axon. Apparently the Ca^{2+} conductance increases slowly in response to depolarization induced by the action poten-

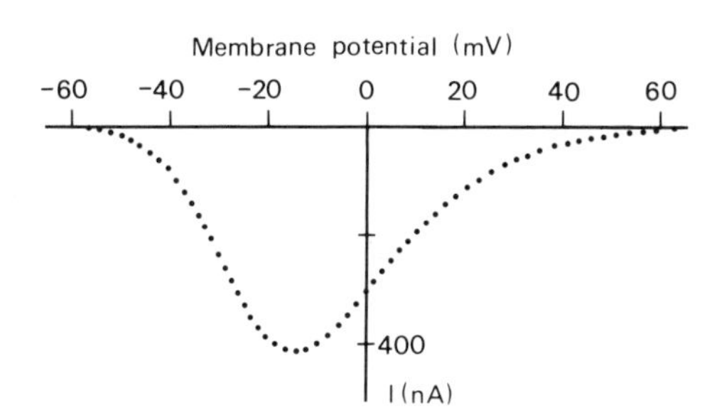

Fig. 2.9. Intensities of Ca^{2+} currents (ordinate) as a function of the membrane potential (abscissa) in a presynaptic terminal of the squid giant synapse. Depolarization was from a holding potential of −70 mV. The external Ca^{2+} concentrations, 10 mM. (Adapted from Llinás *et al.* 1981.)

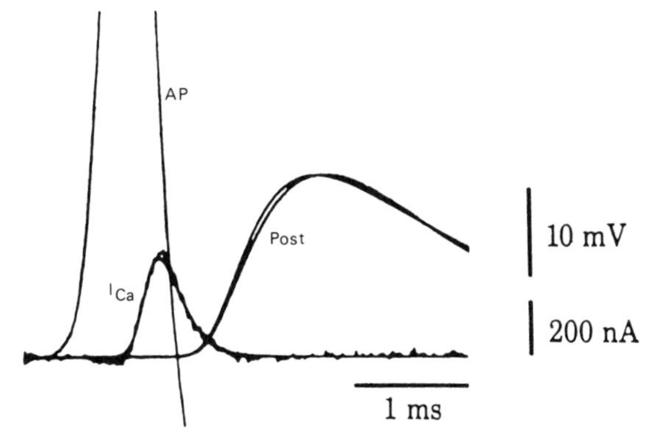

Fig. 2.10. Calcium currents (I_{Ca}) generated in a presynaptic terminal of the squid giant synapse by a replicated action potential. The presynaptic terminal was voltage-clamped with a waveform identical to the normal presynaptic action potential (AP). Post, synaptic potential simultaneously recorded from the postsynaptic axon. (From Llinás *et al.* 1982.)

tial, so that the Ca^{2+} current flows during the falling phase of the action potential. This slow onset of the Ca^{2+} current provides a favourable condition in that the large driving force for Ca^{2+} currents comes at the peak of increased Ca^{2+} conductance. The Ca^{2+} current evoked in the presynaptic terminal by the action potential is thus equivalent to the 'off' Ca^{2+} response in Fig. 2.7B. Were the Ca^{2+} conductance to increase earlier soon after the onset of the action potential, the Ca^{2+} current would be much smaller, because of the small driving force for Ca^{2+} at the peak of the action potential (Llinás *et al.* 1982).

Other attempts to record Ca^{2+} currents from presynaptic terminals have focused on the ciliary ganglion. Its presynaptic terminal forms a cup-shaped calyx, enveloping the ciliary neurone in a sheet-like structure (Carpenter 1911; De Lorenzo 1960). This terminal is large enough to admit an intracellular electrode (Martin and Pilar 1963, 1964b) but not for two electrodes to fit inside the terminal. Membrane currents from the presynaptic terminal were therefore recorded either with a single-electrode voltage clamp system (Martin *et al.* 1989) or with the *patch clamp technique* (Stanley 1989; Stanley and Goping 1991; Yawo and Momiyama 1993). Figure 2.11A illustrates the technique for recording membrane currents with the patch clamp method (Hamill *et al.* 1981). First, firmly pressing the microelectrode tip against the cell membrane surface (Fig. 2.11Aa) forms a seal so tight that its resistance exceeds 1 GΩ($10^9\Omega$). Then, the membrane patch under the electrode is ruptured by applying negative pressure to the pipette interior (Fig. 2.11Ab). As a consequence, the total membrane currents of the cell can be recorded through the electrode with virtually complete insulation from the bath (*tight-seal whole-cell recording*; Hamill *et al.* 1981). With this technique, Yawo and Momiyama (1993) made whole-cell recordings simultaneously from a presynaptic calyx and its postsynaptic neurone in the chick ciliary ganglion (Fig. 2.11B). The application of a depolarizing pulse to the presynaptic terminal induced Ca^{2+} currents in the terminal (Fig. 2.11Cb), which in turn elicited synaptic currents in its postsynaptic neurone (Fig. 2.11Cc). The Ca^{2+} current recorded from the presynaptic terminal was activated when the terminal was depolarized to about -30 mV; it reached the maximum level at about 0 mV (Fig. 2.12). This threshold for activating the Ca^{2+} current exceeds that observed in the squid giant synapse (Fig. 2.9). Perhaps, the Ca^{2+} channels present in these two presynaptic terminals differ in nature (see section 2.2.3). Tachibana *et al.* (1993) recorded Ca^{2+} currents from the terminal of bipolar cells

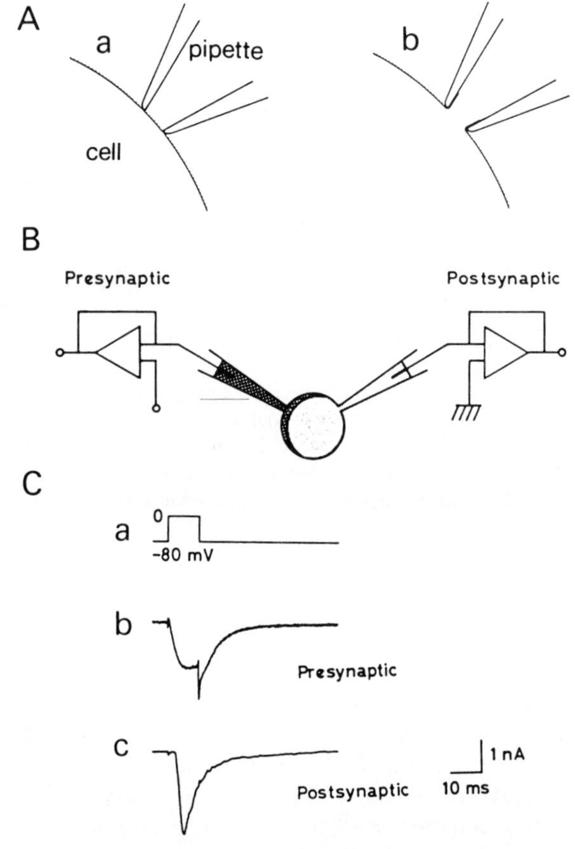

Fig. 2.11. Recording of whole-cell Ca^{2+} currents. A, the technique for recording the whole-cell currents with the patch clamp. The recording electrode is first pressed tightly against the cell surface (a), then the membrane patch under the electrode is ruptured by negative pressure. (From Hamill *et al.* 1981.) B, the experimental procedure for the whole-cell recording technique applied simultaneously to a presynaptic terminal (left) and its postsynaptic neuronee (right) in the chick ciliary ganglion. C, Ca^{2+} currents recorded from the presynaptic terminal (b) and the resultant synaptic current monitored in the postsynaptic neuronee (c) in the ciliary gagnlion when the presynaptic terminal was depolarized from -80 mV to 0 mV (a). (B and C, from Yawo and Momiyama 1993.)

in the goldfish retina. As we will see later (Section 2.2.3), the properties of these Ca^{2+} currents recorded from the bipolar cell terminal are distinctly different from those observed in the squid or ciliary presynaptic terminals.

Although ionic currents at motor nerve terminals cannot be recorded directly with intracellular electrodes, local electrical events can be monitored with a microelectrode placed in close proximity to the motor nerve

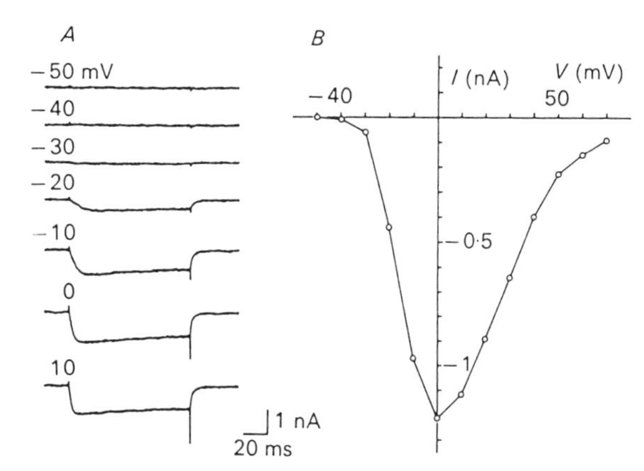

Fig. 2.12. Ca^{2+} currents recorded from a presynaptic terminal of the chick ciliary ganglion at different membrane potentials. *A*, sample records. *B*, the relationship between the intensity of Ca^{2+} currents and the membrane potential. (From Yawo and Momiyama 1993.)

terminal (Katz and Miledi 1965b) or by inserting a microelectrode under the perineural sheath of an intramuscular nerve bundle within 50–500 μm of the terminal arborizations (Gundersen *et al.* 1982; Mallart 1985; Penner and Dreyer 1986). Intracellular potentials have also been recorded from motor nerve fibres near their terminals (Wojtowicz and Atwood 1988; Morita and Barrett 1989). All these techniques enable the potentials produced by Ca^{2+} currents generated in motor nerve terminals to be recorded with K^{+} conductances blocked. Judging from several similarities in behaviour of the monitored Ca^{2+} currents, the concept established for transmitter release at the squid giant synapse seems fully applicable to neuromuscular junctions and other synapses such as the ciliary ganglion. It is reasonable to conclude that the generation of a regenerative inward Ca^{2+} current, induced by an action potential reaching the nerve terminal, is essential for the transmitter release in both neuronal and neuromuscular synapses. However, it is not certain whether the Ca^{2+} currents produced in motor nerve terminals are similar in nature to those observed in neuronal synapses. Ca^{2+} channels are present ubiquitously in almost every excitable cell, including its cell body, axon, and terminal. Might the nature of Ca^{2+} channel be distinct at different regions in a given cell? Might the nature of Ca^{2+} channel be specific for cell type? Is there a particular type of Ca^{2+} channel which is involved specifically in transmitter release? Let us now examine the diversity of Ca^{2+} channels in excitable cells.

2.2 THE DIVERSITY OF CALCIUM CHANNELS

According to Hagiwara and Byerly (1981), 'Ca^{2+} currents recorded from different preparations differ in almost every measurable property, whereas the basic kinetics and selectivity in Na^{+} channels remain unchanged from preparation to preparation'. Does the diversity of Ca^{2+} channels stem from differences in cell types? Not necessarily. For example, Purkinje cells in the mammalian cerebellum show Ca^{2+} spike potentials and long-lasting Ca^{2+} plateau potentials, both of which are involved in excitation of the dendrites (Llinás and Sugimori 1980). Similarly, mammalian inferior olivary neurones can produce low- and high-threshold Ca^{2+} spikes in the dendrites (Llinás and Yarom 1981). Low- and high-threshold Ca^{2+} components also occur in dorsal horn neurones of the rat spinal cord (Murase and Randić 1983). Moreover, multiple types of Ca^{2+} channel coexist in dorsal root ganglion cells of the vertebrate (Carbone and Lux 1984, 1987a,b; Nowycky *et al.* 1985; Fox *et al.* 1987a,b). How many types of Ca^{2+} channel are there?

2.2.1 Classification of calcium channels

The most thorough characterization of Ca^{2+} channels has been carried out in sensory neurones. Figure 2.13

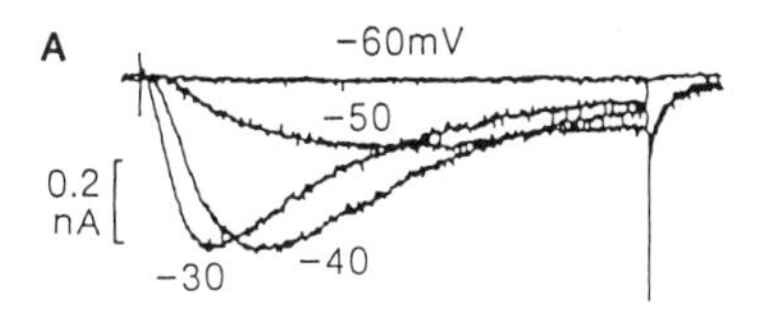

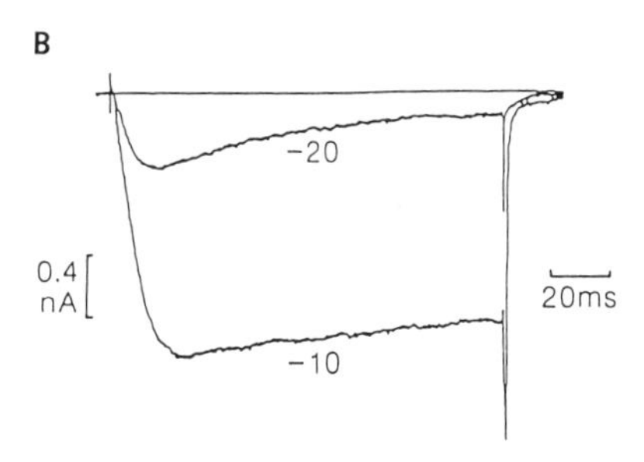

Fig. 2.13. Whole-cell Ca^{2+} currents recorded from a sensory neurone. A, when the cell was depolarized from a holding potential of −80 mV, transient (fast inactivating) Ca^{2+} currents were activated by relatively small depolarization (−50, −40, and −30 mV). B, long-lasting (slow inactivating) Ca^{2+} currents required large depolarization (−20 mV and −10 mV) for activation. (From Swandulla *et al.* 1991a.)

shows Ca^{2+} currents recorded from a rat embryonic sensory neurone (Swandulla *et al.* 1991a) with the whole-cell mode of the patch clamp (Fig. 2.11A). When the cell was depolarized from a holding potential of −80 mV, two different types of Ca^{2+} current were recorded (Fig. 2.13A,B). One type of Ca^{2+} current could be activated by a relatively small depolarization (at about −50 mV; Fig. 2.13A), whereas another type required a greater depolarization for activation (at −20 mV; Fig. 2.13B). Furthermore, the two types differ regarding inactivation. The former type is inactivated rapidly (hence, *transient*; Fig. 2.13A), whereas the latter is only slowly inactivated (hence, *long-lasting*; Fig. 2.13B). The low-threshold transient type of Ca^{2+} current was termed *low-voltage-activated* type (Carbone and Lux 1984) or *T*-type (Nowycky *et al.* 1985). The high-threshold long-lasting Ca^{2+} current was termed *high-voltage-activated* type (Carbone and Lux 1984) or *L*-type (Nowycky *et al.* 1985).

An additional type of Ca^{2+} current, *N*-type (*neither* T nor L), was recorded from chick sensory neurones (Nowycky *et al.* 1985). The N-type Ca^{2+} current requires a relatively large depolarization for activation like the L-type, but unlike the L-type, the N-type Ca^{2+} current is rapidly inactivated. The three types of Ca^{2+} current obtained from chick sensory neurones are illustrated in Fig. 2.14 (Nowycky *et al.* 1985). Superimposed Ca^{2+} currents (Fig. 2.14*a*) were elicited by depolarization to various membrane potentials (−50 mV to +20 mV) from two initial holding potentials (−40 mV and −100 mV). The current–voltage (I–V) relationships under these conditions are revealing. Depolarization from −100 mV (Fig. 2.14*b*, circles) elicits the Ca^{2+} current at a relatively low threshold (about −60 mV), but the I–V profile shows an additional component at more depolarized levels beyond about −10 mV. With depolarization from −40 mV (Fig. 2.14*b*, squares), the low-threshold component disappears, leaving only the high-voltage-activated Ca^{2+} current. Hence, the low-threshold component appears to be already inactivated at a membrane potential of −40 mV. As shown in the inset of Fig. 2.14*a* (+10 and

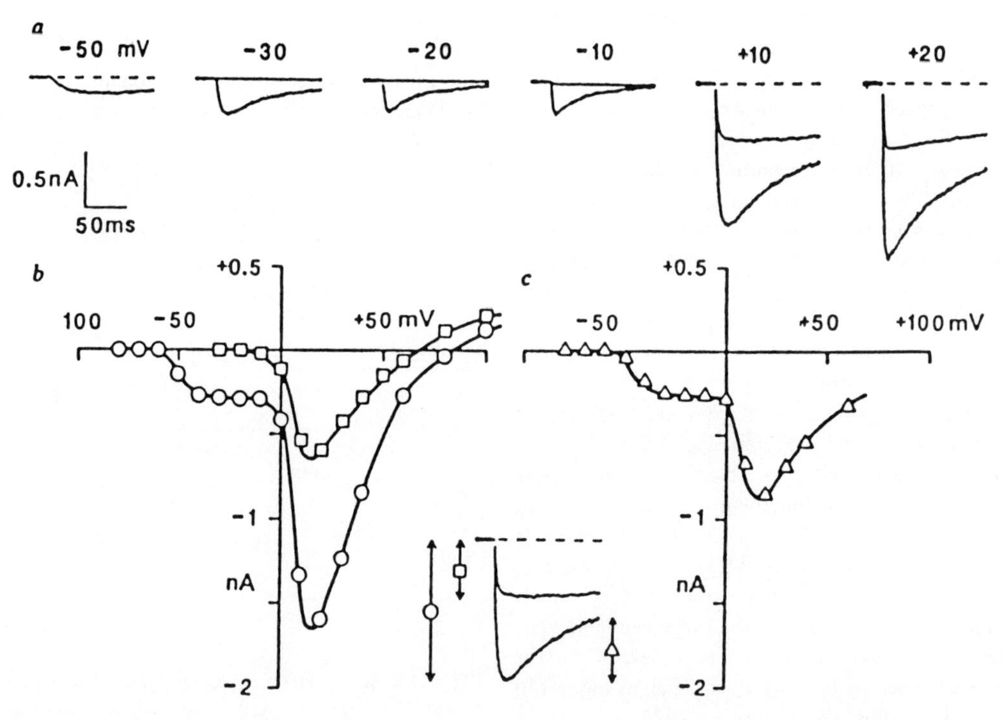

Fig. 2.14. Three components of Ca^{2+} currents recorded from a chick dorsal root ganglion cell with the whole-cell recording technique. *a*, superimposed current traces evoked by depolarization to the levels indicated on each record from holding potentials of −40 mV and −100 mV. *b*, current-voltage (I–V) relations for the records obtained from a holding potential of −100 mV (circles) and from −40 mV (squares). *c*, I–V relation for the peak relaxing (transient, inactivating) part (indicated by triangle with arrows in inset) of the records obtained from a holding potential of −100 mV. (From Nowycky *et al.* 1985.)

+20 mV), the high-voltage-activated Ca^{2+} current remaining after inactivation of the low-threshold component is a long-lasting component (L-type), whereas the low-threshold component (−50 mV to −10 mV) is transient in nature (T-type). If only the T-and L-types of Ca^{2+} current exist, the difference between the curves shown by circles and squares in Fig. 2.14*b* (equivalent to 'transient' components in Fig. 2.14*a*) should give the I–V curve for the T-type. The inflection in the I–V curve reconstructed for the difference (Fig. 2.14*c*) suggests the presence of a third component (N-type). This N-type component is apparently high-voltage-activated channels (Fig. 2.14*c*) and transient (or fast-inactivating) in nature (Fig. 2.14, inset, triangle with arrows).

These procedures enabled the three types of Ca^{2+} current to be distinguished and their individual I–V relations determined (Fox *et al.* 1987a; Tsien *et al.* 1988). Other types may exist as well. For example, the Ca^{2+} currents recorded from presynaptic terminals in the squid giant synapse resemble the L-type in lacking fast inactivation (Katz and Miledi 1971; Llinás *et al.* 1981) but have a markedly different I–V relationship (compare Figs. 2.9 and 2.14*b* squares). Yet, the general classification of Ca^{2+} channels into three types remains a useful framework, even if modification and additions occur (see below).

Selective blockade of Na^+ and K^+ currents with TTX and TEA, respectively, provides strong evidence for the existence of separate Na^+ and K^+ channels. Might the three types of Ca^{2+} current be distinguishable by some selective blockers? We have seen that Ca^{2+} currents in presynaptic terminals of the squid synapse are blocked by Cd^{2+} (Fig. 2.8Bb). The low-threshold Ca^{2+} current (T-type) was more resistant to Cd^{2+} than the high-threshold Ca^{2+} currents (L-and N-types; Fox *et al.* 1987a), but Cd^{2+} could not distinguish between the L- and N-types. Fortunately, however, dihydropyridine compounds (DHPs), developed in 1969 for therapeutic use of coronary disease (Fleckenstein 1983), act selectively on the L-type Ca^{2+} channel. The L-type channel is blocked by DHP-antagonists (e.g., nifedipine; Fox *et al.* 1987a) and enhanced by DHP-agonists (e.g., BAY K 8644; Nowycky *et al.* 1985; Fox *et al.* 1987b), whereas the T-and N-types are insensitive to DHP compounds.

The peptide toxin ω-conotoxin (ω-CgTx), purified from the venom of a marine mollusc (Olivera *et al.* 1985), is another useful chemical, because it blocks the N-type channel selectively (Fox *et al.* 1987a; Aosaki and Kasai 1989; Plummer *et al.* 1989). Thus, the two types (L- and N-types) of the high-threshold Ca^{2+} current can be distinguished by applying DHPs and ω-CgTx. As shown in Fig. 2.12, presynaptic terminals of the chick ciliary ganglion possess only the high-threshold Ca^{2+} channels. The Ca^{2+} currents recorded from this presynaptic terminal were affected by neither DHP-agonist (BAY K 8644) nor DHP-antagonist (nifedipine) but reduced markedly by ω-CgTx (Fig. 2.15A; Yawo and Momiyama 1993). Clearly, the major component of Ca^{2+} currents in this presynaptic terminal is the N-type, and this terminal appears to lack the L-type of Ca^{2+} channel. However, a small fraction of the Ca^{2+} currents (about 10 per cent) was resistant to both DHPs and ω-CgTx and blocked only by Cd^{2+} (Fig. 2.15A). This fraction must be generated by Ca^{2+} channels other than the three classified in the chick sensory neurones. A similar DHP- and ω-CgTx-resistant fraction of Ca^{2+} currents (about 20 per cent) also occurred in the ciliary postganglionic cell (Fig. 2.15B). Unlike the presynaptic terminal, the postsynaptic cell has both nifedipine-sensitive (L-type; 25 per cent) and ω-CgTx-sensitive (N-type; 55 per cent) components, in additon to the unidentified fraction (Fig. 2.15B).

In a variety of rat central and peripheral neurones, a substantial fraction of high-threshold Ca^{2+} currents is resistant to both DHPs and ω-CgTx (Regan *et al.* 1991). Similarly, synaptic transmission in the ciliary ganglion is blocked by ω-CgTx (Yoshikami *et al.* 1989; Stanley and Atrakchi 1990), whereas the synaptic potential in the squid giant synapse is not affected by ω-CgTx or DHPs (Charlton and Augustine 1990). Llinás *et al.* (1989) found that both the presynaptic Ca^{2+} current and synaptic transmission in the squid giant synapse can be blocked by a toxin purified from the venom of funnel-web spiders (FTX). The Ca^{2+} channel in the squid presynaptic terminal thus comprises an additional independent class of channel.

FTX also blocks Ca^{2+}-dependent spikes and Ca^{2+}-dependent plateau potentials recorded from Purkinje cells of the mammalian cerebellum (Llinás *et al.* 1989). Once the toxin-binding proteins were isolated from the membrane fractions of the guinea-pig cerebellum and of the squid optic lobe, the purified proteins could be reconstituted in a lipid bilayer and Ca^{2+} currents recorded. Amazingly, the characteristics of the Ca^{2+} currents recorded from these reconstituted proteins derived from the cerebellum and from the squid tissue proved markedly similar: these currents were blocked by FTX but insensitive to DHPs or ω-CgTx (Llinás *et al.* 1989). Since the presence of this type of Ca^{2+} current was first

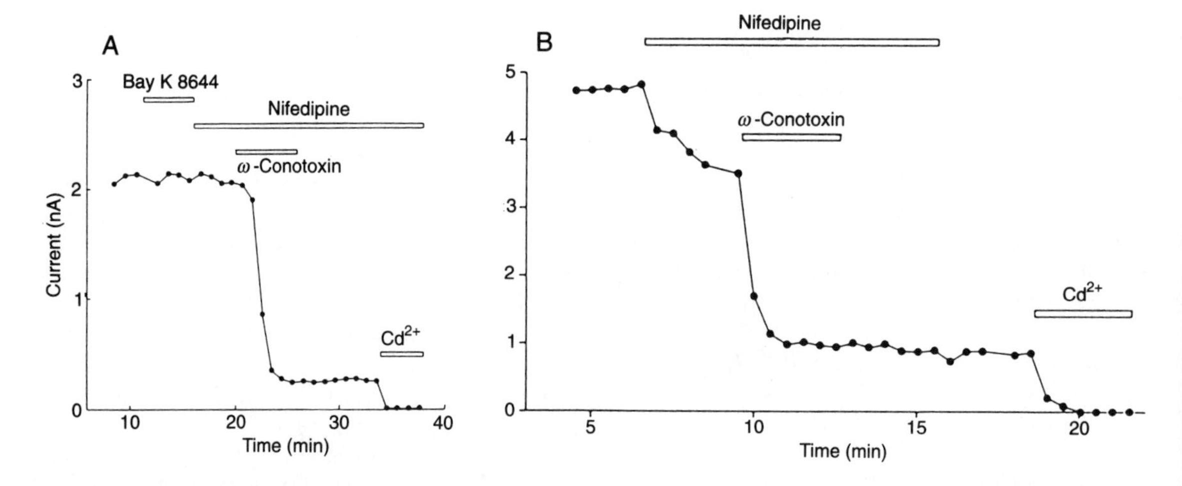

Fig. 2.15. DHP- and ω-CgTx-sensitive components of Ca^{2+} currents recorded from a presynaptic terminal (A) and a postganglionic cell (B) of the chick ciliary ganglion. Open bars indicate the period of time during which each chemical is applied. Note the presence of DHP- and ω-CgTx-resistant component which can be blocked only by Cd^{2+}. (From Yawo and Momiyama 1993.)

observed in Purkinje cells of the mammalian cerebellum (Llinás and Sugimori 1980), it was termed the *P-channel* (Llinás *et al.* 1989). Mintz *et al.* (1992) found that the P-channel is blocked also by a peptide toxin purified from funnel-web spider venom, ω-Aga-IVA. This peptide toxin is distinct from FTX, a non-peptide toxin. It should be noted, however, that FTX is a selective blocker for the P-channel, whereas high doses of ω-Agfa-IVA appear to block another type of Ca^{2+} channel (*Q-type*) in addition to the P-channel (Wheeler *et al.* 1994; see section 2.2.3).

Inconveniently, the term *P-channel* has also been used for potassium-preferring cationic channels observed in the synaptic vesicular membrane (Rahamimoff *et al.* 1988). Here, we use the P-channel to refer only to the above type of Ca^{2+} channel, not to the potassium-preferring channel in the vesicular membrane.

Undoubtedly, there are at least four different classes of Ca^{2+} channel: T-, L-, N-, and P-channels. Two additional novel types of Ca^{2+} channel are observed in bovine chromaffin cells. Chromaffin cells are secretory cells, innervated by sympathetic preganglionic fibres, and release catecholamines. Catecholamine release from chromaffin cells requires the presence of Ca^{2+} in the external solution (Douglas and Rubin 1963). As expected, Ca^{2+} channels were found in chromaffin cells. In addition, the channel activity was facilitated by a preceding depolarizing pulse (Fenwick *et al.* 1982; Hoshi *et al.* 1984). Figure 2.16B shows the current–voltage (I–V) relations of the Ca^{2+} currents elicited with and without a conditioning depolarizing pulse (to +120 mV for 50 ms; Fig. 2.16A). Although the Ca^{2+} current was clearly facilitated by the depolarizing pulse (Fig. 2.16B, triangles), the profile of the I–V relation was essentially the same as that for the control Ca^{2+} current (open circles). Surprisingly, however, the control Ca^{2+} current showed no inactivation during a sustained depolarization (to + 10 mV for 1.5 sec), whereas the Ca^{2+} current facilitated by a conditioning depolarizing pulse was progressively inactivated during the sustained depolarization (Fig. 2.17). Evidently, the Ca^{2+} channels facilitated or recruited by the depolarizing pulse are different in nature from those present under the control or standard condition. These two currents were termed *facilitation* and *standard* currents, respectively (Artalejo *et al.* 1991a).

The standard and facilitation Ca^{2+} currents are also pharmacologically distinct. The standard Ca^{2+} current was largely blocked by ω-CgTx but insensitive to DHPs (Artalejo *et al.* 1992a), whereas the facilitation Ca^{2+} current was sensitive to DHPs but not to ω-CgTx (Artalejo *et al.* 1991a,b). Thus, pharmacologically, the standard channel is similar to the N-type, however, unlike the N-type, the standard current shows no inactivation; hence, it represents a novel type of Ca^{2+} channel (Artalejo *et al.* 1991b, 1992a). Pharmacologically, the facilitation channel resembles the L-type, but unlike the

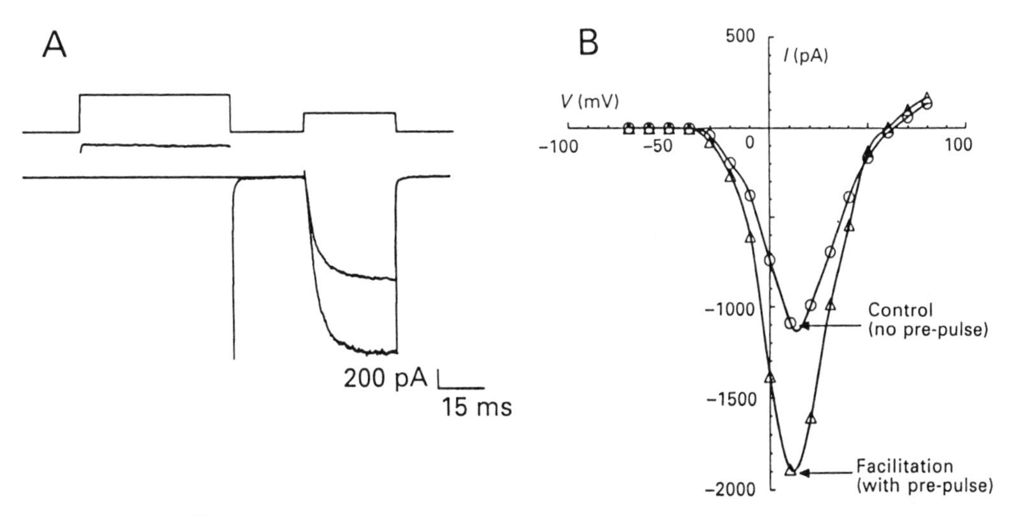

Fig. 2.16. Facilitation of Ca^{2+} currents by a preceding depolarizing pulse (pre-pulse) in bovine chromaffin cells. A, Ca^{2+} currents were induced by a 30 ms test depolarizing pulse to +10 mV from −80 mV with and without a 50 ms pre-pulse to +120 mV from −80 mV. The test pulse was applied 20 ms after the pre-pulse. Note an increase in the inward Ca^{2+} current produced by the pre-pulse. B, the current-voltage relationships for the Ca^{2+} currents induced with (triangles; facilitation) and without (circles; control) the 50 ms pre-pulse. The current-voltage relation was obtained by changing the test pulse in a range from −70 mV to +80 mV. (From Artalejo *et al.* 1991a.)

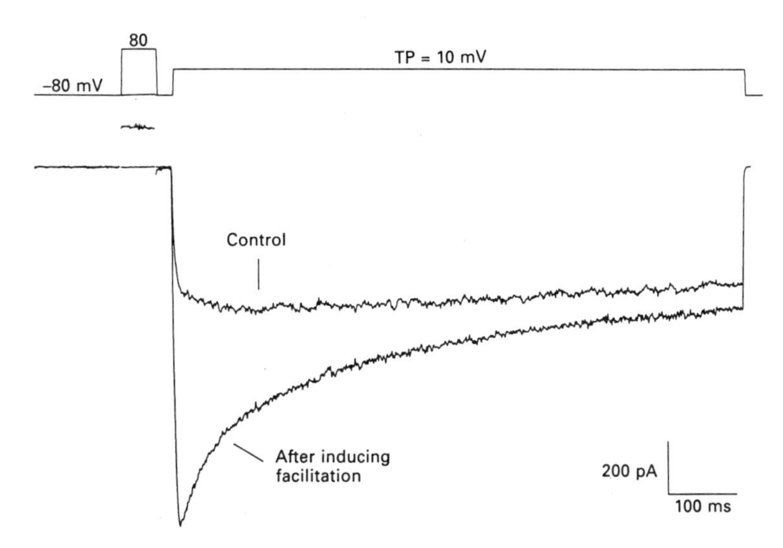

Fig. 2.17. Inactivation kinetics in control versus facilitated Ca^{2+} currents in a chromaffin cell. The Ca^{2+} currents were induced by a long (1.5 sec) depolarizing pulse from −80 mV to +10 mV with (after inducing facilitation) and without (control) a preceding depolarizing pulse (from −80 mV to +80 mV for 50 ms). (From Artalejo *et al.* 1991a.)

L-type, the facilitation Ca^{2+} current cannot be activated even by large depolarizations (−20 to +30 mV) unless it is primed with a preceding depolarizing pulse. Hence, it too represents a novel type of Ca^{2+} channel (Artalejo *et al.* 1991b). As we will discuss later (7.1.2), activity of Ca^{2+} channels can be modulated under a variety of conditions. A large depolarizing pulse is one of the modulators.

So far we have classified four major types of Ca^{2+} channel and two additional classes unique to chromaffin cells. However, each type of Ca^{2+} channel classified by its pharmacological and biophysical characteristics may not be identical in different tissues. For example, a mutant mouse with muscular dysgenesis lacks the DHP-sensitive L-type Ca^{2+} current in skeletal muscle (Tanabe *et al.* 1988). If the animal had only a single type of gene encoding the L-type Ca^{2+} channel, the L-type Ca^{2+} current would not be expressed in any tissues of this mutant. Yet, this mutant mouse shows normal DHP-sensitive L-type Ca^{2+} currents in cardiac cells and sensory neurones (Tanabe *et al.* 1988). Thus, Ca^{2+} channel diversity must be greater than that expected from the number of channel types classified pharmacologically or biophysically. The cloning and protien purification of Ca^{2+} channel molecules are now beginning to define the properties of each type of channel on the molecular basis. Let us now trace this remarkable progress in molecular biology of Ca^{2+} channels.

2.2.2 The molecular structure of calcium channels

The ionic channel is a protein (or an aggregate of proteins) embedded in the cell membrane. As illustrated in Fig. 2.18A, the cell membrane itself consists of a phospholipid bilayer in contact with two aqueous compartments: the extracellular medium and the intracellular or cytoplasmic fluid. It is the *hydrophilic* (water-loving) region of each molecule in the lipid bilayer that faces outward, jutting into the extracellular or intracellular fluid, whereas the *hydrophobic* (water-hating) tails of the lipid molecules reach inward from both surfaces to form a hydrocarbon core in the middle of the cell membrane. Thus, hydrophilic and hydrophobic interactions draw some regions of molecules to an aqueous environment and drive other regions away from contact with water. In other words, the ionic channel proteins embedded in the cell membrane must be structurally asymmetric (*amphipathic*) so as to accomodate both the aqueous surfaces and the membrane's hydrophobic core (Singer and Nicolson 1972). Singer (1975) also suggested that certain proteins may exist within the membrane as a subunit aggregate (*polymer*) which has a water-filled pore (channel) in its centre, spanning the entire membrane (Fig. 2.18B). These protein subunits would have amphipathic sidedness, for each subunit needs a hydrophilic lining on one side to face its water-filled central pore and a hydrophobic lining on the other side to face the hydrocarbon core of phospholipids. In effect, the subunit must resemble the inner and outer surfaces of a barrel (Fig. 2.18B). This early suggestion has proved remarkably consistent with our current view of the ionic channel.

DHP-sensitive Ca^{2+} channels are present in a variety of tissues, but they are highly concentrated in the transverse tubules of skeletal muscle (Fosset *et al.* 1983). DHP-binding proteins purified from the transverse tubules form functional L-type Ca^{2+} channels when reconstituted into phospholipid bilayer membranes (Flockerzi *et al.* 1986). The muscle DHP-sensitive Ca^{2+} channel comprises five subunits; α_1, α_2, β, γ, and δ (Catterall 1988). These subunits were assumed to assemble at the cell membrane, as illustrated in Fig. 2.19. The largest subunit, α_1, contains the DHP-binding site (Vandaele *et al.* 1987) and substantial hydrophobic domains; hence, it must be the principal component of Ca^{2+} channel (Catterall 1988).

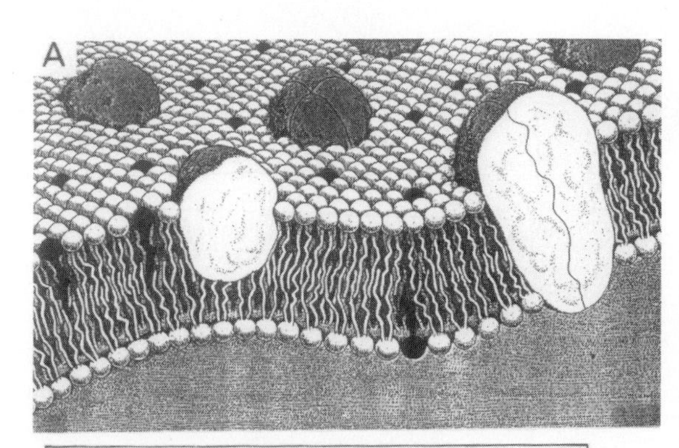

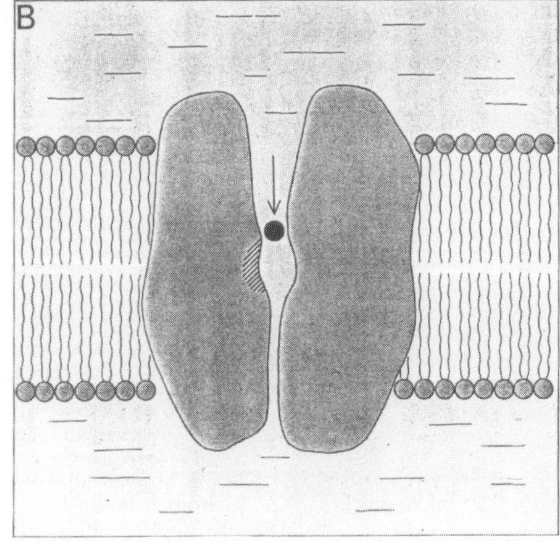

Fig. 2.18. Hypothetical topography of proteins embedded in the cell membrane. A, a phospholipid bilayer (small balls with hairpin-like tails) forming the cell membrane. Some globular proteins are simply embedded on one or the other side, while others pass entirely through the bilayer. B, an ionic channel formed by membrane-spanning proteins. Circle with arrow indicates an ion travelling through the central pore of the channel. (From Singer 1975.)

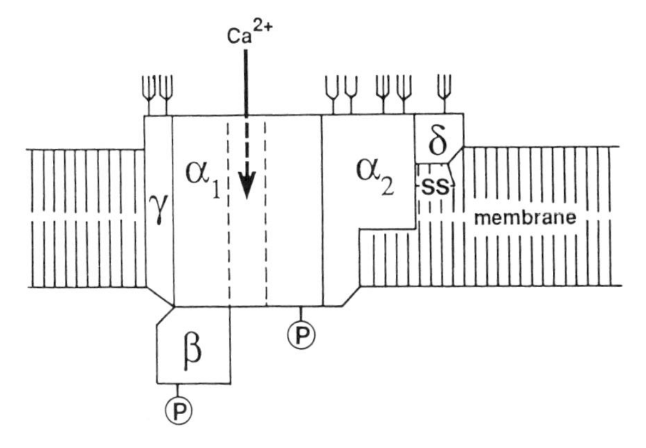

Fig. 2.19. Hypothetical subunit structure of the rabbit skeletal muscle Ca^{2+} channel. Top, the extracellular fluid. Glycosylation sites are indicated by forks on the extracellular side. The α_1 subunit is assumed to form the channel through which Ca^{2+} flows. The α_2 and δ-subunits are a disulphide-linked complex (-S-S-). P in a circle indicates phosphorylation sites. (From Catterall 1988.)

The primary structure of the α_1-subunit of rabbit skeletal muscle was elucidated by the cloning and sequencing of the DNA complementary (cDNA) to its messenger RNA (mRNA; Tanabe *et al.* 1987). The α_1-subunit consists of 1,873 amino acids and shows two features in its structure. First, there are four repeated units (domains), each of which contains 250–300 amino acid residues with homologous sequences (I, II, III, and IV in Fig. 2.20A). Second, the hydropathicity profile analysis (Box B) suggests that each domain has six segments (S1 to S6 in Fig. 2.20A) composed of about 20 predominantly hydrophobic amino acids. From the membrane model shown in Fig. 2.18, these hydrophobic segments are assumed to be the regions that traverse the hydrophobic core of the cell membrane with α-helical formations (Fig. 2.20A). In an α-helix, each turn is made of 3.6 amino acids and represents an advance of 5.4 Å down the long axis of the helix. A helix formed by about 20 amino acid residues would be just long enough to span the hydrophobic core of the cell membrane, which is about 40 Å or less. The amino terminus (N-terminus) of the α_1-subunit must be located in the intracellular side, because its N-terminus does not contain the sequence characteristics of the *signal peptide* (a stretch of 15–60 hydrophobic amino acids) essential for translocating the N-terminus of a protein from the cytosolic side. Because of the even number of transmembrane segments (24 in total), the

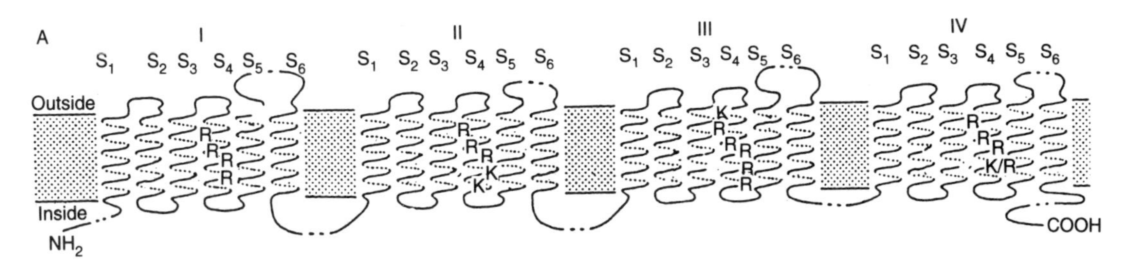

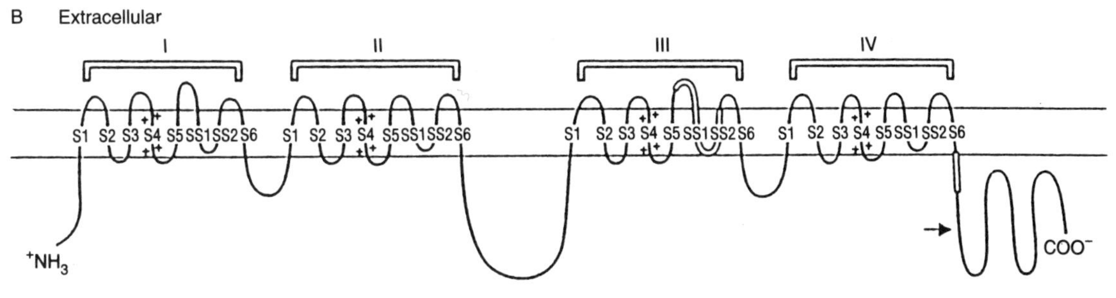

Fig. 2.20. Hypothetical structure of the α_1-subunit of the rabbit skeletal muscle Ca^{2+} channel. A, it consists of four domains or repeats (I–IV), and each domain has six transmembrane segments (S_1–S_6). The S_4 segment in each repeat has positively charged amino acids (R, arginine; K, lysine). (Adapted from Tanabe *et al.* 1987.) B, an alternative model of the α_1-subunit. The region separating S5 and S6 segments of each domain is assumed to contain two additional short segments (SS1 and SS2). (Adapted from Snutch and Reiner 1992.)

Box B The hydropathy profile of a protein

A protein is an array of amino acids connected by peptide bonds. The peptide bond is formed between the carboxyl group (—COOH) of one amino acid and the amino group ($-NH_2$) of the next amino acid. Thus a protein has its amino (N) terminus on one end and its carobxyl (C) terminus on the other end. The side chains of some amino acids are charged (ionized) or uncharged but polar (having an electric dipole moment), allowing these amino acids to interact with water (hydrophilic). The side chains of other amino acids consist only of hydrocarbons, and such amino acids are hydrophobic. Where hydrophobic amino acids predominate in one region or segment of the protein, it becomes locally hydrophobic, and this segment of the protein can then be incorporated into the hydrophobic core of the cell membrane. It is not so much that the hydrophobic regions can be included in hydrophobic lipids, but rather that they are excluded from the aqueous environment; i.e., they are pushed, not pulled, into the membrane. Identifying such hydrophobic segments in a given protein indicates which protein segments interact with or reside within the cell membrane.

Hopp and Woods (1981) and Kyte and Doolittle (1982) developed a method for displaying the hydropathicity profile of a protein. Each of the 20 amino acids is scored numerically according to its hydrophilic or hydrophobic properties. Given the amino acid sequence of any protein, the average hydropathy score is then calculated for the first seven amino acids (or any appropriate number) counting from the N-terminus of the protein. Then, the next segment of seven amino acids (from the second amino acid to the eighth amino acid from the N-terminus) is counted, and so on. Finally, the entire hydropathy profile is assembled from the scores for every segment of seven contiguous amino acids in the protein.

The hydropathy profile in Figure (A) describes the γ-subunit of the nicotinic ACh receptor (Claudio *et al.* 1983), which will be discussed in section 4.1.1. The locus of the four hydrophobic segments implies the subunit's structure in relation to the cell membrane, as indicated in the Figure (B).

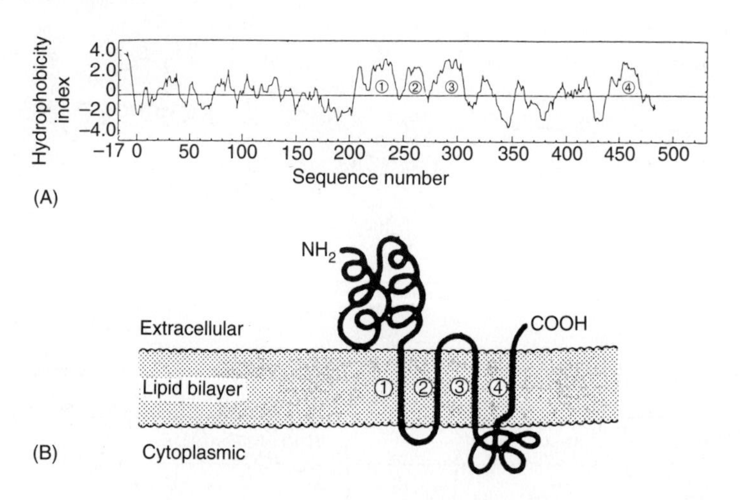

carboxyl terminus (C-terminus) of the channel protein should be also in the cytoplasmic side (Fig. 2.20A; Numa 1989).

Despite the presence of other subunits (Fig. 2.19), the α_1-subunit alone can express a functional L-type channel (Tanabe *et al.* 1988; Mikami *et al*, 1989; Perez-Reyes *et al.* 1989). Therefore, this subunit must be involved directly in Ca^{2+} permeation. The molecular size and its structural features of the α_1-subunit are remarkably similar to those of the α-subunit of Na^+ channel (Fig. 2.21A; Noda *et al.* 1984; Catterall 1988). The α-subunit of Na^+ channel alone also expresses a functional TTX-sensitive channel (Noda *et al.* 1986). A similar structural motif can be seen also in a family of

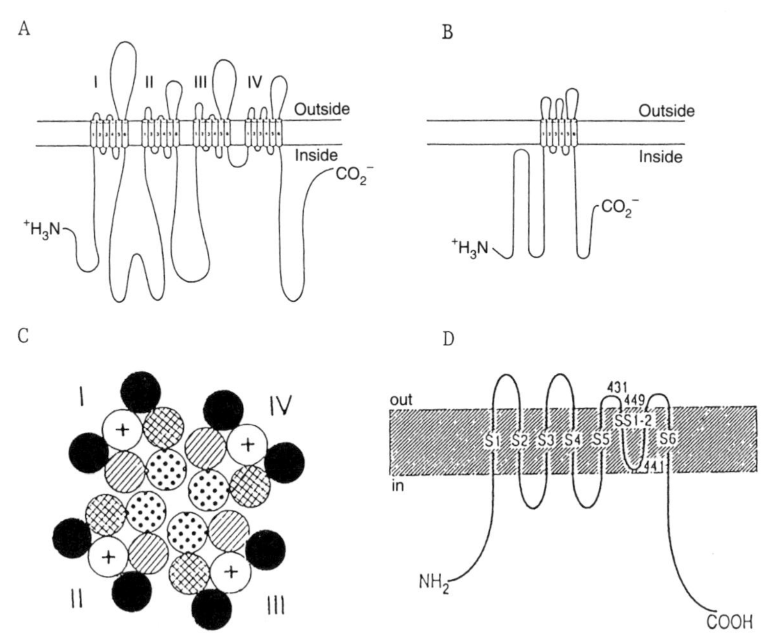

Fig. 2.21. Schematic transmembrane arrangements of Na^+ and K^+ channel molecules. A, Na^+ channel α-subunit having four domains (I–IV), each of which has six transmembrane segments (1–6). (From Catterall 1988.) B, K^+ channel, resembling one of the four domains of the Na^+ channel. (From Catterall 1988.) C, hypothetical arrangment of the transmembrane segments in a Na^+ channel viewed in the direction perpendicular to the cell membrane. I–IV, four domains. Eight filled circles in the outer surface represent segments 5 and 6. Four stippled circles in the inner surface represent segment 2. Four circles with + represent segment 4. (From Numa 1987.) D, an alternative model for transmembrane topology of K^+ channels. Short segments (SS1–2) are located between segments 5 and 6. (From Yellen *et al.* 1991.)

voltage-gated K^+ channels, as illustrated in Fig. 2.21B (Pongs *et al.* 1988; Catterall 1988). The K^+ channel protein, however, differs from the Ca^{2+} or Na^+ channel in that it is relatively small (consisting of 500–800 amino acids) and contains only a single domain. In effect, the K^+ channel resembles one of the four domains of Ca^{2+} or Na^+ channels. The four homologous domains of Ca^{2+} or Na^+ channels may be oriented in a quasi-symmetrical fashion across the cell membrane, presumably forming an ionic channel in the centre (Fig. 2.21C). Likewise, a single K^+ channel may be formed by the assembly of multiple (probably four) subunits. Individual K^+ channels can be formed by the assembly of different subunits (heteromultimer), and this heteromeric combination may account for the diversity of functional K^+ channels (Christie *et al.* 1990; Isacoff *et al.* 1990; Ruppersberg *et al.* 1990). The striking similarity in the basic molecular motif suggests that all these ionic channels are members of the same gene superfamily (Numa 1989, 1991).

Since Ca^{2+}, Na^+, and K^+ channels are activated by depolarization (voltage-gated), some region of the channel protein is expected to act as a voltage sensor. The segment S4 has several positively charged amino acids (arginine or lysine; Fig. 2.20A) and is conserved in all the voltage-gated channel proteins. The S4 segment was then suggested to be the voltage sensor (Numa 1987). This suggestion was strongly supported by *site-directed mutagenesis*. A genetic engineering technique now permits specific individual amino acid residues to be altered in a given protein. This can be achieved by deleting, inserting or altering a particular set of three nucleotides (codon that specifies an amino acid) in the cloned cDNA. The altered cDNA is an artificial (site-directed) mutant gene. After transcription of the altered cDNA (making mRNA from the cDNA) and translation (making a protein from the mRNA), a mutant protein is produced. When the positively charged amino acid residues in the S4 segment of the Na^+ channel α-subunit were replaced by neutral or negatively charged residues, the voltage-dependence of channel activation was altered (Stümer *et al.* 1989). It is likely that the S4 segment in the Ca^{2+} or K^+ channel has a similar function (Numa 1989, 1991).

Numa (1987) proposed that the segment S2 in each of

the four homologous domains forms the wall of the ionic pore of the Na^+ channel (Fig. 2.21C). In K^+ channels, however, a short stretch of amino acids between segments S5 and S6 (SS1–2 in Fig. 2.21D) has been suggested to form the ionic pore (Yellen *et al.* 1991; Yool and Schwarz 1991). This stretch (a hairpin loop with 8–10 amino acids for each limb; Fig. 2.21D) may form β-strands, because it is too short to span the membrane with an α-helical structure. The presence of similar short transmembrane segments is also found in the Na^+ channel (Guy and Seetharamulu 1986; Guy and Conti 1990). The equivalent positions in Ca^{2+} channel are indicated by SS1 and SS2 in Fig. 2.20B. Heinemann *et al.* (1992) showed that the replacement of lysine or alanine in the SS2 segment of the sodium channel by glutamic acid converts its ion-selectivity for Na^+ to the Ca^{2+}-selective form. Since glutamic acid is present in the equivalent postion in the Ca^{2+} channel, these results imply that the SS1–SS2 region forms the pore wall in both Na^+ and Ca^{2+} channels, as in the K^+ channel. A recently cloned family of K^+ channels (ATP-regulated, inward rectifier, and muscarinic K^+ channels) are characterized by a structure with only two transmembrane regions (M_1 and M_2 corresponding to S_5 and S_6) and a putative pore region (SS1–SS2 region; Ho *et al.* 1993; Kubo *et al.* 1993a,b). Thus, the basic molecular motif of ionic channels illustrated in Fig. 2.21C,D cannot be generalized.

We have seen that Purkinje cells in the mammalian cerebellum express the P-channel which is resistant to DHP and ω-CgTx but blocked by FTX (Llinás *et al.* 1989). Therefore, specific mRNAs encoding the P-channel must be abundant in the cerebellum or in the mammalian central nervous system. If, then, mRNAs are extracted from these tissues and transferred into an appropriate expression system, the P-channel protein would be newly synthesized (translated) by the exogenous mRNAs. Gurdon *et al.* (1971) were the first to show how well the amphibian oocyte serves as an expression system, demonstrating the synthesis of haemoglobin by *Xenopus* oocytes injected with mRNA purified from rabbit reticulocytes. This technique is now widely used for functional expression of ionic channels and transmitter receptors (Sumikawa *et al.* 1986; Snutch 1988).

When mRNAs extracted from rat brain were injected into *Xenopus* oocytes, the Ca^{2+} currents expressed in the oocytes were activated by a relatively large depolarization (high threshold); these Ca^{2+} currents were slowly inactivated and resistant to both DHP and ω-CgTx (Leonard *et al.* 1987) but blocked by FTX (Lin *et al.* 1990a). Mori *et al.* (1991) cloned the cDNA encoding a brain Ca^{2+} channel (α_1-subunit) that has a high threshold for activation and is resistant to both DHP and ω-CgTx. The current produced by this brain Ca^{2+} channel, termed *brain I* (BI), differs from the P-channel current in that the BI current is rapidly inactivated and resitant to FTX. The BI-channel, however, was expressed predominantly in the cerebellum and blocked by crude venom from the funnel-web spider (Mori *et al.* 1991). Therefore, it is likely that the BI-channel is a member of the P-channel family. The BI-channel α_1-subunit consists of about 2 300 amino acids and again shows four repeating domains (I–IV), each of which has the same transmembrane topology (S1–S6) as the DHP-sensitive Ca^{2+} channel.

The ω-CgTx-sensitive N-type Ca^{2+} channel α_1-subunit was also cloned (Williams *et al.* 1992). This subunit again consists of about 2 300 amino acids and has the same transmembrane topology as described for the DHP-sensitive Ca^{2+} channel. The amino acid sequence of the α_1-subunit of the N-type channel was 64 per cent identical to that of the BI-channel. At present, no clue has as yet been obtained about the structure–function correlation in the three cloned Ca^{2+} channels.

What might be the functional significance of other subunits of the Ca^{2+} channel? The α_2- and β-subunits markedly enhance the functional expression of the α_1-subunit for all the L-type (Mikami *et al.* 1989), BI-type (Mori *et al.* 1991), and N-type Ca^{2+} channels (Williams *et al.* 1992). The γ-subunit affects the voltage-dependence of inactivation of the expressed Ca^{2+} channel (Singer *et al.* 1991). The α_2- and δ-subunits are now known to be derived from the same gene (De Jongh *et al.* 1990); the gene product is then cleaved into the two subunits which are linked by a disulphide bond (Fig. 2.18). At this stage, it is premature to suggest how other subunits modulate the function of the principal subunit, α_1. However, the full complement of subunits seems to be essential for the physiological function of every Ca^{2+} channel (Catterall 1991). The α_1- and β-subunits are each encoded by several distinct genes. The heterogeneity of these subunits may account for the diversity of Ca^{2+} channels (Williams *et al.* 1992). We have seen that multiple types of Ca^{2+} channel coexist in a given nerve terminal (Fig. 2.15A). Might any type of Ca^{2+} channel contribute to transmitter release? Let us now examine the types of Ca^{2+} channel that are involved in transmitter release at different synapses and see whether there is any rule governing the relationship between the Ca^{2+} channel type and transmitter release.

2.2.3 Calcium channels involved in transmitter release

The presynaptic terminal of the squid giant synapse seems to possess only the P-type of Ca^{2+} channel. Since both the presynaptic Ca^{2+} current and synaptic transmission are completely blocked by FTX, the P-channel must be responsible for transmitter release at the giant synapse (Llinás *et al.* 1989). The toxin, ω-Aga-IVA reduces markedly (> 50 per cent) the voltage-dependent Ca^{2+} influx in *synaptosomes* (isolated presynaptic terminals) prepared from the rat brain, whereas ω-CgTx has no effect (Mintz *et al.* 1992). This suggests that the ω-Aga-IVA-sensitive channel is involved in transmitter release at central synapses. However, synaptosomes prepared from chick brain showed ω-CgTx-sensitive and ω-Aga-IVA-resistant Ca^{2+} influx (Mintz *et al.* 1992). Clearly, there is a species difference in the nature of Ca^{2+} channels in the central nervous system. Takahashi and Momiyama (1993) examined the effect of different Ca^{2+} channel blockers on synaptic currents recorded from rat central neurones. These synaptic currents were markedly suppressed by ω-Aga-IVA and to a less extent by ω-CgTx, whereas the L-type channel blocker (nicardipine) had no effect. The functional significance of the involvement of two types of Ca^{2+} channel in transmitter release at a given synapse is not clear. As noted previously, the P-channel is blocked selectively by low doses of ω-Aga-IVA, whereas high doses of this toxin can also block another type of Ca^{2+} channel, termed Q-channel (Wheeler *et al.* 1994). Whether the ω-Aga-IVA sensitive channel responsible for transmitter release at central synapses is the P- or Q-channel remains to be identified.

Might the release of ACh at neuromuscular junctions require a particular type of Ca^{2+} channel? The first clue came from ω-CgTx, which blocks neuromuscular transmission in the frog (Kerr and Yoshikami 1984). Oddly, this toxin did not block the mouse neuromuscular junction (Anderson and Harvey 1987; Sano *et al.* 1987; Yoshikami *et al.* 1989), again suggesting species differences of the Ca^{2+} channel involved in transmitter release. Uchitel *et al.* (1992) showed that neuromuscular transmission in the mouse is resistant to ω-CgTx but blocked by FTX. It is likely that ACh release from mammalian motor nerve terminals is mediated by the P-channel.

The involvement of another type of Ca^{2+} channel in transmitter release at mammalian neuromuscular junctions is suggested by studies involving *Lambert–Eaton myasthenic syndrome* (LEMS), an autoimmune disease (Vincent *et al.* 1989). The muscular weakness results from a decrease in ACh release (Lambert and Elmqvist 1971). Mice injected with immunoglobulin G (IgG) prepared from patients with LEMS develop an LEMS-like disorder (passive transfer; Lang *et al.* 1983), and their immunoreactivity to this LEMS IgG is distinctly localized at the *active zones* of motor nerve terminals (Fukuoka *et al.* 1987b). These mice also exhibit a significant reduction in the number of the large membrane particles associated with active zones (Fukuoka *et al.* 1987a). Under the general assumption that active zones are the sites of transmitter release (Couteaux and Pecot-

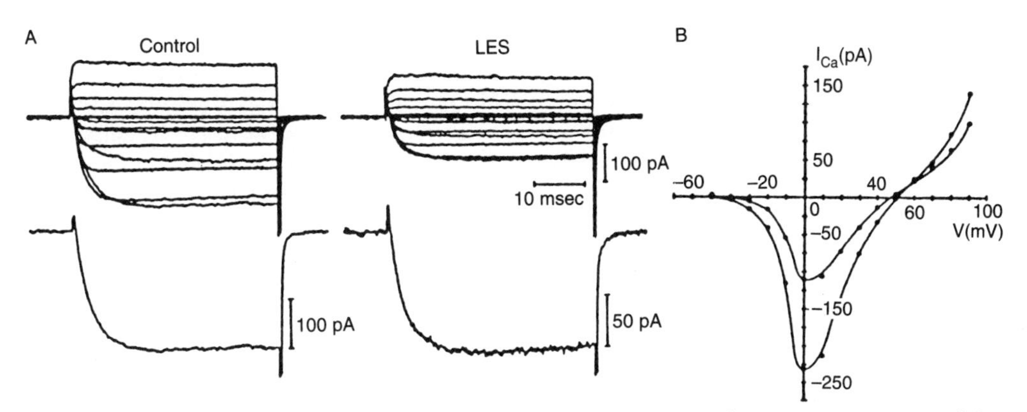

Fig. 2.22. Ca^{2+} currents recorded from bovine adrenal chromaffin cells treated for 24 hours with IgG prepared from normal human subjects (Control) and from patients with Lambert–Eaton myasthenic syndrome (LES). A, upper traces, whole-cell Ca^{2+} currents evoked by depolarization from −80 mV to levels ranging from −60 mV to +90 mV. Lower traces, Ca^{2+} currents recorded at 0 mV. B, current–voltage relationships of Ca^{2+} currents (open circles, control; filled circles, LES). (From Kim and Neher 1988.)

Dechavassine 1970; Heuser *et al.* 1979) and that the large membrane particles reflect voltage-gated Ca^{2+} channels (Pumplin *et al.* 1981), it appears that the LEMS IgG specifically binds Ca^{2+} channels involved in transmitter release. The obvious question is: what type of Ca^{2+} channel are they?

Kim and Neher (1988) applied the LEMS IgG to bovine adrenal chromaffin cells. As shown in Fig. 2.22, the antibodies reduced the magnitude of depolarization-induced Ca^{2+} currents. This change was not associated with any alteration of the kinetics of the Ca^{2+} channel. Presumably, the LEMS IgG blocks the voltage-gated Ca^{2+} channels in an all-or-nothing fashion, leaving fewer functional Ca^{2+} channels. The Ca^{2+} currents blocked by the antibodies were characterized as high-threshold and long-lasting, thus resembling the L-type channel (Fig. 2.22). Might this Ca^{2+} channel blocked by the LEMS IgG represent the Ca^{2+} channel involved in transmitter release from the mouse motor nerve terminal? This possibility is unlikely. Recall the standard Ca^{2+} current recorded from bovine chromaffin cells (Figs 2.16 and 2.17). In fact, it is the standard Ca^{2+} current that is blocked by the LEMS IgG (see I–V curves in Figs 2.16 and 2.22). Unlike neuromuscular transmission in the mouse, the standard Ca^{2+} current is blocked by ω-CgTx (Artalejo *et al.* 1992a; section 2.2.1). How, then, does the LEMS IgG block both neuromuscular transmission in the mouse and the standard Ca^{2+} current? Presumably, it is due to non-specific cross-reaction of the antibody. Alternatively, the antigen to which the LEMS antibody binds may not necessarily be the Ca^{2+} channel protein itself at the neuromuscular junction (see section 3.1.2).

Let us now turn to other examples in the peripheral nervous system. Sympathetic neurones feature N-and L-type Ca^{2+} channels, but display virtually no T-type channel (Hirning *et al.* 1988; Tsien *et al.* 1988). The release of norepinephrine when cultured sympathetic neurones are depolarized with K^+-rich solutions is blocked by ω-CgTx but not by DHP-antagonist (Hirning *et al.* 1988). In sympathetic neurones, then, the N-type Ca^{2+} channel appears to be responsible for transmitter release. Interestingly, the DHP-antagonist substantially reduces the voltage-gated Ca^{2+} influx in sympathetic neurones, despite its failure to block the norepinephrine release (Hirning *et al.* 1988). Might the transmitter in sympathetic neurones discriminate the Ca^{2+} channel type for its release? This question invites the investigation of neuropeptides.

In a variety of neurones, many peptides coexist with a classical transmitter such as acetylcholine or norepinephrine (Hökfelt *et al.* 1986; Hökfelt 1991). Neuropeptides are present only in large (about 70–130 nm in diameter), dense-core synaptic vesicles, whereas small vesicles (about 50 nm in diameter; clear or dense-core) contain only classical transmitters (Lundberg and Hökfelt 1983; Hökfelt *et al.* 1986). Sympathetic nerve terminals of the rat, for example, have norepinephrine in both the small and large vesicles, whereas neuropeptide Y is found only in the large vesicles (Fried *et al.* 1985). Similarly, substance P and calcitonin gene-related peptide (CGRP) occur in large vesicles of synaptosomes isolated from the rat spinal cord (Fried *et al.* 1989). CGRP in the frog motor nerve terminal is also confined within large, dense-core vesicles (Matteoli *et al.* 1988). Furthermore, depleting ACh quanta and small, clear synaptic vesicles by α-latrotoxin isolated from the black widow spider venom leaves both CGRP immunoreactivity and large, dense-core vesicles in the motor nerve terminal (Matteoli *et al.* 1988). Therefore, the release of neuropeptides that coexist with a classical transmitter in presynaptic terminals may employ another mechanism than that which releases the classical transmitter.

Does the release of both the classical transmitter and neuropeptides involve activating the same type of Ca^{2+} channel? The evidence is mixed. When dorsal root ganglion cells cultured from neonatal rats are depolarized by K^+-rich solutions, substance P is released, and this release is completely suppressed by DHP-antagonist (Perney *et al.* 1986). Yet, substance P release from dorsal root ganglion cells induced by electrical stimulation is not affected by DHP-antagonists (Rane *et al.* 1987). Furthermore, quantal release of ACh at the rat neuromuscular junction is enhanced by DHP-agonist (BAY K 8644), an increase blocked by DHP-antagonist (nimodipine), whereas the DHP-antagonist has no effect on the e.p.p.s evoked by nerve stimulation (Atchison and O'Leary 1987). These results suggest that although DHP-sensitive Ca^{2+} channels occur in sensory and sympathetic neurones, as well as in the rat motor nerve terminal, they do not normally participate in the release of peptides or the classical transmitter.

To account for this behaviour, Miller (1987) proposed the hypothesis illustrated in Fig. 2.23. In this scheme, DHP-sensitive Ca^{2+} channels are distributed in the presynaptic terminal membranes distant from the sites of transmitter release, whereas the DHP-insensitive Ca^{2+} channels normally involved in transmitter release are preferentially located near the release sites. If, then, enough Ca^{2+} ions enter through the remote DHP-

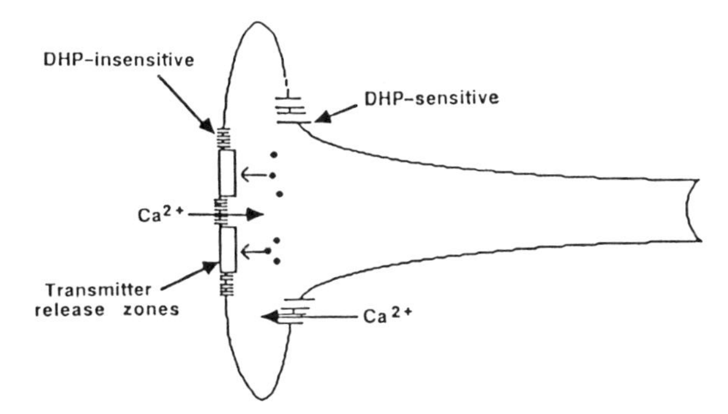

Fig. 2.23. Hypothetical distributions of dihydropyridine (DHP)-sensitive and DHP-insensitive Ca^{2+} channels in a presynaptic terminal. DHP-insensitive channels may cluster near the transmitter release sites. DHP-sensitive channels may preferentially distribute in regions distant from (but not necessarily opposite to) the release sites. (From Miller 1987.)

sensitive channels following either the application of DHP-agonist (BAY K 8644) or K^+-induced prolonged depolarization, the transmitter or peptides may be released by the spillover effect of the excessive Ca^{2+} (Miller 1987). Under normal conditions, on the other hand, the rise of the intracellular Ca^{2+} is restricted to 'microdomains' close to the Ca^{2+} channels (Augustine and Neher 1992a; Llinás *et al.* 1992), so that transmitter release would be triggered only by Ca^{2+} entering through the DHP-insensitive channels clustered near release sites (active zones).

Miller's hypothesis is intriguing but does not explain why the release of substance P induced by K^+-rich solutions can be blocked by DHP-antagonists, but norepinephrine release under similar conditions is not blocked by the DHP antagonist (Perney *et al.* 1986). Hirning *et al.* (1988) suggest that a Ca^{2+} influx through DHP-sensitive L-type channels may favour the exocytosis of large, dense-core vesicles, whereas Ca^{2+} entry through DHP-insensitive channels may lead to exocytosis of small synaptic vesicles. How would these two types of synaptic vesicle disinguish the source of a Ca^{2+}-influx? Clear synaptic vesicles containing classical transmitters accumulate near the active zones thought to harbour voltage-gated Ca^{2+} channel particles (Pumplin *et al.* 1981). In contrast, large, dense-core vesicles are widely dispersed within the presynaptic terminal and are released at ectopic sites outside the active zones (Zhu *et al.* 1986). Thus, if DHP-insensitive Ca^{2+} channels are preferentially located near active zones, these channels would favour the release of clear vesicles. Consistent with this view, Ca^{2+} channels labelled with fluorescently tagged ω-CgTx in the frog motor nerve terminal were found to cluster exclusively at discrete sites corresponding in distribution to active zones (Robitaille *et al.* 1990; Cohen *et al.* 1991; Torri Tarelli *et al.* 1991). However, in retinal bipolar cells, DHP-sensitive L-type channels are responsible for the release of classical transmitters. Large terminals of bipolar cells possess ribbon-type release sites, each with many small clear vesicles which presumably contain glutamate as the transmitter (Tachibana and Okada 1991). The bipolar cell terminal shows DHP-sensitive Ca^{2+} currents, and the synaptic response recorded from a horizontal cell in response to depolarization of a bipolar cell is blocked by DHPs, but not by ω-CgTx (Tachibana *et al.* 1993). It should be noted that bipolar-horizontal synaptic function is continuously regulated by changes in light intensity. The involvement of non-inactivating DHP-sensitive (L-type) Ca^{2+} channels in transmitter release at retinal cells might reflect their functional requirements.

Does the Ca^{2+} dependence of peptide release differ from that for classical transmitters? At neuromuscular junctions, the release of ACh occurs as a fourth power function of the external Ca^{2+} concentration (Dodge and Rahamimoff 1967; see section 3.1.1). In contrast, the Ca^{2+} dependence of CGRP release from nerve terminals in muscle is less than unity (Sakaguchi *et al.* 1991). Why so low? The Ca^{2+} entering the nerve terminal on depolarization may be greatly diluted before reaching the large dense-core vesciles that are dispersed within the terminal. This would account for the relatively low external Ca^{2+} dependence of CGRP release. When the intraterminal Ca^{2+} concentration is measured, exocytosis of large, dense-core vesicles increases also with the fourth power of the cytosolic Ca^{2+} (Thomas *et al.* 1990). Thus, peptide release depends on the average cytosolic Ca^{2+} concentration in the whole presynaptic terminal, whereas the release of classical transmitters depends on the local Ca^{2+} concentration at the active zones. Attaining an intraterminal Ca^{2+} concentration sufficient to release peptides may thus require an accumulation of residual Ca^{2+} achieved by prolonged terminal depolarization or by repetitive nerve stimulation (Lundberg and Hökfelt 1983; Smith and Augustine 1988; De Camilli and Jahn 1990). However, when the intracellular Ca^{2+} concentration of isolated nerve terminals was uniformly increased with ionomycin (Ca^{2+} ionophore), the Ca^{2+} level required for the release was lower for peptide release than for classical transmitters (Verhage *et al.* 1991). In fact, Von Gersdoff and Mathews (1994) found that transmitter release from bipolar cell terminals of the goldfish retina requires an increase of the intraterminal Ca^{2+} level to about 50 μM (see also Adler *et al.* 1991). This is about 50-fold higher than the Ca^{2+} level (1.4 μM) sufficient for the exocytosis of dense-core vesicles in chromaffin cells (Augustine and Neher 1992b). This suggests that the release of classical transmitters at the active zone is triggered by activation of a low-affinity Ca^{2+} binding site, whereas a high-affinity Ca^{2+} binding site is responsible for exocytosis of large, dense-core vesicles. How would Ca^{2+} trigger transmitter release? This is the subject of the next chapter.

Summary and prospects

Ca^{2+} is an essential co-factor for transmitter release. In the absence of external Ca^{2+}, a presynaptic impulse reaches the nerve terminal but does not release the transmitter. Intracellular injections of Ca^{2+} into a presynaptic terminal enhances the release of individual transmitter quanta. It is thus the entry of Ca^{2+} into the nerve terminal that is required for transmitter release. This Ca^{2+} entry is triggered by depolarization due to the action potential reaching the terminal. It is the Ca^{2+} channel embedded in the cell membrane that opens a gate selectively for Ca^{2+} when the cell is depolarized. When the channel opens by depolarization, Ca^{2+} enters the cell according to its electrochemical gradient between the external and internal solutions. When the depolarization is very large, the entry of positively charged Ca^{2+} is obstructed by electrical repulsion. As a consequence, Ca^{2+} enters the nerve terminal during the falling phase of the action potential. The Ca^{2+} hypothesis for transmitter release seems fully applicable to every chemical synapse.

Ca^{2+} channels are not homogeneous. There are at least four basic types of Ca^{2+} channel: T–, L–, N–, and P-types. The T-type Ca^{2+} channel requires relatively small depolarizations for its activation (low threshold), whereas the three other channels have a high threshold for activation. The Ca^{2+} currents induced by activation of the three high-threshold channels can be distingushied by their biophysical and pharmacological properties. Pharmacologically, the L-channel is blocked by DHP-antagonists; the N-channel is blocked by ω-CgTx, and the P-channel by FTX or ω-Aga-IVA.

The Ca^{2+} channel is formed by multiple subunit molecules. The molecular cloning elucidated the amino acid sequence of the principal α_1-subunit for different types of Ca^{2+} channel. The molecular motif of the principal subunit of Ca^{2+} channels is characterized by four homologous domains, each of which has six transmembrane segments. Since this basic motif is conserved in Ca^{2+}, Na^+, and K^+ channels, all these voltage-gated channel molecules must be members of the same gene superfamily. By analogy with Na^+-channels, the possible sites of the voltage senosor and the ion pore lining can be located in the Ca^{2+} channel molecule. Yet, how the functional channel properties of the three high threshold Ca^{2+} channels are specified in terms of the molecular structure still remains elusive.

The classical transmitters, such as acetylcholine or norepinephrine, often coexist with neuropeptides in a given nerve terminal. The classical transmitter is contained in small, clear synaptic vesicles, whereas peptides occur exclusively in large, dense-core vesicles. Both the classical transmitters and peptides require Ca^{2+} for their release, but the Ca^{2+} dependence and the type of Ca^{2+} channel involved in the release are different between classical transmitters and peptides. These differences are presumably accounted for by the preferential localization of a particular type of Ca^{2+} channel near the active zones where small, clear synaptic vesicles are clustered. The unequivocal test of this view requires precise information about the intraterminal concentration of Ca^{2+} and its spatial and temporal distributions in the terminal following an action potential.

Suggested reading

Calcium influx and calcium currents in nerve terminals

Augustine, G. J., Charlton, M. P., and Smith, S. J. (1985). Calcium entry into voltage-clamped presynaptic terminals of squid. *Journal of Physiology (London)*, **367**, 143–62.

Katz, B. (1969). *The release of neural transmitter substances*. Charles C. Thomas, Springfield.

Llinás, R., Steinberg, I. Z., and Walton, K. (1981). Presynaptic calcium currents in squid giant synapse. *Biophysical Journal*, **33**, 289–322.

Llinás, R., Sugimori, M., and Simon, S. M. (1982). Transmission by presynaptic spike-like depolarization in the squid giant synapse. *Proceedings of the National Academy of Sciences of the USA*, **79**, 2415–19.

Miledi, R. (1973). Transmitter release induced by injection of calcium ions into nerve terminals. *Proceedings of the Royal Society B*, **209**, 447–52.

Yawo, H. and Momiyama, A. (1993). Re-evaluation of calcium currents in pre-and postsynaptic neurones of the chick ciliary ganglion. *Journal of Physiology (London)*, **460**, 153–72.

Classification of calcium channels

Aosaki, T. and Kasai, H. (1989). Characterization of two kinds of high-voltage-activated Ca-channel currents in chick sensory neurones. Differential sensitivity to dihydropyridines and ω-conotoxin GVIA. *Pflügers Archiv*, 414, 150–6.

Llinás, R., Sugimor, M., Lin, J. W., and Cherksey, B. (1989). Blocking and isolation of a calcium channel from neurones in mammalian and cephalopods utilizing a toxin fraction (FTX) from funnel-web spider poison. *Proceedings of the National Academy of Sciences in the USA*, **86**, 1689–93.

Mintz, I. M., Adams, M. E., and T. D., Bean (1992). P-type calcium channels in rat central and peripheral neurones. Neuron, **9**, 85–95.

Regan, L. J., Sah, D. W. Y., and Bean, B. P. (1991). Ca^{2+} channels in rat central and peripheral neurones: high-threshold current resistant to dihydropyridine blockers and ω-conotoxin. *Neuron*, **6**, 269–80.

Tsien, R. W., Lipscombe, D., Madison, D. V., Bley, K. R., and Fox, A. P. (1988). Multiple types of neuronal calcium channels and their selective modulation. *Trends in Neurosciences*, **11**, 431–8.

The molecular structure of calcium channels

Catterall, W. A. (1988). Structure and function of voltage-sensitive ion channels. *Science*, **242**, 50–61.

Mori, Y., Friedrich, T., Kim, M. S., Mikami, A., Nakai, J., Ruth, P., Bosse, E., Hofmann, F., Flockerzi, V., Furuichi, T., Mikoshiba, K., Imoto, K., Tanabe, T., and Numa, S. (1991). Primary structure and functional expression from complementary DNA of a brain calcium channel. *Nature*, **350**, 398–402.

Snutch, T. P. and Reiner, P. B. (1992). Ca^{2+} channels: diversity of form and function. *Current Opinion in Neurobiology*, **2**, 247–53.

Tanabe, T., Takeshima, H., Mikami, A., Flockerzi, V., Takahashi, H., Kangawa, K., Kojima, M., Matsuo, H., Hirose, T., and Numa, S. (1987). Primary structure of the receptor for calcium channel blockers from skeletal muscle. *Nature*, **328**, 313–18.

Tanabe, T., Beam, K. G., Powell, J. A., and Numa, S. (1988). Restoration of excitation-contraction coupling and slow calcium current in dysgenic muscle by dihydropyridine receptor complementary DNA. *Nature*, **336**, 134–9.

Tsien, R. W., Ellinor, P. T., and Horne, W. A. (1991). Molecular diversity of voltage-dependent Ca^{2+} channels. *Trends in Pharmacological Sciences*, **12**, 349–54.

Williams, M. E., Brust, P. F., Feldman, D. H. Patthi, S., Simerson, S., Maroufi, A., McCue, A. F., Velicelebi, G., Ellis, S. B. and Harpold, M. M. (1992). Structure and functional expression of an ω–conotoxin-sensitive human N-type calcium channel. *Science*, **257**, 389–95.

Calcium channels involved in transmitter release

Hirning, L. D., Fox, A. P., McCleskey, E. W., Olivera, B. M., Thayer, S. A., Miller, R. J., and Tsien, R. W. (1988). Dominant role of N-type Ca^{2+} channels in evoked release of norepinephrine from sympathetic neurones. *Science*. **239**, 57–61.

Kim, Y. I. and Neher, E. (1988). IgG from patients with Lambert-Eaton syndrome blocks voltage-dependent calcium channels. *Science*, **239**, 405–8.

Llinás, R., Sugimori, M., and Silver, R. B. (1992). Microdomains of high calcium concentration in a presynaptic terminal. *Science*, **256**, 677–9.

Lundberg, J. M. and Hökfelt, T. (1983). Coexistence of peptides and classical neurotransmitters. *Trends in Neurosciences*, **6**, 352–33.

Matteoli, M., Haimann, C., Torri-Tarelli, F., Polas, J. M., Ceccarelli, B., and De Camilli, P. (1988). Differential effect of α–latrotoxin on exocytosis from small synaptic vesicles and from large dense-core vesicles containing calcitonin gene-related peptide at the frog neuromuscular junction. *Proceedings of the National Academy of Sciences of the USA*, **85**, 7366–70.

Miller, R. J. (1987). Multiple calcium channels and neuronal function. *Science*, **235**, 46–52.

Robitaille, R., Adler, E. M., and Charlton, M. P. (1990). Strategic location of calcium channels at transmitter release sites of frog neuromuscular synapses. *Neuron*, **5**, 773–9.

Takahashi, T. and Momiyama, A. (1993). Multiple types of calcium channel mediate central synaptic transmission. *Nature*, **366**, 156–8.

Wheeler, D. B., Randall, A., and Tsien, R. W. (1994). Roles of N-type and Q-type Ca^{2+} channels in supporting hippocampal synaptic transmission. *Science*, **264**, 107–11.

3

How transmitter release is triggered

More transmitter is released as more Ca^{2+} is added to the external solution, but the relation between the two parameters is not linear. Dodge and Rahamimoff (1967) have shown that the increase in amplitude of end-plate potentials (e.p.p.s) at the frog neuromuscular junction is approximately a fourth power function of the extracellular Ca^{2+} concentration. Similarly, the amount of transmitter release at the squid giant synapse is related to the third power of the external Ca^{2+} concentration (Augustine and Charlton 1986). On the other hand, the magnitude of Ca^{2+} currents recorded from the squid presynaptic terminal is roughly proportional to the external Ca^{2+} level (Augustine and Charlton 1986). Evidently, the high-order Ca^{2+} dependence concerns the process of transmitter release rather than the process of Ca^{2+} entry in the presynaptic membrane. To Dodge and Rahamimoff (1967) the steep Ca^{2+} dependence suggested a cooperative action, with three or four Ca^{2+} ions jointly triggering the release of one quantum of the transmitter. Thus, if transmitter quanta originate from individual synaptic vesicles, three or four Ca^{2+} ions may act in concert to induce exocytosis of one synaptic vesicle.

Exocytosis of the content of a synaptic vesicle requires the vesicular membrane and the presynaptic terminal membrane to contact, then fuse. We have seen that small, clear synaptic vesicles containing the classical transmitter are clustered at the active zones in presynaptic terminals. How would these synaptic vesicles fuse with the nerve terminal membrane following Ca^{2+} influx?

3.1 FUSION OF SYNAPTIC VESICLES WITH THE NERVE TERMINAL MEMBRANE

Stanley and Ehrenstein (1985) postulate that the vesicular membrane possesses chloride (Cl^-) channels and Ca^{2+}-activated K^+ channels. The Ca^{2+} entry triggered by an action potential in the presynaptic terminal would activate the K^+ channels in the vesicular membrane, causing an influx of K^+ and Cl^- ions into the vesicle. Water would also enter because of increased salt concentration in the vesicle, and the resultant osmotic swelling could act as a driving force for fusion between the vesicular and terminal membranes (Stanley and Ehrenstein 1985).

Calcium-activated K^+ channels, originally discovered in nerve cells by Meech and Strumwasser (1970), are now known to exist in the membranes of a variety of excitable and nonexcitable cells (Meech 1978; Schwarz and Passow 1983). In cultured muscle cells, about a third power relation was found between the intracellular Ca^{2+} concentration and the probability that Ca^{2+}-activated K^+ channels are open (Barrett *et al.* 1982). This steep dependence of the opening of the ionic channels on the Ca^{2+} concentration is consistent with the above hypothesis that Ca^{2+}-activated K^+ channels are involved in vesicular exocytosis. In the squid giant synapse, however, applying blockers of Ca^{2+}-activated K^+ channels to the presynaptic terminal did not affect transmitter release (Augustine *et al.* 1988). Thus, membrane fusion between synaptic vesicles and the presynaptic terminal seems to require some other mechanisms.

3.1.1 How synaptic vesicles become releasable

Figure 3.1A is a hypothetical scheme originally proposed by Katz (1962) to explain how the transmitter molecules in a vesicle may escape into the synaptic cleft across barriers formed by the vesicular and terminal membranes. It is assumed that both membranes have special reactive sites and that when the two reactive sites touch one another in a critical collision (docking), the colliding membranes become leaky enough to release the vesicular contents.

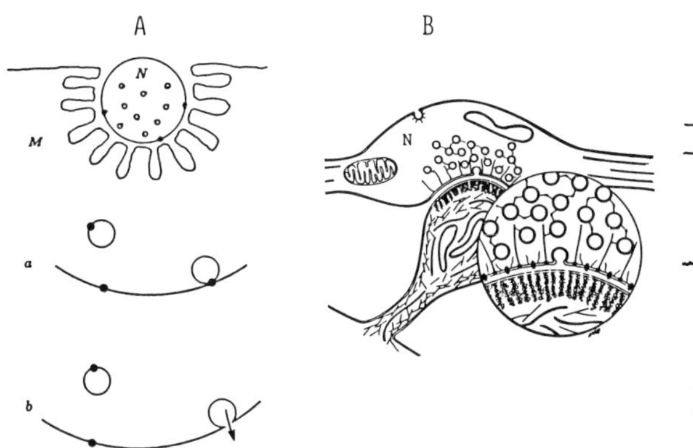

Fig. 3.1. Synaptic vesicles in presynaptic terminals. A, exocytosis in the vesicular hypothesis. *N*, a motor nerve terminal containing vesicles. *M*, a muscle fibre with junctional folds. Dots indicate reactive sites on the vesicular surface and in the terminal membrane. *a*, the two reactive sites meet; *b*, exocytosis occurs at arrow. (After Katz 1962.) B, schematic diagram of filamentous proteins. N, presynaptic terminal on the the postsynaptic spine. Short filaments (presumably synapsin I) are attached to synaptic vesicles, and larger filaments extend from the terminal membrane and are contacted by synapsin I-like filaments. The release site including exocytosis of a synaptic vesicle is enlarged in a circle. (From Landis *et al.* 1988.)

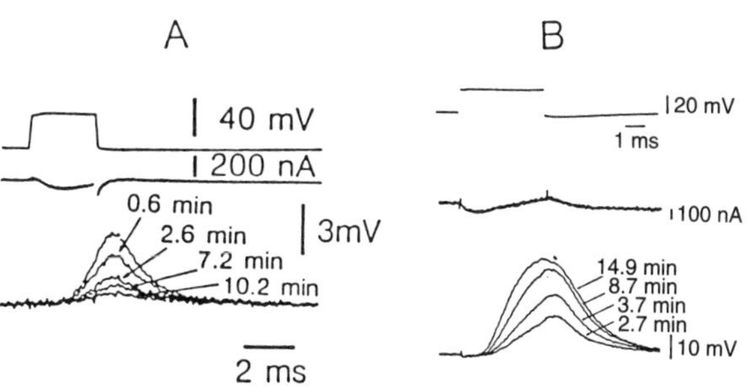

Fig. 3.2. Effects of dephosphorylated synapsin I (A) and calmodulin kinase II (B) on transmitter release at the squid giant synapse. Synaptic potentials (bottom traces) were evoked by constant depolarizing pulses (top traces) applied to the presynaptic terminal. Middle traces, membrane currents recorded from the presynaptic terminal in response to presynaptic depolarizing pulses. Time (in minutes) after the intracellular injection is indicated in the bottom traces. (From Llinás *et al.* 1985.)

This hypothesis invites several questions: How do synaptic vesicles collide with the terminal membrane? Can synaptic vesicles indeed move freely in the cytoplasm? The ultrastructure of a meshwork of cytoskeletal filaments intermingled with synaptic vesicles has been shown by Landis *et al.* (1988) with the freeze-etch technique. Figure 3.1B diagrams this structure, which was observed in the vicinity of the active zone at excitatory synapses between parallel fibres and Purkinje cell spines in the mouse cerebellum. At least two sets of filaments can be seen: 'lollipop'-shaped filaments extending radially from the surface of synaptic vesicles, and larger filaments anchoring in the plasma membrane of the presynaptic terminal. Individual synaptic vesicles thus appear to be immobilized by linking to one another with 'lollipop'-shaped smaller filaments and by anchoring at the terminal membrane with larger filaments (Fig. 3.7A). Similarities in shape and size suggested that the 'lollipop'-shaped filaments are *synapsin I* (Landis *et al.* 1988), a vesicle-associated protein. Indeed, immunocytochemistry with anti-synapsin I antibodies subsequently confirmed this suggestion both in motor nerve terminals and in nerve terminals of the cerebellum (Hirokawa *et al.* 1989). Let us look closer at the possible role of synapsin I.

Synapsin I occurs only in nerve cells, and it is associated with the cytoplasmic surface of small synaptic vesicles but not in large, dense-core vesicles (Navone *et al.* 1984). Its binding affinity with the small synaptic vesicles decreases when synapsin I is phosphorylated (De Camilli and Greengard 1986). *Phosphorylation* is a mechanism for switching the state or conformation of a protein, thereby regulating or modulating biological function operated by the protein (Box C). Synapsin I is phosphorylated by Ca^{2+} via *calmodulin-dependent kinase II* (CaM kinase II; De Camilli and Greengard 1986; Trimble and Scheller 1988; Burgoyne 1990). How synapsin I affects transmitter release was examined at the presynaptic terminal of the squid giant synapse (Llinás *et al.* 1985). Injecting phosphorylated synapsin I (phosphosynapsin) into the terminal left synaptic transmission unaffected. Injecting dephosphorylated synapsin I (dephosphosynapsin), however, progressively reduced synaptic transmission, as shown in Fig. 3.2A. The decreased postsynaptic response to constant depolarizing pulses applied to the presynaptic terminal occurred even though the depolarization-induced Ca^{2+} current in the presynaptic terminal was unchanged. Therefore, this synaptic depression must be due to a decrease in the amount of transmitter released by a given level of Ca^{2+} influx. When CaM kinase II was injected into the terminal, the postsynaptic responses to constant presynaptic depolarizing pulses increased markedly (Fig. 3.2B). Again, this change occurred without affecting Ca^{2+} currents in the terminal. These results suggest

Box C Alteration of biological response by protein phosphorylation

The covalent transfer of phosphate (PO_4^-) from adenosine triphosphate (ATP) to a protein is called *protein phosphorylation*. This reaction is catalysed by a specific protein kinase, and removal of the phosphate, *dephosphorylation*, is catalysed by another enzyme, *phosphatase*. Thus, we have the reversible reaction:

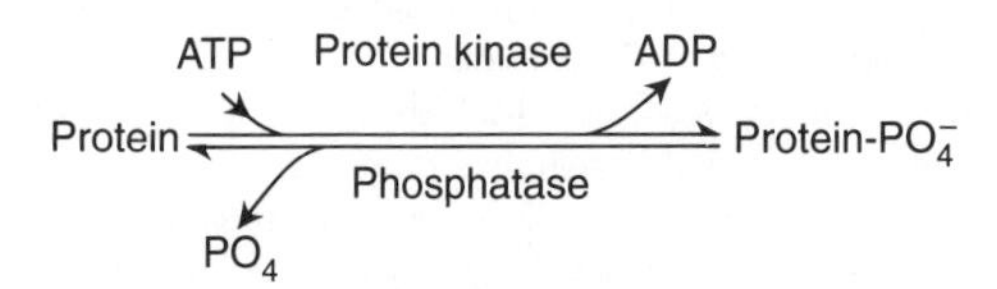

The key issue here is that the protein gains a negative charge at the site of phosphorylation. This develops electric interactions, attraction or repulsion, with pre-existing charges in other regions of the protein. In consequence, the protein undergoes conformational changes. The conformational change of the protein would alter its function. Imagine, for example, the phosphorylation of an ionic channel molecule and the resultant conformational change. The domains of the channel molecule lining the ionic pore might be tilted or displaced, so that the kinetics or the opening rate of the channel could be altered. In fact, we will see later many examples of channel modulation by the phosphorylation. Thus, protein phosphorylation is the molecular basis for regulation and modulation of a wide variety of neurobiological phenomena (Nestler and Greengard 1983).

The conformational changes of a given protein by the phosphorylation occur in a streotypical manner, because each protein has specific sites for phosphorylation. These sites are specific not only for phosphorylation but also for the protein kinase that catalyses the phosphorylation. In synapsin I, for example, it is the serine residues that are phosphorylated by CaM kinase II or by cyclic AMP-dependent kinase A (protein kinase A). However, the serine (Ser) residue phosphorylated by protein kinase A is located in a sequence, Arg-Arg-Lue-Ser, at the N-terminus, whereas the Ser residues phophorylated by CaM kinase II reside in a sequence, Arg-Gln-Thr(or Ala)-Ser at the C-terminus (Czernik *et al.* 1987). Thus, each phosphorylation site is located in a consensus sequence specific for the protien kinase involved in the phosphorylation. Therefore, the conformational change or functional alteration of the protein depends on which protein kinase is activated.

---

the following scheme for interactions between synapsin I and synaptic vesicles. At rest, dephosphosynapsin I binds to synaptic vesicles, restraining their mobility; next, CaM kinase II, activated by the Ca^{2+} influx associated with an action potential in the presynaptic terminal, phosphorylates the synapsin I; phosphosynapsin I then dissociates from the vesicle surface as its binding affinity weakens, and the vesicle is released from its restraint (Llinás *et al.* 1985). The 'lollipop'-shaped synapsin I attaches to synaptic vesicles with its C-terminus in the tail region and to large filaments with its N-terminus in the globular head (Fig. 3.7A). The phosphorylation sites for CaM kinase II are located in the C-terminus of synapsin I (Box C). Presumably, conformational changes in the C-terminus of synapsin I by the phosphorylation would reduce its binding affinity to synaptic vesicles.

Hirokawa *et al.* (1989) found that most synaptic vesicles located within 30 nm (less than one vesicle diameter) of the terminal membrane are devoid of synapsin I. They inferred that the mobility of synaptic vesicles in a reservoir pool, distant from the terminal membrane, is restrained by synapsin I, whereas the vesicles released from synapsin I move to the active zone and become ready for exocytosis (see Fig. 3.7A,B). Electrophysiological studies on the squid giant synapse also suggest that CaM kinase II increases the availability of synaptic vesicles for exocytosis without affecting the mechanism of transmitter release (Lin *et al.* 1990b).

Ca^{2+}-dependent phosphorylation of synapsin I thus appears to regulate the number of synaptic vesicles available for release. But how the transmitter is released from these vesicles remains unclear. In the hypothesis illustrated in Fig. 3.1A, Katz (1962) assumed that exocytosis of the vesicular content occurs as the vesicle fuses with the terminal membrane. Let us first examine the experimental basis for this assumption.

3.1.2 Exocytosis of the vesicular content

If exocytosis were achieved by fusion of synaptic vesicles with the terminal membrane, the surface area of the presynaptic membrane would enlarge as the vesicular membranes incorporated. Electrical correlates of such an increase in the terminal surface area would be an increase in the electrical capacitance of the terminal membrane. In fact, monitoring the membrane capacitance in chromaffin cells with the whole-cell recording technique revealed stepwise, discrete increments in the membrane capacitance on electrical stimulation (Neher and Marty 1982). These fluctuations of the membrane capacitance occurred when the intracellular Ca^{2+} concentration ranged from 100 to 1000 nM but not when the internal Ca^{2+} was buffered to low levels (10 nM). Such discrete, or quantal, changes in the membrane capacitance are consistent with the notion of Ca^{2+}-dependent exocytosis of chromaffin secretory vesicles. Indeed, each step of capacitance change corresponded well to the surface area of a single chromaffin granule. Neher and Marty (1982) also found quantal decrements in the membrane capacitance, implying vesicle retrieval or recycling.

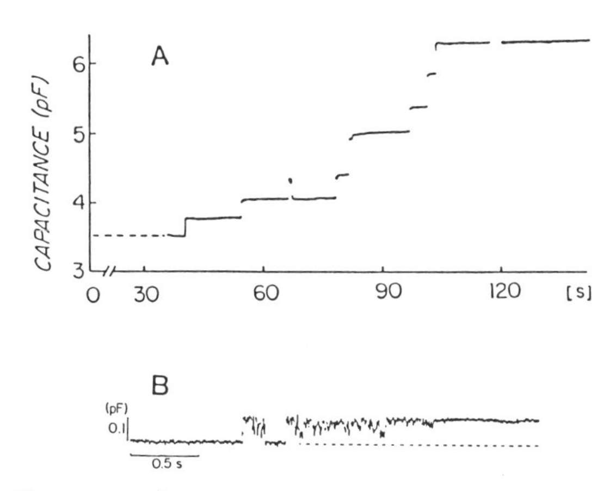

Fig. 3.3. Membrane capacitance changes in a single mast cell of the beige mouse. A, stepwise increments of the membrane capacitance during degranulation induced by applying GTP-γS, a non-hydrolyzable analogue of GTP, into the mast cell; time-scale, seconds. B, flickering of the membrane capacitance between two states. (From Breckenridge and Almers 1987a.)

Figure 3.3A shows stepwise increments of the membrane capacitance observed in a mast cell of the beige mouse in response to an agent that stimulates secretion. Beige mice are useful for this study, because their mast cells contain giant secretory vesicles (Breckenridge and Almers 1987a; Zimmerberg *et al.* 1987). Occasionally, the membrane capacitance flickered rapidly between two levels (Fig. 3.3B). These observations suggest a sequence of events for exocytosis as outlined in Fig. 3.4A. First, the vesicle binds to the cell membrane at specific sites (a and b in Fig. 3.4A), then the vesicle interior connects to the cell exterior when a low resistance channel, the *fusion pore*, forms (Fig. 3.4Ac). Through this electric path of the fusion pore, the vesicle membrane raises the capacitance of the cell membrane. As the fusion pore opens and closes repeatedly in the course of establishing a stable pore (b and c in Fig. 3.4A), the membrane capacitance flickers. Finally, the fusion pore is replaced by a wide opening which leads to exocytosis (Fig. 3.4Ad). Breckenridge and Almers (1987b) also recorded a current transient probably generated through the fusion pore in mast cells of the beige mouse (see below).

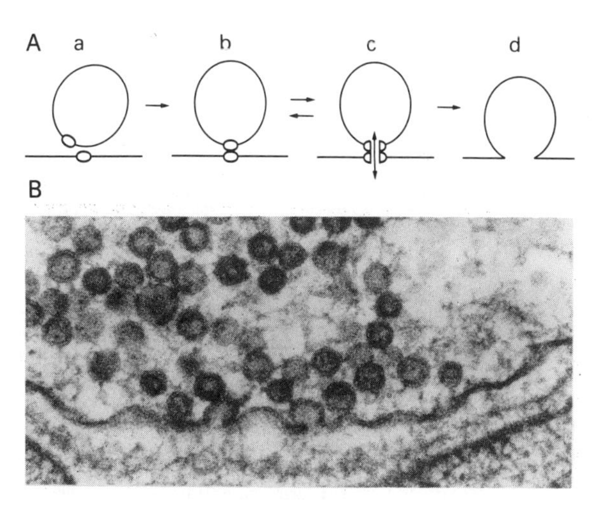

Fig. 3.4. A, hypothetical sequence of events during exocytosis of a secretory vesicle in the mast cell. A fusion pore is formed (c), then replaced by a wide opening (d), resulting in exocytosis. Before the formation of a stable wide opening, the fusion pore repeats the opening (c) and closing (b). (After Breckenridge and Almers 1987a.) B, an electron photomicrograph of a frog motor nerve terminal following nerve stimulation. Two synaptic vesicles communicate freely with the junctional cleft through structures resembling fusion pore or dilatated pore. (From Heuser and Reese 1981.)

Would the capacitance increase be indeed associated with exocytosis of the vesicular content? Chow *et al.* (1992) monitored catecholamines secreted from a bovine chromaffin cell with a *carbon-fibre microelectrode*. The carbon-fibre electrode records electrical signals when catecholamines are electrically oxidized. The magnitude of the electrical signal is proportional to the catecholamine concentration. By placing the electrode in close

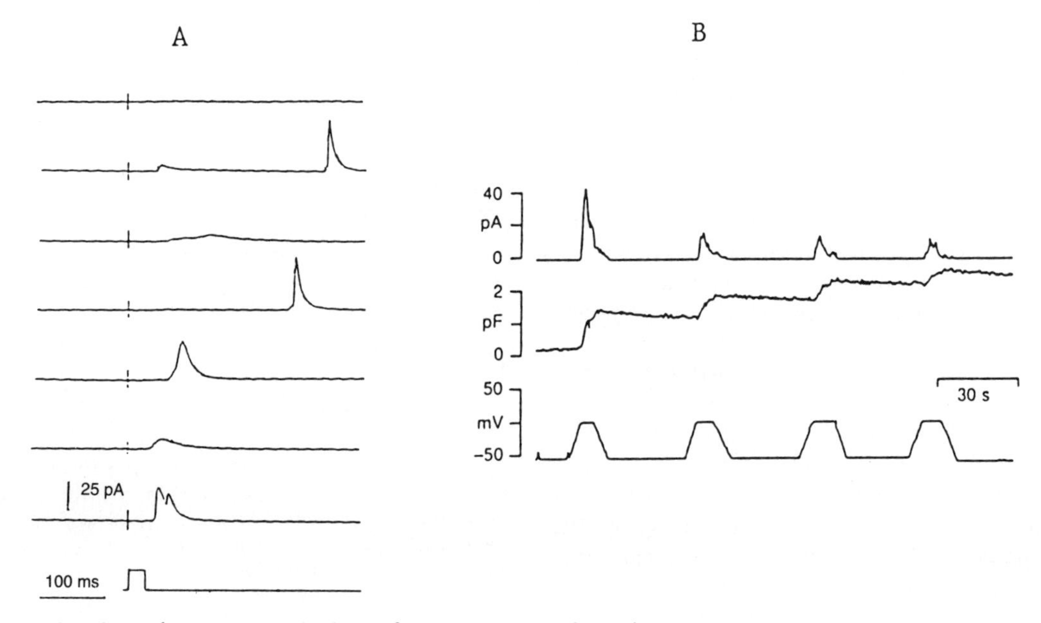

Fig. 3.5. Correlation between quantal release of catecholamines and membrane capacitance increments in bovine chromaffin cells. A, the release of catecholamines monitored with a carbon-fibre electrode in response to a depolarizing pulse applied to a chromaffin cell. The vertical bar on each record represents the onset of the depolarizing pulse. The electrical transient reflecting catecholamine secretion occurs with a latency of about 50 ms. B, simultaneous recordings of electrical transients associated with catecholamine secretion (top) and changes in the membrane capacitance (middle) when the cell was repeatedly depolarized (bottom). (Adapted from Chow *et al.* 1992.)

proximity of a single chromaffin cell, the release of catecholamines can be measured when the cell is depolarized by an electrical pulse (Fig. 3.5A). Occasionally, the depolarizing pulse failed to release catecholamines, and in other trials two or more signals occurred in response to the depolarizing pulse (Fig. 3.5A). This behaviour suggests that catecholamines are released in a quantal fashion. When the electrical signal of the catecholamine release (Fig. 3.5B, top) and changes in the membrane capacitance (middle) were recorded simultaneously in response to repeatedly applied depolarizations (bottom), a clear correlation was observed between the magnitude of the signal and the size of the capacitance change. Therefore, there is little doubt that the increase in the membrane capacitance reflects exocytosis or quantal release of the vesicular content (Chow *et al.* 1992).

The scheme illustrated in Fig. 3.4A is essentially that of Katz (1962; Fig. 3.1A), buttressed by new evidence. Structures resembling fusion pores can be seen in synaptic vesicles at the motor nerve terminal membrane (Fig. 3.4B; Couteaux and Pecot-Dechavassine 1970; Heuser and Reese, 1981). Furthermore, increments of membrane capacitance were observed in association with Ca^{2+} entry in nerve terminals isolated from the mammalian neurohypophysis (Lim *et al.* 1990) and in the terminal membrane of retinal bipolar cells (Von Gersdorff and Mathews 1994). Therefore, the basic mechanism of quantal release of neurotransmitters would be similar to that of exocytosis of granular vesicles in chromaffin or mast cells.

Katz (1962) postulated that both the vesicular and terminal membranes have 'reactive sites' for fusion (Fig. 3.1A). This implies that it is not the membranes *per se* but particular structures in the membranes that contact and fuse. What would be the molecular basis for the structures involved in fusion?

3.1.3 Vesicle-associated fusion molecules

In chromaffin or mast cells, the presence of particular membrane proteins involved in the fusion, *fusion proteins*, was suggested but not proven (Almers 1990). For the exocytosis of synaptic vesicles, on the other hand, a strong candidate of a fusion protein has emerged: *synaptophysin*. The action and locus of synaptophysin provide important insights into the possible mechanism of exocytosis.

Synaptophysin was first isolated from the synaptic vesicle fraction of the brain and called *p38*, because it is a protein with an estimated molecular weight of 38 000 (Jahn *et al.* 1985; Wiedenmann and Franke 1985). The primary structure deduced by the cloning and sequencing of the cDNA encoding synaptophysin shows that this protein contains about 300 amino acids and has four hydrophobic segments (see Box B), presumably transmembrane segments (Buckley *et al.* 1987; Leube *et al.* 1987; Südhof *et al.* 1987). Synaptophysin lacks a hydrophobic signal sequence at its N-terminus (section 2.2.2). Thus, synaptophysin appears to be incorporated into the vesicular membrane, having both its N- and C-termini in the cytoplasm (the exterior of the vesicle; Fig. 3.6A; Trimble and Scheller 1988). Synaptophysin purified from synaptic vesicles, however, was a large complex with an estimated molecular weight of 230 000, which exhibited a rosette-like structure under the electron microscope (Thomas *et al.* 1988). This suggests that synaptophysin *in situ* is a homomeric hexamer, made of six p38 subunit proteins.

The molecular motif and subunit configuration of synaptophysin resemble those of *gap junction proteins* or *connexins* (Leube *et al.* 1987; Südhof *et al.* 1987; Thomas *et al.* 1988). The gap junction is a pathway formed between two closely apposed cells, which admits passage of small molecules. The gap junction protein is also a hexamer (Fig. 3.6C), and its subunit is characterized by four transmembrane segments (Fig. 3.6B; M. V. L. Bennett *et al.* 1991; Südhof and Jahn 1991). The gap junction proteins were isolated from different tissues (liver, lens) and incorporated into lipid bilayers to measure electric currents through the gap junction channel (Zampighi *et al.* 1985; Young *et al.* 1987). The *single-channel conductance* was 140–280 pS (pico siemens), being more than 10-fold greater than that of ordinary ionic channels. This large conductance may account for cell-to-cell passage of small molecules through gap junctions. Synaptophysin was also reconstituted into lipid bilayers to record its channel currents; its single-channel conductance (150 pS) was similar to that found in gap junction channels (Thomas *et al.* 1988). Interestingly, the conductance estimated for the fusion pore in mast cells was also high (230 pS; Breckenridge and Almers 1987b). Plausibly, synaptophysin may act as a fusion pore during the exocytosis of synaptic vesicles rather like the gap junction channel.

A single gap junction channel consists of two half channels united in series, each being located in the membrane of the two apposed cells (Fig. 3.6C; M. V. L. Bennett *et al.* 1991; Hille 1992a). If a similar mechanism were involved in the exocytosis of synaptic vesicles, half a synaptophysin channel complex (a hexameric complex) should be present in the terminal membrane, and another in the vesicular membrane, so that the two half channel complexes can dock (Fig. 3.7C). Why, then, is synaptophysin present in the vesicular membrane but not in the terminal membrane? Two possibilities may be considered. First, a fully assembled hexameric complex (two-half-channel complex) of synaptophysin may exist on the vesicular membrane and simply penetrate the terminal membrane on collision; this single synaptophysin complex would be capable of forming a fusion pore by itself (Almers 1990). However, this possibility is unlikely, because it is difficult to assume that synaptophysin exposes its hydrophobic segments to the cyto-

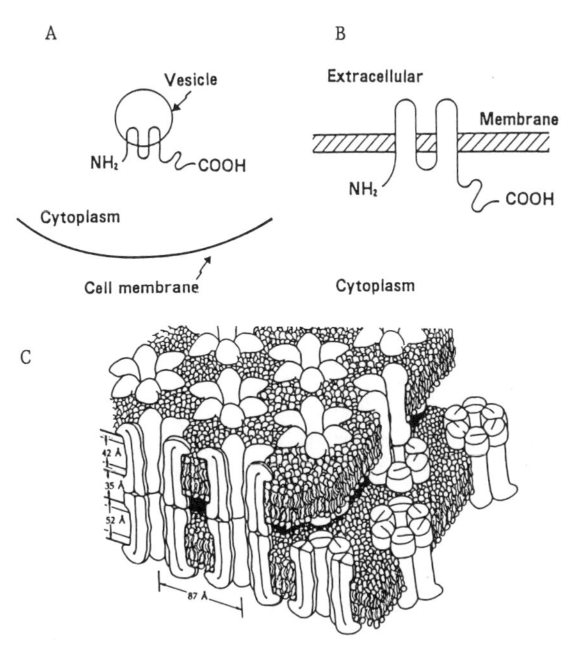

Fig. 3.6. Similarities between synaptophysin and gap junction protein. A, schematic illustration of a synaptophysin subunit (p38) incorporated in the vesicular membrane. Both the N-and C-termini of synaptophysin are located in the exterior of the vesicle (in the cytoplasm). B, schematic illustration of a gap junction protein which is incorporated in the cell membrane. Both the N-and C-termini of the gap junction protein subunit are located in the intracellular space. C, scheme of the assembly of gap junction subunits (connexins) to form channels (connexons) in the lipid bilayer membrane of two apposed cells. Six subunits form half a gap junction channel in each of the two membranes, and two half channels are united in series to make a gap junction channel. (From Makowski *et al.* 1977.)

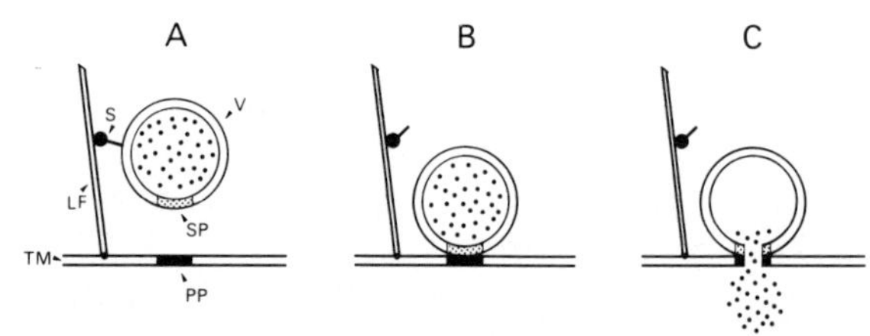

Fig. 3.7. Hypothetical models for exocytosis of a synaptic vesicle. A, at rest, synaptic vesicles (V) are immobilized by synapsins (S) that extend from the vesicular surface and link to large filaments (LF) anchored in the terminal membrane (TM). Synaptophysin (SP) in the vesicular membrane and physophilin (PP) in the terminal membrane are assumed to join for fusion. B, synaptic vesicles move to the active zone when the phosphorylation detaches the synapsin. C, exocytosis of the vesicular content occurs when SP and PP form a dilated fusion pore that is similar to a gap junction channel.

plasm. Second, the half fusion protein in the terminal membrane, which joins the synaptophysin half channel of the vesicular membrane, may simply be a different molecule (Fig. 3.7A, PP). If this were the case, the complementary fusion protein in the terminal membrane would have to bind to synaptophysin (Fig. 3.7A, SP). There is a protein, termed ***physophilin*** ('vesicle adorer'), which Thomas and Betz (1990) have isolated from rat brain synaptosomes. Physophilin does bind specifically to synaptophysin. Moreover, physophilin was found in the terminal membrane but not in the vesicular membrane. Physophilin seems to form an oligomeric complex, but its primary structure is not yet known. Should its amino acid sequence and transmembrane motif prove similar to those of synaptophysin, physophilin would be a good candidate for the complementary fusion protein in the terminal membrane. Figure 3.7C is a speculative diagram of exocytosis of the vesicular content, based on the assumption that synaptophysin and physophilin join to form the fusion pore.

In the proposed model of synaptophysin fusion pores, one crucial uncertainty persists. We have seen that synaptophysin has four transmembrane segments with its N- and C-termini in the cytoplasm (Fig. 3.6A) like the gap junction protein (Fig. 3.6B). A gap junction channel is formed by interactions between the extracellular regions of the proteins from two apposed cells. If synaptophysin on the vesicular membrane forms a channel with its complementary protein in the terminal membrane, the synaptophysin would have to do it only via its cytoplasmic regions (M. V. L. Bennett *et al.* 1991; Südhof and Jahn 1991). It remains uncertain whether this opposite orientation can form a functional channel. This uncertainty raises two questions: Is synaptophysin indeed essential for transmitter release, and does synaptophysin indeed merge with the terminal membrane during transmitter release?

Alder *et al.* (1992) showed that transmitter release is suppressed when synaptophysin is blocked by its antibody. Since synaptophysin resides in the vesicular membrane, its antibody must be applied into the terminal to observe the effect. Because of difficulties in injecting antibodies into nerve terminals, the approach was made by way of *Xenopus* oocytes, which are large, about 1 mm in diameter. Recall that the injection of mRNAs into oocytes results in new protein synthesis by translation of the exogenous mRNAs (section 2.2.2). The oocytes injected with mRNAs extracted from the rat cerebellum expressed synaptophysin and Ca^{2+}-dependent glutamate secretion (Alder *et al.* 1992). When the mRNAs were injected with an antibody against synaptophysin, the Ca^{2+}-dependent glutamate release from the oocyte was significantly reduced. This observation was complemented further by coinjection of the mRNAs and ***antisense oligonucleotides*** to synaptophysin mRNA. Antisense oligonucleotides would hybridize to their complementary mRNA (the synaptophysin mRNA) in the oocyte, so that translation of synaptophysin would be suppressed. Indeed, the expression of synaptophysin was consistently diminished in the oocytes co-injected with the antisense oligonucleotides; this was associated with marked reduction of Ca^{2+}-dependent glutamate release (Alder *et al.* 1992). These results strongly suggest that synaptophysin plays a crucial role in transmitter release.

Valtorta *et al.* (1988) stained synaptophysin with its antibodies at frog neuromuscular junctions. To stain cytoplasmic proteins with their antibodies, the cell membrane must be made permeable (permeabilization) with detergents. In fact, synaptophysin could be stained with the externally applied antibodies only after permeabilization of the motor nerve terminal with detergents. However, it was possible to induce a massive transmitter release by applying the spider toxin, α-latrotoxin (section 2.2.3) at neuromuscular junctions, while preventing recycling of synaptic vesicles (endocytosis), and thereby disclose synaptophysin immunoreactivity along the nerve terminal membrane without detergent treatment (Valtorta *et al.* 1988). This indicates that synaptophysin is exposed to the extracellular surface during transmitter release, presumably by fusion of synaptic vesicles with the nerve terminal membrane.

How does the spider toxin induce transmitter release?

We have seen that α-latrotoxin (LTX) purified from the black widow spider venom depletes ACh and small, clear vesicles in motor nerve terminals (section 2.2.3). Interestingly, transmitter release induced by LTX does not require Ca^{2+} influx (Valtorta *et al.* 1988). Then, the structure to which LTX binds might provide a clue about the mechanism of transmitter release. The LTX receptor was found to be *neurexins*, a family of neural membrane proteins (Ushkaryov *et al.* 1992). The cytoplasmic domain of the neurexins appears to interact with a vesicle-associated protein, *synaptotagmin* (Petrenko *et al.* 1991; Hata *et al.* 1993). Synaptotagmin comprises about 400 amino acids and has a long C-terminus in the cytoplasm, which binds to membrane phospholipids in a Ca^{2+}-dependent manner (Brose *et al.* 1992). Moreover, immunoglobulin prepared from patients with Lambert–Eaton myasthenic syndrome (LEMS; section 2.2.3) recognizes synaptotagmin (Leveque *et al.* 1992). These results imply that synaptotagmin is associated with Ca^{2+} channels and anchors the synaptic vesicle to phospholipids of the terminal membrane in a Ca^{2+}-dependent manner immediately before or during exocytosis of the vesicular content (Brose *et al.* 1992; Leveque *et al.* 1992). At this stage, however, it would be still premature to elaborate this hypothesis further (Bennett and Scheller 1993).

3.2 THE ORIGIN OF TRANSMITTER QUANTA

So far we have assumed that the transmitter is released by exocytosis of individual synaptic vesicles and that the vesicular release represents one quantum of the transmitter. Let us now examine whether the quantal release and the vesicular hypothesis are well-established concepts or whether there are still some discrepancies.

Del Castillo and Katz (1954a) showed that the release of ACh from the presynapic terminal occurs with a certain amount as a unit (quantum). Because this observation was made just prior to or at about the same time as the discovery of synaptic vesicles (De Robertis and Bennett 1954; Palade 1954; Palay 1954), it suggested a mechanism of quantal release without considering the presence of preformed packets of transmitter (synaptic vesicles). In the hypothesis schematically illustrated in Fig. 3.8A, for example, it is assumed that free ACh molecules are distributed uniformly within the motor nerve terminal (Del Castillo and Katz 1956). As a nerve impulse arrives in the terminal, individual release sites in the terminal membrane open briefly with a certain probability (p). The amount of ACh leaked at each release site would be approximately constant (quantum) if each gate stayed open for the same period of time. If the number of release sites is n, the mean number of quanta of ACh released by a nerve impulse (m; mean quantum content) can be given by $m = np$, the equation used for the quantum hypothesis (Del Castillo and Katz 1954a). Assuming that each release gate occasionally opens at rest would account for the generation of spontaneously occurring miniature end-plate potentials (m.e.p.p.s). It is unlikely, however, that the open time of the release gate remains unaltered regardless of the presence or absence of a large depolarization produced by the action potential. For this reason this possibility was considered untenable (Del Castillo and Katz 1956).

The vesicular hypothesis, diagrammed in Fig. 3.8B, assumes that the quantal release of transmitter results from exocytosis of individual synaptic vesicles, each containing about the same amount of the transmitter. If the total number of preformed packets available for release in a presynaptic terminal is n, and each packet is released with a mean probability of p, in response to a nerve impulse, the mean quantum content is again $m = np$. This vesicular hypothesis has now won overwhelming support, and numerous experimental results are consistent with it. Intriguingly, however, some results still cannot be explained by the vesicular hypothesis. We shall examine these discrepancies.

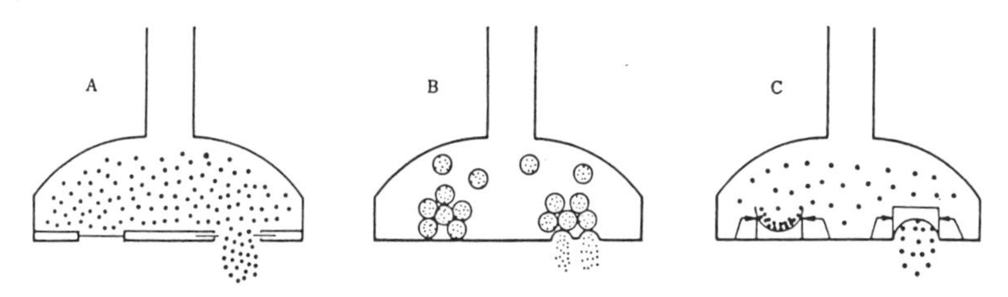

Fig. 3.8. Three hypothetical models for quantal release of transmitter. A, gate or shutter hypothesis. B, vesicular hypothesis. C, operator hypothesis. In A and C, the transmitter is assumed to be released from the terminal cytoplasm.

3.2.1 Unsolved problems in the vesicular hypothesis

The vesicular hypothesis has been challenged by several investigators. In their view, ACh is released from the cytoplasmic pool of a presynaptic terminal rather than from synaptic vesicles. After all, ACh is present in the terminal cytoplasm as well as in synaptic vesicles (Whittaker *et al.* 1972; Wagner *et al.* 1978). Moreover, since the ACh synthesizing enzyme, choline acetyltransferase, exists only in the terminal cytoplasm, not in synaptic vesicles (Fonnum 1967; Marchbanks and Israel 1972), the vesicular ACh must be initially synthesized in the cytoplasmic compartment before it is taken up by the vesicles (Carpenter *et al.* 1980). Dunant *et al.* (1982) developed a rapid freezing technique to measure changes in the ACh level of the *Torpedo* electric organ during repetitive nerve stimulation. Although nerve stimulation significantly reduced the total ACh in this terminal, the total amount of vesicular ACh remained unchanged (Dunant *et al.* 1982) even when the ACh release was markedly enhanced by applying K^+ channel blockers (Corthay *et al.* 1982). Furthermore, there was no evidence for ACh transfer from the cytoplasmic compartment to the vesicular compartment during or after nerve stimulation (Corthay *et al.* 1982). These results and other similar observations (see Israel *et al.* 1979; Tauc 1982) cast some doubt on the vesicular hypothesis.

The second discrepancy concerns preferential release of newly synthesized ACh by nerve stimulation (Collier 1969; Potter 1970). After the sympathetic ganglion is loaded with radioactively labelled ACh, stimulating its presynaptic nerve causes the release of ACh that has a relatively high specific radioactivity (70–80per cent) in the first few minutes; five minutes later the specific activity of the ACh released by nerve stimuli is only 35–45per cent if the ganglion is superfused with unlabelled choline (Collier 1969). That is, at least half of the ACh released by the second trial must be newly synthesized by uptake of the unlabelled choline from the extracellular medium. If the released ACh originated from vesicular ACh, the newly synthesized ACh must have been transferred from the cytoplasmic compartment to the synaptic vesicles within five minutes. Again, there is no evidence that cytoplasmic ACh is taken up by the vesicles so rapidly.

The third conflict arises from the effect of the ACh hydrolytic enzyme, acetylcholinesterase (AChE). When the *Torpedo* electric organ is homogenized, cytoplasmic ACh cannot be recovered unless the endogenous AChE is inhibited, whereas the inhibitor is not required for the recovery of vesicular ACh (Marchbanks and Israel 1972; Morel *et al.* 1978). This indicates that, unlike the cytoplasmic ACh, ACh contained in synaptic vesicles is not accessible to AChE. In other words, the vesicular membrane is not permeable to AChE. Injecting AChE into cholinergic neurones of *Aplysia* causes synaptic transmission produced by stimulating the cholinergic neurones to decline (Tauc *et al.* 1974). Since the injected AChE cannot enter the synaptic vesicles, ACh destroyed by the AChE must be only the cytoplasmic ACh. Therefore, this observation further supports the idea that ACh released by nerve stimulation originates from the cytoplasmic compartment rather than from synaptic vesicles. Figure 3.8C diagrams this alternative hypothesis of transmitter release, which is variously termed the *operator* (Israel *et al.* 1979), *vesigate* (Tauc 1982) or *mediatophore* (Dunant 1986) hypothesis. Without specifying detailed mechanisms, it simply postulates the release of cytoplasmic ACh with a large number of molecules as a unit. As Tauc (1982) himself admits, the evidence in favour of the vesicular hypothesis is now overwhelming. Still, several results cannot be accounted for by the present form of the vesicular hypothesis and certainly should not be ignored. Should a rapid transfer and equilibrium of ACh prove to occur between the cytoplasm and synaptic vesicles, the three discrepancies discussed above would be removed.

The quantal release of cytoplasmic ACh postulated in Fig. 3.8C should not be confused with steady leakage of cytoplasmic ACh. That non-quantal ACh leakage from the cytoplasmic pool of a nerve terminal occurs is undeniable, although its functional significance remains unclear. The local application of curare to the neuromuscular junction in frog (Katz and Miledi 1977) or mouse (Vyskocil and Illes 1977), in the presence of AChE inhibitors, induces a small hyperpolarization in the muscle fibre. This suggests that each muscle fibre is steadily depolarized presumably by the leakage of ACh from the motor nerve terminal. The amount of hyperpolarization produced by curare was not related to the frequency of m.e.p.p.s (Katz and Miledi 1977); it was not affected by changes in the external Ca^{2+} level nor by nerve stimulation (Katz and Miledi 1981). Thus, ACh appears to leak steadily in a non-quantal fashion from the cytoplasmic compartment of motor nerve terminals. This is a phenomenon entirely independent of transmitter release induced by nerve impulses. The rate of non-quantal ACh leakage is estimated to be about 10^6 molecules per second at each end-plate in the frog muscle (Katz and Miledi 1977)—around 100 times greater than the rate of ACh release by spontaneous quantal discharge. A steady ACh leakage would explain the amount of ACh released from a nerve-muscle preparation at rest, which is disproportionately larger than would be expected from spontaneous m.e.p.p.s (Mitchell and Silver 1963; Fletcher and Forrester 1975). How this leakage occurs or whether it plays any functional role remains unknown. In any case, it would all presumably be hydrolysed by AChE.

3.2.2 Unsolved problems in the quantum hypothesis

In the quantum hypothesis, n was defined as the number of releasable units within a nerve terminal, and p as the average probability that each unit will respond to a nerve impulse (Del Castillo and Katz 1954a). However, because the test of this hypothesis employed Poisson's statistics, which requires measuring only the m parameter, the physical correlates of n and p remained undefined (Box D).

Both the mean quantum content (m) and the frequency of m.e.p.p.s at a neuromuscular junction are proportional to the total length of its nerve terminal (Kuno *et al.* 1971; Grinnell and Herrera 1981). That is, the motor nerve terminal appears to be homogeneous with respect to the density of release sites. This situation is illustrated schematically in Fig. 3.9. If the mean number of units capable of responding to a nerve impulse (i.e., the mean number of releasable vesicles) at each release site is k, and if a nerve terminal has n' release sites (i.e., the number of active zones), we can describe the mean quantum content at this synapse as $m = np = kn'p$. In fact, Propst and Ko (1987) found the m value at the frog neuromuscular junction to be better correlated with the total length of active zones than with the terminal length.

How many units (k) are available for release at each release site? This question was addressed in inhibitory synapses formed by a single presynaptic neurone on Mauthner cell of the goldfish. The quantal analysis of fluctuations of the inhibitory synaptic potentials was made, using binomial statistics (Box D). The parameter n (where $n = kn'$) estimated by binomial equation appeared to coincide with the total number of synaptic boutons formed by the presynaptic neurone (Korn *et al.* 1981, 1982). Furthermore, each synaptic bouton arising from this neurone contained only one release site (active zone or presynaptic grid; Triller and Korn 1982).

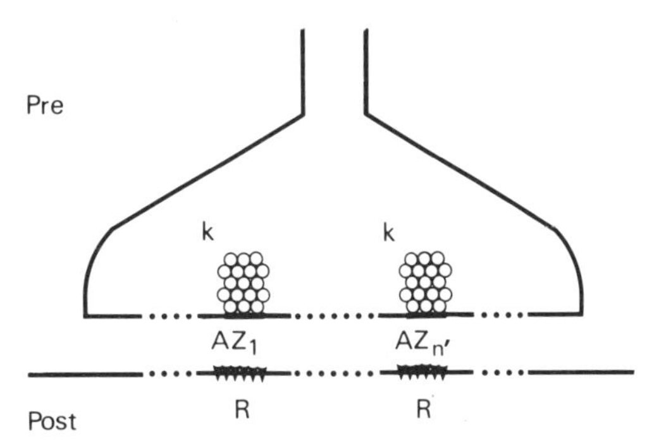

Fig. 3.9. Hypothetical distribution of transmitter quanta in a presynaptic terminal (Pre). Each terminal has n' release sites or active zones (AZ) distributed at approximately equal distances. Each release site has k quanta or synaptic vesicles as releasable units. In the postsynaptic membrane (Post), transmitter receptor molecules (R) accumulate in each active zone.

The results imply that n reflects the total number of release sites ($n = n'$) and that the number of quanta released by a nerve impulse at each synaptic site does not exceed unity ($kp < 1$). Similar results have now been reported for a variety of synapses, including excitatory synapses on central neurones (Redman and Walmsley 1983; Walmsley *et al.* 1988), the crayfish neuromuscular junction (Wojtowicz and Atwood 1986) and excitatory junctions of smooth muscle (Brock and Cunnane 1987; Astrand and Stjarne 1989; Nickolas *et al.* 1992).

Taken together, these give us a picture of transmitter release as occurring at each release site in an all-or-none fashion, as if only one releasable vesicle waits at each active zone. Perhaps, this notion is also applicable to neuromuscular junctions. In a typical neuromuscular junction of the frog, the number of active zones (n') is about 700 (Katz and Miledi 1979), and the m value is normally about 300 (Martin 1955; Katz and Miledi 1979). Thus, even at neuromuscular junctions the number of quanta liberated from each active zone may not exceed one. Katz and Miledi (1979) measured the m value in the frog end-plate in the presence of K^+ channel blockers which enhance the transmitter release. The m value estimated by the ratio of average end-plate current to miniature end-plate current peak amplitudes exceeded 1 000—more than the number of active zones. In frog motor nerve terminals, however, each active zone spans almost the entire terminal width perpendicular to its long axis. Presumably, some active zones in frog may contain more than one release site.

Del Castillo and Katz (1954a) recognized the possibility that p, the probablity of each unit being released

Box D Binomial and Poisson's statistics in the quantal analysis

The amplitude of end-plate potentials (e.p.p.s) evoked by consecutive nerve impulses shows a random fluctuation, including occasional failures of response, in a low Ca^{2+} solution. This fluctuation occurs in a stepwise fashion as multiples of a certain amplitude (v; see Figure). The mean smallest response (v) coincides with the mean amplitude of spontaneous miniature end-plate potentials (m.e.p.p.s; see inset in the Figure). The basic assumption in the quantum hypothesis is that the transmitter is stored in the presynaptic terminal in the form of n releasable units or quanta and that each unit is released with a certain probability, p, in response to a nerve impulse (Del Castillo and Katz 1954a). The mean number of quanta released by one impulse (m, mean quantum content) would be then $m = np$, and the mean amplitude of e.p.p.s (E) is $E = mv$. By recording a large number of e.p.p.s and m.e.p.p.s, we can determine the E and the v; hence, the m can be estimated.

The Figure is an example of the amplitude distribution of e.p.p.s (Boyd and Martin 1956). The total number of stimuli (N) applied to the nerve was 198, and the number of failures of response (F) was 18. Then, the chance of total failures of transmitter release (P_0) should be $P_0 = (1 - p)^n = F/N = 0.091$. Similarly, the chance of observing e.p.p.s evoked by the release of only one unit (P_1) can be given by $P_1 = np(1 - p)^{n-1}$. Mathematically, the general form for the chance of observing e.p.p.s evoked by the release of x quanta (Px) can be given by

$$Px = \frac{n!}{(n-1)!x!} p^x (1-p)^{n-x}$$

This is the prediction based on binomial statistics. Although n and p are the concepts used to form the quantum hypothesis, these two parameters cannot be measured independently.

If n is sufficiently large and if p is very small, the above binomial equation can be approximated by Poisson's equation

$$Px = \frac{m^x}{x!} e^{-m}$$

Here, the only parameter is m which is measurable. Del Castillo and Katz (1954a) used this approximation. The average amplitude of e.p.p.s composed of x units (quanta) will equal xv. The predicted amplitude distribution of e.p.p.s on the basis of the Poisson's equation (curves in Figure) was in good agreement with the observed distribution (histogram in Figure). Thus, the results were consistent with the quantum hypothesis, but the concepts of n and p remained undefined.

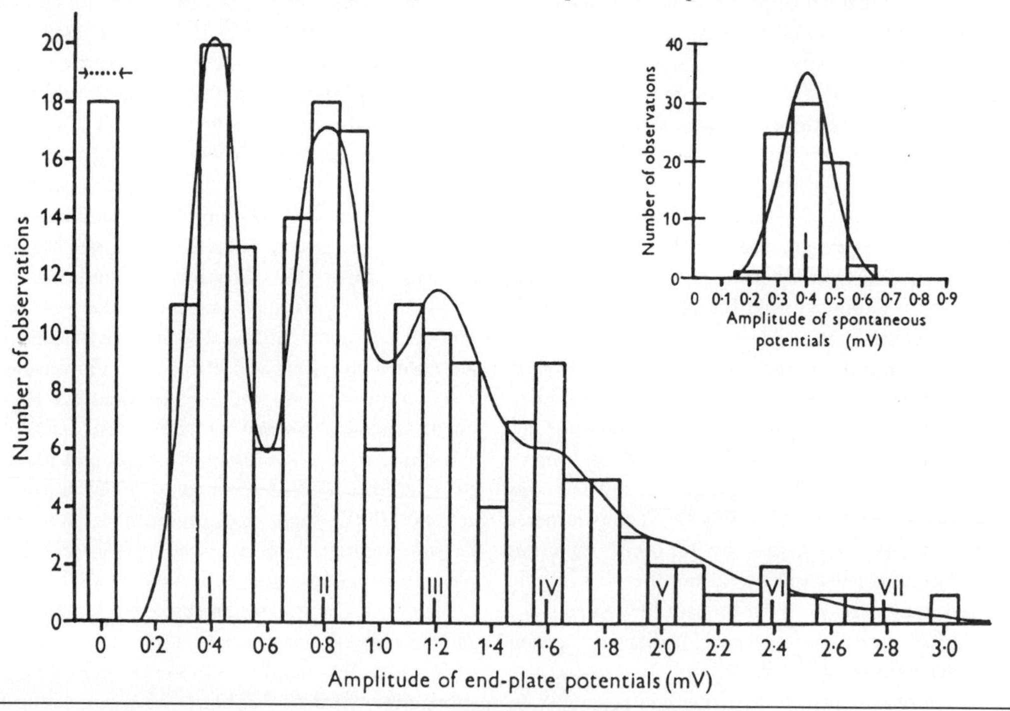

by a nerve impulse, may vary considerably for individual units or at individual release sites. Non-uniform probabilities at different release sites have now been suggested for central synapses (Jack *et al.* 1981; Redman and Walmsley 1983; Walmsley *et al.* 1988) as well as for neuromuscular junctions (D'Alonzo and Grinnell 1985; Bennett *et al.* 1986; Nickolas *et al.* 1992). If the variance of p at individual release sites is large, then the amplitude fluctuations of synaptic responses may not satisfy the distributions predicted by simple statistics based on the averaged p.

Despite the complexities in analysing the fluctuations of synaptic responses in central neurones, quantal analysis remains a potentially useful approach to identify the site of plastic changes at synapses. Long-term potentiation, for example, is a phenomenon implicated in learning and memory (section 7.2.2). Whether this potentiation is due to increased amount of transmitter release or to increased sensitivity of the receptor for the transmitter is a principal question for elucidating its mechanisms. Theoretically, changes in synaptic efficacy can be distinguished in terms of the presynaptic or postsynaptic level by quantal analysis. As explained in Box D, potentiation of the mean synaptic response (E) must be due either to an increase in m (presynaptic factor) or to an increase in v (postsynaptic factor). For this analysis, however, the quantal size of synaptic responses (v) must be evaulated unequivocally. This evaluation is unfortunately difficult at central synapses. What would be the source of such difficulties?

The quantum hypothesis of transmitter release at central synapses was first proposed on the basis of statistital fluctuations of the amplitude of excitatory postsynaptic potentials (e.p.s.p.s) evoked in spinal motor neurones by stimulation of single sensory fibres (Kuno 1964a, 1971; Kuno and Miyahara 1969). The amplitude fluctuations of these e.p.s.p.s conformed to Poisson's law at least in several experiments, although Poisson's or binomial statistics consistently failed to apply when the analysis was extended to large e.p.s.p.s with high m values (> 3). This analysis, particularly its large variability in the m value (less than one to about 15; Kuno and Miyahara, 1969), was questioned, because the poor signal-to-noise ratio in the records might make the quantal size of e.p.s.p.s hard to discern (Edwards *et al.* 1976; Redman 1979, 1990). Subsequent analyses of the e.p.s.p.s evoked by single presynaptic impulses subtracted the noise component from the noise-contaminated records to compute the 'true' e.p.s.p. signal. This deconvolution method initially contradicted the quatum hypothesis previously proposed by the *old method*, finding no quantal e.p.s.p. fluctuations at central synapses (Edwards *et al.* 1976). Why the initial deconvolution analysis anomalously failed to demonstrate the quantal nature of e.p.s.p.s is not clear (Korn and Faber 1987). When the quantal size of e.p.s.p.s was again estimated by the deconvolution method (0.08–0.20 mV; Jack *et al.* 1981), the value proved similar to that found with the old method (0.12–0.24 mV; Kuno 1964a). Moreover the mean number of quanta released by a single presynaptic impulse (m value) at central inhibitory synapses estimated by the old method (2–5; Kuno and Weakly 1972) was consistent with the value (about 4) measured recently (Takahashi 1992). Thus, despite some criticism, the conceptual framework built on the old method remains unchanged.

Establishing non-uniform release probabilities at individual release sites in central neurones was a major achievement of the deconvolution method. This finding now accounts for the previous puzzle that the amplitude fluctuations of large e.p.s.p.s in central neurones cannot be described by Poisson's or binomial law (Kuno 1964a; Kuno and Miyahara 1969; Mendell and Weiner 1976). Another new concept proposed by the deconvolution method was *quasi-invariability in quantal size* of e.p.s.p.s. It is this concept that contradicts distinctly between the findings made by the old method and by the deconvolution method. At neuromuscular junctions, the unit e.p.p.s (or m.e.p.p.s) evoked by the release of individual quanta of ACh in a given muscle fibre vary in size, their standard deviation being about 30per cent of their mean amplitude (i.e., coefficient of variation = 0.3; Martin 1965). The coefficient of variation of unit e.p.s.p.s measured in a given spinal motor neurone with the old method was similar (0.3; Kuno 1964a; Blankenship and Kuno 1968). By contrast, the amplitude of unit e.p.s.p.s estimated by the deconvolution method in spinal motor neurones is virtually constant, its coefficient of variation being less than 0.05 (Jack *et al.* 1981; Redman 1990; Walmsley 1991). This discrepancy must be either due to some errorneous measurements made by the old method because of the low signal-to-noise ratio (Kuno 1964a, 1971) or to some misalignment in the deconvolution procedure (Edwards *et al.* 1976; Redman 1990). The deconvolution method must optimize multiple factors to find the best fit in the observed amplitude distribution of e.p.s.p.s. This calculation yields a somewhat dubious picture. For example, in selecting the optimal quantal size, 0,186 mV and

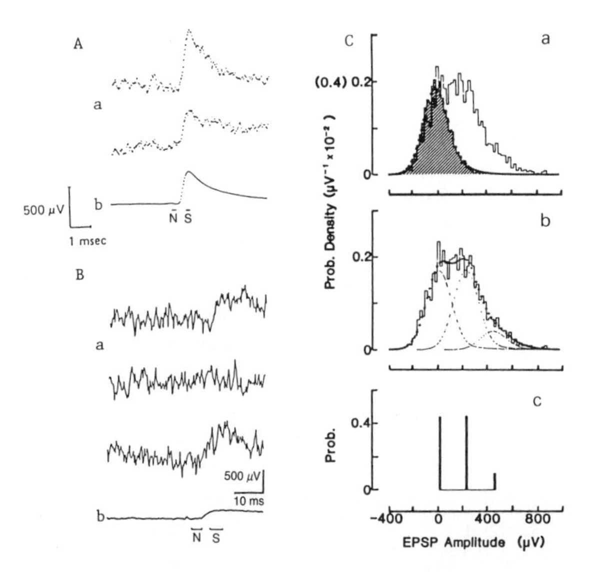

Fig. 3.10. The quantal analysis of synaptic potentials with the deconvolution method. Excitatory postsynaptic potentials (e.p.s.p.s) evoked by single presynaptic impulses in a spinal motor neurone (A) and a hippocampal pyramidal neurone (B). a, sample records of individual e.p.s.p.s; b, averaged e.p.s.p.s. The baseline noise (N) and signal (S) were sampled during the periods indicated by horizontal bars. C, deconvolution analysis of the records in B. Top, noise-contaminated e.p.s.p. histogram and noise (shaded) histogram. Middle, reconstruction of the noise-contaminated e.p.s.p. histogram by three normal curves with the same standard deviation as measured noise distribution. Bottom, deconvolution. (A, adapted from Jack *et al.* 1981. B and C, adapted from Redman 1990.)

0.202 mV appear satisfactory, but not 0.194 mV (Walmsley *et al.* 1988). Although such misalignment is not surprising computationally, it defies logical interpretation. The quantal analysis cannot be applied to central synapses without resolving this uncertainty.

Another shortcoming in the deconvolution procedure stems from a lack of visual control of individual records. As shown in Fig. 3.10 (A,B), the noise-contaminated peak amplitude of e.p.s.p.s evoked by single presynaptic impulses in central neurones is estimated by averaging individual records in a fixed time window corresponding to the peak (S) and the baseline noise (N) and taking the difference for a large number of responses (Jack *et al.* 1981; Redman 1990). The 'pure' e.p.s.p. amplitude distribution (Fig. 3.10Cc) is then constructed by subtracting the noise component (Fig. 3.10Cb). The period used for averaging was 0.25 ms in A (for a spinal motor neurone) and 3 ms in B (Fig. 3.10). Cope and Mendell (1982) found that the onset of e.p.s.p.s evoked in a spinal motor neurone by consecutive single presynaptic impulses fluctuates in a range from 0.2 to 0.6 ms. Therefore, if the time window for averaging was too narrow (e.g., 0.25 ms), the peak of some e.p.s.p.s might have been out of phase. If the window was too wide (e.g., 3 ms), the procedure would give an average around the peak of e.p.s.p.s, including points in their rising and falling phases. The variability of the estimated quantal size might be reduced under such conditions. Furthermore, it is not clear whether the variability of the quantal size estimated by subtracting the noise component (Fig. 3.10Cc) is entirely independent of the background noise level (Fig. 3.10Cb, dotted-line curves). Even at neuromuscular junctions, the variability of unit e.p.p.s was unusually small when the deconvolution method was applied after addition of extra noise to simulate the analysis at central synapses (Edwards *et al.* 1976). This possibiity should be examined more systematically.

The signal-to-noise ratio can be improved substantially by recording central synaptic responses in the tight-seal whole-cell recording mode of the patch clamp technique. With this technique, Takahashi (1992) analysed the unit or minimal inhibitory postsynaptic currents (i.p.s.c.s) recorded from spinal motor neurones in neonatal rats. When i.p.s.c.s were evoked by stimulating a neighbouring interneurone, the magnitude of i.p.s.c.s showed fluctuations, including occasional failures (six failures out of 299 trials; Fig. 3.11A). When the external Ca^{2+} concentration was reduced from 2 mM to 0.7 mM, the number of failures markedly increased (217 failures out of 299 trials), and the amplitude distribution followed a Gaussian curve (Fig. 3.11B). A further reduction in the external Ca^{2+} level (to 0.5 mM) increased the number of failures (562 failures out of 600 trials), but the amplitude distribution remained unchanged. Therefore, the amplitude distribution shown in Fig. 3.11B must represent that of the minimal or quantal i.p.s.c.s. Their coefficient of variation (0.42) was much larger than that of the background noise level (Fig. 3.11C). A similar coefficient of variation (0.41) was also observed for the minimal i.p.s.c.s recorded from hippocampal neurones with the whole-cell recording technique (see Fig. 7 in Edwards *et al.* 1990). What would the variability of the quantal size of synaptic responses signify?

In principle, two factors may contribute to the degree of variability in quantal size. One factor concerns the amount of transmitter contained in each quantal packet; if the number of transmitter molecules varies consider-

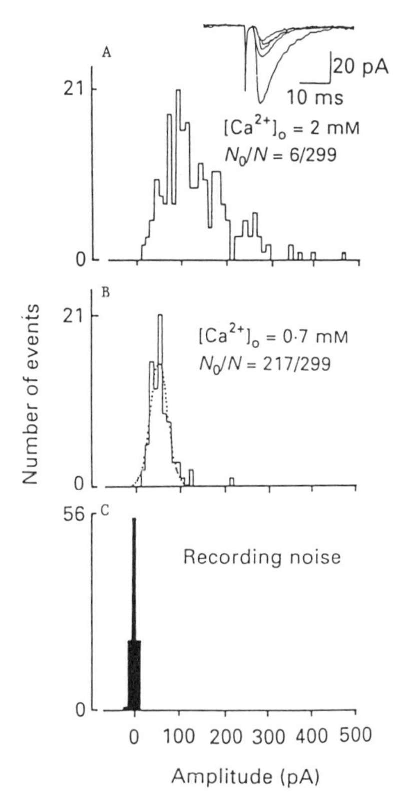

Fig. 3.11. The minimal inhibitory postsynaptic currents (i.p.s.c.s) recorded from a spinal motor neurone. A, sample i.p.s.c.s recorded in 2 mM Ca^{2+} (inset), and their amplitude distribution. B, similar to A, but in 0.7 mM Ca^{2+}. C, baseline noise amplitude histogram. (Adapted from Takahashi 1992.)

ably from one vesicle to another, the amplitude of quantal synaptic responses would show a large variability (Del Castillo and Katz 1956; Elmqvist and Quastel 1965; Bekkers *et al.* 1990). Another factor is the number of the postsynaptic receptors for the transmitter. If the number of the receptors is limited (e.g., 50), compared with the number of transmitter molecules released by a quantal discharge (e.g., 5 000), all the receptors would be saturated, so that there might be no variability of quantal size. At neuromuscular junctions, a quantum is estimated to comprise 1 000 to 10 000 ACh molecules (Katz and Miledi 1972; Kuffler and Yoshikami 1975b), and the density of ACh receptors in the end-plate region is about 10 000/μm^2 (Fambrough and Hartzell 1972; Fertuck and Salpeter 1974). Thus, the variability of quantal size of e.p.p.s would be due mainly to that in the number of ACh molecules in individual quanta, because the ACh receptors clearly outnumber the ACh molecules in each quantum. The release of one quantum of ACh at neuromuscular junctions is estimated to open 1 000–3 000 ACh receptor channels (Katz and Miledi 1972; Gage and McBurney 1972). Similarly, about 1 500 receptor channels are opened by a quantal discharge at glycine-mediated inhibitory synapses in the lamprey (Gold and Martin 1983) and in the goldfish Mauthner cell (Korn *et al.* 1987). At inhibitory synapses mediated by gamma-aminobutyric acid (GABA) in the rat hippocampus (Edwards *et al.* 1990) or by glycine in rat motor neurones (Takahashi 1992), each quantum appears to activate only 15–30 receptor channels. At excitatory synapses in central neurones, the number of glutamate receptor channels opened by the release of one quantum is estimated to be only 10–20 (Hesterin 1992; Stern *et al.* 1992; Traynelis *et al.* 1993). This limited number of receptor channels activated at individual mammalian central synapses suggests that each quantum saturates virtually all the receptor molecules at the release site; this may cause the quantal size to remain virtually invariable (Edwards *et al.* 1990; Redman 1990; Walmsley 1991). However, each presynaptic neurone forms multiple synaptic sites on a given postsynaptic neurone (Fig. 3.13). Then, the number of receptor channels at every synaptic site must be almost identical to show the quasi-invariability of quantal size. For instance, if the coefficient of variation of the quantal size at the GABA-inhibitory synapse is less than 0.05, the number of GABA receptors and its standard deviation at every synaptic site would have to be 30 ± 1.5, a figure somewhat unrealistic. The variability of quantal size at central synapses may reflect the variability in the number of receptors at individual synaptic sites formed by a given presynaptic fibre (see below; Fig. 3.13).

At both GABA- and glycine-mediated inhibitory synapses in mammalian central neurones, the amplitude distribution of spontaneously occurring i.p.s.c.s is positively skewed, displaying large responses corresponding to multiple quantal discharge, even in the presence of tetrodotoxin (Fig. 3.12). This behaviour is similar to that of spontaneous miniature e.p.s.p.s observed in the chick ciliary ganglion (Martin and Pilar 1964a). Martin and Pilar (1964a) suggested that when a unit is released spontaneously, each of the remaining units has a small probability of being 'dragged' with it. Figure 3.13 shows an alternative view to account for the variability of quantal size and the spontaneous multiple quantal event at central synapses. Here, a presynaptic axon forms four synaptic sites on a postsynaptic neurone, having non-uniform probabilities of transmitter release (P_a

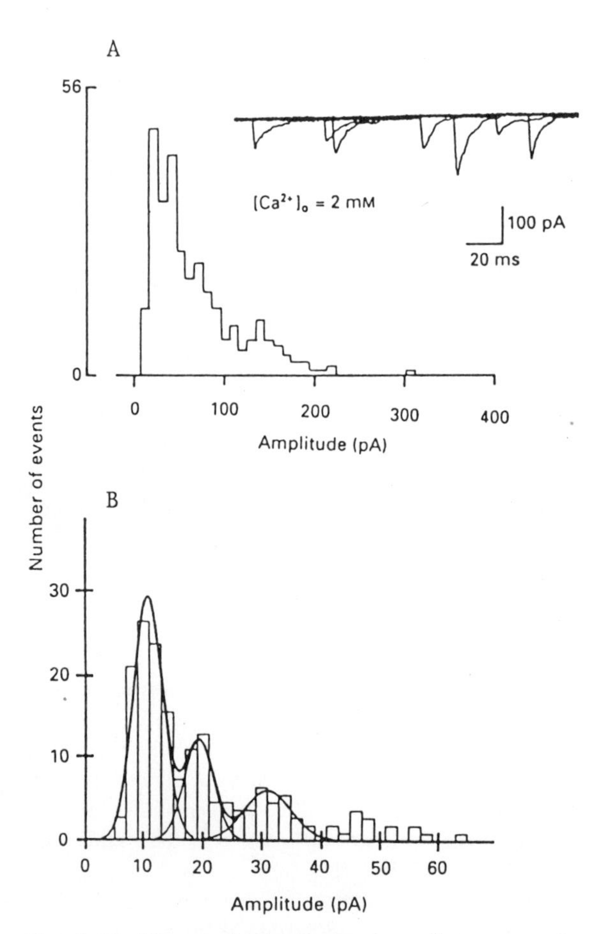

Fig. 3.12. The amplitude distributions of spontaneously occurring inhibitory postsynaptic currents (i.p.s.p.s) recorded from mammalian central neurones. A, sample records from a spinal motor neurone (inset) and their amplitude distribution (From Takahashi 1992.) B, similar to A, but recorded from a hippocampal neurone in the dentate gyrus. (From Edwards *et al.* 1990.)

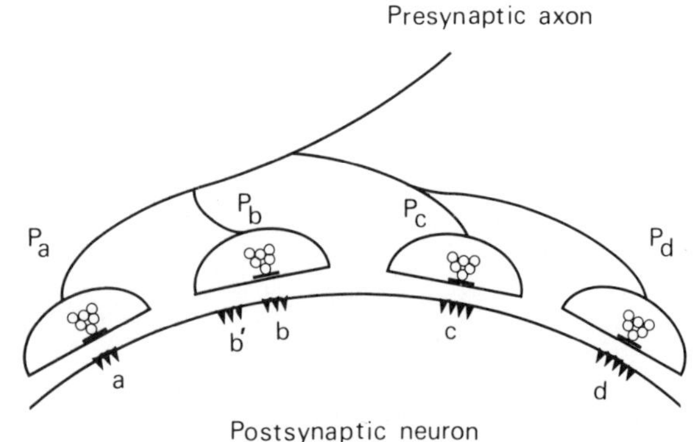

Fig. 3.13. Hypothetical distribution of terminals arising from a single presynaptic fibre (P_a to P_d) and receptor modules (a–d) at the postsynaptic sites in a central neurone. Although the release probability may differ from one terminal to another, a single presynaptic impulse is assumed to be incapable of releasing more than one quantum of the transmitter from each terminal. There is a cluster of transmitter receptors (receptor module) at each postsynaptic site. The number of receptor molecules in each receptor module may not be identical, which presumably accounts for the variability of quantal size. The subsynaptic membrane beneath some presynaptic terminal may contain multiple receptor modules (e.g., b and b′), which may explain spontaneously occurring multiple quantal events.

to P_d). The postsynaptic site has a cluster of transmitter receptor molecules, forming a receptor module (a–d). The number of receptor molecules contained in each module differs; this accounts for the variability of quantal size. Because of the relatively small number of receptor molecules in each module (10–30; see above), the area occupied by the module is much smaller than that covered by each presynaptic terminal. More than one module may then exist aberrantly at some of the postsynaptic sites (b′). This would result in the generation of a multiple quantal event by the release of one quantum of transmitter. The basic assumption in this model is that the transmitter molecules in individual quanta outnumber the receptor molecules at each synaptic site. There are, of course, several other possible models, all equally speculative. Clearly, more information is needed before the notion of receptor saturation by quantal release of transmitter is accepted. At present, the conclusion inferred by quantal analysis of synaptic responses depends on the interpretation of the amplitude fluctuations of the synaptic response (Korn and Faber 1991). Evidently, the application of quantal analysis to central synapses is unreliable without resolving the uncertainties of the quantal size and its variability.

Summary and prospects

Transmitter release occurs in a quantal fashion. According to the current view, the amount of transmitter contained in individual synaptic vesicles is the quantal unit (the vesicular hypothesis). Exocytosis of the vesicular content requires the vesicular membrane and the nerve terminal membrane to contact, then fuse. Specialized vesicle proteins are likely to be involved in this process. Two vesicle-associated proteins, synapsin I and synaptophysin, appear to be particularly relevant to transmitter release. Synapsin I restrains the mobility of synaptic vesicles, at rest, by binding to large filaments within the terminal. The Ca^{2+} influx associated with an action potential in the presynaptic terminal activates calmodulin (CaM)-dependent protein kinase II which in turn phosphorylates synapsin I. Under this condition, synapsin I dissociates from the vesicular surface, and the vesicle becomes available for exocytosis. Synaptophysin is similar to the gap junction protein involved in the formation of a pathway between two apposed cells, which permits transfer of small molecules. Based on this similarity, synaptophysin is assumed to be the vesicular protein involved in fusion with the terminal membrane, thereby forming a fusion pore for exocytosis of the vesicular content.

Exocytosis of granular vesicles in chromaffin or mast cells gives us a model for the molecular mechanisms of transmitter release. Fusion of granular vesicles in these cells is evidenced by stepwise increments in electrical capacitance of the cell membrane. In fact, quantal release of catecholamines from chromaffin cells is associated with quantal increments of the membrane capacitance. Unfortunately, fusion proteins involved in granular vesicles of chromaffin cells have not yet been identified.

Despite its overwhelming support, several results still cannot be accounted for by the present form of the vesicular hypothesis. The major shortcoming in this hypothesis is a lack of information about the transfer of the transmitter molecules between the cytoplasm and synaptic vesicles. Similarly, the details of the mechanisms of transmitter release proposed by the quantum hypothesis still remain uncertain. The probability of transmitter release appears to vary at different release sites. Yet, it is not known what determines the release probability. The mechanism of quantal transmitter release is ambiguous particularly at central synapses. In some cells the synaptic responses induced by the release of individual quanta appear to be relatively invariable, whereas a large variability occurs in others. Theoretically, quantal analysis of synaptic responses can predict whether alterations of synaptic efficacy originate from changes at the presynaptic or postsynaptic level. At present, the quantal analysis of central synaptic responses cannot make this prediction reliably.

Suggested reading

Vesicle-associated proteins: synapsin I and synaptophysin

Bennett, M. K. and Scheller, R. H. (1993). The molecular machinery for secretion is conserved from yeast to neurons. *Proceedings of the National Academy of Sciences of the USA*, **90**, 2559–63.

De Camilli, P. and Greengard, P. (1986). Synapsin I: a synaptic vesicle-associated neuronal phosphoprotein. *Biochemical Pharmacology*, 35, 4349–57.

Landis, D. M. D., Hall, A. K., Weinstein, L. A., and Reese, T. S. (1988). The organization of cytoplasm at the presynaptic active zone of a central nervous system synapse. *Neuron*, 1, 201–9.

Llinás, R., McGuinness, T. L., Leonard, C. S., Sugimori, M., and Greengard, P. (1985). Intraterminal injection of synapsin I or calcium/calmodulin-dependent protein kinase II alters neurotransmitter release at the squid giant synapse. *Proceedings of the National Academy of Sciences of the USA*, **82**, 3035–9.

Thomas, L., Hartung, K., Langosch, D., Rehm, H., Bamberg, E., Franke, W. W., and Betz, H. (1988). Identification of synaptophysin as a hexameric channel protein of the synaptic vesicle membrane. *Science*, **242**, 1050–3.

Südhof, T. C. and Jahn, R. (1991). Proteins of synaptic vesicles involved in exocytosis and membrane recycling. *Neuron*, 6, 665–77.

Ushkaryov, Y. A., Petrenko, A. G., Geppert, M., and Südhof, T. C. (1992). Neurexins: Synaptic cell surface proteins related to α-latrotoxin receptor and laminin. *Science*, 257, 50–6.

Valtorta, F., Jahan, R., Fesce, R., Greengard, P., and Ceccarelli, B. (1988). Synaptophysin (p38) at the frog neuromuscular junction: its incorporation into the axolemma and recycling after intense quantal secretion. *Journal of Cell Biology*, **107**, 2717–27.

Vesicular exocytosis and fusion pore

Bennett, M. V. L., Barrio, L. C., Bargiello, T. A., Spray, D. C., Hertzberg, E., and Saez, J. C. (1991). Gap junctions: new tools, new answers, new questions. *Neuron*, 6, 305–20.

Breckenridge, L. J. and Almers, W. (1987a). Final steps in exocytosis observed in a cell with giant secretory granules. *Proceedings of the National Academy of Sciences of the USA*, **84**, 1945–9.

Breckenridge, L. J. and Almers, W. (1987b). Currents through the fusion pore that forms during exocytosis of a secretory vesicle. *Nature*, **328**, 814–7.

Chow, R. H., von Rudden, L., and Neher, E. (1992). Delay in vesicle fusion revealed by electrochemical monitoring of single secretory events in adrenal chromaffin cells. *Nature*, **356**, 60–3.

Neher, E. and Marty, A. (1982). Discrete changes of cell membrane capacitance observed under conditions of enhanced secretion in bovine adrenal chromaffin cells. *Proceedings of the National Academy of Sciences of the USA*, 79, 6712–6.

Spruce, A. E., Breckenridge, L. J., Lee, A. K., and Almers, W. (1990). Properties of the fusion pore that forms during exocytosis of a mast cell secretory vesicle. *Neuron*, 4, 643–54.

The vesicular hypothesis

Dunant, Y. (1986). On the mechanism of acetylcholine release. *Progress in Neurobiology*, 26, 55–92.

Heuser, J. E. (1989). Review of electronmicroscopic evidence favouring vesicle exocytosis as the structural basis for quantal release during synaptic transmission. *Quarterly Journal of Experimental Physiology*, 74, 1051–69.

Israel, M., Dunant, Y., and Manaranche, R. (1979). The present status of the vesicular hypothesis. *Progress in Neurobiology*, 13, 237–75.

Katz, B. and Miledi, R. (1965). The quantal release of transmitter substance. In *Studies in physiology*, (ed. D. R. Curtis and A. K. McIntyre), pp. 118–25. Springer-Verlag, New York.

Tauc, L. (1982). Nonvesicular release of neurotransmitter. *Physiological Reviews*, 62, 857–93.

Quantal analysis of synaptic responses

Korn, H. and Faber, D. S. (1987). Regulation and significance of probabilistic release mechanisms at central synapses. In *Synaptic function* (ed. G. M. Edelman, W. E. Gall, and W. M. Cowan), pp. 57–108. John Wiley, New York.

Korn, H. and Faber, D. S. (1991). Quantal analysis and synaptic efficacy in the CNS. *Trends in Neurosciences*, 14, 439–45.

Kuno, M. (1971). Quantum aspects of central and ganglionic synaptic transmission in vertebrates. *Physiological Reviews*, 51, 647–78.

Martin, A. R. (1965). Quantal nature of synaptic transmission. *Physiological Reviews*, 46, 51–66.

Martin, A. R. (1977). Junctional transmission II. Presynaptic mechanisms. In *Handbook of the nervous system*, (ed. E. Kandel) Vol. 1, pp. 329–55. American Physiological Society, Baltimore.

Redman, S. (1979). Junctional mechanisms at group Ia synapses. *Progress in Neurobiology*, 12, 33–83.

Takahashi, T. (1992). The minimal inhibitory synaptic currents evoked in neonatal rat motoneurones. *Journal of Physiology (London)*, 450, 593–611.

Walmsley, B. (1991). Central synaptic transmission: studies at the connections between primary afferent fibres and dorsal spinocerebellar tract (DSCT) neurones in Clarke's column of the spinal cord. *Progress in Neurobiology*, 36, 391–423.

4

Ionotropic receptors mediate fast synaptic responses

The transmitter released from a presynaptic terminal diffuses across the synaptic cleft and binds to its postsynaptic receptor. The ensuing synaptic response consists of a change in the membrane potential of the postsynaptic cell, which is caused by the opening or closing of transmitter-operated ionic channels. Elucidation of the mechanisms underlying this transmitter-induced response has been advanced enormously by three major developments during the past two decades.

The first area of progress was in analysing the elementary voltages (Katz and Miledi 1972) or currents (Anderson and Stevens 1973) produced by individual transmitter-operated channels. Characteristic properties of individual channels could now be discerned by fluctuations in voltage or current (*noise analysis*). A further technical refinement allowed direct recording of the single-channel currents using the *patch clamp technique* (Neher and Sakmann 1976; Hamill *et al.* 1981).

The second major advance was in determining the primary structure of receptor molecules by *cloning and sequencing cDNAs* encoding the receptors (Numa *et al.* 1983). It is now feasible to delineate the structure–function relationship of receptor molecules. The analysis of receptor function at the molecular level has been further strengthened by joining recombinant DNA technology with the patch clamp recording technique (Methfessel *et al.* 1986).

The third advance was in finding a set of intracellular messengers involved in the opening or closing of ionic channels in response to the transmitter. These intracellular factors are called the *second messengers* to distinguish them from the first messenger, the transmitter that binds to the receptor itself. A clue to the existence of this second type of message system came from the well-known differences in the pattern of the synaptic response. As mentioned earlier (section 2.1.2), the e.p.p.s are initiated at neuromuscular junctions within 1 ms of an impulse's arrival in the motor nerve terminal. In the first class of receptor, typified by the nicotinic ACh receptor at the neuromuscular junction, the ionic permeability change induced by the transmitter is instantaneous and short-lasting. By contrast, in the second class of synapse, typified by those in the autonomic nervous system (e.g., muscarinic ACh receptors), a single presynaptic impulse arouses a long-lasting synaptic response after a relatively long delay (Kuba and Koketus 1978; Kuffler 1980). The features of both types of *ligand-operated channel* are illustrated in Fig. 4.1. We use term *ligand* (to tie or bind) to refer to specific molecules that bind the receptor, including the transmitter.

The channel shown in Fig. 4.1A is a part of the receptor molecule, and conformational changes produced in the molecule by the transmitter binding cause the channel to open. Here, receptor activation affects the ionic channel directly, hence this class of receptor is called the *ionotropic receptor* (Eccles and McGeer 1979) or *intrinsic sensor receptor* (Hille 1992a). In the second class of receptor (Fig. 4.1B), the binding of the transmitter to the receptor activates intracellular second messengers, which in turn cause the ionic channel, located at a distance, to open or close. The initial step in dispatching the messengers is the production of some metabolite, hence such a receptor is called *metabotropic receptor* (Eccles and McGeer 1979) or *remote sensor receptor* (Hille 1992a). As we will see later (section 5.1.1), all metabotropic receptors are now known to be linked to a family of special proteins called the *guanine nucleotide-binding protein* (G-protein). Therefore, the metabotropic receptor is also termed the *G-protein-linked receptor*.

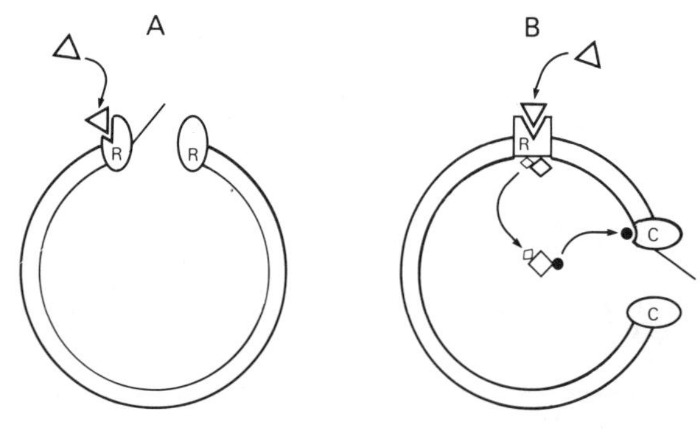

Fig. 4.1. Schematic illustrations of ionotropic (A) and metabotropic (B) receptors. A, the transmitter (triangle) binds to the ionotropic receptor (R) in the cell membrane (large ring) and opens the channel formed by the receptor subunit molecules. B, the transmitter binds to the metabotropic receptor, which uses a series of intracellular second messengers (diamonds and dots) to open or close the target ionic channel (C), located at a distance.

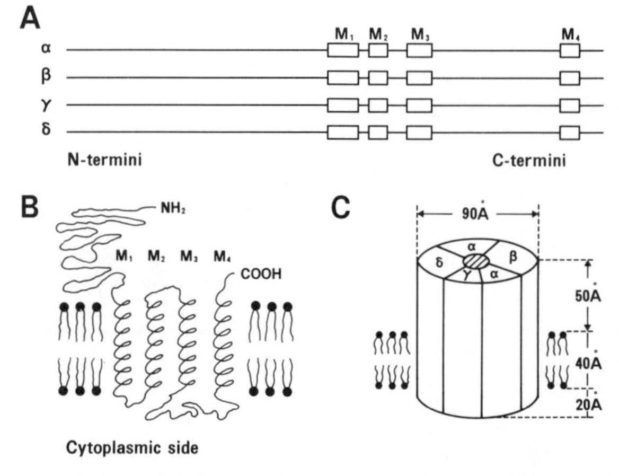

Fig. 4.2. A schematic model of the nicotinic acetylcholine receptor. A, each of the receptor's four subunits (α, β, γ, δ) contains four hydrophobic transmembrane segments (M_1–M_4) at equivalent positions. Amino (N) and carboxyl (C) termini are at opposite ends of each subunit. B, the hypothetical incorporation of each subunit in the hydrophobic lipid bilayer of the cell membrane. The array of hairpin structures with black circles represents the lipid bilayer. Each transmembrane segment of the subunit is assumed to traverse the cell membrane with an α-helix. C, the hypothetical assempbly of five subunits projecting through the cell membrane.

The information acquired on these three fronts has vastly enlarged our understanding of receptor mechanisms in synaptic response. We shall examine the structure and function of ionotropic receptors first.

4.1 THE MOLECULAR STRUCTURE OF IONOTROPIC RECEPTORS

We have seen that an ionic channel is a protein or an aggregate of proteins embedded in the cell membrane (section 2.2.2; Fig. 2.18). Receptor molecules are also integral proteins in the postsynaptic cell, having hydrophilic regions in contact with the extracellular and cytoplasmic fluids and a hydrophobic region intercalated within the inner hydrocarbon core of the cell membrane's lipid bilayer. The ionotropic receptor must have the binding site for its ligand (transmitter) and a pore (channel) through which ions permeate on receptor activation (Fig. 4.1A). How might an ionic pore be formed in the receptor molecule?

4.1.1 Muscular nicotinic acetylcholine receptors

The nicotinic ACh receptor of the *Torpedo* electric organ is an ionotropic receptor, the first to have its primary structure elucidated by the cloning and sequencing of the cDNAs (Numa *et al.* 1983). The molecular cloning, together with other findings described below, has disclosed many structural details indicative of how such receptors perform their function in synaptic transmission.

The nicotinic ACh receptor is a pentameric protein, consisting of two α-subunits, a β-, a γ-, and a δ-subunit. Each subunit is made of about 430–500 amino acids whose sequence is markedly homologous among all four subunits (Raftery *et al.* 1980; Numa *et al.* 1983). Before deducing the complete amino acid sequence, Raftery *et al.* (1980) sequenced 54 amino acid residues at one end (the N-terminus) of each of the four subunits from the purified mature ACh receptor. Every subunit precursor identified by recombinant DNA technology showed about 20 additional hydrophobic amino acids preceding the N-terminus. Evidently, these are the signal peptides (section 2.2.2) which translocate the protein across the membrane and are subsequently cleaved. The N-terminus of every subunit must then be located on the extracellular side, as illustrated in Fig. 4.2B.

The hydropathicity profile analysis (Box B; section 2.2.2) provides further details in subunit organization.

Each subunit seems to have four segments composed of 18–27 predominantly hydrophobic amino acids (Numa *et al.* 1983; Claudio *et al.* 1983). These segments occupy equivalent positions (M_1 to M_4 in Fig. 4.2A) in all four subunits. The four hydrophobic regions are assumed to be the transmembrane segments that traverse the hydrophobic core of the cell membrane with α-helical structures (Fig. 4.2B; but see Unwin 1993b). Thus, every subunit has both its N-and C-termini in the extracellular side (Fig. 4.2B). Both the high homology in amino acid sequence and the equivalent positions of transmembrane segments among the four subunits suggest that all the subunits are oriented vertically and symmetrically. This alignment would then form an ionic channel down the centre of the receptor (Fig. 4.2C). Such a scheme accords with Singer's (1975) prediction (Fig. 2.18B) and indeed with electron microscopic studies of purified or membrane-bound ACh receptors, viewed *en face* and from the side. In these studies, the receptor was found to form a rosette about 90 Å in diameter, around a distinct, central pit. This axial, rod-like structure not only spans the cell membrane but projects beyond the extracellular and cytoplasmic surfaces (Cartaud *et al.* 1978; Zingsheim *et al.* 1980; Kistler *et al.* 1982).

Improved resolution of these morphological details has been achieved by computer-generated image processing of electron micrographs of crystallized receptors (Brisson and Unwin 1985; Toyoshima and Unwin 1988). The resulting three-dimensional maps of the ACh receptor are shown in Fig. 4.3. The *en face* image viewed from the synaptic cleft (Fig. 4.3A) shows a central hole, about 25 Å in diameter, surrounded by five homologous subunits. The side view of the receptor (Fig. 4.3B) shows the vertical wall that the subunits form; it extends from the synaptic cleft (top) to the cytoplasmic side (bottom). The cytoplasmic ends of the subunits appear to attach to another protein—presumably the receptor-associated 43 K protein (a protein with a molecular weight of 43 000; Sealock *et al.* 1984; Froehner 1986). The 43 K protein helps anchor and cluster ACh receptors at postsynaptic sites (Froehner 1986; Froehner *et al.* 1990). The vertical wall lines a hole that narrows near the lipid bilayer to a diameter less than 10 Å which is unfortunately beyond the resolution (Fig. 4.3B). It is quite likely that this narrow tubular structure represents the ionic channel of the ACh receptor.

Some years ago, each subunit of the nicotinic ACh

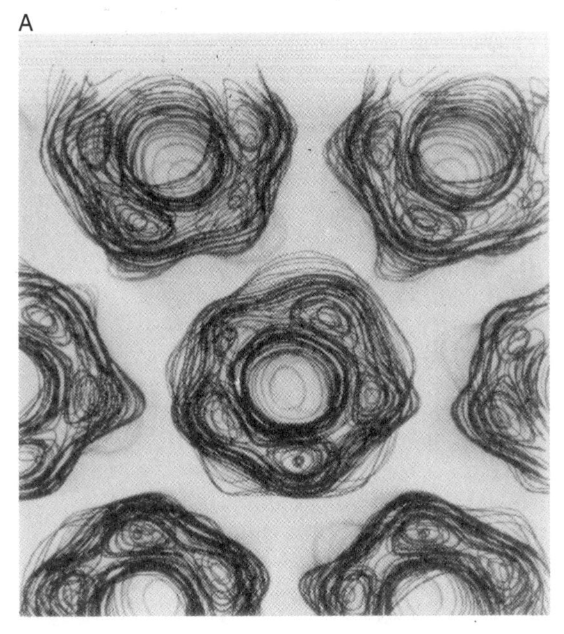

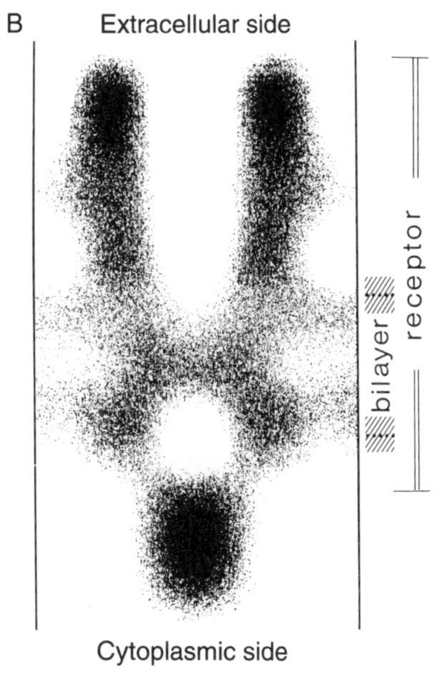

Fig. 4.3. Three-dimensional map of the nicotinic ACh receptor in the crystal lattice. A, *en face* view of the receptor from the extracellular side. (From Brisson and Unwin 1985.) Note the central hole symmetrically surrounded by five subunits. B, side view of the receptor. (From Toyoshima and Unwin 1988 and courtesy of N. Unwin.) Note the central hole narrowing in the lipid bilayer region. A molecule attached to the receptor in the cytoplasmic side is a receptor-associated protein, presumably the 43 K protein.

receptor was thought to comprise five transmembrane segments (Finer-Moore and Stroud 1984; Guy 1984). Although this possibility is not completely excluded, the molecular motif of other ionotropic receptors (see below) suggests that each ionotropic receptor subunit possesses four transmembrane segments. Furthermore, the ACh receptor structure recently viewed at high resolution suggests that the M_2 of each subunit is an α-helical segment, whereas other segments (M_1, M_3, M_4) have β-sheet configuration (Unwin 1993b). Thus, the detailed structure of the receptor has yet to be resolved (section 4.3.2).

As we will discuss later, the ligand, ACh, appears to bind to the extracellular N-terminus segment of each of the two α-subunits.

4.1.2 Neuronal nicotinic acetylcholine receptors

The nicotinic ACh receptor has been studied most extensively on the material purified from *Torpedo* electric organ. The nicotinic ACh receptor is also present at neuromuscular junctions and at neuronal synapses. The ACh receptors at neuromuscular junctions and in the *Torpedo* electric organ are basically identical in molecular structure, having four distinct subunits, whereas those at neuronal synapses exhibit only two types of subunit: α and β, or α and non-α (Boulter *et al.* 1987; Deneris *et al.* 1988).

The α-subunit of neuronal nicotinic receptors contains the ACh-binding site, and each subunit has four putative hydrophobic transmembrane segments (Duvoisin *et al.* 1989; Patrick *et al.* 1989). Interestingly, the α-and β-subunits of neuronal ACh receptors have multiple subtypes or variants (Goldman *et al.* 1987; Nef *et al.* 1988; Duvoisin *et al.* 1989). Such structural diversity also implies a functional diversity among the neuronal ACh receptors. Which neurones express which α and β variants remains an open question, however. In overall structure, the neuronal ACh receptor appears to be also pentameric, presumably with a stoichiometry of $\alpha_2\ \beta_3$ (Boulter *et al.* 1987; Anad *et al.* 1991; Cooper *et al.* 1991).

The neuronal ACh receptor subunits show high homology in amino acid sequence with the muscular receptor subunits, particularly in their putative transmembrane stretches. Both the neuronal and the muscular ACh receptors are made of different but homologous subunits, which presumably form the ionic channel within the receptor barrel. In basic molecular motif, therefore, both the muscular and the neuronal nicotinic ACh receptors are similar. Evidently, the two types of nicotinic ACh receptor must evolve from a common ancestral sequence of closely related genes.

4.1.3 Glycine receptors

Besides nicotinic ACh receptors, other ionotropic receptors generate rapid synaptic responses, including both excitatory and inhibitory synaptic potentials. Two major inhibitory transmitter substances are glycine and gamma-aminobutyric acid (GABA). The primary structures of their receptor subunits have been elucidated (Grenningloh *et al.* 1987; Schofield *et al.* 1987), and in both cases there is marked sequence homology with the nicotinic ACh receptor subunits.

The glycine receptor is schematically diagrammed in Fig. 4.4, in which receptor peptides are described in terms of their molecular weight. The purified glycine receptor contains peptides of 48 K, 58 K and 93 K (Graham *et al.* 1985; Betz 1987). The 48 K (α) and 58 K (β) subunits span the postsynaptic membrane. The 93 K peptide is located in the cytoplasm and may function to anchor the receptor the way the 43 K protein seems to anchor the muscular nicotinic ACh receptor (section 4.1.1; Fig. 4.3B).

The primary structure of the α-(Grenningloh *et al.* 1987) and β-subunits (Grenningloh *et al.* 1990) has been deduced. Each subunit has four hydrophobic segments. Again, both the N-and C-termini of each subunit are located on the extracellular side. The glycine receptor may be a pentameric assembly of both subunits (Langosch *et al.* 1988, 1990), which share a high degree

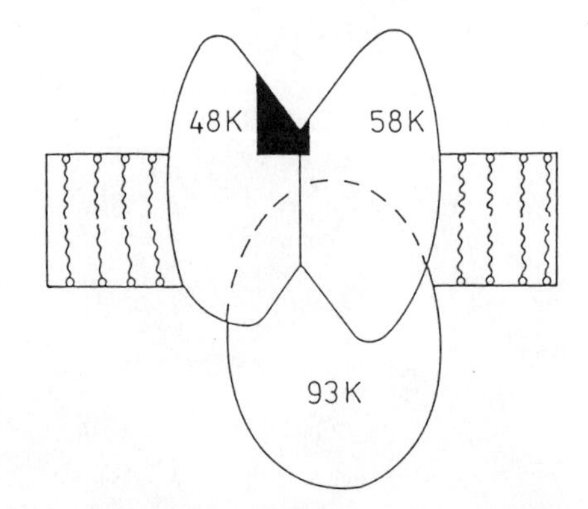

Fig. 4.4. A model of the glycine receptor complex. The membrane-spanning receptor consists of the small (48 K or α) and large (58 K or β) subunits. The stoichiometry is assumed to be $\alpha_3\beta_2$. On the cytoplasmic side (bottom), the receptor subunits may be conjoined to a receptor-associated protein (93 K). (From Betz 1987.)

Box E **The Hill coefficient**

If the reaction of oxygenation of haemoglobin (Hb) follows the equation, $Hb + O_2 \rightleftharpoons HbO_2$, the dissociation curve should be hyperbolic, but actually this curve is sigmoid. Hill (1910) assumed that Hb molecules aggregate, so that the reaction can be shown by $Hb_n + nO_2 \rightleftharpoons (HbO_2)_n$, where n represents the average number of Hb molecules per aggregate (corresponding to the number of subunits of a Hb molecule in the current term). Then, the ratio of the percentage of oxyhaemoglobin (y) to the percentage of deoxyhemoglobin ($1 - y$) at O_2 concentration (x) can be given by:

$$y/(1 - y) = Kx^n$$

where K is the equilibrium constant. Brown and Hill (1923) plotted log $y/(1 - y)$ against log x. This plot (*Hill plot*) gives a straight line with a slope of n, the *Hill coefficient*. The n was found to be 2.3. This result is consistent with the current notion of *cooperative interaction* of the multiple O_2-binding sites (or subunits) of a Hb molecule.

The same idea can be applied to the binding of a ligand to its receptor. Perhaps Clark (1926) was the first to apply the Hill plot to the dose-response curve of the transmitter action. This application has since been used widely. The Figure shows the Hill plot applied to the dose-response curve for gamma-aminobutyric acid (GABA) in the locust muscle fibres (Brookes and Werman 1973). The Hill coefficient was 2.8. This suggests that three GABA molecules are required for activation of each GABA receptor and that the GABA-binding sites are not independent of each other but positively cooperative. Thus, the binding of GABA at one site increases the affinity of the GABA binding at a second site, which in turn increases the binding affinity at a third site (Koshland *et al.* 1966).

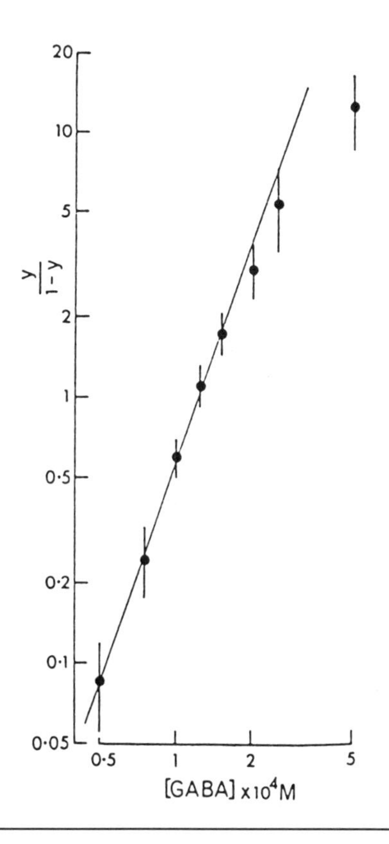

of sequence identity (58 per cent; Grenningloh *et al.* 1990).

We have seen that an injection of the exogenous mRNA specific for a Ca^{2+} channel protein into *Xenopus* oocytes results in synthesis of the channel protein; the injected oocytes then generate Ca^{2+} currents on depolarization, revealing functional expression of the channel protein (section 2.2.2). Glycine operates as an inhibitory transmitter mainly in the spinal cord (Krnjević 1974). When mRNAs extracted from the rat spinal cord were injected into oocytes, the injected oocytes responded to glycine (Akagi and Miledi 1988). This response increased in a dose-dependent manner as more glycine was applied. From the slope of this dose-response curve (the *Hill coefficient*; Box E), the number of glycine molecules required to activate each glycine receptor can be estimated. The Hill coefficient was close to three (Akagi and Miledi 1988), suggesting the presence of three ligand-binding sites for each receptor (Box E). The Hill coefficient evaluated for the glycine response recorded from spinal neurones *in situ* was similar (Werman 1969). The oocytes injected with the mRNA specific for the glycine receptor α-subunit responded to glycine (Schmieden *et al.* 1989), whereas the injection of its β-subunit specific mRNA failed to induce functional glycine receptors (Grenningloh *et al.*

1990). Thus, the α-subunit must have the glycine-binding site. The observed Hill coefficient then suggests that each glycine receptor has three α-subunits, its stoichiometry being $\alpha_3\beta_2$ (Langosch *et al.* 1988; Schmieden *et al.* 1989). That is, the basic molecular motif of the glycine receptor echoes that of the muscular and neuronal nicotinic ACh receptors.

4.1.4 Gamma-aminobutyric acid ($GABA_A$) receptors

Gamma-aminobutyric acid (GABA) ia also an inhibtory transmitter substance found in the central nervous system (Krnjević 1974). Two types of GABA receptor are now distiniguished by their pharmacological properties (Dunlap 1981; Hill and Bowery 1981), the $GABA_A$ and $GABA_B$ receptors (Hill and Bowery 1981). The $GABA_A$ receptor proves to be an ionotropic receptor, whereas the $GABA_B$ receptor is a metabotropic receptor (Bormann 1988; Nicoll 1988). This section examines the $GABA_A$ receptor.

Figure 4.5 predicts the topology of the $GABA_A$ receptor, given its amino acid sequences (Schofield *et al.* 1987). Note that only two types of subunit are depicted: α (53 K) and β (58 K). Both subunits show a high homology (57per cent), and each appears to have four transmembrane segments. In its basic motif the $GABA_A$ receptor thus conforms to the nicotinic ACh receptor.

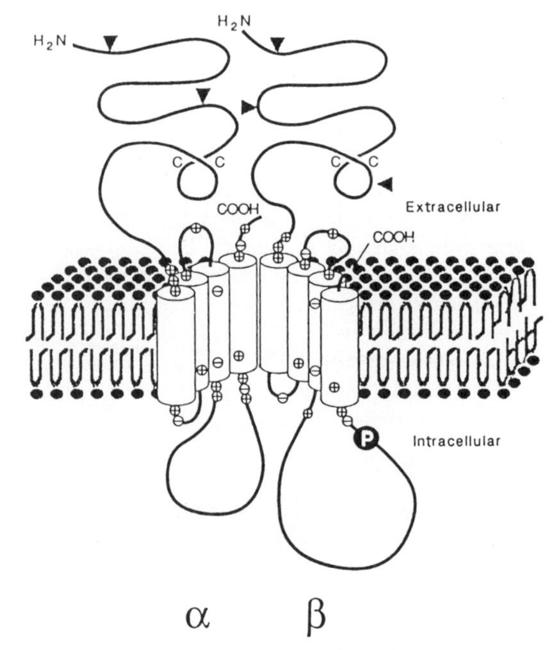

Fig. 4.5. A topological model of the $GABA_A$ receptor, comprising the α-and β-subunits. The receptor stoichiometery was assumed to be $\alpha_2\beta_2$ here (but see text). Each subunit has four transmembrane segments. Triangles show potential N-glycosylation sites. C, cysteines presumably involved in the disulphide bond. P, phosphorylation site. (From Schofield *et al.* 1987.)

Initially, the subunit structure of the $GABA_A$ receptor complex seemed to be $\alpha_2\beta_2$ (Mamalaki *et al.* 1987; Schofield *et al.* 1987). Subsequent studies found unexpected complexities, however, in structurally interpreting these $GABA_A$ receptors.

A bewildering number of distinct $GABA_A$ receptor subunit isoforms has been derived from cDNA cloning, including α-, β-, γ-, and δ-subunits as well as their variants (Levitan *et al.* 1988b; Olsen and Tobin 1990). The scheme in Fig. 4.5 suggests that the α- and β-subunits are essential for expressing the normal function of the $GABA_A$ receptor. This prediction was addressed by injecting the mRNAs specific for both the α- and β-subunits into *Xenopus* oocytes in order to *coexpress* the two subunits. The injected oocytes responded to GABA, and this response was inhibited by bicuculline, a specific blocker of the $GABA_A$ receptor (Levitan *et al.* 1988a, b). However, the dose-response curve examined on these oocytes showed no evidence for the cooperativity of GABA action, the Hill coefficient being less than or equal to unity, in contrast with the behaviour of native $GABA_A$ receptors (Box E). Moreover, the GABA response recorded from the oocytes showed no potentiation by benzodiazepine (Levitan *et al.* 1988a). The $GABA_A$ receptor is known to have the benzodiazepine-binding site, having been originally purified, in fact, by benzodiazepine affinity chromatography (Sigel and Barnard 1984). Thus, additional factors may be required in order to attain expression of a fully functional $GABA_A$ receptor. It now seems that a newly cloned $GABA_A$ receptor subunit (γ_2) is essential for benzodiazepine sensitivity (Pritchett *et al.* 1989; Sigel *et al.* 1990). The cooperativity of GABA action was also found when the three subunits (α, β, and γ) were coexpressed (Sigel *et al.* 1990; Verdoorn *et al.* 1990). Some $GABA_A$ receptors in central neurones are known to be blocked reversibly by zinc (Zn^{2+}; Westbrook and Mayer 1987). Interestingly, the presence of a γ-subunit in any combination with the other subunits reduces Zn^{2+} sensitivity of the $GABA_A$ receptor (Draguhn *et al.* 1990). It is possible that the native $GABA_A$ receptors are heterogenous and consist of both α,β and α,β,γ subclasses (Smart 1992; Moss *et al.* 1992). For now, the exact molecular stoichiometry of the $GABA_A$ remains unknown, so that the diagram in Fig. 4.5 is admittedly speculative and probably over-simplified.

4.1.5 Glutamate receptors

L-Glutamate has long been proposed as an excitatory transmitter in the mammalian central nervous system (Krnjević 1974). Pharmacological studies point to at least three types of ionotropic glutamate receptor (Watkins and Evans 1981; Fagg 1985; Mayer and Westbrook 1987b): *NMDA* (*N*-methyl-D-aspartate) receptor, kainate receptor, and *AMPA* (α-amino-3–hydroxy-5–methyl-isoxazole-4–propionic acid) receptor. Moreover, metabotropic glutamate receptors have also been found (Sladeczek *et al.* 1985; Sugiyama *et al.* 1987; Masu *et al.* 1991). The presence of a specific receptor blocker confirms the NMDA receptor as a distinct subtype, but how many other ionotropic glutamate receptor subtypes exist is less certain. It is even difficult to distinguish between kainate and AMPA receptors (Watkins *et al.* 1990). For this reason ionotropic glutamate receptors are often classified into two classes: NMDA receptor and *non-NMDA* (or AMPA/kainate) receptor.

A non-NMDA glutamate receptor subunit, GluR1, was first cloned and sequenced by Hollmann *et al.* (1989). This molecule comprises 889 amino acids and has four putative transmembrane segments, again having its N-and C-termini on the extracellular side. The similarity in the basic structure accords with the view that all ionotropic receptors are members of a gene superfamily (Barnard *et al.* 1987; Betz 1990), although the amino acid sequence of the non-NMDA receptor reveals little homology with the ACh, glycine or $GABA_A$ receptor subunits.

Six other non-NMDA receptor subunits subsequently cloned (GluR2 to GluR7) showed a high homology with the GluR1 and the same transmembrane topology (Sommer and Seeburg 1992; Seeburg 1993). A combination of different subunits would form a large variety of non-NMDA receptors. Activation of the ionotropic glutamate receptor opens its channel selectively for cations. The non-NMDA receptors have been considered virtually impermeable to Ca^{2+}; yet, some non-NMDA receptor channels now prove permeable to Ca^{2+} (Ogura *et al.* 1990; Iino *et al.* 1990). This observation was followed by three exciting findings. First, the expression of various combinations of the receptor subunits (GluR1, GluR2 and GluR3) in *Xenopus* oocytes showed that the presence of GluR2 (GluR1 plus GluR2, or GluR2 plus GluR3) eliminates Ca^{2+} permeability in the expressed receptor (Hollmann *et al.* 1991). Second, this disparate receptor channel property depends on the positively charged amino acid, arginine (Arg) located in the second transmembrane segment (M_2) of the GluR2 subunit (Hume *et al.* 1991; Verdoorn *et al.* 1991). The Arg residue in this position is found only in the GluR2 subunit, whereas an uncharged glutamine (Gln) residue is present instead at the equivalent position in other subunits. The replacement of the Arg in the GluR2 with Gln causes the expressed receptor to become permeable to Ca^{2+}; similarly, the replacement of the Gln residue in other subunits with Arg eliminates Ca^{2+} permeability. A further mutational analysis showed that both the positive charge and the size of the side chain of the amino acid located at the particular site (Arg/Gln site) of the M_2 segment regulate the divalent permeability of the glutamate receptor (Burnashev *et al.* 1992a). This finding also implies that the M_2 segment is involved in the formation of the wall lining the ionic pore (section 4.3.2).

The third finding was a novel mechanism for altering the sequence of transcripts for the glutamate receptor subunits. Sommer *et al.* (1991) analysed the sequence of the genomic DNA encoding the M_2 segment of different GluR subunits. As expected, the genomic DNA encoding the Gln residue in the M_2 segment of GluR3 or GluR4 shows a set of three nucleotides (codon) specifying Gln (the codon being CAA or CAG). Surprisingly, the genomic DNA expected to encode the Arg residue in the M_2 segment of the GluR2 also showed a Gln codon (CAG), instead of an Arg codon. The primary RNA transcript from the GluR2 gene must then undergo changes in its sequence, giving rise to an mRNA coding for the Arg residue (Sommer *et al.* 1991). This process of *RNA editing*, originally found in the protozoa mitochondria (Benne *et al.* 1986), now seems to be employed also for regulation of receptor molecules in the vertebrate (Sommer and Seeburg 1992).

The long-awaited cloning of the NMDA receptor was finally achieved. (Moriyoshi *et al.* 1991). The NMDA receptor has unique properties and plays a crucial role in long-term potentiation of central synaptic function (see Part II, 7.2.2). One way it is known to differ from non-NMDA receptors is that the activated NMDA receptor is much more permeable to Ca^{2+} (Mayer and Westbrook 1987a; Ascher and Nowak 1988) and blocked by extracellular Mg^{2+} (Mayer *et al.* 1984; Nowak *et al.* 1984; section 7.2.2). Presumably, the resultant increase in the intracellular Ca^{2+} level triggers activation of intracellular messengers and protein kinase, thereby inducing synaptic plasticity. The cloned NMDA receptor subtype (NMDAR1) consists of 938 amino acids and unexpectedly reveals marked structural similarities with non-NMDA receptors (Moriyoshi *et al.* 1991). Interestingly, the NMDAR1 has an uncharged asparagine (Asn) at the Arg/Gln site of the M_2 segment. The replacement of the Asn in the NMDAR1 by Gln decreases Ca^{2+} permeability as well as the blocking effect by Mg^{2+} (Burnashev *et al.* 1992b; Mori *et al.* 1992). Therefore, like non-NMDA receptors, the M_2 segment of the NMDA receptor also seems to line the ionic channel of the receptor.

Although the NMDAR1 expressed in *Xenopus* oocytes fulfilled all the properties characteristic of the native NMDA receptor, the magnitude of currents evoked by receptor activation was much smaller than that observed in oocytes injected with non-specific mRNAs extracted from the brain (Moriyoshi *et al.* 1991). This suggests that the formation of a fully functional NMDA receptor requires additional complementary subunits. Newly cloned NMDA receptor subtypes (NMDAR2-A, -B, -C, and -D) did not express functional NMDA receptors by themselves but potentiated the receptor function when they were coexpressed with the NMDAR1 subunit (Meguro *et al.* 1992; Monyer *et al.* 1992). It is likely that the native NMDA receptors require the NMDAR1 as their fundamental subunit and form a heteromeric configuration with different NMDAR2 subtypes (Kutsuwada *et al.* 1992; Nakanishi 1992).

Figure 4.6 gives a list of 14 glutamate receptor subunits with their preferential ligand (see reviews by Sommer and Seeburg 1992 and Seeburg 1993). These receptor subunits were found in different laboratories and given different designations. Numbering the subunits from GluR1 to GluR7 is simple but ignores the degree of their sequence similarities. Sakimura *et al.* (1992) classified the receptor subunits into several subfamilies, designated with Greek letters, according to the sequence homology. A unification of nomenclature is desirable, but the pace of current research on glutamate receptors seems to be too rapid to correct this inconvenient problem of nomenclature.

As we will discuss later (Section 7.2), NMDA receptor channels play a crucial role in inducing functional and developmental plasticity at central synapses. Distinct properties of different NMDA receptors appear to depend on the $NMDAR_2$ subunit composition. In fact, different $NMDAR_2$ subunits show regionally and developmentally specific expression in the central nervous system (M. Watanabe *et al.* 1992, 1993; Monyer *et al.* 1994).

Receptor type	Preferential ligand	Other nomenclature
GluR1	AMPA	GluR–A or α1
GluR2	AMPA	GluR–B or α2
GluR3	AMPA	GluR–C or α3
GluR4	AMPA	GluR–D or α4
GluR5	Kainate	β1
GluR6	Kainate	β2
GluR7	Kainate	β3
KA–1	Kainate	γ1
KA–2	Kainate	γ2
NR1	NMDA	ζ1
NR2A	NMDA	ε1
NR2B	NMDA	ε2
NR2C	NMDA	ε3
NR2D	NMDA	ε4

Fig. 4.6. Classification of ionotropic glutamate receptors. Note that each receptor subtype has different designations. See text.

4.1.6 Serotonin (5HT-3) receptors

Serotonin (5–hydroxytriptamine; 5HT) is a biogenic amine and produces diverse pharmacological actions in both central and peripheral neurones (Krnjević 1974). The physiological and pharmacological studies now classify the serotonin receptor into multiple subtypes: 5HT-1a, -1b, -1c, -1d, 5HT-2, 5HT-3, and 5HT-4 (Bradley *et al.* 1986). The actions of 5HT are generally slow and long-lasting, characteristic of the metabotropic receptor (Peroutka 1988; Julius 1991). At present, only the 5HT-3 subtype proves to be an ionotropic serotonin receptor (Derkach *et al.* 1989; Sugita *et al.* 1992).

The 5HT-3 receptor was cloned and sequenced (Maricq *et al.* 1991). This receptor consists of 487 amino acids, and its molecular structure is again characterized by four putative transmembrane segments. The 5HT-3 receptor expressed in oocytes generated cation-selective currents in response to 5HT (Maricq *et al.* 1991). The mRNA encoding 5HT-3 receptor is expressed in the brain as well as in the heart. The 5HT-3 receptor is clearly involved in excitatory synaptic transmission in central neurones (Sugita *et al.* 1992), but which neuronal pathways employ 5HT as a rapid excitatory transmitter remains to be clarified.

Rapid synaptic responses are mediated by a variety of transmitters, including ACh, amino acids, and amines. All these ligand-gated ionotropic receptor channels have two structural features in common: first, they comprise multiple homologous subunits; second, each subunit has four transmembrane segments. Adenosine triphosphate (ATP) is also shown to act as a rapid excitatory transmitter for central and peripheral neurones (Edwards *et al.* 1992; Evans *et al.* 1992). Although the structure of the ionotropic ATP receptor is as yet unknown, probably its molecular motif would again conform to other ionotropic receptors. Such structural similarity also implies a similarity in functional mechanisms among different ionotropic receptors. We now turn to functional aspects of the ionotropic receptor.

4.2 HOW IONOTROPIC RECEPTORS FUNCTION

Locally applying ACh to a neuromuscular junction produces a depolarization (*ACh potential*) in the muscle fibre, the amount of depolarization being proportional to the amount of ACh applied. A constant dose of ACh induces a significant increase in voltage fluctuations or 'noise' around the average level of depolarization (Katz and Miledi 1972). What causes the ACh responses to fluctuate? Katz and Miledi (1972) postulated that probabilistic variations in how many ACh receptor channels are open from moment to moment would account for this phenomenon. Thus, each elementary 'blip' represents the opening of a single ACh receptor, and the number of the open channels varies from moment to moment. Fluctuation analysis of the ACh noise unveiled the characteristics of the elementary ACh current (Katz and Miledi 1972; Anderson and Stevens 1973). Ultimately, the electric current flowing through a single ACh receptor channel (*single-channel current*) was recorded with the patch clamp technique (Neher and Sakmann 1976).

The membrane current recorded from a muscle cell in response to ACh (whole-cell current) is the sum of the single-channel currents produced by all the individual ACh receptors activated. We will now discuss the basic principles governing biophysical aspects of the ionotropic receptor channel. Nicotinic ACh receptors expressed in *Xenopus* oocytes illustrate them well. Mishina *et al.* (1984) showed that functional ACh receptors can be expressed in oocytes injected with the specific mRNAs prepared from the cDNAs encoding the four subunits (α, β, γ, and δ) in the *Torpedo* electric organ. With this procedure, functional ACh receptors were also expressed, using the specific mRNAs for the four subunits in the calf muscle (Sakmann *et al.* 1985). The properties of the expressed receptors were distinctly different between the *Torpedo* ACh receptor and the calf muscle receptor. Let us examine how different they are, in terms of the whole-cell current and the single-channel current.

4.2.1 Relationship between the whole-cell current and single-channel currents

Transmission at the neuromuscular junction is mediated by selective permeation of cations through the ACh receptor channel. As expected, the ACh receptors expressed in oocytes are also cationic-permeable channels. A *Xenopus* oocyte is about 1 mm in diameter, so that two microelectrodes can be inserted easily to record its whole-cell current by one electrode while its membrane potential is clamped at desired levels by the second electrode. Figure 4.7 shows inward currents recorded from oocytes injected with the mRNAs specific for the four subunits of ACh receptors in *Torpedo* electric organ (A), and in calf muscle (B) in response to the application of ACh. Although each oocyte synthesized comparable densities of the receptor, responses were consistently greater in the calf than in the *Torpedo* ACh receptor. Moreover, when the membrane potential of the oocyte was altered, the ACh current of the *Torpedo* receptor was linearly related to the membrane potential (Fig. 4.7*c*), unlike the non-linear relation for the calf ACh receptor (Fig. 4.7*d*). In the latter, the current induced by ACh flows inward more easily than outward, a phenomenon called *inward rectification*. Nonetheless, in both types of receptor the ACh current switches from inward to outward at the same level of membrane potential, a reversal potential of about −10 mV (Box A). This indicates that the species as well as the permeability ratio of cations involved in the generation of ACh currents is identical in both types of ACh receptor. Furthermore, the Hill coefficient calculated from the slope of the dose-response relation (Fig. 4.7*e*, *f*) was essentially the same for both receptors (1.7 and 1.8). In other words, both types of receptor seem to open when two ACh molecules bind simultaneously to a receptor molecule (Box E).

Single-channel currents can be recorded by pressing the tip of a patch clamp electrode tightly against the cell membrane surface, because ionic currents through the small membrane patch flow into the pipette rather than outward through the tight pipette–membrane seal (*cell-attached patch clamp recording*; Hamill *et al.* 1981). To induce ACh currents under this condition, however, ACh must be present in the solution of the recording electrode. It is then difficult to control the dose or the timing of ACh application. This difficulty can be overcome by the *outside-out patch technique* (Hamill *et al.* 1981). When the electrode is pulled away slowly from the cell surface after rupturing the membrane patch under the electrode (the whole-cell mode), the electrode

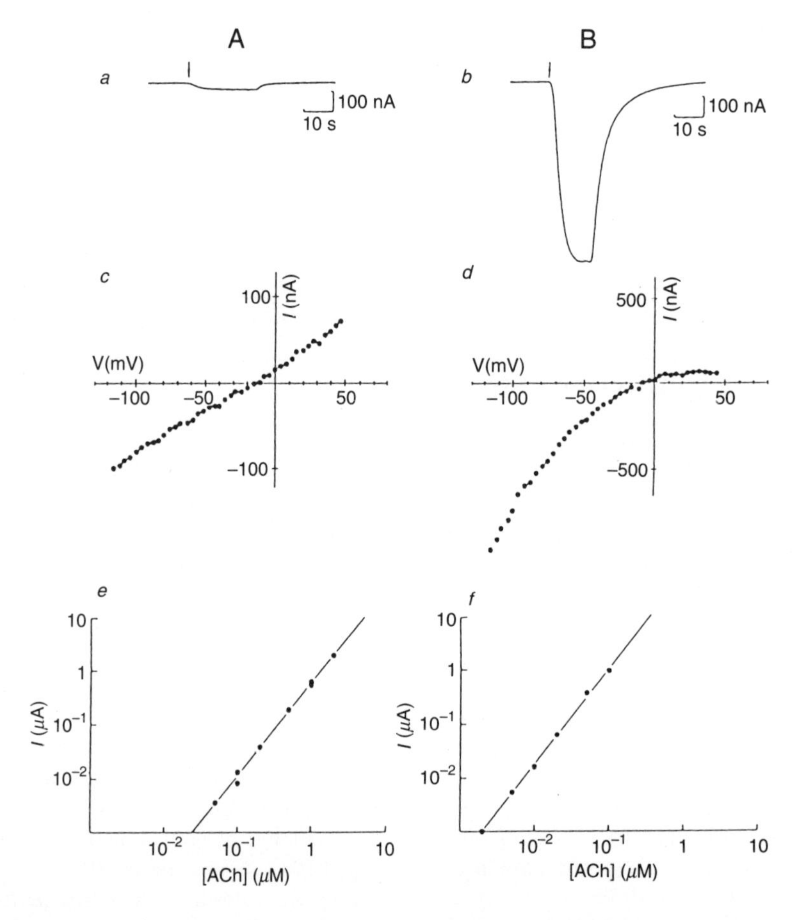

Fig. 4.7. Functional expression of nicotinic ACh receptors in Xenopus oocytes injected with mRNAs specific for receptor subunits in the *Torpedo* electric organ (A) and the calf muscle (B). *a* and *b*, inward currents recorded under voltage clamp in response to ACh at a membrane potential of −70 mV. *c* and *d*, relationship between the ACh current and the membrane potential of the injected oocytes. *e* and *f*, the dose-response curves for ACh concentrations versus ACh currents plotted on double logarithmic cordinates. The slope (Hill coefficient) is 1.7 for *e* and 1.8 for *f*. (From Sakmann *et al.* 1985.)

tip will be sealed by a new membrane patch which is separated from the cell and oriented with its extracellular surface facing the bath solution (*cell-free outside-out patch*). Therefore, single-channel currents can be induced by applying ACh to the bath. With this technique, single-channel currents were recorded from oocytes injected with the mRNAs encoding the *Torpedo* (Fig. 4.8*A*) and calf (B) ACh receptors. ACh applied to the outside-out patch induced elementary current pulses approximately constant in magnitude. Thus, the single-channel current flow is either *on* or *off*, with either an open or closed state, as previously predicted (Magleby and Stevens 1972; Anderson and Stevens 1973).

By analogy with the enzyme–substrate reaction, we can describe the change in ACh receptor state (Del Castillo and Katz 1957; Katz and Miledi 1972; Magleby and Stevens 1972; Colquhoun and Hawks 1981):

$$A + R \underset{k_2}{\overset{k_1}{\rightleftharpoons}} AR \underset{\alpha}{\overset{\beta}{\rightleftharpoons}} AR^* \qquad (1)$$

That is to say, ACh (A) binds to its receptor (R) to form a ligand–receptor complex (AR) whose initial closed state (AR) changes to the open state (AR*). These reactions are reversible, and the shift from one state to another occurs at certain reaction speeds (*rate constants*), indicated by k_1, k_2, α and β.

As illustrated in Fig. 4.8, the single channels stay open much longer in the calf muscle ACh receptor (B)

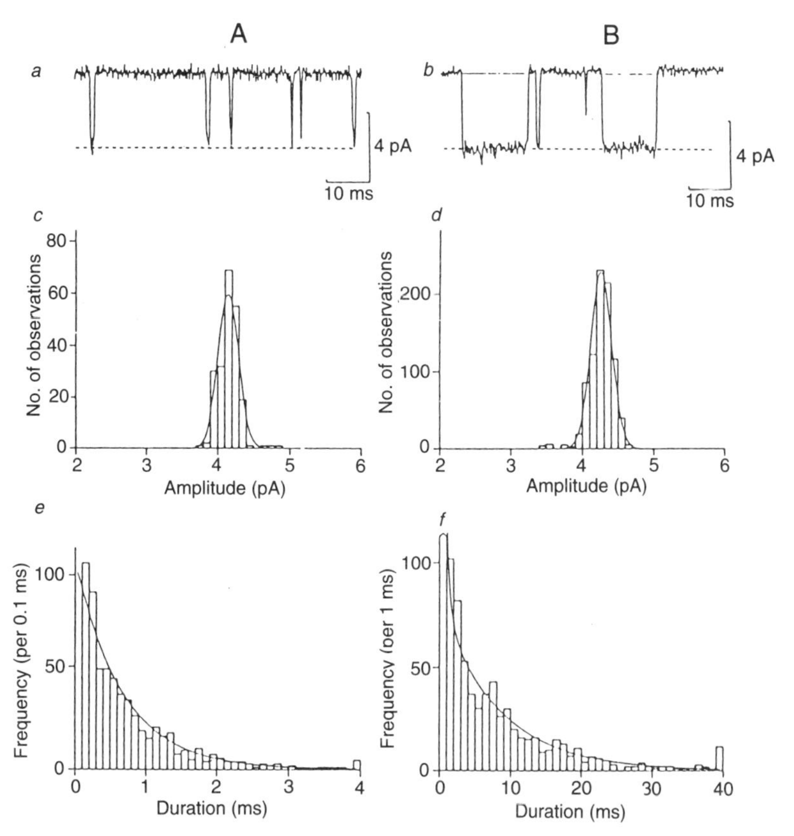

Fig. 4.8. Single-channel currents recorded from outside-out patches from oocytes injected with mRNAs specific for ACh receptor subunits of *Torpedo* (A) and calf (B). *a* and *b*, sample records in response to ACh; the holding membrane potential was −100 mV. *c* and *d*, amplitude histogram of single-channel currents recorded at −100 mV. *e* and *f*, histograms of open times. (From Sakmann *et al.* 1985.)

than in *Torpedo* (A). The mean duration of open times measured for a number of events is only 0.6 ms in the *Torpedo* receptor (Fig. 4.8*e*) versus 8.6 ms in calf (*f*). On the other hand, in mean magnitude their single-channel currents are comparable (Fig. 4.8*a–d*; measured at a membrane potential of −100 mV).

An ACh receptor may open immediately whenever two ACh molecules bind to the receptor, whereas its closure may result from the dissociation of the ACh molecules from the receptor. If this were the case, there would be no need to postulate two states for the ligand–receptor complex. The mean open time for a given ACh receptor, however, is independent of the dose of ACh applied. In other words, unlike the ACh-binding step (AR), the activation step (AR*) is not directly related to k_1; hence, AR and AR* must be two separate states (Magleby and Stevens 1972). The mean open time of individual channels can be given by the time constant of the exponential curve describing the distribution of their open times (Box F). The lower the rate constant α, the longer the mean open time, because the reaction speed of the transition from AR* (open state) to AR (closed state) is slower. The mean open time is thus the reciprocal of α. Then, the ACh receptor must have a lower α in calf muscle than in the *Torpedo* electric organ.

The whole-cell currents in Fig. 4.7*a*, *b* represent the sum of the single-channel currents produced by all the individual ACh receptors functionally expressed in the oocyte. Therefore, the whole-cell current (I) is:

$$I = N \cdot i \cdot p(\mathrm{o}) \qquad (2)$$

where N is the number of functional receptors expressed, i the magnitude of the single-channel current, and $p(\mathrm{o})$ is the mean probability of individual channels being open (Hille 1992a). As noted, each oocyte expresses the same density of functional ACh receptors, so N should

Box F The distribution of single-channel open times

The single-channel current amplitudes show a normal distribution around the mean (Fig. 4.8*c*,*d*), whereas the distribution of open times follows an exponential curve (Fig. 4.8*e*,*f*). This is because each open channel closes at random with a certain probability. Its probability of being closed is, in fact, invariant at any given moment after opening. Let p be the mean probability of closing during a short interval of time, Δt, after a channel opens. Of course, the channel may fail to close during this time interval. This probability of failure is $(1 - p)$. The probability of the channel's closing during the next interval, between Δt and $2\Delta t$, is $p \cdot (1 - p)$. It may again fail to close during this time interval, and the probability of the 'double fault' is $(1 - p)^2$. During the third interval, between $2\Delta t$ and $3\Delta t$, the probability of its closing is $p\ (1 - p)^2$. Generally, the probability that the channel will close during the nth Δt interval is $p \cdot (1 - p)^{n-1}$. In other words, the shortest class interval of time has the highest likelihood of a closing, and the chance of remaining open progressively decreases at longer durations. The probability that the channel will not close until the nth Δt interval is virtually nil, if n is sufficiently large. Mathematically, the entire distribution of actual open times follows an exponential curve whose time constant is the mean open time, as shown in Fig. 4.8e,f.

p: Probability that the channel will close during a time interval, Δt. $p < 1$.

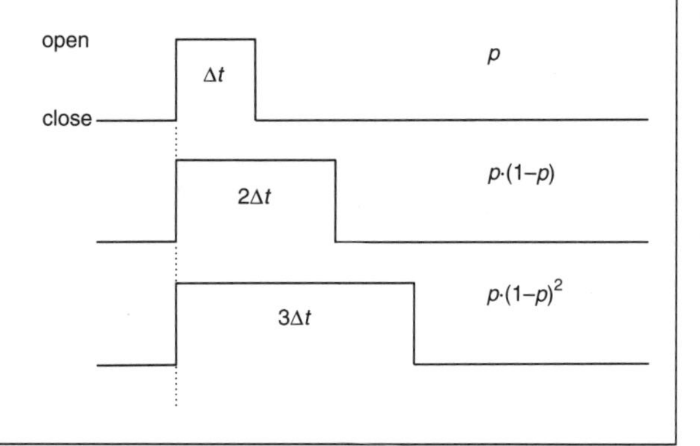

be the same for the two ACh receptors. Likewise, the magnitude of their single-channel currents (i) is the same (Fig. 4.8*c*,*d*). Clearly, the difference in responsiveness (I) between the two types of receptor reflects a difference in the mean probability that their individual channels are open, p(o).

The *mean open probability*, p(o) derives from the fraction of time the channel spends open during the total sampling time. In other words, p(o) depends on the rate of channel opening (how often the channel opens) and the mean open time (how long the channel spends open, on average, once it opens). Thus, the longer the mean open time ($1/\alpha$) the greater the p(o). Therefore, the difference in the magnitude of their whole-cell currents (I) results largely, if not entirely, from the difference in their mean open times.

The inward rectification in the calf ACh receptor (Fig. 4.7*d*) magnifies the difference in the whole-cell current as the two receptors are compared at more negative membrane potentials of the oocytes. This phenomenon in the calf receptor stems from prolongation of the receptor's mean open time by hyperpolarization. From these results it is clear that the *Torpedo* and the calf muscle ACh receptors distinctly differ in behaviour, primarily in their mean open times. Which molecular substrate might be responsible for this difference?

4.2.2 Subunit specificity in receptor function

The different behaviour of ACh receptors in the two species was analysed by constructing *hybrid receptors* using subunits from the *Torpedo* and the calf muscle. Thus, Sakmann *et al.* (1985) injected oocytes with mRNAs specific for the calf α–subunit and the *Torpedo* β–, γ–, and δ–subunits. Although this procedure generated a hybrid ACh receptor that had the three *Torpedo* subunits plus the α–subunit native to the calf muscle, it behaved like the original *Torpedo* ACh receptor. The α–subunits in the two species thus appear to be functionally similar. On the other hand, when only the δ–subunit of *Torpedo* was replaced by the calf δ–subunit, the resulting hybrid ACh receptor had a long mean open time (8.6 ms), like the original calf ACh receptor. The δ–subunit thus appears to affect the gating or channel closing process. Unfortunately, other hybrid ACh receptors substituting β–or γ–subunits of calf could not be tested, because with these combinations the injected oocytes failed to express functional

ACh receptors. Still, this study on hybrid receptors strongly suggests that a particular subunit can specify the receptor properties.

Another approach along this line revealed that the developmental change of ACh receptors is mediated by switching the expression of genes encoding different subunits. The e.p.p.s recorded from neuromuscular junctions in fetal or neonatal rats are known to be slower in time course than in adult rats (Diamond and Miledi 1962). Consistent with this finding, the single-channel currents of ACh receptors also differed between adult and neonatal mammalian muscles, the latter having a longer mean open time and a smaller current magnitude (Sakmann and Brenner 1978; Fischbach and Schuetze 1980; Vicini and Schuetze 1985). Moreover, these ACh receptors in adult and neonatal muscles were distinguished immunologically using appropriate antibodies (Hall *et al.* 1985). Therefore, the ACh receptors expressed in adult and neonatal muscles must have distinct molecular structures which can be recognized by different antibodies. Which molecular substrate distinguishes the adult and neonatal types of ACh receptor? The answer lies in a novel subunit, designated the ε-subunit, discovered in the ACh receptor of calf muscle (Takai *et al.* 1985; Mishina *et al.* 1986). In amino acid sequence, the calf ε-subunit was more homologous with the calf γ-subunit (53 per cent) than with other subunits (30–40 per cent; Takai *et al.* 1985).

Figure 4.9A illustrates the single-channel currents of ACh receptors expressed in oocytes injected with the calf α, β, γ, and δ mRNAs (a), and with calf α, β, δ, and ε mRNAs (b). When the γ-subunit was present, in the place of the ε-subunit (Fig. 4.9Aa), the single-channel current was smaller and its open time was longer, compared with the presence of the ε-subunit (Ab; Mishina *et al.* 1986). These features correlated well with recordings from ACh receptor channels in bovine muscle tissue cultured from the fetus (Fig. 4.9Ba) and the adult (Bb), respectively (Mishina *et al.* 1986). Moreover, hybridization of RNA extracted from muscle tissues by the subunit cDNA probe (*Northern blot*) showed that the α-subunit mRNA is abundantly expressed at fetal stages, whereas the expression of the ε-subunit mRNA is enhanced only at postnatal stages (Mishina *et al.* 1986). Thus, the properties of ACh receptors expressed in oocytes reflect well those expressed *in situ*.

It seems likely that the γ-subunit of the ACh receptor initially expressed at early developmental stages is gradually replaced by the ε-subunit. This subunit switch would explain why the receptor channel (or e.p.p.) properties differ between immature and adult animals. This is analogous to the developmental change in haemoglobin subunit molecules, which switch from the fetal form, $\alpha_2\gamma_2$, to the adult form, $\alpha_2\beta_2$. Switching from the γ-subunit to the β-subunit renders the adult haemoglobin more efficient in unloading oxygen into the tissues. Similarly, at glycine-mediating inhibitory synapses the developmental switching from the fetal (α_2) to adult (α_1) isoform appears to lead to a significant accelerating of the functional kinetics of the glycine receptor (Takahashi *et al.* 1992).

How might the expression switching from the γ-to ε-subunit be regulated? The expression of the ε-subunit may be triggered by nerve–muscle contact at the early developmental stage (Brenner *et al.* 1990) or by some *trophic factor* originating from the motor neurones

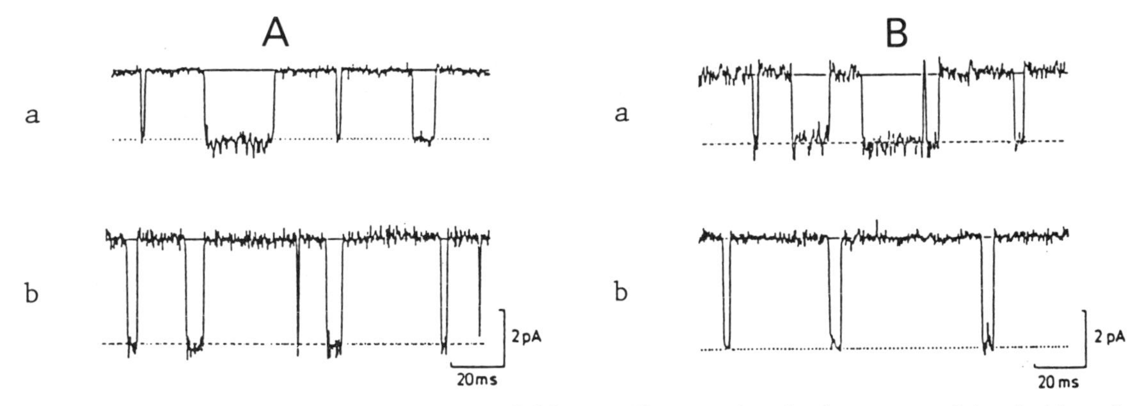

Fig. 4.9. Single-channel currents of ACh receptors recorded from outside-out patches. Aa, from an oocyte injected with α-, β-, γ-and δ-subunit specific mRNAs of the bovine ACh receptor. Ab, similar to Aa, but using α-, β-, δ-, and ε-subunit-specific mRNAs. Ba, from a fetal bovine muscle cell. Bb, from an adult bovine muscle cell. All responses were elicited by 0.5–1 μM ACh at a membrane potential of −60 mV. (From Mishina *et al.* 1986.)

(Martinou *et al.* 1991). The γ-subunit suppressed in adult rat muscle is expressed anew after denervation of the muscle (Gu and Hall 1988). The synthesis of the γ- and ε-subunits may thus be controlled neurally or by muscle activity. These topics relate to neurotrophic action and will be discussed in Part III (section 12.1.2).

4.3 THE STRUCTURE–FUNCTION CORRELATION IN IONOTROPIC RECEPTORS

How ionotropic receptors operate is steadily unfolding: information about their molecular structure is available, and their functional properties are being analysed at the single-channel level. By assembling the available data, we can now begin to delineate the molecular basis for the biophysical characteristics of ionotropic receptors.

Several questions spring to mind. Where does the ligand (transmitter) bind to its receptor molecule? How does this ligand binding cause the receptor channel to open? Where is the ionic channel located in an ionotropic receptor? How does the receptor channel restrict permeation to cations or anions selectively? Although comprehensive answers are still lacking, new experimental approaches have yielded significant insights. This section traces the remarkable progress molecular neurobiology has made on these key issues.

4.3.1 Ligand-binding site

The site at which a receptor binds its ligand (transmitter) must be exposed to the extracellular synaptic cleft. For the muscular and *Torpedo* nicotinic ACh receptor, the ligand-binding site apparently lies on the long, extracellular N-terminus segment (containing about 200 amino acids) of the two α-subunits (Karlin 1980; Conti-Tronconi and Raftery 1982). An ACh molecule has a positively charged quaternary ammonium head. This positively charged head may interact with negatively charged sites of the receptor (Karlin 1969), although the scheme discussed below points to another basis than simple charge–charge interactions.

Some time ago, Karlin (1969) predicted the ACh-binding site from the action of a quaternary ammonium derivative, MBTA (4-(*N*-maleimido)benzyl-trimethyl-ammonium), which acts as a competitive inhibitor against cholinergic agonists. The maleimido moiety of MBTA attaches covalently to a sulphhydril (–SH) group of cysteines when the disulphide bridge (S–S bonds) between two cysteine residues in the ACh receptors is cleaved. Because the quaternary ammonium head of MBTA, which would interact with the ACh binding site, lies about 10 Å from its maleimido moiety, the ACh binding site would likewise be about 10 Å from the cysteine residues in the α-subunit, with which the maleimido moiety interacts (Karlin 1969). Where are the cysteine residues?

The extracellular, N-terminus region of the α-subunit has four cysteines, Cys-128, -142, -192, and -193; their positions are indicated by the number of amino acid residues counted from the N-terminus (Noda *et al.* 1982). Kao *et al.* (1984) found that the cysteines labelled by radioactive MBTA are Cys-192 and Cys-193. To test their role, Mishina *et al.* (1985) replaced Cys-192 or Cys-193 with serine (Ser) by site-directed mutagenesis (see 2.2.2). When this mutant α-subunit mRNA was injected into oocytes together with the mRNAs encoding three other subunits (β, γ, and δ), ACh elicited no response. This failure suggests that the integrity of the structure near Cys-192 and Cys-193 is essential for functional expression of the receptor.

The extracellular N-terminus domain of the α-subunit labelled with a photo-affinity ligand also included Cys-192 and Cys-193 (Dennis *et al.* 1988). Moreover, this region contained three negatively charged aspartic acid residues, Asp-180, -195, and

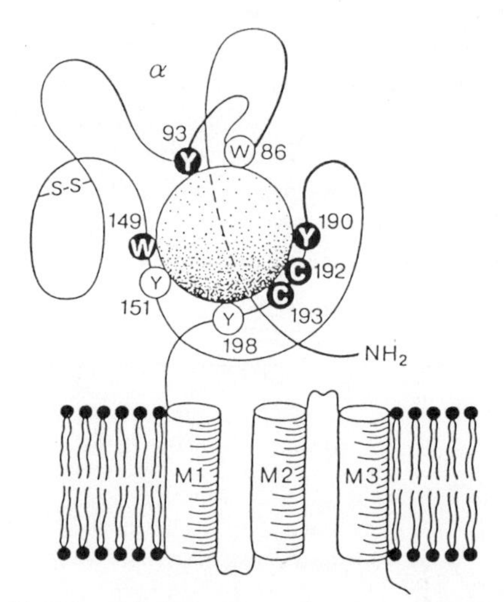

Fig. 4.10. A hypothetical scheme of the ligand-binding site in the extracellular N-terminus of the α-subunit of the nicotinic ACh receptor. A pocket ringed by aromatic amino acids (W, tryptophan; Y, tyrosine) and cysteine residues (C) is the binding site for an ACh molecule (shaded circle). The position of each residue is indicated by the number of amino acids counted from the N-terminus. M1 to M3, transmembrane segments. The M4 segment is omitted for simplicity. (Adapted from Changeux 1990.)

-200. Interestingly, none of these anionic residues was labelled. Instead, the photo-affinity ligand labelled six aromatic amino acids (tryptophan and tyrosine residues) in addition to the two cysteines (Fig. 4.10; Dennis *et al.* 1988; Changeux 1990). The electron-rich aromatic residues possess π electrons which distribute perpendicularly to the axis of double bonds in carbon atoms (Dougherty and Stauffer 1990). Then, the binding of ACh's positively charged quaternary ammonium group to the receptor may employ the cation–π interaction, rather than the charge–charge interaction with negatively charged residues that Karlin assumed (Changeux 1990; Dougherty and Stauffer 1990). The extracellular N-terminus stretch of the α-subunit has, in total, 26 negatively charged amino acids (Noda *et al.* 1982). To examine their role, each anionic residue in this stretch was replaced, one by one, with a non-charged amino acid; the ACh response was not impaired when these mutant α-subunits were expressed together with three other subunits (β, γ, and δ) in oocytes (M. Mishina, T. Takahashi, M. Kuno, and S. Numa, unpublished observations). This is also consistent with the view that negatively charged amino acids do not act as the ACh-binding site.

Figure 4.10 illustrates how cycteines and aromatic π systems in the α-subunit could surround a pocket that holds the quaternary ammonium group of ACh (Changeux 1990). If the disulphide bond formed by Cys-192 and Cys-193 is cleaved, this pocket would no longer be formed. Perhaps, this is why functional ACh receptors are not expressed when one of these two residues is replaced by Ser (Mishina *et al.* 1985; see above).

4.3.2 Ionic permeation in receptor channels

The conductance change produced by ACh in the muscle membrane at the end-plate region is confined to cations, mainly Na^+ and K^+ (A. Takeuchi and N. Takeuchi 1960). The end-plate current is thus the sum of Na^+ and K^+ currents. If Na^+ and K^+ flowed through separate channels with independent conductance changes, the ionic flow through the Na^+ channel would cease when the muscle membrane is held at the Na^+ equilibrium potential (about +50 mV); similarly, the ionic flow through the K^+ channel should halt at the K^+ equilibrium potential (about −100 mV). Consequently, if the channel activity of ACh receptors were recorded as current fluctuations (*noise*), the least fluctuation should occur at these two potential levels. Yet Dionne and Ruff (1977) demonstrated that the fluctuations reach a minimum at the reversal potential of the macroscopic end-plate current (about 0 mV), without having two separate minima. This undifferentiated activity indicates that both Na^+ and K^+ flow through the same ACh receptor channel.

Apparently, the ACh receptor channel cannot distinguish between Na^+ and K^+ ions for permeation. Might other large ions also be peremeable through this receptor channel? Maeno *et al.* (1977) and Dwyer *et al.* (1980) examined whether and how well many organic cations permeate the ACh receptor channel at the frog neuromuscular junction. Their relative permeability (*selectivity sequence*) was inversely related to their size. From this relation, the cross-sectional area of the receptor channel was estimated to be 6.5 Å × 6.5 Å (Dwyer *et al.* 1980; Hille 1992a). These results give us a picture of the ACh receptor channel as a water-filled pore with a relatively large diameter (Huang *et al.* 1978; Adams *et al.* 1980).

We have seen that the transmembrane segment of every subunit presumably lines an ionic channel located in the centre of the ACh receptor molecule (Fig. 4.3). Each subunit has four transmembrane segments (M_1 to M_4). Several lines of evidence have now singled out the M_2 segment as being crucial in forming the ionic channel. One approach labelled the ACh receptor by non-competitive blockers, for example, the local anaesthetic, QX-222 (Fig. 4.11A). Because QX-222 blocks ACh receptor activity but does not compete with ACh, we know that its blocking effect is not exerted at the level of the ligand-binding site. Where then does it act? QX-222 produces *flickering*, so that the normal open duration of single ACh receptor channel activity is chopped into many short pulses (Fig. 4.11B). This pulsation suggests that QX-222 enters the ionic channel when the ACh receptor opens, thereby blocking the channel directly (Neher and Steinbach 1978; Leonard *et al.* 1988). In other words, non-competitive blockers can occupy some site within the channel. Then, the structure labelled by non-competitive blockers must represent the transmembrane segments that actually constitute the ionic channel.

Two non-competitive cholinergic blockers, TPMP (triphenyl-methylphosphonium; Hucho *et al.* 1986) and chlorpromazine (Giraudat *et al.* 1987) labelled the M_2 segments of the ACh receptor. The binding site for the open-channel blocker, QX-222 also lies in the M_2 segment (Leonard *et al.* 1988; Charnet *et al.* 1990). Figure 4.12 aligns the amino acid sequences of the M_2 segment and its extracellular and cytoplasmic vicinities for the

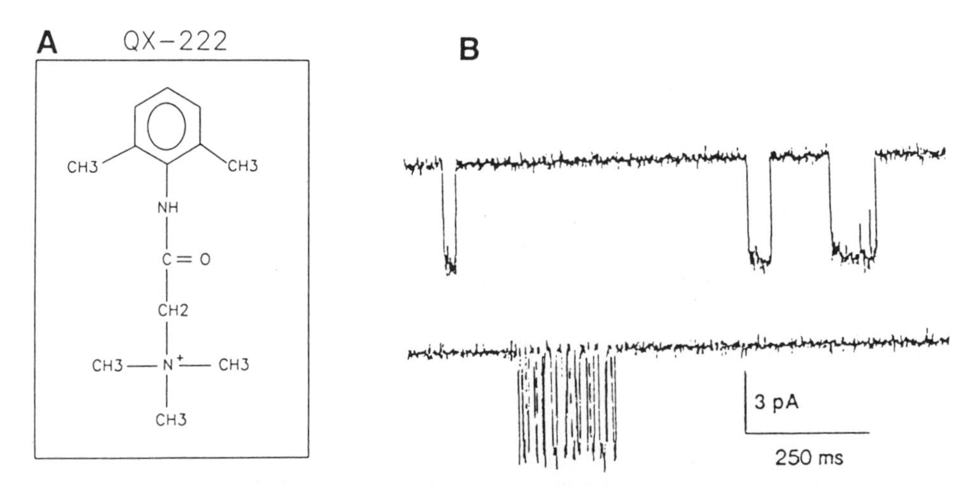

Fig. 4.11. The structure (A) and action of QX-222 (B). QX-222 has a positively charged quaternary ammonium head and an aromatic end (A; from Charnet *et al.* 1990). B, single-channel currents of ACh receptors recorded from an outside-out patch membrane of an oocyte injected with the mouse ACh receptor subunit mRNAs before (upper trace) and after the application of QX-222 (lower trace). (From Leonard *et al.* 1988.)

four subunits of the *Torpedo* ACh receptor. Each membrane-spanning M_2 segment has 19 amino acids numbered from the cytoplasmic side toward the extracellular side (its N-terminus being number 1; Miller 1989). The amino acids labelled by non-competitive blockers were the Ser residue located in position 6 of each subunit (Hucho *et al.* 1986; Giraudat *et al.* 1987). QX-222 has a positively charged quaternary ammonium head and a hydrophobic aromatic group 5–6 Å apart (Fig. 4.11A). This distance corresponds with the distance down the long axis of each turn of α-helix (5.4 Å) made by 3.6 amino acids (section 2.2.2). QX-222 presumably binds to the OH side chain of the polar residues in position 6 (Ser) with its quaternary head and to the non-polar residues in position 10 (alanine, Ala) with its hydrophobic group (Leonard *et al.* 1988; Charnet *et al.* 1990). Of the 19 amino acids in each subunit, then, at least the residues in positions 6 and 10 appear to be directly exposed to the ionic channel of the ACh receptor.

As noted, substituting a single amino acid in the M_2 segment of NMDA (Burnashev *et al.* 1992b; Mori *et al.* 1992) and non-NMDA receptors (Hume *et al.* 1991; Verdoorn *et al.* 1991) clearly alters the channel permeability to Ca^{2+} (section 4.1.5). Therefore, it is likely that the M_2 segments of glutamate receptor subunits are also responsible for the channel formation. By analogy, the M_2 segments in glycine and $GABA_A$ receptor subunits are also suggested that they line the receptor ionic chan-

		α	β	γ	δ	
Outside		Ile	Val	Val	Leu	
		Leu	Lys(+)	Lys(+)	Arg(+)	
		Glu(−)	Asp(−)	Gln	Gln	Extra. ring
Transmembrane segment (M_2)	19	Val	Ala	Ala	Ser	
	18	Ile	Leu	Ile	Thr	
	17	Val	Leu	Leu	Leu	
	16	Leu	Leu	Phe	Leu	
	15	Leu	Leu	Leu	Leu	
	14	Phe	Pne	Phe	Phe	
	13	Val	Val	Ile	Val	
	12	Thr	Thr	Thr	Ala	
	11	Leu	Val	Gln	Gln	
	10	Ser	Ala	Ala	Ala	
	9	Leu	Leu	Leu	Leu	
	8	Leu	Leu	Leu	Leu	
	7	Val	Ala	Val	Val	
	6	Ser	Ser	Ser	Ser	
	5	Ile	Ile	Ile	Ile	
	4	Ser	Ser	Ser	Ala	
	3	Leu	Leu	Leu	Thr	
	2	Thr	Ser	Thr	Ser	
	1	Met	Met	Cys	Met	
Cytoplasmic		Lys(+)	Lys(+)	Lys(+)	Lys(+)	
		Glu(−)	Glu(−)	Glu(−)	Glu(−)	Inter. ring
		Gly	Gly	Gly	Gly	
		Ser	Ala	Gly	Ser	
		Asp(−)	Asp(−)	Ala	Glu(−)	Cyto. ring
		Thr	Pro	Gln	Ala	

Fig. 4.12. The aligned amino acid sequences in and near the M_2 segments of the four subunits (α, β, γ, δ) of the *Torpedo* ACh receptor. The amino acid residues are numbered from the cytoplasmic side (see inset). The serine (Ser) residues in position 6 (circled) are the residues labelled with non-competitive ACh receptor blockers. Note the three rings of negative charges at equivalent positions.

nel (Langosch *et al.* 1990; Olsen and Tobin 1990; Stroud *et al.* 1990).

By comparing the δ-subunits of ACh receptors between the calf muscle and the *Torpedo* electric organ, Imoto *et al.* (1986) found that the M_2 segment is not only the structure lining the ionic channel but also the site at which the rate of ionic transport is set. Since the rate of ionic flow is a question of access to the channel, let us examine the amino acid residues around the channel entries. As shown in Fig. 4.12, in the cytoplasmic and extracellular vicinities of each M_2 segment there are clusters of negatively charged amino acid residues. Their equivalent positions from one subunit to another create three rings (extracellular, intermediate, and cytoplasmic) of negative charges in the channel (Iomoto *et al.* 1988; Miller 1989; Numa 1989, 1991). Removing the extracellular ring by site-directed mutagenesis reduced the inward current, whereas removing the cytoplasmic ring decreased the outward current (Imoto *et al.* 1988). These negative charges clustered in the transition zones between the bulk solution (the extracellular or cytoplasmic solution) and the channel are well positioned to determine the rate of cation transport through the ACh receptor channel if that is indeed their function (Imoto *et al.* 1988; Miller 1989; Numa 1991).

Figure 4.13 is a schematic orientation of the three negative rings in the ACh receptor channel (Numa 1991). Recall that the ACh receptor embedded in the cell membrane reaches from the extracellular space to the cytoplasmic side, thereby forming a vestibule at each end of the narrow receptor channel (Fig. 4.3B; Toyoshima and Unwin 1988; Unwin 1989). Thus, the two negative rings in the extracellular and cytoplasmic vestibules, by attracting cations from the bulk solutions, would raise the local concentration of cations (Dani and Eisenman 1987; Imoto *et al.* 1988; Miller 1989; Numa 1991; Hille 1992a). These high local concentrations of cations at the entrance of the ACh receptor channel would play a significant role in its cation selective permeation. If selective permeation of the glycine or $GABA_A$ receptor by anions is achieved by similar mechanisms, a cluster of positively charged amino acids would be expected at the entries of these inhibitory receptor channels. Indeed, a positively charged arginine (Arg) lies on the extracellular side of the M_2 segment of the glycine and GABA subunits (Grenningloh *et al.* 1987; Schofield *et al.* 1987; Langosch *et al.* 1990). These Arg residues would form a ring of positive charges, thereby increasing the local concentration of anions in the extracellular vestibule.

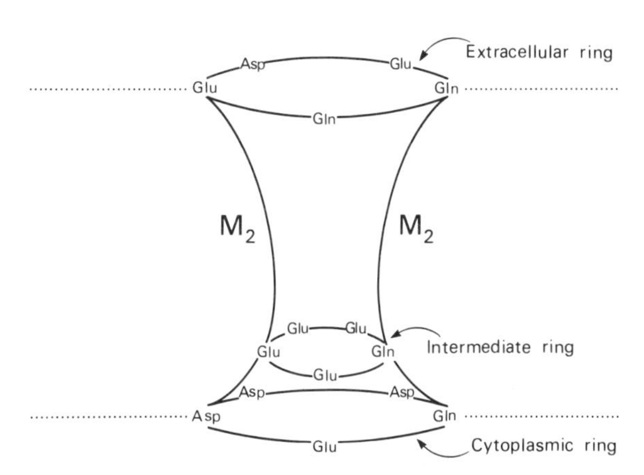

Fig. 4.13. A hypothetical location of the three rings of negative charges formed by the M_2 segments of the ACh receptor subunits. The cell membrane is shown by two dotted lines. The 'hour-glass' structure represents the receptor channel. Top, the extracellular side. Bottom, the cytoplasmic side. (Adapted from Numa 1991.)

These anion-selective glycine and $GABA_A$ receptor subunits lack counterparts of the cytoplasmic and intermediate rings, however.

Unwin (1989) considered the presence of cylindrical vestibules at both ends of the channel to be important for the charge rings to exert a strong, local electrostatic effect (see also Hille 1992a). Yet a synthetic peptide comprising only 23 amino acids that mimic the M_2 segment of the ACh receptor δ-subunit, can form a cation-selective channel in the lipid bilayer, despite being too short to form a cylindrical vestibule on each side (Oiki *et al.* 1988). Recall too that mutagenic removal of the extracellular or cytoplasmic negative ring of the M_2 segments in the ACh receptor reduces the inward or outward currents, respectively (Imoto *et al.* 1988); even so, the mutant receptor still bars Cl^- from permeation. Evidently, the electrostatic effect provided by negative rings cannot fully explain the cation-selective permeation of the ACh receptor.

The intermediate ring (Figs 4.12 and 4.13) was assumed to represent a selective filter, because mutation of the intermediate ring altered ion-selectivity among different monovalent cations. Another ring, termed the *central ring* (Imoto *et al.* 1991), is formed by uncharged, polar residues, serine (Ser) or threonine (Thr) in position 2 (Fig. 4.12). When these polar residues were replaced with non-polar (hydrophobic) residues or with residues having a larger side chain, the single-channel conductance of the receptor was reduced (Imoto *et al.* 1991;

Villarroel *et al.* 1991). It was postulated that the intermediate ring attracts cations into the channel constriction, whereas the central ring contributes to dehydration of permeating cations, by interactions of the –OH group of Ser or Thr with water molecules that hydrate a permeant ion (Imoto *et al.* 1991). However, synthetic peptides consisting of leucine (Leu) and Ser residues alone, without any negatively charged amino acids, can form a cation-selective channel when incorporated into a planar bilayer (Lear *et al.* 1988). There must be some as yet unidentified molecular mechanisms that discriminate between cations and anions for permeation through the ionotropic receptor channel.

Recently, Galzi *et al.* (1992) could convert a cation-selective ACh receptor into an anionic-selective ACh receptor channel by site-directed mutagenesis of the $\alpha7$ neuronal ACh receptor. A critical mutation was the insertion of an uncharged residue (proline or alanine) at the N-terminus of the M_2 segment, the cytoplasmic side of the position equivalent to the intermediate ring. Mutation of the negatively charged amino acid (glutamic acid) at the intermediate ring to a neutral residue (alanine) did not alter the cationic selectivity of the ACh receptor. Thus, it seems to be the geometry of the M_2 segment that is crucial for the maintenance of the cationic or anionic selectivity (Galzi *et al.* 1992). Although the multiple ring model is an attractive hypothesis, it must be viewed as a tentative guide for further evaluation.

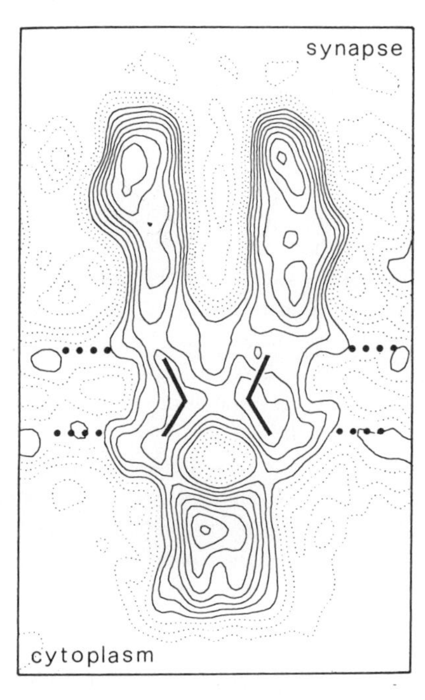

Fig. 4.14. The three-dimensional profile of the nicotinic ACh receptor in the closed conformation. Horizontal double-dotted lines indicate the membrane lipid bilayer 30 Å apart. The apexes of the two heavy lines bent towards the centre of the pore represents the conserved leucine (Leu) residue in the M_2 helix of every subunit. (From Unwin 1993b.)

The three-dimensional images of the *Torpedo* ACh receptor at high resolution (9 Å; Unwin 1993b) showed that amino acid residues in the M_2 segments of the contributory subunits come closest to the axis of the pore just below the middle of the lipid bilayer when the channel is closed (Fig. 4.14). These residues appear to be the leusines (Leu) conserved at position 9 in all the subunits (Fig. 4.12). Interestingly, this hydrophobic Leu ring is conserved at the same position in all subunits of glycine, GABA, and 5HT-3 receptors as well as muscular and neuronal nicotinic ACh receptors (Unwin 1989, 1993b). Unwin suggested that these Leu residues project towards the centre of the pore, interacting with their side chains. This side-to-side association of Leu residues forms the 'leucine zipper' (O'Shea et al. 1991), presumably contributes to the switch between open and closed states of the receptor channel (Unwin 1993a,b). The importance of these conserved Leu residues was further supported by mutation of the neuronal ACh receptor $\alpha7$. When the $\alpha7$ mRNA is injected into oocytes, the oocyte can express functional, homoeric $\alpha7$ ACh receptors. When the Leu residue in the M_2 segment of the $\alpha7$ receptor was replaced with a polar residue (serine or threonine), the mutated receptor showed a second 80 pS conductance state, in addition to the 45 pS conductance state observed in the wild-type receptor (Revah *et al.* 1991). The mutant also showed reduced desensitization and was activated by the receptor antagonists (Bertrand *et al.* 1992). In other words, if the Leu residue is removed, the ACh receptor can take an open channel conformation even when the receptor is in a desensitized state or in a blocked state induced by the competitive antagonists. The mutant receptor takes a closed conformation at rest (without the ligand), however. This implies that the Leu residue is not required for the closed state. Undoubtedly, the leucine hypothesis provides important insights into the molecular mechanisms of gating in the receptor channel. At least, the desensitization state appears to be different from the closed state at rest. Let us now examine how receptor desensitization develops.

4.3.3 Receptor desensitization

Applying ACh to a neuromuscular junction depolarizes the muscle fibre, but this response subsides when the ACh application is prolonged (Kuffler 1943; Fatt 1950). The neuromuscular junction also becomes refractory following an initial brief exposure to highly concentrated ACh (Thesleff 1955). Both alterations reflect the phenomenon of *desensitization* (Katz and Thesleff 1957). Once ACh receptor desensitization develops, it takes from about 10 seconds to several minutes to recover at the frog neuromuscular junction (Feltz and Trautman 1982).

In principle, desensitization could occur in two ways: first, reduced affinity at the transmitter binding site of the receptor (a decrease of k_1 in relationship (1) described in section 4.2.1); second, reduced ionic permeation through the receptor channel by a decrease in the single-channel current or the open probability of the channel (i or $p(o)$ in relationship (2) described in section 4.2.1). Clearly, the first possibility is not the case, because prolonged applications of ACh actually shift the ACh receptors from a low-affinity binding state to a high-affinity state (Heidmann and Changeux 1980; Neubig and Cohen 1980). At the single-channel level, desensitization entailed less frequent openings (or more silent periods), but no significant changes occurred in the mean open time or in the mean current amplitude (Sakmann *et al.* 1980; Mulle *et al.* 1988). In other words, the two major changes induced in the ACh receptor during desensitization are an increased affinity for ACh binding and a lowered rate of openings in its ionic channel.

Miledi (1980) found that desensitization is due to modulation of the ACh receptor by some factors from the cytoplasmic side. Thus, desensitization of ACh receptors in muscle fibres could be enhanced by Ca^{2+} influx and diminished by injecting a Ca^{2+} chelator. A variety of studies have now traced receptor desensitization to phosphorylation of the receptor molecule by protein kinases (Huganir and Greengard 1983; Box C). In the *Torpedo* ACh receptor, protein kinase A, protein kinase C and tyrosine kinase all phosphorylate different subunits (Huganir and Greengard 1990). Phorbol ester, which activates protein kinase C, accelerates the desensitization of ACh receptors (Eusebi *et al.* 1985). Forskolin or cyclic AMP (adenosine monophosphate; see section 5.1.1), which activates protein kinase A, enhances the desensitization of ACh receptors (Albuquerque *et al.* 1986; Middleton *et al.* 1986). Phosphorylation is further implicated in desensitization by the action of calcitonin gene-related peptide (CGRP), which coexists with ACh in motor nerve terminals (section 2.2.3). CGRP both increases intracellular cyclic AMP in the muscle (Takami *et al.* 1986; Laufer and Changeux 1987) and enhances the desensitization of ACh receptors (Mulle *et al.* 1988). CGRP also causes phosphorylation of the α- and δ-subunits of the ACh receptor in rat cultured muscle (Miles *et al.* 1989). Thus, the phosphorylation of particular subunits appears to lead to the desensitization of the receptor.

How does the phosphorylation of some subunits of the ACh receptor induce desensitization? As noted above, desensitization is associated with a decreased rate of opening in the receptor channel. The most plausible explanation then involves conformational changes of the receptor molecule, which may immobilize the receptor configuration. Cryoelectron microscopy (Fig. 4.15) shows distortion in the *Torpedo* ACh receptor subunit assembly following exposure to carbamylcholine, an ACh analogue (Unwin *et al.* 1988). Compared with the normal configuration (Fig. 4.15N), the δ-subunit is quite tilted and the γ-subunit is slightly displaced after exposure to the ACh agonist (Fig. 4.15C). This asymmetrical distortion in the subunit configuration around the ionic pore may stabilize or freeze the channel in its closed state, inducing desensitization (Unwin *et al.* 1988). As noted previously (section 4.3.2), however, the neuronal ACh receptor mutated by replacing the leucine with threonine fails to show normal desensitization but can take a closed state in the absence of the ligand. Thus, the 'closed state' at rest appears to differ from the 'closed state' in desensitization. Unwin (1989) further speculates that for normal opening or closing (*gating*) of the receptor channel, the encircling subunits also switch from one symmetrical configuration to another. How is the receptor configuration distorted in an asymmetrical fashion by prolonged exposure to ACh? How do prolonged applications of ACh increase the affinity of the receptor binding site? These questions remain to be answered. Moreover, the crucial question of how the binding of the ligand to its receptor opens the receptor channel is still elusive.

Summary and prospects

The molecular cloning of transmitter receptors and the recording of single-channel currents have greatly increased our understanding of interactions between transmitters and their receptors. A variety of transmitter

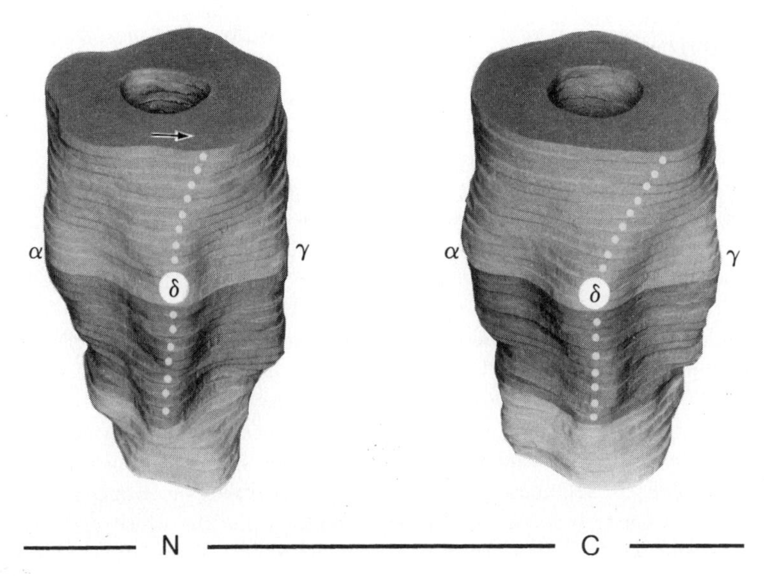

Fig. 4.15. Three-dimensional features of the subunit assembly reconstructed from data obtained by cryoelectron microscopy of crystallized *Torpedo* ACh receptors. N, normal receptor. C, receptor exposed to carbamylcholine. Note inclination of the ridge formed by the δ-subunit in C. (From Unwin *et al.* 1988 and courtesy of N. Unwin.)

substances can produce rapid excitatory or inhibitory synaptic responses. Amazingly, all the ionotropic receptors that mediate fast synaptic responses have the same molecular motif, despite the different natures of the transmitters. Each ionotropic receptor is composed of multiple homologous subunits, and each subunit has four transmembrane segments. Site-directed mutagenesis and substituting receptor subunits into other species revealed which membrane segments contribute to the formation of the receptor channel and which subunit specifies the receptor properties. A possible mechanism for selective permeation of cations or anions through the receptor channel has been proposed. Yet many questions remain unanswered. How does the binding of the transmitter to its receptor lead to the opening of the receptor channel? How do prolonged applications of the transmitter cause its receptor to desensitize? What determines the magnitude and the mean open time of the single-channel current? The molecular basis of biophysical properties of transmitter channels still awaits exploration.

Suggested reading

The primary structure of ionotropic receptors

Langosch, D., Becker, C. M., and Betz, H. (1990). The inhibitory glycine receptor: a ligand-gated chloride channel of the central nervous system. *European Journal of Biochemistry*, **194**, 1–8.

Nakanishi, S. (1992). Molecular diversity of glutamate receptors and implication for brain function. *Science*, **258**, 597–603.

Numa, S., Noda, M., Takahashi, H., Tanabe, T., Toyosato, M., Furutani, Y., and Kikyotani, S. (1983). Molecular structure of the nicotinic acetylcholine receptor. *Cold Spring Harbor Symposia on Quantitative Biology*, **48**, 57–69.

Olsen, R. W. and Tobin, A. J. (1990). Molecular biology of $GABA_A$ receptors. *FASEB Journal*, **4**, 1469–80.

The molecular basis of ionotropic receptor function

Bertrand, D., Devillers-Thiery, A., Revah, F., Galzi, J. L., Hussy, N., Mulle, C., Bertrand, S., Ballivet, M., and Changeux, J. P. (1992). Unconventional pharmacology of a neuronal nicotinic receptor mutated in the channel domain. *Proceedings of the National Academy of Sciences of the USA*, **89**, 1261–5.

Changeux, J. P. (1990). The nicotinic acetylcholine receptor: an allosteric protein prototype of ligand-gated ion channels. *Trends in Pharmacological Sciences*, **11**, 485–92.

Charnet, P., Labarca, C., Leonard, R. J., Vogelaar, N. J., Czyzyk, L., Gouin, A., Davidson, N., and Lester, H. A. (1990). An open-channel blocker interacts with adjacent turns of α-helices in the nicotinic acetylcholine receptor. *Neuron*, **4**, 87–95.

Galzi, J. L., Devillers-Thiery, A., Hussy, N., Bertrand,

S., Changeux, J. P., and Bertrand, D. (1992). Mutations in the channel domain of a neuronal nicotinic receptor convert ion selectivity from cationic to anionic. *Nature*, 359, 500–5.

Imoto, K., Busch, C., Sakmann, B., Mishina, M., Konno, T., Nakai, J., Bujo, H., Mori, Y., Fukuda, K., and Numa, S. (1988). Rings of negatively charged amino acids determine the acetylcholine receptor conductance. *Nature*, 335, 645–8.

Miller, C. (1989). Genetic manipulation of ion channels: a new approach to structure and mechanism. *Neuron*, 2, 1195–205.

Numa, S. (1991). Neurotransmitter receptors and ionic channels: from structure to function. *Fidia Research Foundation Neuroscience Award Lectures*, 5, 23–44.

Sommer, B. and Seeburg, P. H. (1992). Glutamate receptor channels: novel properties and new clones. *Trends in Pharmacological Sciences*, 13, 291–6.

Unwin, N. (1989). The structure of ion channels in membranes of excitable cells. *Neuron*, 3, 665–76.

Unwin, N. (1993). Neurotransmitter action: opening of ligand-gated ion channels. *Neuron*, 10 (Supplement), 31–41.

Macroscopic and single-channel currents of ionotoropic receptors

Anderson, C. R. and Stevens, C. F. (1973). Voltage clamp analysis of acetylcholine produced end-plate current fluctuations at frog neuromuscular junction. *Journal of Physiology (London)*, 235, 655–91.

Edwards, F. R., Konnerth, A., Sakmann, B., and Takahashi, T. (1989). A thin slice preparation for patch clamp recordings from neurones of the mammalian central nervous system. *Pflügers Archiv*, 414, 600–12.

Katz, B. and Miledi, R. (1972). The statistical nature of the acetylcholine potential and its molecular components. *Journal of Physiology (London)*, 224, 665–99.

Methfessel, C., Witzemann, V., Takahashi, T., Mishina, M., Numa, S., and Sakmann, B. (1986). Patch clamp measurements on *Xenoupus lavis* oocytes: currents through endogenous channels and implanted acetylcholine receptor and sodium channels. *Pflügers Archiv*, 407, 577–88.

Mishina, M., Takai, T., Imoto, K., Noda, M., Takahashi, T., and Numa, S. (1986). Molecular distinction between fetal and adult forms of muscle acetylcholine receptor. *Nature*, 321, 406–11.

Sakmann, B., Mathfessel, C., Mishina, M., Takahashi, T., Takai, T., Kurasaki, M., Fukuda, K., and Numa, S. (1985). Role of acetylcholine receptor subunits in gating of the channel. *Nature* 318, 538–543.

Sakmann, B. (1992). Elementary steps in synaptic transmission revealed by currents through single ion channels. *Neuron*, 8, 613–29.

Silver, R. A., Traynelis, S. F., and Cull-Candy, S. G. (1992). Rapid-time-course of miniature and evoked excitatory currents at cerebellar synapses *in situ*. *Nature*, 355, 163–166.

5

Metabotropic receptors mediate slow synaptic responses

Activation of metabotropic receptors causes their target ionic channels, located at a distance, to open or close (Fig. 4.1B). On activation, the receptor must signal the target ionic channel via the intracellular milieu. The factors involved in conveying this signal are called the *second intracellular messengers*. Thus, the metabotropic receptor differs from the ionotropic receptor in two respects: first, the metabotropic receptor molecule must have a domain specific for activating the second intracellular messengers; second, the ionic channel domain must be lacking in metabotropic receptor molecules.

We have seen that all the ionotropic receptors have the same basic molecular motif. We will now examine structural characteristics of the metabotropic receptor family.

5.1 MOLECULAR STRUCTURE OF METABOTROPIC RECEPTORS

How might the metabotropic receptor dispatch the second messengers? The binding of a specific ligand to the receptor activates an intermediary regulatory component attached to the receptor's cytoplasmic domain. As we shall see, this regulatory component becomes active when it binds guanosine triphosphate (GTP) and inactive when it binds guanosine diphosphate (GDP), hence it is called the guanine nucleotide-binding protein (or G-protein). All the metabotropic receptors are linked to G-proteins; hence, as noted previously, the metabotropic receptor is also termed the *G-protein-linked receptor*. Let us first examine how the metabotropic receptor is linked to the G-protein.

5.1.1 The linkage to G-proteins

The concept of a second messenger arose in the mid-1960s when Sutherland and his colleagues showed that the interaction of a hormone with its receptor stimulates *adenylate cyclase* which catalyses the synthesis of cyclic AMP from ATP. Cyclic AMP then acts as a second messenger, inducing a variety of physiological responses (Sutherland 1972). The hormone receptor was widely thought to interact directly with the catalytic unit (i.e., adenylate cyclase molecule) until Rodbell *et al.* (1971a,b) proved otherwise. The activation of the catalytic unit by the hormone receptor is actually indirect, being mediated through the G-protein. Moreover, one type of G-protein (G_s) was found to mediate stimulation of the adenylate cyclase activity, whereas another type of G-protein (G_i) inhibited its activity (Rodbell 1980; Gilman 1984). Indeed, adenylate cyclase is not the only enzyme to be regulated by G-proteins: a variety of enzymes are (e.g., phosphodiesterase or phospholipase). In accordance with the multiplicity of effector enzymes, there are also multiple subtypes of G-proteins (Gilman 1987). Hence, the activation or inhibition of different enzymes by the G-proteins yields a wide array of second messengers.

Figure 5.1 outlines the information processing network operated by G-protein-linked receptors. As will be seen, the target ionic channels can be modulated in three ways: (1) by phosphorylation of the channel (e.g., K^+ channel) via activation of protein kinases (e.g., protein kinase A) through the second messengers (e.g., cyclic AMP), by the second messengers themselves (e.g., cyclic AMP or cyclic guanosine monophosphate, cyclic GMP) or directly by the subunit (α-subunit or $\beta\gamma$-subunits) of the G-protein. Inconveniently, this

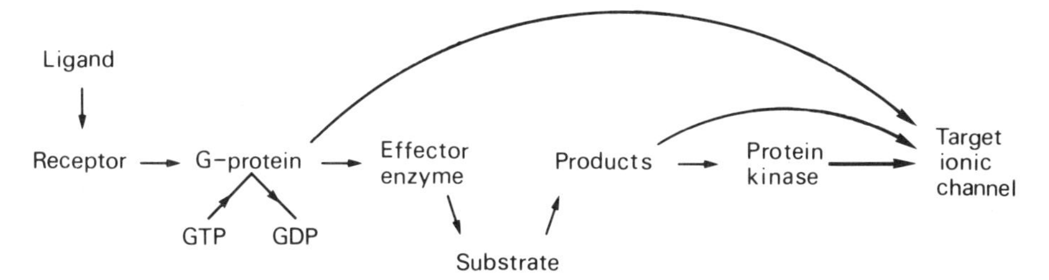

Fig. 5.1. Outline of information processing by G-protein-linked receptors. Regulation of the target ionic channel can follow three paths: I, phophorylation of the channel protein by protein kinases; II, products of the substrate catalysed by effector enzymes; III, subunits of the G-protein.

summary renders the term, *second messengers* rather ambiguous. Originally it meant only the product that is synthesized from the substrate catalysed by the effector enzyme and that induces various cellular responses (e.g., cyclic AMP). Evidently, however, factors besides the enzyme-induced products also regulate the target channel. We now call them second messengers as well, because they too serve as signal conveyers.

A given transmitter substance can activate its metabotropic receptor as well as its ionotropic receptor. Therefore, the metabotropic and ionotropic receptors cannot be distinguished by the nature of their specific ligands. In skeletal muscle, for example, ACh directly opens the nicotinic receptor channel at neuromuscular junctions, whereas in cardiac muscle ACh opens K^+ channels by activating the metabotropic muscarinic receptor; moreover, activation of the metabotropic muscarinic receptor in sympathetic neurones causes particular K^+ channels to close. Similarly, GABA increases the Cl^- conductance by activating ionotropic $GABA_A$ receptors but opens K^+ channels by activating metabotropic $GABA_B$ receptors. Glutamate also activates its metabotropic receptors as well as its ionotropic receptors. The same applies to serotonin as described in section 4.1.6. However, the responses of the ionotropic and metabotropic receptors to the same ligand can be blocked differentially by selective antagonists. Each class of receptor must therefore have structurally different binding sites for the same ligand. As we have seen in section 4.1, none of the ionotropic receptors is linked to the G-protein. Obviously, the two classes of receptor must differ in their molecular motif of receptor structure.

5.1.2 The seven-transmembrane segment superfamily

Figure 5.2 shows the structures of four subtypes of the muscarinic ACh receptor deduced from their cloned cDNAs. We have seen that the muscular nicotinic ACh receptor is a pentameric protein, formed by four different subunits. Unlike the nicotinic ACh receptor, the muscarinic ACh receptor is a single peptide made of 500–600 amino acids. This receptor is characterized by seven hydrophobic segments that presumably traverse the cell membrane with α-helical structures. The receptor protein has its N-terminus on the extracellular side and its C-terminus in the cytoplasm.

This membrane spanning organization with seven segments, dubbed 'the magnificent seven' by Hanley and Jackson (1987), is characteristic of all G-protein-linked receptor molecules (Dohlman *et al.* 1987; Lefkowitz and Caron 1988; Haga 1989; Numa 1989). The deduced molecular structure of a wide variety of receptors now places them in the seven-transmembrane segment superfamily, including muscarinic ACh receptors, adrenergic receptors, dopamine receptors, histamine receptors, metabotropic glutamate receptors, metabotropic serotonin receptors, $GABA_B$ receptors, and the receptors for neuropeptides (Hille 1992b).

The structural motif of the seven transmembrane segments was initially documented in bacteriorhodopsin (Engelman *et al.* 1982; Ovchinnikov 1982), a membrane protein found in bacteria that live in salty environments (e.g. *Halobacterium halobium*). This protein is linked to a light-absorbing element, *retinal*, and converts energy from light into stored energy by pumping protons across the membrane (the *light-driven proton pump*). Early on, an electron diffraction analysis (Henderson and Unwin 1975) predicted that each bacteriorhodopsin molecule is folded into seven α-helices which span the lipid bilayer in the cell membrane (Fig. 5.3). This prediction accords remarkably well with the structure now indicated by the deduced amino acid sequence. Strikingly similar molecular structures also exist in the vertebrate and inver-

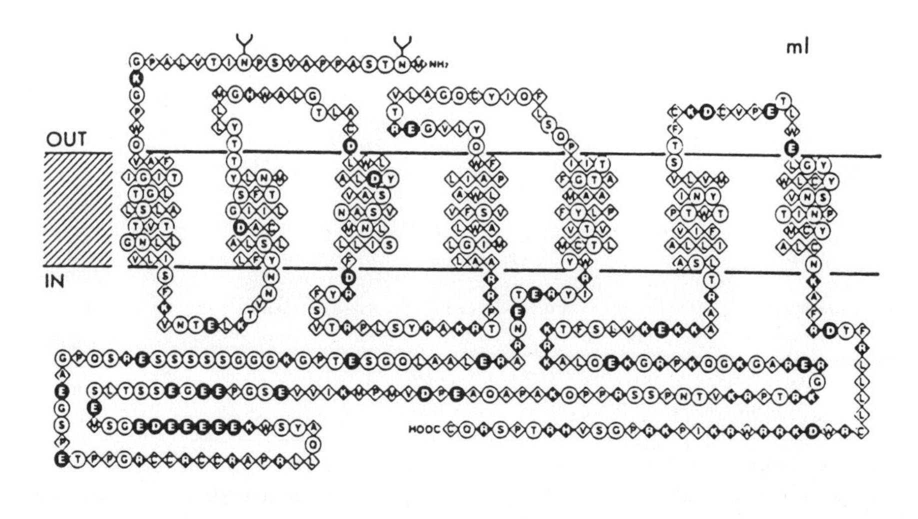

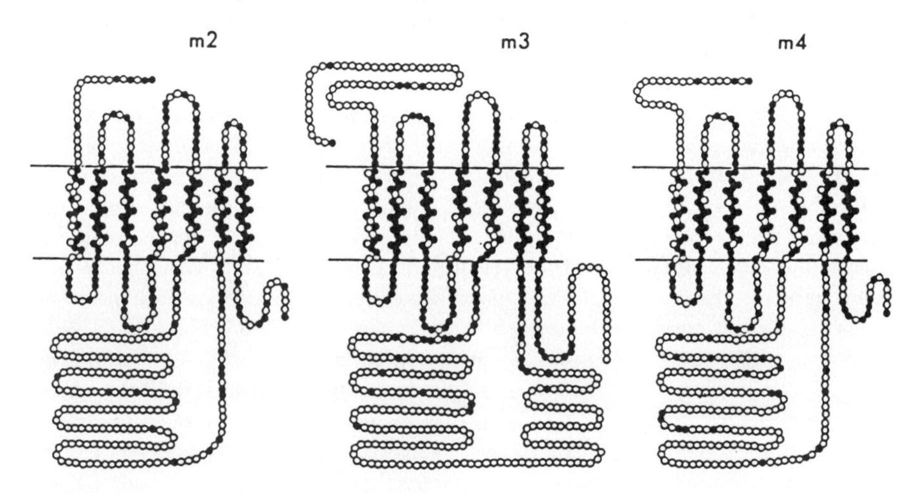

Fig. 5.2. Predicted structure of four muscarinic ACh receptor subtypes (m1, m2, m3, m4) incorporated in the cell membrane (two horizontal lines). Out (or top), the extracellular side. In (or bottom), the cytoplasmic side. The amino acid sequence of the m1 subtype is shown by single letter symbols. Filled circles in m2, m3, and m4 indicate the same amino acid residues as in m1. (Adapted from Haga 1989.)

tebrate visual pigment, rhodopsin (Ovchinnikov 1982; Baehr and Applebury 1986). Rhodopsin is of special interest for the clues it holds about the site of ligand binding in the metabotropic receptor.

Unlike the ligand-binding site in the ionotropic receptor, which lies in the extracellular N-terminal segment, the ligand-binding site of the metabotropic receptors appears to be in a hydrophobic transmembrane segment. Evidence for this locus comes from the visual pigment. Rhodopsin consists of a receptor membrane protein, *opsin*, and a small vitamin A aldehyde, 11–*cis* retinal. The latter acts as a light-activated switch to change the conformation of the opsin protein. Thus, specifically, photo-isomerization of the 11–*cis* retinal to the all-trans retinal activates the G-protein (*transducin*) coupled to the opsin (Baehr and Applebury 1986). In this process, the 11–*cis* retinal is in effect a ligand leading to signal transduction through the receptor molecule.

The 11–*cis* retinal is attached to a positively charged lysine (Lys) located about midway in the seventh transmembrane helix (Thomas and Stryer 1982; Dratz and Hargrave 1983). Dixon *et al.* (1987) proposed that the ligand-binding site in the β-adrenergic receptor

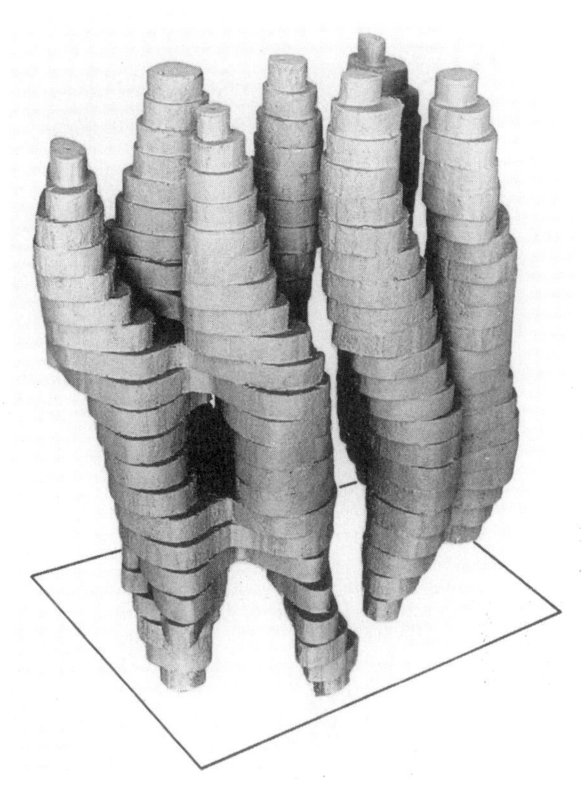

Fig. 5.3. A three-dimensional model of the bacteriorhodopsin molecule predicted by electron diffraction analysis. The seven rods are assumed to be α-helices which span the cell membrane. Each rod is about 35–40 Å long. (From Henderson and Unwin 1975 and courtesy of N. Unwin.)

is likewise located in a membrane segment. Although the seventh membrane segment of the adrenergic receptor lacks the Lys residue, negatively charged amino acid residues in the second and third segments have been implicated as a binding site for positively charged adrenaline (Dixon *et al.* 1987; Dohlman *et al.* 1987, 1988). Where is the binding site in the muscarinic ACh receptor? Again, the second and possibly the third membrane segments are strong candidates (Dohlman *et al.* 1987; Peralta *et al.* 1987). However, an exception to this general rule was found for metabotropic glutamate receptors; both metabotopic and ionotropic receptors have a large (400–500 amino acids), extracellular N-terminus, and this N-terminus appears to contain the ligand-binding site (O'Hara *et al.* 1993; K. Takahashi *et al.* 1993).

Which second messengers are derived by activation of a metabotropic receptor depends on the type of G-protein that is coupled to the receptor molecule. Therefore, the specificity of the receptor action must be determined, at least in part, by the G-protein that mediates the signal transduction. Because G-proteins are anchored in the cytoplasmic face of the plasma membrane, they probably interact with the receptor molecule through its cytoplasmic loop that is formed by the 'tails' of the membrane segments protruding into the cytoplasm. Site-directed mutagenesis has shown that deleting the third cytoplasmic loop (connecting the fifth and sixth membrane segments) uncouples the β-adrenergic receptor from the G-protein (Strader *et al.* 1987). This third cytoplasmic loop shows exceptional diversity of form among different metabotropic receptors. Each of these receptors couples a different G-protein, so that the idiosyncratic loops may reflect specificity in the respective receptor-G-protein complexes (Dohlman *et al.* 1987; Peralta *et al.* 1987; Strader *et al.* 1987). This view was strongly supported by mutagenic studies on muscarinic ACh receptors. Kubo *et al.* (1988) expressed the M_1 or M_2 muscarinic ACh receptors in *Xenopus* oocytes by injecting their specific mRNA. The functional expression of M_1 and M_2 receptors could be distinguished by the presence or absence of Ca^{2+} dependency of the recorded inward currents. They also produced chimeric receptors by replacing the third cytoplasmic loop of the M_2 receptor with the corresponding segment of the the M_1 receptor; this chimeric receptor, when expressed in oocytes, was functionally indistinguishable from the M_1 receptor. Similarly, the M_1 receptor whose third cytoplamic segment was replaced by the third loop of the M_2 receptor expressed the functional properties indistinguishable from those of the M_2 receptor. Thus, the third cytoplasmic loop of each metabotropic receptor seems to have a specific amino acid sequence for activation of a particular G-protein. Surprisingly, however, this third loop specificity is not rigid but rather compatible with multiple types of G-protein. Several examples now show that activation of a single metabotropic receptor produces versatile actions, suggesting its linkage to G-proteins of more than one family or the presence of 'crosstalk' (Taylor 1990). How a single metabotropic receptor can activate multiple effector enzymes remains unclear, but two possibilities are proposed. First, the phosphorylation of a G-protein activator region of the receptor may alter the G-protein specificity of the activator region (Okamoto *et al.* 1991). Second, activation of one type of G-protein may lead to stimulation of one enzyme by its α-subunit, whereas its βγ-subunits stimulate the other enzyme (Baumgold 1992; Birnbaumer 1992; Bourne and Nicoll 1993). Whatever the underlying mechanism, the response of metabotropic receptors appears to be more flexible and versatile than that in ionotropic receptors. We will now examine physiological consequences of activation of metabotropic receptors.

5.2 SIGNAL TRANSDUCTION BY METABOTROPIC RECEPTORS

The diversity of function in G-protein-linked receptors results from the high number of distinct G-proteins—perphaps, as many as 20 (Ross 1989; Simon *et al.* 1991; Hepler and Gilman 1992). Each G-protein regulates a specific effector enzymes, resulting in the generation of second intracellular messengers. Signal transduction mediated by a large number of distinct G-proteins thus opens different avenues leading to a variety of cellular responses. In other words, each G-protein serves the function of signal distribution. We will now examine how G-proteins distribute the signal as a result of activation of the metabotropic receptor.

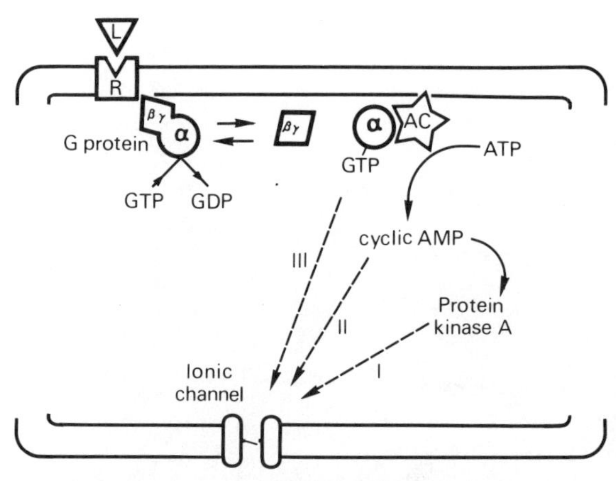

Fig. 5.4. A scheme for regulating an ionic channel by activation of a G-protein-linked receptor stimulating adenylate cyclase. L, ligand. R, G-protein-linked receptor. G-protein consists of α-, β- and γ-subunits. On activation, GTP binds to the α-subunit, replacing GDP. The G-protein then dissociates into the GTP-bound α-subunit and the βγ complex. The α-subunit activates adenylate cyclase (AC), catalysing the synthesis of cyclic AMP from ATP. Cyclic AMP activates protein kinase A. The target ionic channel can be regulated by three routes, I, II, and III.

5.2.1 The role of G-proteins

We have seen that the muscarinic ACh receptor can be expressed in *Xenopus* oocytes injected with the mRNA specific for the receptor molecule. The oocyte that displayed functional expression of the metabotropic receptor did not receive the exogenous mRNA encoding G-proteins. It is then likely that the seven-transmembrane receptor molecule, newly synthesized in the oocyte by injecting the exogenous receptor-specific mRNA, becomes functional by coupling with a particular type of G-protein native to the oocyte. Thus, although the cytoplasmic loop of a metabotropic receptor can recognize a specific G-protein, the receptor and the G-protein are not preassembled. The receptor molecule and the G-protein encounter in 'a random walk' only by chance, but this probability is very high presumably because of the high number of available G-proteins.

Figure 5.4 outlines the role of G-proteins in signal transduction (Gilman 1987; Ross 1989; Taylor 1990), taking the adenylate cyclase-stimulating system as an example. The G-protein is a heterotrimeric protein, consisting of α-, β-, and γ-subunits. At rest, GDP is bound to the α-subunit. When the receptor (β-adrenergic receptor) is activated by the ligand (adrenaline), the GDP bound to the α-subunit is replaced with GTP. This causes the G-protein to dissociate into a βγ complex and an α-GTP complex. The activated α-subunit (α-GTP) then stimulates the effector enzyme (adenylate cyclase), thereby increasing the intracellular level of the second messenger (cyclic AMP).

The dissociation of the G-protein into the α-GTP and βγ complexes is a reversible process. Reversal terminates the signal transduction. This termination occurs when the GTP bound to the α-subunit is hydrolysed to GDP by GTPase. Substituting non-hydrolyzable GTP analogues (e.g., GTPγS) for GTP allows the activation of the effector enzyme to persist. This procedure is widely used to test the presence of G-protein-dependent signal induction. It should be noted that as long as GTP or its analogue is bound to the α-subunit, both the α- and βγ-subunits remain active. In fact, besides the α-subunit, the βγ-complex can be involved in activation of the effector enzyme (Clapham and Neer 1993).

As described previously, the G-protein that stimulates adenylate cyclase is called G_s, and G_i refers to the G-protein that inhibits adenylate cyclase. Similarly, the G-protein coupled to the opsin (rhodopsin), transducin, is often described as G_t. Figure 5.5 summarizes the major G-proteins in relation to their effector enzymes and second messenger pathways. The four effector enzymes regulated by G-proteins include adenylate cyclase

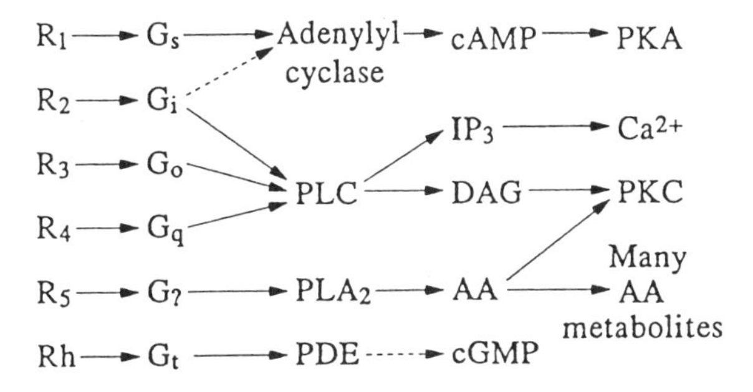

Fig. 5.5. A list of the major second messenger pathways initiated by activation of G-protein-linked receptors. R, receptors. Rh, rhodopsin. PLC, phospholipase C. PLA_2, phospholipase A_2. PDE, phosphodiesterase. cAMP, cyclic AMP. IP_3, inositol trisphosphate. DAG, diacylglycerol. AA, arachidonic acid. cGMP, cyclic GMP. PKA, protein kinase A. PKC, protein kinase C. (From Hille 1992b.)

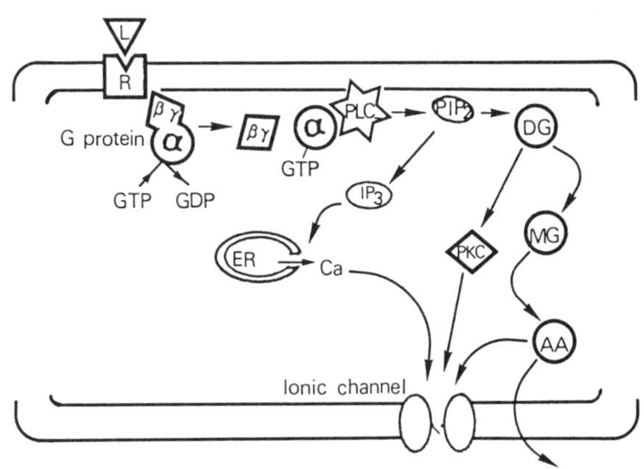

Fig. 5.6 Scheme of signal transduction via phospholipase after ligand (L) activation of G-protein-linked receptors (R). Activation of phosphoinositide-specific phospholipase C (PLC) yields PIP_2 (phosphatidylinositol bisphosphate). IP_3, inositol trisphosphate. ER, endoplasmic reticulum. DG, diacylglycerol. PKC, protein kinase C. MG, monoacylglycerol. AA, arachidonic acid.

(Krupinski *et al.* 1989), phospholipase C (PLC; Ashkenazi *et al.* 1989), phospholipase A_2 (PLA_2; Burch *et al.* 1986), and phosphodiestrase (PDE; Stryer 1986). Again, these effector enzymes are regulated by the G-α subunit, the G-βγ complex, or both (Clapham and Neer 1993). In addition, ionic channels can be regulated directly by G-proteins, so that these ionic channels themselves become the effectors (Fig. 5.4, route III; Brown and Birnbaumer 1990; Dolphin 1991; Clapham and Neer 1993).

The response induced by G-protein-linked receptors is complicated by the multiple steps in the signal pathway. As Fig. 5.4 depicts, the product of each step not only acts as a second messenger, mediating between the receptor and the target ionic channel, but also may spawn an additional second messenger. The result is multiple regulation of the target ionic channel along with signal amplification. Should the product initiate a *bifurcating signalling system*, the outcome would be even more amplified and more diverse. This happens in some muscarinic ACh receptors, α_1-adrenergic receptors, some metabotropic glutamate receptors, and the receptors activated by neuropeptides. All these receptors share one feature: their G-proteins activate PLC (Fig. 5.5), and its signal transduction is schematically illustrated in Fig. 5.6. PLC, the effector anzyme, acts on phosphatidylinositol bisphosphate (PIP_2) which is located in inner surface of the cell membrane. The cleavage of PIP_2 creates the bifurcating signalling system, by generating two products; inositol trisphophate (IP_3) and diacylglycerol (DAG or DG). Each product then modulates the ionic channel differently.

The molecular structure of the IP_3 receptor (Furuichi *et al.* 1989) is very similar to that of the ryanodine receptor in skeletal muscle (Takeshima *et al.* 1989). The IP_3 receptor is located on the endoplasmic reticulum (ER; Otsu *et al.* 1990; Satoh *et al.* 1990) and functions as a calcium-release channel (Berridge 1987; Meyer *et al.* 1988), as does the ryanodine receptor in the muscle sarocplasmic reticulum (Inui *et al.* 1987; Lai *et al.* 1988). Elevated intracellular Ca^{2+} concentrations in turn produce modulation of different ionic channels in the cell membrane (Miledi 1982; Yellen 1982; Marty *et al.* 1984).

The other bifurcation product, diacylglycerol (DG), is converted to monoacylglycerol (MG) and arachidonic acid (AA) by the action of diacylglycerol lipase. In addition, DG activates protein kinase C (PKC). Protein kinase C mediates a variety of cell responses (Nishizuka 1986), including regulation of ionic channels (Baraban *et al.* 1985; Strong *et al.* 1987). Arachidonic acid yields several products; one of them, prostaglandin, produces many cellular responses, and other products modulate ionic channels (Belardetti and Siegelbaum 1988). Moreover, arachidonic acid, a lipids soluble substance, is able to leave the cell, thereby inducing signal transduction in neighbouring neurones.

Activation of metabotropic receptors can thus dispatch a variety of intracellular second messengers which in turn modulate ionic channels. We have seen many

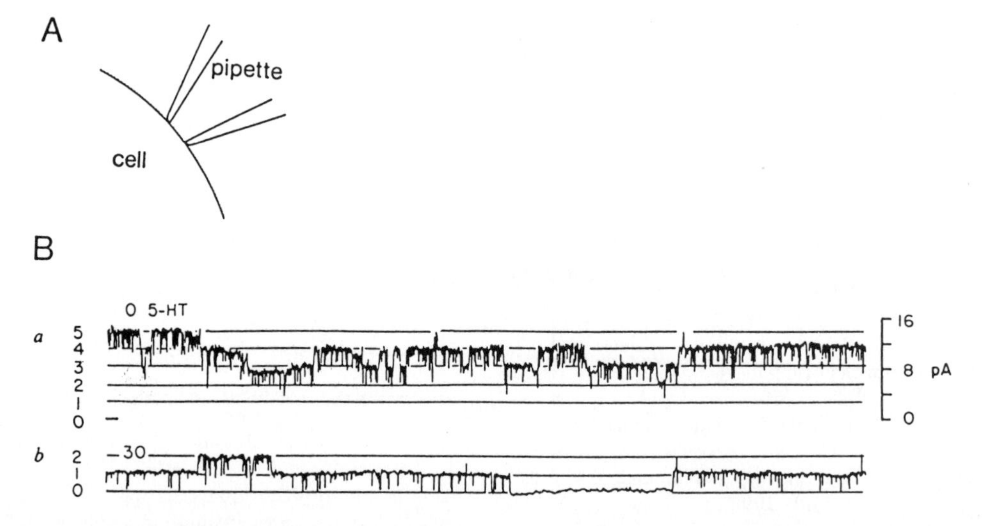

Fig. 5.7. Modulation of single K^+ channel activity by serotonin (5HT) in an *Aplysia* sensory neurone. A, in the cell-attached recording mode a patch recording electrode (pipette) is pressed firmly against the cell membrane surface. From Hamill *et al.* 1981. B, K^+ currents with individual channel steps of about 2.6 pA. Numbers at the left indicate the number of open channels. *a*, before serotonin. *b*, 2 min after adding serotonin to the bathing solution (outside the recording electrode). (From Siegelbaum *et al.* 1982.)

examples of voltage-gated ionic channels or ligand-gated ionotropic receptor channels. How can ionic channels be regulated from the cytoplasmic side by second messengers?

5.2.2 Regulation of the target ionic channel

So far we have assumed that second messengers regulate the target ionic channels without adducing any experimental evidence. Can we in fact prove that the target channel is located in the cell membrane some distance from the receptor, rather than directly associated with the receptor? Yes, by recording single-channel activity with the cell-attached patch clamp technique (Hamill *et al.* 1985).

An example is illustrated in Fig. 5.7 for K^+ channels of a sensory neurone in *Aplysia* (Siegelbaum *et al.* 1982). Firmly pressing the patch clamp pipette against the cell membrane (Fig. 5.7A) forms a tight seal. In consequence, most of the K^+ currents originating from the membrane patch flow into the pipette. The tight seal also prevents any substance applied to the bathing solution from reaching the membrane patch under the recording pipette. In the case of nicotinic ACh receptors on the muscle membrane, for example, single ACh receptor channel actvity can be recorded in the cell-attached mode only when ACh or its analogue is added to the pipette solution; adding the ligand to the external solution will not work. The K^+ channels recorded from the *Aplysia* neurone were sensitive to serotonin (5HT), and a few minutes after 5HT was added to the bath the channel activity significantly declined (Fig. 5.7B*b*). This change indicates that the serotonin receptors lie on the cell membrane ouside the pipette tip, whereas the K^+ channels under study are within the membrane patch beneath the pipette tip. The activity of the K^+ channels must therefore be modulated by intracellular signals sent by the distant receptors. This corroborates the notion of intracellular second messengers being directed to distant target channels.

This type of K^+ channel sensitive to serotonin is now known as the *S-channel* (Klein *et al.* 1982). The sensory neurones with S-channels are involved in plastic changes of the *gill-withdrawal* reflex in *Aplysia*. The studies of plasticity in the gill-withdrawal reflex have provided many important insights into the possible mechanisms of learning, discussed in Part II, 8.1. These plastic changes can be induced pharmacologically in two ways: by applying 5HT to the sensory neurones or by intracellularly injecting cyclic AMP into the neurones (Brunelli *et al.* 1976). This second method suggests that the serotonin-sensitive K^+ channels (S-channels) may likewise be affected by cyclic AMP. In fact, Siegelbaum *et al.* (1982) observed that activity of the S-channel is suppressed by intracellular injections of cyclic AMP into *Aplysia* sensory neurone.

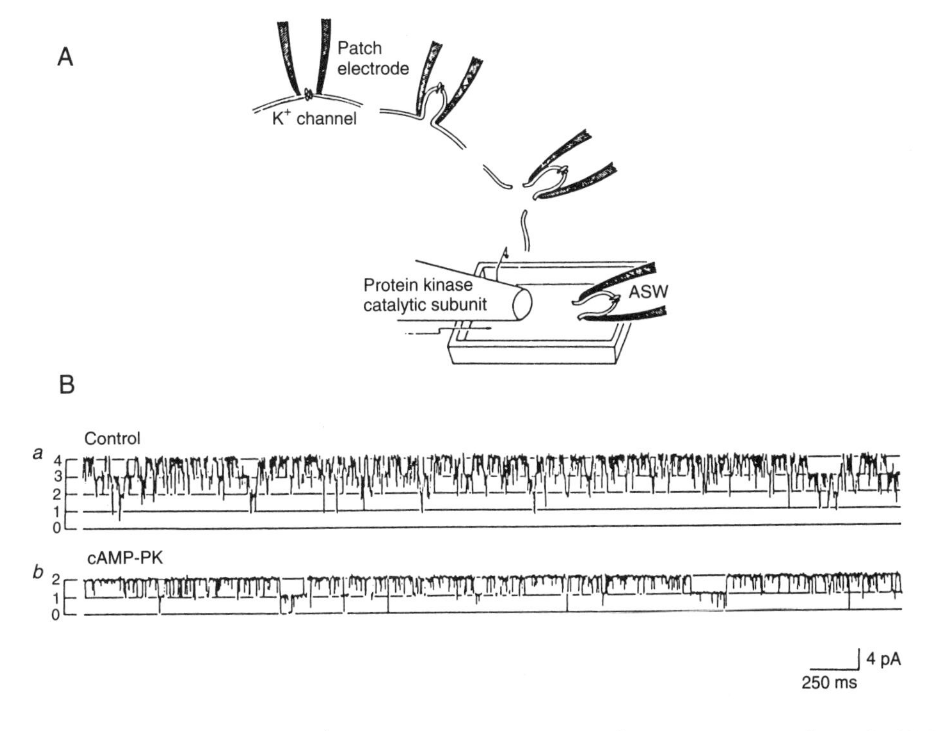

Fig. 5.8. Modulation of serotonin-sensitive K^+ channels (S-channels) in an *Aplysia* sensory neurone by cyclic AMP-dependent protein kinase. A, the procedure for isolating an inside-out patch containing S-channels. (From Kandel *et al.* 1987.) B, activity of single S-channels as described in Fig. 5.7B. *a*, control activity. *b*, 3 min after adding cyclic AMP-dependent protein kinase plus Mg-ATP. (From Shuster *et al.* 1985.)

These results accord well with the scheme illustrated in Fig. 5.1. Initial activation of the G-protein-linked serotonin receptor stimulates adenylate cyclase (the effector enzyme), which in turn increases the level of cyclic AMP (the product, or second messenger) for modulation (down-regulation or suppression) of the S-channel (the target ionic channel). As Fig. 5.1 shows, the target channel may be regulated by cyclic AMP itself or by cyclic AMP-dependent protein kinase (protein kinase A). To discover which process is involved, the experiment illustrated in Fig. 5.8 was performed, employing the *inside-out patch* configuration (Hamill *et al.* 1981). First, the electrode for recording the S-channel activity from an *Aplysia* sensory neurone in the cell-attached mode was carefully pulled away, thereby detaching the membrane patch containing the S-channels (Fig. 5.8A). The cytoplasmic side of the membrane patch now faced the bath. Next, the purified protein kinase A was applied to the cytoplasmic surface of the membrane patch. As before, the S-channel activity declined (Fig. 5.8B; Shuster *et al.* 1985), showing that the target channel is modulated by cyclic AMP-dependent kinase, rather than by cyclic AMP itself. Thus, modulation of the S-channel activity results from phosphorylation of the channel protein by protein kinase A.

However, some target ionic channels are regulated directly by an effector enzyme-induced product (cyclic AMP or cyclic GMP; Fig. 5.1, route II). A clear example occurs in the cation-selective channels of photoreceptor cells (Stryer 1986; Yau and Baylor 1989). As we shall see, the study of the photoreceptors has unveiled many details of this modulatory process.

In photoreceptor cells rhodopsin embedded in the intracelluar retinal disk membrane has seven transmembrane segments and is coupled to the G-protein transducin at its cytoplasmic loop. The effector enzyme that transducin activates in response to light is cyclic GMP phosphodiesterase (Fig. 5.5), which in turn reduces the cyclic GMP levels by hydrolysis. This cyclic GMP cascade directly targets cation-selective channels, thereby modulating the membrane potential. In the dark, photoreceptors display persistent inward currents ('dark currents'), mainly a Na^+ influx, and they maintain depolarization. A flash of light reduces the levels of cyclic

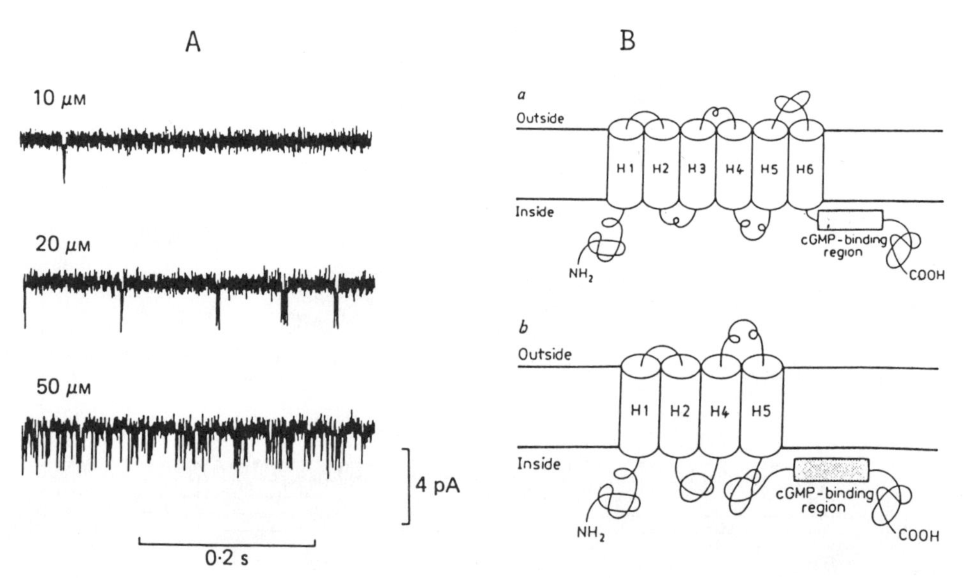

Fig. 5.9. Cyclic GMP-gated channels in photoreceptors. A, recorded single-channel activity increases in a dose-dependent fashion when cyclic GMP is applied to the cytoplasmic side of the inside-out membrane patch. (From Matthews and Watanabe 1988.) B, membrane topology of the cyclic GMP-gated channel protein based on its deduced amino acid sequence. Its hydropathicity profile implies six (*a*) or four (*b*) transmembrane segments. The cytoplasmic C-terminus displays a cyclic GMP-binding region. (From Kaupp *et al.* 1989.)

GMP in the photoreceptor, suppressing the inward currents and thereby hyperpolarizing the cell.

As Fig. 5.9A illustrates, the cation-selective channel activity can be recorded from the plasma membrane of a photoreceptor's inner segment in the inside-out mode. Activity increases in proportion to the concentration of cyclic GMP applied to the cytoplasmic surface of the membrane patch (Matthews and Watanabe 1988). Besides its action, the primary structure of the cyclic GMP-gated channel has also been determined. After cloning and sequencing, its cDNA was functionally expressed in *Xenopus* oocytes (Kaupp *et al.* 1989). Figure 5.9B sketches the transmembrane domain structure of the cyclic GMP-gated channel suggested from the deduced amino acid sequence (Kaupp *et al.* 1989). The C-terminus of the channel protein definitely has cyclic GMP-binding domains, which must be the site for modulation of the ionic channel (Fig. 5.9B). Initially, the number of the channel transmembrane segments was estimated to be four or six; a recent study suggests that the transmembrane topography comprises eight segments, including positively charged S4-like segment and short SS_1-SS_2 pore segments (Bönigk *et al.* 1993).

Olfactory transduction exhibits a cyclic AMP cascade that strongly resembles the cyclic GMP cascade in visual transduction. Exposing olfactory sensory neurones to various odours elevates cyclic AMP levels only in the presence of GTP, suggesting the involvement of G-proteins (Lancet 1986; Takagi 1989). These changes are associated with depolarization of the olfactory neurones caused by the opening of cation-selective channels (Kurahashi 1989). These channels are activated directly by cyclic nucleotides, including cyclic AMP (Nakamura and Gold 1987). The cDNA encoding the cyclic nucleotide-gating channel protein has been cloned and sequenced (Dhallan *et al.* 1990). This channel protein had a putative cyclic nucleotide-binding site and, strikingly, was highly homologous with the cyclic GMP-gated channel found in photoreceptors. An even more exciting finding was that a large family of proteins specifically expressed in olfactory epithelium (presumably a family of odour receptors) is characterized by seven transmembrane segments (Buck and Axel 1991), again like the opsin protein.

Historically, metabotropic receptors were considered to mediate slow and long-lasting synaptic responses, thereby regulating excitability of the neurone according to changes in the membrane potential produced by the lingering synaptic potential. Yet, activation of metabotropic receptors initiates cascades of intracellular signalling, so that the neurone not only alters its excitability by changes in its membrane potential but also modulates its responsiveness to other stimuli. Then, a presynaptic signal directed to a metabotropic receptor in the postsynaptic neurone may display more diverse synaptic functions than those expected from changes in the postsynaptic membrane potential. We shall examine this possibility next.

5.3 MODULATION OF SYNAPTIC FUNCTION

We have seen that the second messengers, cyclic AMP or cyclic GMP, modulated by activation of metabotropic receptors target only those ionic channels which have cyclic nucleotide-binding domains (Fig. 5.9B). Some other ionic channels, on the other hand, can be regulated by protein kinases, as exemplified by the S-channel in *Aplysia* sensory neurones (Fig. 5.8). Recall that the phosphorylation of proteins by the protein kinase can alter markedly the state of the proteins, hence biological function of the proteins (Box C). The ionotropic receptor has phosphorylation sites. We have seen, in fact, that phosphorylation by different protein kinases causes the nicotinic ACh receptor to enhance its desensitization (section 4.3.3). Each neurone responds to different transmitters and possesses both ionotropic and metabotropic receptors. The protein kinases activated as a result of signal transduction initiated by some metabotropic receptors in a given neurone might then phosphorylate the ionotropic receptors in the same neurone in additon to the target ionic channel. In other words, intracellular signals dispatched from metabotropic receptors might modulate the responsiveness of the ionotropic receptor in the same neurone Thus, the ionotropic receptor might interact with the metabotropic receptor.

5.3.1 Interactions between ionotropic and metabotropic receptors

All the ionotropic receptors have several structural features in common (section 4.1): Each receptor comprises multiple homologous subunits, and each subunit contains four transmembrane segments. Moreover, each subunit has a long cytoplasmic loop (the second cytoplasmic loop) between the M_3 and M_4 transmembrane segments (Fig. 5.10A). It is this second cytoplasmic loop that contains phosphorylation sites (Swope *et al.* 1992). Might phosphorylation of the ionotropic receptor significantly alter its function? If so, how are the ionotropic receptors phsophorylated?

Glutamate is the major excitatory transmitter at the central synapses. We have seen that the ionotropic glu-

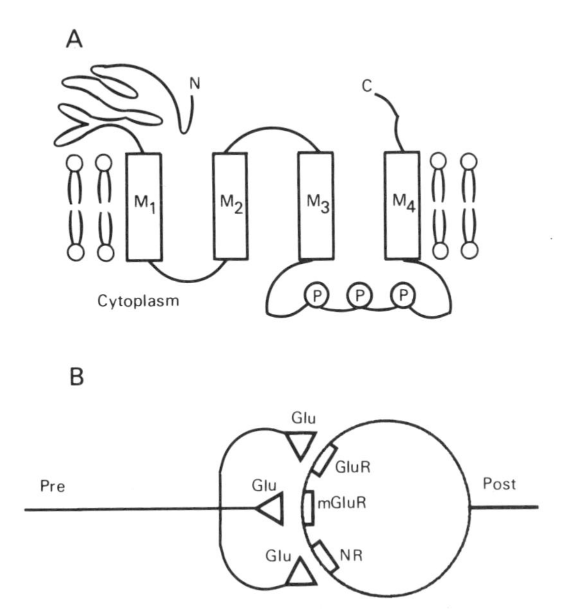

Fig. 5.10. Hypothetical interactions between metabotropic and ionotropic receptors by phosphorylation of the ionotropic receptor through protein kinases stimulated by activation of metabotropic receptors. A, membrane topology of ionotropic receptor subunits, indicating phosphorylation sites (P) in the second cytoplasmic loop between the M_3 and M_4 segments. (Adapted from Swope *et al.* 1992.) B, hypothetical multiple glutamate receptors located on a single postsynaptic neurone. Pre, a single presynaptic fibre, excitation of which causes the release of glutamate. Postsynaptic glutamate receptors: GluR, ionotorpic non-NMDA receptors; NR, ionotropic NMDA receptors; mGluR, metabotropic glutamate receptors.

tamate receptor can be classified into NMDA and non-NMDA receptors (section 4.1.5). These two types of ionotropic receptor not only coexist in a given central neurone but are also co-localized at individual excitatory synapses (Bekkers and Stevens 1989; Johns and Baughman 1991). The metabotropic glutamate receptor also coexists with the ionotropic glutamate receptor in central neurones. In principle, then, glutamate released from presynaptic fibres can activate three different types of glutamate receptor in the same postsynaptic neurone, as illustrated in Fig. 5.10B. Kelso *et al.* (1992) replicated this situation by expressing different glutamate receptors in *Xenopus* oocytes. Thus, the oocytes injected with crude rat brain RNA responded to NMDA through the NMDA receptor, to kainate or AMPA through the non-NMDA receptor as well as to *trans*-ACPD (*trans*-1-aminocyclopentane-1,3-dicarboxylic acid), a metabotropic glutamate receptor agonist. The

NMDA-induced inward current in the oocyte was significantly enhanced following activation of metabotropic glutamate receptors by their agonist (*trans*-ACPD) or by activation of protein kinase C with phorbol esters. By contrast, the inward current induced by non-NMDA receptor agonists was not affected by activation of metabotropic glutamate receptors or by phorbol esters. Kelso *et al.* (1992) thus suggested that the responsiveness of NMDA receptors can be enhanced by phosphorylation via protein kinase C which is stimulated by activation of metabotropic glutamate receptors. According to Kutsuwada *et al.* (1992), the NMDA receptors potentiated by activation of protein kinase C are the heteromeric receptors formed by NR1 and NR2A or NR1 and NR2B, but not those formed by NR1 and NR2C (for nomenclature of the receptor subunits, see Fig. 4.6 in section 4.1.5). Thus, some modulation or plastic change in the NMDA receptor appears to be specific for the subunit. Some metabotropic glutamate receptors (mGluR1 and mGluR5) are known to link to stimulation of phosphatidylinositol (PI) hydrolysis (Masu *et al.* 1991; Abe *et al.* 1992), whereas other metabotropic glutamate receptors (e.g., mGluR2 and mGluR3) mediate inhibition of the cyclic AMP cascade (Tanabe *et al.* 1992). The metabotropic glutamate receptors expressed in the oocytes thus appear to be one of the metabotropic glutamate receptor types which couple to the PI turnover leading to stimulation of protein kinase C (Fig. 5.6).

Consistent with this finding in oocytes, potentiation of NMDA responses through activation of protein kinase C was found in spinal trigeminal neurones of the rat. These trigeminal neurones are involved in signalling nociceptive information. The opioid receptor agonist was found to potentiate NMDA-induced responses in trigeminal neurones; this action was again mediated by activation of protein kinase C (Chen and Huang 1991). As we shall see (section 7.2.1), the NMDA receptor channel is blocked by extracellular Mg^{2+} in a voltage-dependent manner. Chen and Huang (1992) found that protein kinase C potentiates the NMDA response in trigeminal neurones by reducing the voltage-dependent Mg^{2+} block of the NMDA channel. Although the underlying mechanism remains unclear, reduction of the Mg^{2+} block would facilitate Ca^{2+} influx through the NMDA receptor channel, which in turn may increase protein kinase C activity further. Thus, activation of metabotropic receptors can interact with the ionotropic NMDA receptor-mediated response.

Regulation of the ionotropic non-NMDA receptor by protein kinases was also examined in *Xenopus* oocytes injected with mRNAs specific for ionotropic GluR1 and GluR3 receptor subunits (Keller *et al.* 1992). Recall that the glutamate receptor formed by GluR1 and GluR3 subunits is relatively permeable to Ca^{2+}, despite the non-NMDA receptor (section 4.1.5). The kainate-induced currents as well as the kainate-induced Ca^{2+} influx in the oocytes were potentiated by cyclic AMP, and this potentiation was blocked by protein kinase A inhibitors (Keller *et al.* 1992). Similarly, non-NMDA responses recorded from cultured hippocampal neurones were enhanced by protein kinase A (Greengard *et al.* 1991; Wang *et al.* 1991), and this potentiation was associated with increases in the mean open time and the opening frequency of the single-channel activity (Greengard *et al.* 1991). By contrast, the NMDA response in cultured hippocampal neurones was not affected by protein kinase A.

Modulation by protein kinases can also be seen in the ionotropic receptors involved in synaptic inhibition. Protein kinase A markedly enhanced glycine responses recorded from cultured trigeminal neurones and increased the open probability of their single-channel activity (Song and Huang 1990). Song and Huang (1990) suggested that serotonin or noradrenaline released from the descending fibres projecting to trigeminal neurones stimulates protein kinase A via their metabotropic receptors, thereby potentiating their responsiveness to glycine. $GABA_A$ receptors expressed in cultured cells, on the other hand, showed decreased responsiveness when their particular subunit (β) was phosphorylated by protein kinase A (Moss *et al.* 1992).

All these results point to the interactions between ionotropic and metabotropic receptors. Although which metabotropic receptors regulate which ionotropic receptors must be identified for each group of neurones *in situ*, it is almost certain that the efficacy of ionotropic receptors can be modulated by activation of metabotropic receptors in the same neurone. Greengard *et al.* (1991) found that forskolin, which stimulates protein kinase A, increases both the amplitude and the frequency of spontaneously occurring excitatory synaptic responses. This suggests that second messengers dispatched from metabotropic receptors may regulate not only the postsynaptic ionotropic receptor but also transmitter release from presynaptic terminals. Where might this signal for regulation of transmitter release come from? Might transmitter release be regulated by protein kinases via activation of the postsynaptic transmitter receptors? Let us now examine how transmitter release can be regulated.

5.3.2 Regulation of transmitter release

Transmitter release from presynaptic terminals can be regulated by a transmitter substance in two ways: by *axo-axonal synapses* or by *autocrine signalling*. As illustrated schematically in Fig. 5.11A, the axo-axonal synapse is formed by the synaptic connection of a presynaptic terminal with another presynaptic terminal (Gray 1962; Conradi 1969). Autocrine signaling refers to recurring activation of the receptors in a presynaptic terminal by the transmitter released from the terminal (Fig. 5.11B). Let us see the actual examples of these two processes first, then discuss how these processes involve metabotropic receptors.

GABA is a well-documented inhibitory transmitter substance responsible for the generation of inhibitory synaptic potentials in mammalian central neurones (section 4.1.4; Krnjević 1974; Nicoll 1988). Central synaptic transmission is also inhibited by GABA, based on another mechanism, *presynaptic inhibition* (Frank and Fuortes 1957; Eccles *et al.* 1961). For instance, the amplitude of monosynaptic e.p.s.p.s in spinal motor neurones elicited by stimulation of group Ia sensory fibres arising from the muscle can be reduced by stimulating another sensory input without any changes in the

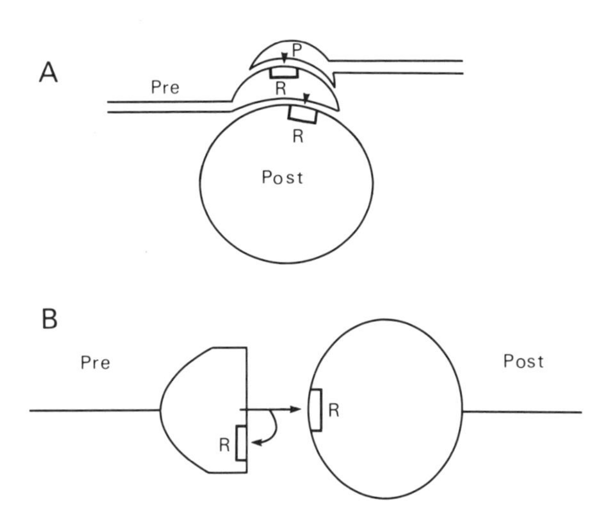

Fig. 5.11. Schematic illustration of presynaptic inhibiton by axo-axonal synapses (A) and autocrine signalling (B). A, a presynaptic terminal, P forms synaptic connections with another presynaptic terminal (Pre) which makes synaptic contacts on a postsynaptic neurone (Post). R, transmitter receptors. B, transmitter released from a presynaptic terminal binds to the postsynaptic receptor and at the same time acts on the receptors located on the presynaptic terminal (autoreceptors).

motor neurones. The presynaptic nature of this inhibition was evidenced by a decrease in the number of quanta of the transmitter released by stimulation of the group Ia sensory fibres projecting to the motor neurones (Kuno 1964b; Clements *et al.* 1987). The morphological basis for presynaptic inhibition is axo-axonal synapses (Gray 1962) which are formed by small *P terminals* (Conradi 1969) on terminals of group Ia sensory fibres in the spinal cord. Pharmacologically, presynaptic inhibition was found to be mediated by GABA (Barker and Nicoll 1972; Davidoff 1972; Nicoll 1988). In fact, the P terminals located on terminals of group Ia sensory fibres contain GABA (Maxwell *et al.* 1990). Recall that GABA binds to its metabotropic $GABA_B$ receptors as well as to its ionotropic $GABA_A$ receptors (section 4.1.4). Pharmacological experiments suggest that presynaptic inhibition in the spinal cord is mediated by both $GABA_A$ receptors (Peng and Frank 1989b; Stuart and Redman 1992) and $GABA_B$ receptors (F. R. Edwards *et al.* 1989; Peng and Frank 1989a), so that the central terminals of the sensory neurones must have both types of receptor.

The presence of transmitter receptors in presynaptic terminals is prerequisite for presynaptic inhibition. If a presynaptic terminal possesses the receptor for the transmitter released from the terminal, autocrine signalling could occur. In 1978, Miledi *et al.* reported a puzzling phenomenon that the amount of ACh released by nerve stimulation at rat neuromuscular junctions shows a two-fold increase when the junctional transmission is blocked by α-bungarotoxin (BUTX), the snake toxin which binds irreversibly to the nicotinic ACh receptor. There was no evidence for BUTX-sensitive nicotinic ACh receptors located on motor nerve terminals (Jones and Salpeter 1983). However, the blockade of neuromuscular transmission markedly reduced the amount of adenosine released from the muscle (Smith 1991). Adnosine is known to suppress the quantal release of ACh from motor nerve terminals (Ginsborg and Hirst 1972; M. R. Bennett *et al.* 1991). Thus, the puzzling phenomenon observed by Miledi *et al.* (1978) can now be accounted for by reduced concentrations of adenosine (hence reduced inhibition of transmitter release by adenosine) after the block of junctional transmission.

Activated muscle is not the only source of adenosine at neuromuscular junctions, however. Synaptic vesicles purified from cholinergic synapses contain both ACh and ATP (Wagner *et al.* 1978), and at neuromuscular junctions ACh and ATP are jointly released by nerve stimulation (Silinsky 1975). The ATP released from motor nerve terminals is rapidly hydrolysed to adenosine which in turn acts on adenosine receptors located on the motor nerve terminals (Silinsky 1980; Hamilton and Smith 1991). In effect, this is autocrine signalling, and such presynaptic receptors are called *autoreceptors*.

How does adenosine inhibit transmitter release from motor nerve terminals? Hamilton and Smith (1991) showed that adenosine inhibits Ca^{2+} currents recorded from rat motor nerve terminals. Suprisingly, the application of ACh also reduced the terminal Ca^{2+} current; this ACh action on Ca^{2+} currents was mimicked by muscarine and blocked by pirenzipine, suggesting that the ACh autoreceptor is the M_1 type of muscarinic receptor (Hamilton and Smith 1991). Since the effects of adenosine and ACh on the terminal Ca^{2+} current are not additive, the two autoreceptor processes appear to share common mechanisms. In fact, both ACh- and adenosine-induced inhibition of the terminal Ca^{2+} current involved G-proteins but was not affected by the application of different second messenger analogues or protein kinases. Hamilton and Smith (1991) suggested that inhibition of the terminal Ca^{2+} current is induced by direct action of the G-protein subunit on the channel (Fig. 5.1, route 3; Brown and Birnbaumer 1990; Dolphin 1991; Clapham and Neer 1993). In sensory neurones an adenosine analogue is known to inhibit the ω-CgTX-sensitive Ca^{2+} channel, and this inhibition is mediated directly by the G-protein (Kasai and Aosaki 1989). Adenosine also preferentially inhibits the ω-CgTx-sensitive Ca^{2+} channel in presynaptic terminals of the ciliary ganglion (Yawo and Chuma 1993). At many sympathetic synapses, norepinephrine and neuropeptide Y (NPY) act as cotransmitters (Lundberg and Hökfelt 1983). Interestingly, NPY also selectively inhibits ω-CgTx-sensitive Ca^{2+} currents in sympathetic neurones (Toth *et al.* 1993).

In cultured hippocampal neurones glutamate or quisqualate inhibits the Ca^{2+} current (Lester and Jahr 1990). This inhibitory effect was markedly potentiated when the intracellular medium was dialyzed with GTPγS, non-hydrolyzable GTP, suggesting the involvement of G-proteins. Inhibitors of several second messengers and protein kinases failed to prevent the inhibition of the Ca^{2+} channel by glutamate. Thus, it was suggested that metabotropic glutamate receptor-activated G-protein directly inhibits the Ca^{2+} channel (Lester and Jahr 1990). Moreover, the presence of glutamate autoreceptors was suggested in the hippocampus (Forsythe and Clements 1990; Baskys and Malenka 1991; Yamamoto *et al.* 1991). Thus, monosynaptic excitatory synaptic responses in hippocampal neurones were depressed by AP4 (2-amino-4-phosphonobutyric acid), a metabotropic glutamate receptor agonist, and this depression was associated with a reduction in the quantal release of transmitter (Forsythe and Clements 1990; Yamamoto *et al.* 1991). Thus, glutamate released from a presynaptic neurone appears to activate some metabotropic glutamate receptors on the presynaptic terminal (autoreceptors), thereby acting as a negative feedback for transmitter release.

Monosynaptic excitatory synaptic currents (e.p.s.c.s) recorded from hippocampal neurones were also depresed by GABA, and this action was mimicked by baclofen, a $GABA_B$ receptor agonist (Forsythe and Clements 1990). The statistical analysis of the changes in the e.p.s.c.s indicated that the action of GABA is at the presynaptic level to reduce the amount of transmitter release. Thus, this presynaptic inhibition involves metabotropic GABA receptors. Interestingly, GABA also reduced the amplitude of inhibitory postsynaptic currents (i.p.s.c.s) in hippocampal neurones (Yoon and Rothman 1991). These i.p.s.p.s were mediated by $GABA_A$ receptors, and the reduction in their amplitudes was due to activation of presynaptic $GABA_B$ receptors (Yoon and Rothman 1991). Therefore, in principle, the GABA released from the inhibitory neurones could induce autocrine signalling, activating $GABA_B$ receptors located on the terminals of the inhibitory neuorns. At present, however, whether autocrine signalling by GABA is present (Davies *et al.* 1991) or not (Yoon and Rothman 1991) in these inhibitory synapses remains unclear.

The physiological significance of metabotropic receptor-mediated synaptic responses appears to be quite different from that of ionotropic receptor-mediated responses. The latter is clearly involved in conveying neuronal signals from one neurone to another. The efficiency at the ionotropic receptor-mediated synapse can be modified by alterations either in the amount of transmitter release from its presynaptic terminal or in the responsiveness of its receptor in the postsynaptic neurone. Activation of metabotropic receptors can produce these changes in the efficiency of the ionotropic receptor-mediated synapse. The synaptic modulation is induced by G-proteins, second messengers or protein kinases, so that the alteration in synaptic efficacy can be a relatively long-lasting process. Thus, the function of metabotropic receptors appears to be dual: not only the transmission of signals from neurone to neurone, but also alterations in synaptic function, that is, synaptic plasticity. This second topic is the subject of Part II.

Summary and prospects

Unlike ionotropic receptors, all the metabotropic receptors consist of a single peptide. Each metabotropic receptor is incorporated in the cell membrane with seven transmembrane segments. The binding of the ligand to the metabotropic receptor causes its cytoplasmic domain to interact with a specific G-protein. It is this G-protein that induces signal transduction in the cell. There are many distinct G-proteins, each of which initiates a cascade of particular cellular signals by regulating a specific effector enzyme. Presumably, the idiosyncratic structure of the cytoplasmic loop in each metabotropic receptor is responsible for interactions with a specific G-protein. However, a given metabotropic receptor can interact with more than one G-protein ('cross-talk'); how this occurs remains unclear.

Following activation of a given metabotropic receptor, the target ionic channel some distance from the receptor is opened or closed. The target ionic channel can be regulated in three ways: by protein kinases activated through the second messengers, by the second messenger themselves or directly by the G-protein. The channel alters its configuration by phosphorylation via protein kinases; the second messenger-gated channel has a binding site specific for the second messsenger; how G-proteins regulate ionic channels directly is not known. However, at least some presynaptic inhibition of transmitter release (heterosynaptic presynaptic inhibition or inhibition via autoreceptors) appears to be based on suppression of presynaptic Ca^{2+} channels by G-proteins.

Unlike the ionotropic receptor, activation of metabotropic receptors results in modulation of a wide variety of intracellular signals by protein phosphorylation and regulation of different enzymes. In consequence, the ionotropic receptors can be phosphorylated by activation of metabotropic receptors present in the same neurone or modulated directly by the G-protein, thereby altering synaptic function mediated by the ionotropic receptors. Such modulation of ionotropic receptors by intracellular signals might be one of the mechansisms of induction of long-term synaptic plasticity.

Suggested reading

The primary structure of metabotropic receptors

Applebury, M. and Hargrave, P. A. (1986). Molecular biology of the visual pigments. *Vision Research*, **26**, 1881–95.

Dohlman, H. G., Caron, M. G., and Lefkowitz, R. J. (1987). A family of receptors coupled to guanine nucleotide regulatory proteins. *Biochemistry*, **26**, 2657–64.

Masu, Y., Nakayama, K., Tamaki, H., Harada, Y., Kuno, M., and Nakanishi, S. (1987). cDNA cloning of bovine substance-K receptor through oocyte expression system. *Nature*, **329**, 836–8.

Tanabe, Y., Masu, M., Ishii, T., Shigemoto, R., and Nakanishi, S. (1992). A family of metabotropic glutamate receptors. *Neuron*, **8**, 169–79.

The role of G-proteins

Baumgold, J. (1992). Muscarinic receptor-mediated stimulation of adenyl cyclase. *Trends in Pharmacological Sciences*, **13**, 339–40.

Birnbaumer, L. (1992). Receptor-to-effector signalling through G proteins: roles for $\beta\gamma$ dimers as well as α subunits. *Cell*, **71**, 1969–72.

Clapham, D. E. and Neer, E. J. (1993). New role for G-protein $\beta\gamma$-dimers in transmembrane signalling. *Nature*, *365*, 403–6

Gilman, A. G. (1987). G proteins: transducers of receptor-generated signals. *Annual Review of Biochemistry*, **56**, 615–49.

Ross, E. M. (1989). Signal sorting and amplification through G protein-coupled receptors. *Neuron*, **3**, 141–52.

Taylor, C. W. (1990). The role of G proteins in transmembrane signalling. *Biochemical Journal*, **272**, 1–13.

Modulations of ionic channels by metabotropic receptors

Brown, A. M. and Birnbaumer, L. (1990). Ion channels as G protein effectors. *Annual Review of Physiology*, **52**, 197–213.

Dolphin, A. C. (1991). Regulation of calcium channel activity by GTP binding proteins and second messengers. *Biochimica et Biophysica Acta*, **1091**, 68–80.

Hille, B. (1992b). G protein-coupled mechanisms and nervous signaling. *Neuron* **9**, 187–95.

Kaupp, U. B., Niidome, T., Tanabe, T., Terada, S., Bonigk, W., Stuhmer, W., Cook, N. J., Kangawa, K., Matsuo, H., Hirose, T., Miyata, T., and Numa, S. (1989). Primary structure and functional expression from complementary DNA of the rod photoreceptor cyclic GMP-gated channel. *Nature*, **342**, 762–6.

Siegelbaum, S. A., Camardo, J. S., and Kandel, E. R. (1982). Serotonin and cyclic AMP close single K channels in *Aplysia* sensory neurones. *Nature*, **299**, 413–18.

Shuster, M. J., Camardo, J. S., Siegelbaum, S. A., and Kandel, E. R. (1985). Cyclic AMP-dependent kinase closes the serotonin-sensitive K channels of *Aplysia* sensory neurones in cell-free membrane patches. *Nature*, **313**, 392–5.

Modulation of synaptic function by metabotropic receptors

Bourne, H. and Nicoll, R. (1993). Molecular machines integrate coincident synaptic signals. *Neuron*, 10 (Supplement), 65–75.

Forsythe, I. D. and Clements, J. D. (1990). Presynaptic glutamate receptors depress excitatory monosynaptic transmission between mouse hippocampal neurones. *Journal of Physiology (London)*, 429, 1–16.

Hamilton, B. R. and Smith, D. O. (1991). Autoreceptor-mediated purinergic and cholinergic inhibition of motor nerve terminal calcium currents in the rat. *Journal of Physiology (London)*, 432, 327–41.

Swope, S. L., Moss, S. J., Blackstone, C. D., and Huganir, R. L. (1992). Phosphorylation of ligand-gated ion channels: a possible mode of synaptic plasticity. *FASEB Journal*, 6, 2514–23.

Part II Plasticity

6

Historical perspective of neuronal plasticity

6.1 EARLY CONCEPTS OF NEURONAL PLASTICITY

If the structure and function of the nervous system were rigid and unmodifiable, one's behavioural alteration would never occur. To describe the enduring modification of behavioural response, the term *plasticity* was introduced by James (1890), whose textbook, *The principles of psychology*, states: 'An acquired habit is nothing but a new pathway of discharge formed in the brain. . . Nervous tissue seems endowed with a very extraordinary degree of plasticity. . . Plasticity means the possession of a structure weak enough to yield to an influence, but strong enough not to yield all at once.' That is, a neuronal pathway can emerge *de novo* or the existing reflex pathways can be reinforced gradually in association with past experience.

How might a reflex pathway be reinforced? Muscle contractions of the rabbit foot, for example, can be evoked by stimulating the cerebral cortex or the skin. However, after lowering intensities until neither stimulation alone elicits a contraction, simultaneous stimulation of the cortex and skin is still effective, as if both inputs mutually helped to pave the way for producing a pooled effect. This phenomenon was termed *Bahnung*, which was translated into English as canalization, augmentation or reinforcement. For Starling (1912), *Bahnung* was the process that promotes particular neuronal pathways to the exclusion of others: 'The law of *Bahnung* is really the law of habit. . . Memory itself has the process of *Bahnung* as its neural basis.' In Schäfer's 1900 *Text-book of physiology*, Sherrington suggested that *Bahnung* reflects the summation of multiple excitatory inputs, so that *facilitation* would be a more appropriate term. Evidently Sherrington did not see *Bahnung* as being related to memory.

Instead, Sherrington (1906) considered that 'projicient stimuli' (stimuli which can influence the state of consciousness, perception or emotion), whether harmful or pleasurable, once branded on the memory might henceforth reinforce the reaction to other stimuli, by remembered association. Indeed, Pavlov (1906) found that dogs salivate at the mere sound of a bell if this conditioned stimulus has been applied repeatedly in association with feeding (unconditioned stimulus). This conditioned reflex is a clear example of learning acquired by association with 'projicient stimuli'.

Although Pavlov (1927) formulated a rule for the acquisition of learning, according to Konorski (1948), he ignored the question of how persistent neuronal changes occur in the conditioned reflex. Konorski (1948) postulated that a stimulus evoking a reflex induces a dual change in the nervous system; both a transient excitability change and a sustained transformation that he called *plasticity* (without referring to James). How might the stimulus elicit sustained neuronal activity? A sustained neuronal response can be maintained if impulse activity continuously circulates through closed self-exciting chains of neurones (*reverberating circuits*; Forbes 1922; Rashevsky 1938). Konorski (1948) objected, however, that plastic changes (e.g., conditioned reflexes) persist even after deep anaesthesia, which should have suppressed such continuously circulating impulses. Konorski (1948) hypothesized that acquiring a conditioned reflex (learning) entails the formation and multiplication of synaptic contacts, whereas its eventual extinction (forgetting) occurs in the absence of reinforcement, as prolonged disuse brings progressive atrophy of the synaptic contacts. There was no experimental evidence for this hypothesis, however.

The long-standing fascination with the mysteries of plasticity was marked by more fervour than fact. Even so, as we shall see, several early insights eventually found experimental support.

6.2 USE AND DISUSE EFFECTS ON SYNAPTIC FUNCTION

Konorski's hypothesis was actually based on Cajal's (1911, 1928) notion that synapses that are not used for a long period (disuse) undergo atrophy. It seemed plausible that excessive usage of synapses would lead to increased synaptic function, and disuse to defective function. During the 1950s these hypotheses received the first experimental tests. When repetitive (tetanic) stimuli were applied to presynaptic fibres at high frequencies for a short period (e.g., 50–500 Hz for 10 sec), the subsequent postsynaptic response to each presynaptic impulse was enhanced for several minutes in the sympathetic ganglion (Larrabee and Bronk 1947) and spinal cord (monosynaptic reflexes; Lloyd 1949; Eccles and Rall 1951). This phenomenon was called *post-tetanic potentiation*. Conversely, Eccles and McIntyre (1953) tested the disuse effect by sectioning the dorsal root just distal to its ganglion on one side in the cat. Under this condition the sensory fibres peripheral to the transection degenerate in a few days, but the sensory neurones in the ganglion and their central processes (dorsal root fibres) projecting to the spinal cord survive. The reflex pathway formed by these dorsal root fibres was assumed to be in total disuse, because the sensory activity from the periphery would not reach the spinal cord through the transected segment. Three to six weeks after this operation, the monosynaptic spinal reflex evoked by stimulating the dorsal root in the sectioned segment was markedly depressed, compared with the corresponding reflex observed on the contralateral, intact side. Similarly, monosynaptic excitatory postsynaptic potentials (e.p.s.p.s) evoked in spinal motor neurones by stimulating sensory fibres arising from a muscle nerve were depressed a few weeks after section of the muscle nerve (Eccles *et al.* 1959). Thus, the function of central synapses seemed to be weakened by prolonged disuse of the synapses.

These experiments by Eccles' group had severed sensory nerve fibres to create disuse at the central synapses. Might the decreased synaptic function be from injury rather than from the absence of sensory input? Beranek and Hnik (1959) approached this question in the cat by severing only the muscle tendon (tenotomy), not the nerve; they assumed that tenotomy would minimize the usage of sensory fibres arising from muscle spindles (stretch sensory receptors) because of the inability to stretch. A few weeks later, monosynaptic spinal reflexes evoked by stimulating sensory fibres from the tenotomized muscle were significantly enhanced (Beranek and Hnik 1959; Kozak and Westerman 1961). It now appeared that prolonged disuse of the presynaptic input had enhanced rather than depressed central synaptic function, contrary to expectation.

One might question whether reduced impulse activity in the sensory fibres of tenotomized muscles was an erroneous assumption. This objection was addressed by blocking the conduction of an intact muscle nerve with locally applied tetrodotoxin for two weeks in the cat, so that the sensory input activity from the periphery was chronically deprived. Again, monosynaptic e.p.s.p.s elicited in motor neurones by stimulating the muscle nerve that had been blocked for two weeks were enhanced (Gallego *et al.* 1979). Enhancement of junctional or synaptic transmission following chronic conduction block of the presynaptic input has since been confirmed at neuromuscular junctions (Snider and Harris 1979) and in the sympathetic ganglion (Gallego and Geijo 1987). Thus, although intuitively compelling, the concept of atrophy or functional weakening of synapses by prolonged disuse still lacks experimental support.

6.3 HEBBIAN SYNAPSES

In behavioural psychology, the properties of learning are inferred from the input–output relationship (stimulus and resulting behavioural response), treating its neuronal mechanisms as a black box (Dudai 1989). According to Hebb (1949), a psychological explanation is ultimately a statement of relationships between observed phenomena, devoid of 'physiologizing'. Hebb (1949) assumed that synaptic function is strengthened by co-activation of a presynaptic fibre and its postsynaptic neurone In a scheme illustrated in Fig. 6.1A, for example, the synaptic transmission from a presynaptic fibre, a, to its postsynaptic neurone, b, increases in efficiency when excitation occurs concurrently and repeatedly in both elements. This postualted learning process by concomitant activation of pre-and postsynaptic elements is now termed the *Hebbian synapse* or *Hebbian conjunction*.

Stent (1973) interpreted plastic changes previously observed in the visual cortex by extrapolation from Hebb's hypothesis. When kittens are reared with monocular deprivation of vision (by suturing one eyelid for more than one month), only a few cells in the visual

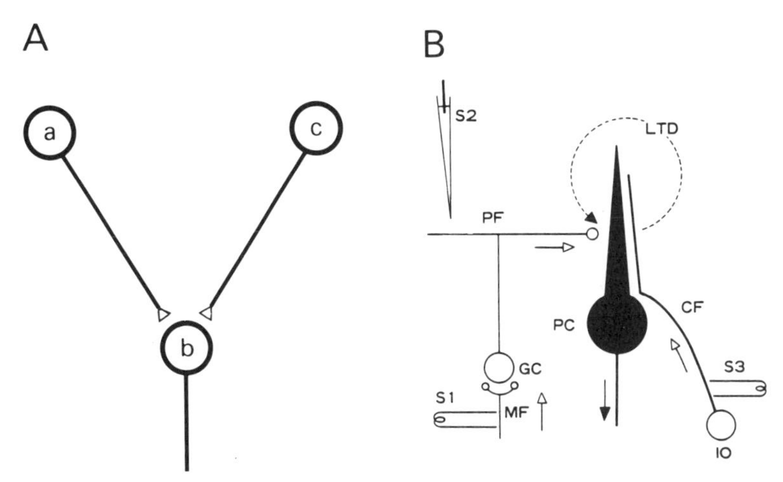

Fig. 6.1. Schematic neuronal networks. A, two excitatory neurones (a, c) converging on a postsynaptic neurone (b). B, two excitatory inputs to a Purkinje cell (PC) in the cerebellar cortex: left, from a mossy fibre (MF) via a granular cell (GC) and its axon, a parallel fibre (PF); right, from a climbing fibre (CF) of the inferior olive neurone (IO). Joint stimulation of mossy fibres (S1) and climbing fibres (S3) induces long-term depression (LTD) of the response of Purkinje cells to stimulation of parallel fibres (S2). (From Ito 1990.)

(striate) cortex respond to visual stimulation of the deprived eye (Wiesel and Hubel 1963). This is not functional waning from disuse, because if both eyes are occluded for the same period of time, many cells in the visual cortex are readily driven by the two eyes as in normal kittens (binocular response; Wiesel and Hubel 1965). Hubel and Wiesel (1965) found that in order to maintain a binocular input, the cells in the visual cortex must receive synchronous activity from both eyes or else be equally devoid of activity from both eyes. It is a lack of synergy or symmetrical activity between the two afferent paths that gives rise to the change in cortical neurones from the binocularly to the monocularly driven state (Hubel and Wiesel 1965; Wiesel 1982). That is, the two inputs from both eyes to a given cortical neurone appear to compete for synaptic connections on the neurone; if the two input activities are not synchronous, the more active input maintains its synaptic connection at the expense of the less active input. In terms of Fig. 6.1A, when a presynaptic input, c, is persistently inactive (or less active) while its postsynaptic neurone, b (visual cortical neurone), is repeatedly activated by another source, a, the synaptic transmission from c to b is weakened (Stent 1973).

Marr (1969) developed a specific hypothesis by applying Hebb's assumption to a known network in the cerebellum. The cerebellar cortex receives two distinct inputs (Fig. 6.1B). One is carried by mossy fibres (MF), which arise from the brainstem and spinal cord and excite Purkinje cells (PC) through granular cells (GC) and their axons, parallel fibres (PF). The other input is carried by climbing fibres (CF) which arise from the inferior olive nucleus (IO) and give a powerful excitatory input directly to Purkinje cells. According to Marr (1969), the efficacy of the synapse formed by a parallel fibre on a Purkinje cell increases if the parallel fibre is active at about the same time as the Purkinje cell is excited by the climbing fibre. A similar mechanism was also proposed by Albus (1971), but he assumed reduced efficiency at the parallel fiber–Purkinje cell synapse by co-activation of the Purkinje cell via the climbing fibre. These contradictory hypotheses were then experimentally tested, and conjunctive stimulation of the parallel fibres and climbing fibres indeed induced long-lasting depression of synaptic transmission from the parallel fibres to Purkinje cells, supporting Albus' assumption (Ito *et al.* 1982; Ito 1990). Thus, Hebbian conjunction appears to be at least one precondition for plastic modulation of synaptic function, but whether its modulation will be potentiation or depression is not readily predictable. Nor do we know how Hebbian conjunction induces prolonged synaptic modulation.

6.4 LONG-TERM POTENTIATION

Mammalian spinal motor neurones receive direct (monosynaptic) excitatory inputs from sensory fibres.

Because of this relatively simple organization, the excitatory synapses on motor neurones have been studied extensively. Eccles and his colleagues employed various approaches to induce plastic changes at these synapses, but the results were unfortunately equivocal (Eccles 1964). As noted (section 6.2), post-tetanic potentiation can occur in the spinal monosynaptic pathway, but such potentiation lasts for only a few minutes and is too short to be related to behavioural plastic changes. This led Eccles (1974) to infer that the function of the spinal cord is rigidly fixed even in young animals, so that plastic changes more readily arise in the brain, at the region closely related to the process of memory.

Where in the brain does memory reside? A first clue that the hippocampus is one crucial site came from clinical observations. Milner and Penfield (1955) reported a total loss of recent memory in patients who underwent bilateral removal of the hippocampus for the control of epilepsy. In the rat, bilateral hippocampectomy impaired maze learning ability (Thomas and Otis 1958), and in monkeys, bilateral ablation of the hippocampus and amygdala resulted in a deficit of memory function (Mishkin 1978).

Based on this localization, several workers investigated the hippocampus for its extrinsic and local circuit connections and cellular phenomenon that might help the understanding of memory. Most influential were studies initiated by Bliss and Lomo (1973). Applying a brief tetanic stimulation (10–100 Hz for 3–20 sec) to the perforant path (presynaptic) fibres in the rabbit potentiated the postsynaptic response recorded from granular cells of the hippocampal formation. Amazingly, this potentiation lasted for more than three days (Bliss and Gardner-Medwin 1973), hence the phenomenon was termed *long-term potentiation* or LTP. LTP has since been found in other ares of the central nervous system (Teyler and DiScenna 1987) and even in peripheral synapses (Kuba and Kumamoto 1990). Today, LTP is one of the most compelling phenomena used for analysing cellular mechanisms of synaptic plasticity in relation to learning and memory (Nicoll *et al.* 1988; Gustafsson and Wigstrom 1988; Malenka *et al.* 1989a). Indeed, we shall return to consider the basis of LTP in some detail later (section 7.2.2).

When the rat is treated with an inhibitor of protein synthesis, tetanic stimulation of the presynaptic fibres fails to maintain the late phase of LTP in the hippocampus (Frey *et al.* 1988; Otani *et al.* 1989). Similarly, the long-term synaptic modulation observed in *Aplysia* requires the synthesis of new macromolecules (Montarolo *et al.* 1986, 1988). Thus, memory of neural information may be retained in the form of newly synthesized molecules (Black *et al.* 1987). Apparently, learning involves the *de novo* formation of molecules, in addition to, or rather than the *de novo* formation of neuronal paths.

7

Functional plasticity at synapses

Currently, *neuronal plasticity* is used as a comprehensive term referring to relatively long-lasting modification in neuronal response (Thompson 1967). Although some common features may exist, a rule describing a given example of plastic changes will not necessarily hold for other examples of plasticity. Hence, plasticity must be defined anew for each individual phenomenon according to its own characteristics. Clearly, an overarching framework for neuronal plasticity has yet to be drawn.

Functional synaptic plasticity is a relatively long-lasting alteration in the efficiency of synaptic transmission. In biophysical terms, the efficiency of synaptic transmission is determined by two parameters: how much transmitter is released from presynaptic terminals, and what size postsynaptic response is produced per unit amount of transmitter. Consequently, plastic changes at synapses must be associated with changes in either one or both parameters. The principal questions concern how these changes are induced (acquisition or learning) and maintained (retention or memory). Depending on how long these changes persist, plasticity can be divided into short-term (minutes to hours) and long-term (days to years) alterations (Kandel and Schwartz 1982; Squire 1982). We will first focus on relatively straightforward forms of plastic changes and see what substrates are involved. For this purpose, we shall begin with short-term plasticity at peripheral synapses.

7.1 SUBSTRATES INVOLVED IN PLASTICITY AT PERIPHERAL SYNAPSES

We described post-tetanic potentiation (PTP) as a simple form of short-term synaptic plasticity (section 6.2). PTP occurs in nearly every chemical synapse studied so far. The enhancement of synaptic responses that follows a brief repetitive stimulation (e.g., 20–500 Hz for 10 sec) of the presynaptic nerve typically lasts only a few minutes. We have seen that changes in synaptic efficacy can be distinguished in terms of the presynaptic or postsynaptic level by quantal analysis (section 3.2.2). PTP is associated with an increase in the mean number of quanta of the transmitter released by presynaptic impulse without any change in the size of the unit synaptic potential (quantal size; Liley 1956; Martin and Pilar 1964b; Kuno 1964b; McLachlan 1975; Hirst *et al.* 1981). Evidently, PTP is due to more transmitter being released by the presynaptic impulse.

How might repetitive presynaptic stimulation for a short period induce a prolonged enhancement of transmitter release? Lloyd (1949) suggested that the action potential in presynaptic terminals may increase in amplitude after tetanic stimulation (see also Eccles and Krnjević 1959; Hubbard and Schmidt 1963). However, in the chick ciliary ganglion, where intracellular potentials can be recorded from its large presynaptic terminal, no change in the amplitude or configuration of the presynaptic action potential was detected during the PTP (Martin and Pilar 1964b). What then is responsible for the PTP?

7.1.1 Residual calcium in post-tetanic potentiation

As early as 1940, Feng found that the amplitude of e.p.p.s at the neuromuscular junction increases in response to the second nerve stimulus applied at various intervals up to 80 ms after the first stimulus. This phenomenon, termed *facilitation*, has since been confirmed in a variety of chemical synapses. Again, this

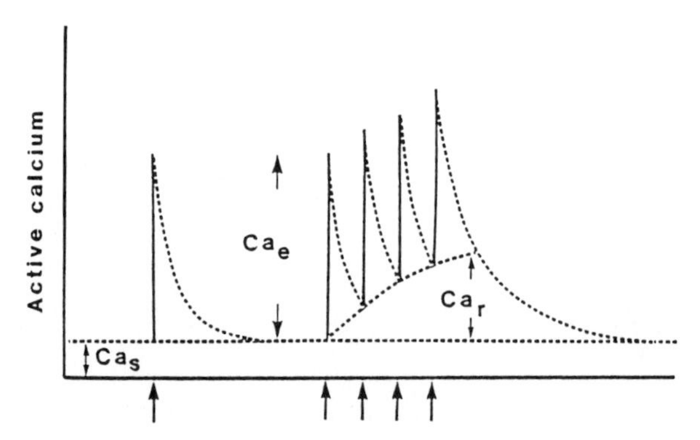

Fig. 7.1. Illustration of the residual calcium hypothesis. The steady level of intracellular Ca^{2+} in a nerve terminal at rest (Ca_s) may increase transiently after a nerve impulse arrives (Ca_e; calcium entry). Repetitive nerve stimulation at short intervals (right four arrows) may accumulate residual Ca^{2+} (Ca_r) in the terminal. Abscissa, time. (From Zucker and Lara-Estrella 1983.)

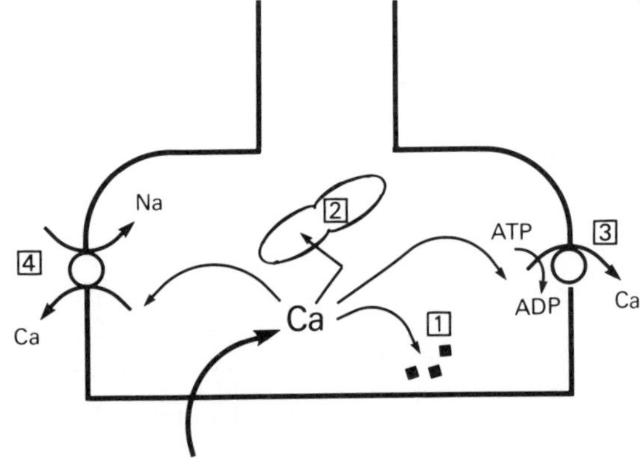

Fig. 7.2. Schematic illustration of four major regulatory processes for cytoplasmic Ca^{2+} levels in a nerve terminal. The bold arrow indicates Ca^{2+} influx on excitation of the terminal, elevating the intraterminal Ca^{2+} level. Ca^{2+} buffering occurs by: (1) binding to calcium-binding proteins, (2) uptake by the intracellular organelles, and extrusion via (3) Ca^{2+} pump and (4) Na^+-Ca^{2+} exchange pump. Ca^{2+} pump operates using energy released by hydrolysis of ATP to ADP. Na^+-Ca^{2+} exchange pump is driven by the concentration gradients of Na^+ and Ca^{2+} between the intracellular and extracellular media.

facilitated synaptic transmission comes from an increase in quantal release of the transmitter (Del Castillo and Katz 1954b; Martin and Pilar 1964b; Kuno 1964b). A single action potential reaching the nerve terminal undoubtedly increases the intraterminal Ca^{2+} concentration to a level sufficient to induce transmitter release. Katz and Miledi (1968) linked the facilitation to a residue of Ca^{2+} that enters the nerve terminal during the preceding impulse (see also Charlton *et al.* 1982). As illustrated in Fig. 7.1, the intracellular Ca^{2+} level in a nerve terminal may increase transiently in association with excitation (Ca_e) and decline to a steady level (Ca_s) over time (Zucker and Lara-Estrella 1983). But if the second presynaptic impulse reaches the terminal before the elevated Ca^{2+} subsides, the second impulse would release more transmitter than the first impulse, thereby producing facilitation. Recall that the amount of transmitter released is related to the third or fourth power of the intraterminal Ca^{2+} concentration (Dodge and Rahamimoff 1967; Augustine and Charlton 1986; Chapter 3). Therefore, the residual Ca^{2+} left behind from the initial influx would contribute substantially to facilitation by the second impulse.

Rosenthal (1969) showed that tetanic stimulation of the motor nerve in a Ca^{2+}-free solution fails to induce PTP in the neuromuscular junction even when Ca^{2+} is added immediately after the tetanization (see also Weinreich 1971). This suggests that the entry of Ca^{2+} into presynaptic terminals during the tetanus is essential for the generation of PTP. Might PTP also rely on residual Ca^{2+} built up in the nerve terminal during tetanic stimulation of the presynaptic nerve? To address this possibility, the residual Ca^{2+} was assumed to rise linearly with each successive stimulus and an exponential decay from each peak (Fig. 7.1, right four arrows). Quantitative analyses of the kinetic behaviour of PTP at the frog neuromuscular junction did not fit the relation predicted from residual Ca^{2+} summating during the tetanization (Magleby 1973, 1987; Zengel and Magleby 1982). Thus, Magleby (1987) inferred that besides residual Ca^{2+} other mechanisms are needed to generate PTP. However, intracellular Ca^{2+} levels can be affected by many processes, so that the precise levels during or after tetanic stimulation are unpredictable in practice.

Figure 7.2 illustrates the four major processes for regulation of intracellular Ca^{2+}: (1) calcium-binding proteins, (2) intracellular organelles, (3) the Ca^{2+} pump, and (4) Na^+–Ca^{2+} exchange transport (Blaustein 1988; Miller 1991; Luther *et al.* 1992). The calcium-binding proteins (calmodulin, parvalbumin, calbindin, and calretinin) trap excessive free (ionized) calcium, thereby reducing intracellular Ca^{2+} level (Ca^{2+} buffering); this process is rapid (within milliseconds) but limited in capacity. The intracellular organelles (mitochondria and endoplasmic reticulum) take up free calcium from the cytosol; this process has a large capacity for Ca^{2+} buffering but operates at a slower pace.

Extrusion of Ca^{2+} from the cell by the Ca^{2+} pump and Na^{+}-Ca^{2+} exchange pump occurs more slowly still (seconds or longer; Blaustein 1988; Miller 1991).

The operation of the Na^{+}–Ca^{2+} exchange depends on the concentration gradient of both Na^{+} and Ca^{2+} between the intracellular and extracellular media. The electrochemical gradients of Na^{+} and Ca^{2+} across the terminal membrane would alter from moment to moment during and after tetanic nerve stimulation as Na^{+} accumulates in the terminal and as intraterminal Ca^{2+} is buffered by Ca^{2+}-binding proteins or by the intracellular organelles. Hence, it is hardly surprising that the time course of PTP diverged from the intracellular Ca^{2+} levels predicted on the simple assumption described above.

Changes in internal Ca^{2+} levels produced by tetanic stimulation were measured directly with Fura-2, a fluorescent Ca^{2+} indicator, in the crayfish motor nerve terminal (Delaney *et al.* 1989) and presynaptic terminals of the squid giant synapse (Swandulla *et al.* 1991b). The intracellular Ca^{2+} level was 0.1–0.2 μM at rest and showed a 10-fold increase after tetanic stimulation of the nerve. The decay rate of the elevated Ca^{2+} level in the presynaptic terminal after tetanic stimulation was roughly correlated with the decay rate of the PTP of synaptic transmission. Thus, residual Ca^{2+} accumulated in the presynaptic terminal following tetanic nerve stimulation indeed appears to be a major factor responsible for PTP.

During tetanic stimulation, Ca^{2+} in the presynaptic terminal first accumulates near the site of transmitter release before spreading throughout the terminal (Swandulla *et al.* 1991b). This spatial compartmentalization must arise from the voltage-gated Ca^{2+} channels localized near the sites of transmitter release (Pumplin *et al.* 1981; Robitaille *et al.* 1990; Cohen *et al.* 1991; Torri Tarelli *et al.* 1991). According to Smith and Augustine (1988), a single action potential may transiently increase the local concentration of Ca^{2+} at the release site to about 100μM or more before it diffuses into the interior of the terminal (see also Roberts *et al.* 1990; Adler *et al.* 1991; Von Gersdoff and Mathews 1994). When a Ca^{2+} chelator, EGTA, was injected into a presynaptic terminal of the squid giant synapse, transmitter release was not affected, but both PTP and the persistent Ca^{2+} accumulation in the presynaptic terminal following tetanic nerve stimulation were blocked (Swandulla *et al.* 1991b). Thus, transmitter release seems to be triggered by a brief, highly localized rise in Ca^{2+} at the release site, whereas the generation of PTP requires excessive Ca^{2+} to be uniformly accumulated in the entire compartment of the nerve terminal. If the Ca^{2+} influx were not sufficient to spread throughout the compartment, the highly localized rise of Ca^{2+} at the relase site would quickly subside to submicromolar levels, and the potentiation of transmitter release would cease. In other words, Ca^{2+} accumulated in the entire compartment of the terminal may act as a reserve source for residual Ca^{2+} at the release site.

Quantitatively, the decay rate of PTP is linearly related to that of residual Ca^{2+} in the crayfish motor nerve terminal (Delaney *et al.* 1989)—which contradicts the widespread notion that the amount of transmitter released is related to a third or fourth power function of the internal Ca^{2+} concentration. In presynaptic terminals of the squid synapse, the Ca^{2+}-dependence of the potentiated synaptic response was also less steep than that of the control response (Swandulla *et al.* 1991b). The explanation for this behaviour is not clear. Nonetheless, the induction and maintenance of PTP can be accounted for largely, if not entirely, by the balance between the Ca^{2+} accumulated in the terminal during the tetanus and its subsequent clearance by Ca^{2+} buffers.

7.1.2 Second messengers in long-lasting potentiation

The intraterminal Ca^{2+} level once elevated by tetanic stimulation declines to the original level within a few minutes. But in the rat sympathetic ganglion, tetanic preganglionic stimulation induces not only PTP but also a synaptic potentiation that lasts for hours (Brown and McAfee 1982). A similar long-lasting potentiation was confirmed in the frog sympathetic ganglion (Fig. 7.3A; Koyano *et al.* 1985). By analogy with the long-term potentiation observed in the hippocampus (Bliss and Lomo 1973), that found in the sympathetic ganglion was also termed *long-term potentiation* or LTP (Brown and McAfee 1982; Koyano *et al.* 1985), although the LTP in the sympathetic ganglion might not last for days.

Figure 7.3A depicts potentiation of e.p.s.p.s in a frog sympathetic neurone following tetanic preganglionic stimulation. The initial phase of potentiation declines with a time constant of about 3 min (Fig. 7.3B), like the PTP observed in other preparations, and the more prolonged second phase (LTP) decays with a time constant of 40 min (Fig. 7.3C). As in PTP, the LTP was associated with an increase in the mean number of quanta of the transmitter released by a presynaptic impulse

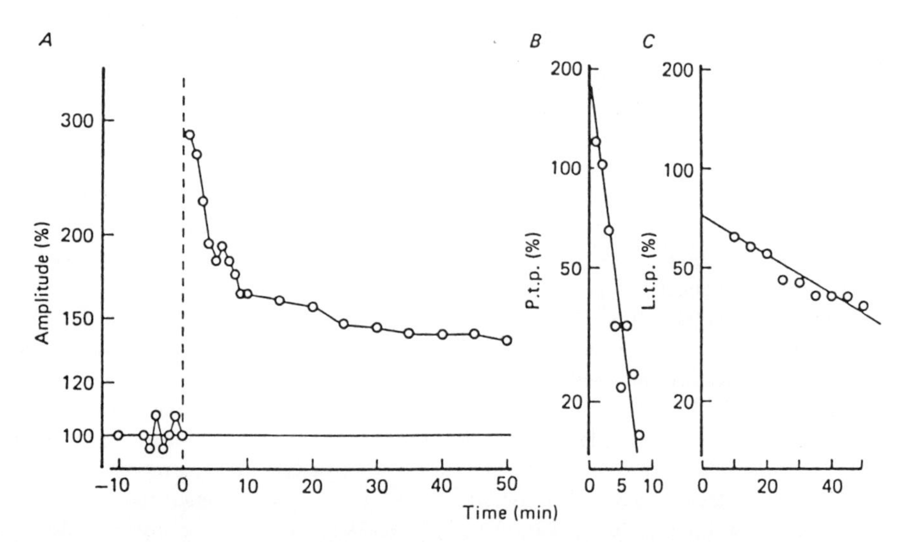

Fig. 7.3. Post-tetanic potentiation (p.t.p) and long-term potentiation (L.t.p.) of synaptic potentials evoked by tetanic stimulation of the preganglionic nerve (33 Hz for 10 sec) in the bullfrog sympathetic ganglion. A, time course of increased amplitude of the synaptic potential after the tetanus applied at time 0. B, relative increases in the amplitude of synaptic potentials during the first 8 min after the tetanus (P.t.p.) plotted on a semilogarithmic scale. C, same as in B, but for the synaptic responses recorded 10 to 50 min after the tetanus (L.t.p.). (From Koyano *et al.* 1985.)

(Koyano *et al.* 1985). Yet tetanically stimulating the preganglionic nerve in a Ca^{2+}-free solution, then immediately adding Ca^{2+} to the external solution, elicited neither PTP nor LTP (Koyano *et al.* 1985). Evidently, both LTP and PTP are triggered by accumulation of Ca^{2+} in the presynaptic terminal during the tetanus. Unlike PTP, however, LTP outlasts the period of overabundant intraterminal Ca^{2+}. Presumably, the Ca^{2+} accumulated in the terminal during the tetanus triggers another process that prompts LTP in sympathetic neurones. To discover what that second process might be, we need to trace Ca^{2+} and its influence in presynaptic terminals.

Calcium is a regulator of diverse cellular processes. This regulation is achieved by activating key enzymes that bind Ca^{2+} when its concentration rises above a certain level (Kennedy 1989). Figure 7.4 illustrates five target enzymes activated by Ca^{2+}: two membrane phospholipases (phospholipase C and phospholipase A_2) and three cytosolic enzymes (calpain, protein kinase C, and calmodulin-dependent kinase). We have seen that activation of some G-protein-linked receptors stimulates phospholipase C (PLC). This results in the production of a second messenger, IP_3 (inositol trisphosphate), which in turn triggers the release of Ca^{2+} from intracellular stores (Fig. 5.6). Ca^{2+} at micromolar ranges can also stimulate PLC directly (Bident *et al.* 1987; Eberhard and Holz 1987). Phospholipase A_2 (PLA_2) is another target enzyme of G-proteins (Axelrod *et al.* 1988) but can also be activated directly by Ca^{2+} when its concentration increases (Van den Bosch 1980), producing a

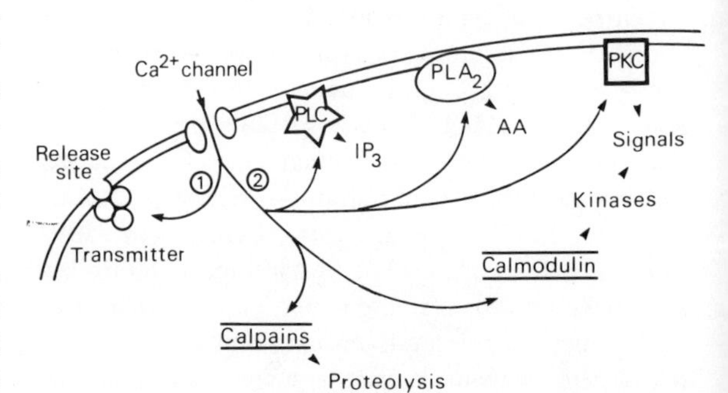

Fig. 7.4. A scheme illustrating cellular responses produced by Ca^{2+} influx (route 2) and transmitter release (route 1). The entry of Ca^{2+} through Ca^{2+} channels in the cell membrane may activate phospholipase C (PLC), phospholipase A_2 (PLA_2), protein kinase C (PKC) or calpains, and may bind to calmodulin. Activation of PLC and PLA_2 produces second messengers, inositol trisphosphate (IP_3) and arachidonic acid (AA), respectively. Calpains hydrolyse proteins (proteolysis). Activation of kinases via calmodulin (or PKC) induces signal transduciton. These Ca^{2+}-dependent cellular responses may lead to plastic changes in transmitter release.

second messenger, arachidonic acid (AA; Fig. 7.4). The *calpains* are Ca^{2+}-activated proteinases (Melloni and Pontremoli 1989; Murachi 1989). By their proteolytic action, the calpains affect a variety of cellular mechanisms, including the modulation of protein kinase C and the modification of the cytoskeletal architecture. Protein kinase C (PKC) also mediates a number of cellular responses, and its activation is facilitated by Ca^{2+} (Nishizuka 1986). Calmodulin (CaM) is a ubiquitous regulatory protein that has Ca^{2+}-binding sites. When it is bound by Ca^{2+}, CaM binds to various kinases, including CaM-dependent kinase II (CaM kinase II; Edelman *et al.* 1987). As we discussed in section 3.1.1 (Fig. 3.2), CaM kinase II phosphorylates synapsin I in presynaptic terminals, thereby enhancing transmitter release (Llinás *et al.* 1985). We have seen that Ca^{2+} entering the nerve terminal during tetanic nerve stimulation first increases its local concentration near the transmitter release site (route 1 in Fig. 7.4), then accumulates inside the terminal as residual Ca^{2+}, which is responsible for PTP (section 7.1.1). As illustrated in Fig. 7.4 (route 2), it is also possible that the Ca^{2+} influx elicits Ca^{2+}-dependent cellular processes, thereby generating long-lasting synaptic plasticity.

In the frog sympathetic ganglion, the generation of LTP was blocked by a calmodulin inhibitor, W-7, but not by a protein kinase C inhibitor, H-7 (Minota *et al.* 1991). This finding favours the involvement of a CaM-dependent process in the LTP. There are, however, no ideal specific inhibitors for each Ca^{2+}-dependent protein kinase. For example, W-7 had been regarded as a selective CaM antagonist (Kanamori *et al.* 1981), but is now known to inhibit other protein kinases (Schatzman *et al.* 1983). Therefore, the results obtained by kinase inhibitors must be interpreted with caution. Further studies are clearly necessary to make a definite conclusion; still, the finding by Minota *et al.* (1991) suggests that at least some protein kinase activated by increased intraterminal Ca^{2+} levels is related to the generation of LTP in sympathetic neurones.

At the crayfish neuromuscular junction, the excitatory junctional potentials are markedly potentiated during tetanic stimulation of the nerve. Although enhanced synaptic response subsides within 10–20 min after the tetanization, some persistent potentiation remains for several hours (Fig. 7.5A; dotted-line curve, Control; Wojtowicz and Atwood 1988). This persistent potentiation is termed *long-term facilitation* or LTF. The LTF is associated with an increase in quantal release of the transmitter (Wojtowicz and Atwood 1986). In this

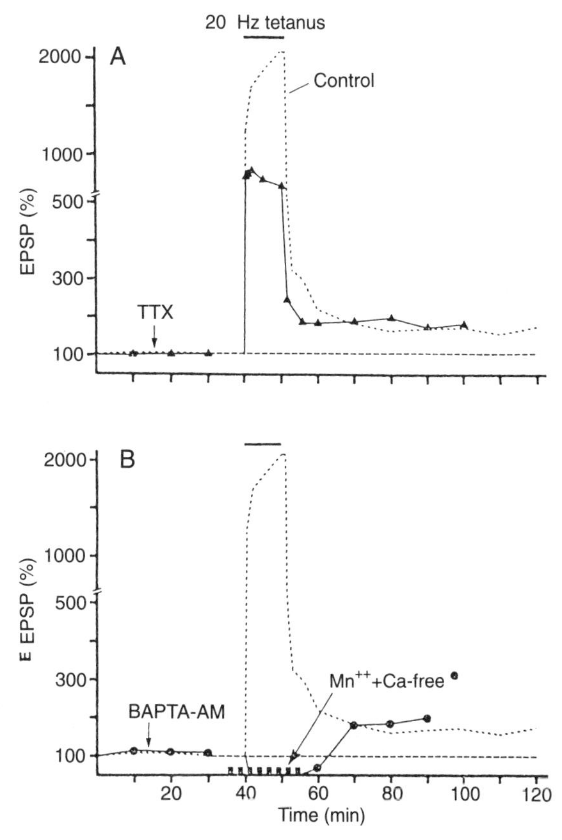

Fig. 7.5. Three phases of potentiation of excitatory junctional potentials (e.j.p.s or EPSP) at the crayfish neuromuscular junction. A, potentiation occurs during tetanic stimulation of the motor nerve, short (10–20 min) and long (hours) periods after the tetanization (dotted line). In the presence of tetrodotoxin (TTX), the short-term potentiation after tetanization disappears (triangles). B, the long-term facilitation can be seen even when the nerve is tetanized in the presence of Ca^{2+} blocker (Mn^{2+}) added to a Ca^{2+}-free solution (squares) in combination with intracellular Ca^{2+} chelator (BAPTA-AM; circles). (From Wojtowicz and Atwood 1988.)

preparation, the motor nerve terminal can be depolarized directly with an intracellular electrode placed near the terminal. Doing so in the presence of tetrodotoxin caused the brief synaptic potentiation immediately after the tetanus to disappear, but the LTF remained (Fig. 7.5A, triangles). This indicates that Na^+ influx into the terminal during the tetanus is essential for the brief potentiation but not for the LTF. Moreover, LTF was observed even when the nerve terminal was tetanized in the absence of external Ca^{2+} (Fig. 7.5B, filled symbols). This experiment was carried out after the addition of

BAPTA-AM to the external solution, which is capable of diffusing into the terminal and buffering intracellular Ca^{2+}. LTF thus appears to be entirely independent of any alterations in the intraterminal Ca^{2+} level, requiring only repetitive depolarizations of the nerve terminal (Wojtowicz and Atwood 1988). Interestingly, the LTF was blocked by a protein kinase A inhibitor or by an adenylate cyclase inhibitor (Dixon and Atwood 1989a). This suggests the involvement of cyclic AMP or cyclic AMP-dependent kinase (protein kinase A) in the LTF. Dixon and Atwood (1989a) postulated that repeated depolarizations of a motor nerve terminal activates adenylate cyclase, which in turn elevates the cyclic AMP level in the terminal, thereby inducing long-lasting synaptic potentiation. But how is adenylate cyclase ativated by membrane depolarization? Although this question remains open, these studies (Wojtowicz and Atwood 1988; Dixon and Atwood 1989a) are perhaps the first demonstration that terminal depolarization itself can directly activate the intracellular second messenger system and induce long-lasting synaptic potentiation, independent of the intraterminal Na^{+} or Ca^{2+} levels. Recall the facilitation Ca^{2+} channel induced in chromaffin cells (section 2.2.1; Figs 2.16 and 2.17) by activation of protein kinase A (Artalejo *et al.* 1990) or other kinases (Artalejo *et al.* 1992b). These protein kinases are also suggested to be activated directly by membrane depolarization.

We have seen that phosphorylation by protein kinases is a mechanism for switching the state or function of a protein (Box C). The protein kinase activated by tetanic stimulation of presynaptic terminals might phosphorylate proteins involved in the enhancement of transmitter, although which target proteins is not known. Synaptic plasticity would be expected to last as long as the key protein kinase was activated. When activation of the protein kinase ceased, the target protein would be dephosphorylated by phosphatase with a time constant of a few minutes (Box C). CaM kinase II, for instance, is activated by increased Ca^{2+} levels in the terminal (Fig. 7.4). However, the LTP in sympathetic neurones clearly outlasts the period of accumulation of excessive Ca^{2+} in the terminal. So how does the target protein stay 'on' without the stimulus for the protein kinase? To solve this dilemma, one must postulate that the signal turned on by Ca^{2+} is preserved in some form. A possible mechanism for such information storage, termed *autophosphorylation*, was first proposed on theoretical grounds (Crick 1984; Lisman 1985) and immediately found experimental support (Miller and Kennedy 1986). CaM kinase II once phosphorylated in the presence of Ca^{2+} and calmodulin is now known to phosphorylate itself even in the absence of Ca^{2+} (Box G). Thus, kinase activity of CaM kinase II can be prolonged beyond the duration of activation by the initial Ca^{2+} signal. Autophosphorylation thus appears to be one of the molecular mechanisms responsible for long-term information storage.

In view of the above scheme, we might expect synaptic plasticity to be induced by activation of some metabotropic receptors, if activation of the receptor yields the same set of second messengers or protein kinases as produced by tetanic stimulation of the presynaptic fibres. Indeed, the LTF at the crayfish neuromuscular junction was mimicked by the application of serotonin. Serotonin, widely distributed in the crustacean nervous system (Livingstone *et al.* 1981; Beltz and Kravitz 1983), enhances neuromuscular transmission by increasing transmitter release (Dudel 1965; Dixon and Atwood 1989b; Goy and Kravitz 1989). The application of serotonin elevates cyclic AMP levels in the crustacean neuromuscular junction (Goy *et al.* 1984). The long-lasting potentiation (hours) induced by a brief application of serotonin to the crustcean neuromuscular junction was found to be mediated by cyclic AMP (Dixon and Atwood 1989b; Goy and Kravitz 1989). Moreover, enhancement of transmitter release by serotonin at the crayfish neuromuscular junction was not due to increased Ca^{2+} levels in the nerve terminal (Delaney *et al.* 1991). It is likely that both the LTF and serotonin action involve increased cyclic AMP levels in the terminal, which presumably activate directly the process of transmitter release (exocytosis). Serotonin-induced enhancement of secretory machinery by increasing its responsiveness to internal Ca^{2+} was also studied in *Aplysia* neurones in detail (Dale and Kandel 1990).

Although information about plasticity at peripheral synapses is still fragmentary, the experiments described in this section provide three important clues about the possible mechanisms underlying long-term alterations in synaptic function. First, the linkage of presynaptic stimulation to the intracellular second messenger system is the key issue for the induction of long-term synaptic plasticity. Second, synaptic plasticity is associated with phosphorylation of the key proteins at the synapse. Third, phosphorylation of the key synaptic proteins can be maintained under some condition (e.g., by autophosphorylation or low abundance of phosphatases) even after removal of the activating stimulus for the protein kinase.

Box G Autophosphorylation and information storage

Long-term synaptic plasticity can last for years, whereas all the molecules in our bodies, except DNA, turn over within days or weeks. How then is the information stored in neurones retained despite molecular turnover? Crick (1984) proposed that the polymeric molecules in the synapse may be replaced with new material one at a time without altering the molecular state. For simplicity, suppose the molecule forms a dimer. The molecule is assumed to be in the active state when both monomers are phosphorylated (symbolized by +, +) and inactive when both are unmodified (−, −). Crick postulated that an enzyme, X, can catalyse phosphorylation of one monomer if the other monomer is already phosphorylated, but not otherwise. If individual monomers are replaced one at a time (hence, +, −), the molecular state will be returned to (+, +) by the enzyme X. If the initial state is (−, −), this state remains unaltered after monomeric turnover. What is this enzyme X?

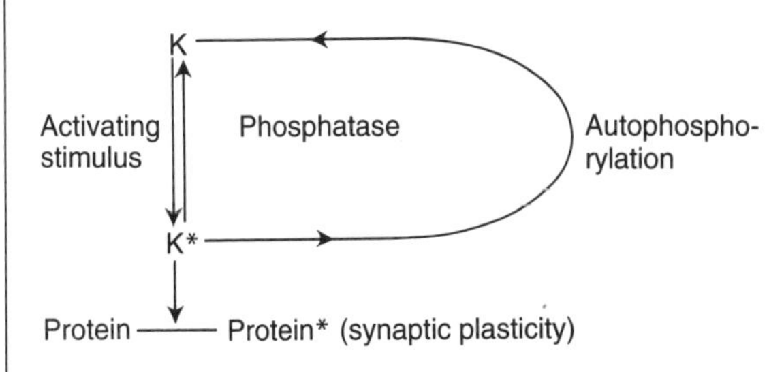

The Figure is a hypothesis independently proposed by Lisman (1985). A protein kinase (K) phosphorylated by a stimulus (e.g., Ca^{2+}) converts the protein in the synapse to its active form by phosphorylation, thereby inducing synaptic plasticity (Protein*). Lisman assumed that the phosphorylated kinase (K*) can catalyse phosphorylation of other non-phosphorylated kinase (K) by autophosphorylation in the absence of an activating stimulus. Miller and Kennedy (1986), in fact, found that CaM kinase II partially phosphorylated in the presence of Ca^{2+} can further phosphorylate the same kinase molecule. There was, however, no evidence that phosphorylated CaM kinase II can phosphorylate other non-phosphorylated kinases. Thus, the autophosphorylation process is intramolecular but not intermolecular. The kinase activity would then survive in the absence of the activating stimulus. The life span of the kinase activity (or synaptic plasticity) would be limited, however, by turnover of the phosphorylated kinase molecule.

A CaM kinase II consists of 12 subunits and has 30 phosphorylation sites. The kinase initiates Ca^{2+}-independent autophosphorylation when three or four of the 30 sites are phosphorylated (Miller and Kennedy 1986). Once more than a few sites in any subunit of the kinase molecule are phosphorylated, other sites on the same molecule will be turned on by autophosphorylation. If each subunit is replaced with new material, one at a time, as proposed by Crick (1984), the enzyme activity would last beyond the life span of the enzyme molecule (Lisman and Goldring 1988, 1989). How long the kinase activity persists would depend on the balance between the rate constant of autophosphorylation and the rate constant of phosphatase for dephosphorylation. In principle, the kinase activity (hence, synaptic plasticity) can outlast the life span of the kinase molecule.

7.2 LONG-TERM PLASTICITY AT CENTRAL SYNAPSES

Traditionally, neurophysiologists have tended to focus on electrical events in neurones or at synapses that last only milliseconds. Young (1951) rightly criticized this failure to investigate enduring reactions at central synapses in connection with learning or memory. In the light of this view, the long-term potentiation (LTP) originally found by Lomo (1966) in the hippocampal formation, which can last for days (Bliss and Gardner-Medwin 1973; Bliss and Lomo 1973), has attracted the attention of many neurophysiologists (Gustafsson and Wigstrom 1988; Nicoll *et al.* 1988; Bliss *et al.* 1990; Malenka and Nicoll 1990). Long-term plasticity in the central nervous system is not confined to synaptic potentiation but also takes the form of synaptic depression, termed *long-term depression* (LTD; Ito 1989, 1990; Sejnowski *et al.* 1990; Linden 1994).

The major excitatory transmitter substance in the central nervous system is L-glutamate. The glutamate receptors are either ionotropic or metabotropic in type, and the ionotropic glutamate receptor is further classified into the NMDA (*N*-methyl-D-aspartate) and non-NMDA subtypes (section 4.1.5). Although a given central neurone possesses both NMDA and non-NMDA receptors, we shall see that the induction and maintenance of long-term plasticity at central synapses occur in a manner specific to the receptor subtype. Obviously these different glutamate receptor subtypes must be functionally distinct. We will begin by discussing the functional properties of different glutamate receptor subtypes, because some of these distinctive attributes seem to hold the key to receptor-specific synaptic plasticity.

7.2.1 Functional features of glutamate receptor subtypes

Long-term synaptic plasticity has now been observed at numerous locations in the central nervous system, but its cellular mechanisms have been explored most extensively in the hippocampus. For this reason, in examining the properties of glutamate receptors we will take the excitatory pathways in the hippocampus as our example. The hippocampus, also called 'Ammon's horn' (or *cornu Ammonis*) because of its unique shape, is divided into four regions, CA1 to CA4 (Lorente de No 1934); for simplicity, only regions CA1 to CA3 are shown in Fig. 7.6, which illustrates the three major excitatory pathways: the fibres from the entorhinal area projecting to the dentate granular cells via the perforant path (pathway 1); the dentate granular cell axons projecting to the CA3 pyramidal cells via mossy fibres (pathway 2); and the CA3 pyramidal cell axons projecting to the CA1 pyramidal cells via Schaffer's collaterals (pathway 3). The CA1 pyramidal cells also receive excitatory inputs from the contralateral hippocampus via commissural fibres. The putative transmitter substance involved in these monosynaptic excitatory pathways is L-glutamate.

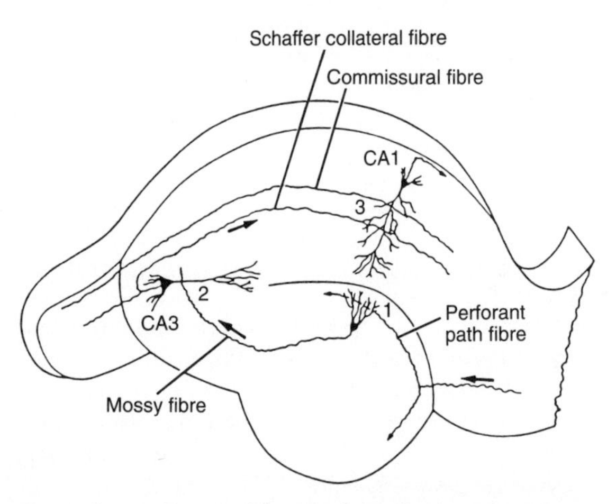

Fig. 7.6. A schematic diagram of the three major excitatory pathways in the hippocampus. 1, from the entorhinal area to dentate granular cells via the perforant path. 2, from dentate granular cells to CA3 pyramidal cells via mossy fibres. 3, from CA3 pyramidal cells to CA1 pyramidal cells via the Schaffer collateral. Commissural fibres project from the contralateral hippocampus to CA1 pyrmaidal cells. (From Nicoll *et al.* 1988.)

Figure 7.7 shows excitatory postsynaptic currents (e.p.s.c.s) recorded from a CA1 pyramidal neurone under voltage clamp following stimulation of the Schaffer collateral and commissural fibres (Hestrin *et al.* 1990). In this experiment, the rat hippocampus was excised and sliced at a thickness of a few hundred μm, and the synaptic currents were recorded in the whole-cell mode of the patch clamp technique (F. A. Edwards *et al.* 1989). When APV (DL-2–amino-5–phosphonovalerate), an NMDA receptor antagonist (Davies and Watkins 1982; Collingridge *et al.* 1983), was applied to the bathing solution, the late phase of the e.p.s.c was markedly depressed, leaving its early phase virtually unaltered (Fig. 7.7A). Evidently, the e.p.s.c. is composed of two phases mediated by the NMDA and non-NMDA receptors.

Figure 7.7B shows changes in the magnitude of the e.p.s.c.s recorded at different membrane potentials of the postsynaptic neurone. Clearly, the APV-insensitive phase measured at its peak and the APV-sensitive phase measured 25 ms after the peak (dotted line in Fig. 7.7A) display different current–voltage relations. The APV-sensitive, NMDA receptor-mediated synaptic response is characterized by nonlinear behaviour (rectification) at membrane potentials more negative than about −20 mV (Fig. 7.7B, filled circles). This rectification is due to a voltage-dependent block of the NMDA receptor channel by extracellular Mg^{2+} (Mayer *et al.* 1984; Nowak *et al.* 1984). With extracellular Mg^{2+} at physiological concentrations (0.5–1 mM), Mg^{2+} enters the NMDA receptor channel when the channel opens, thereby partially blocking the synaptic current. This blocking action is enhanced when the neurone is hyperpolarized, since the entry of positively charged Mg^{2+} into the receptor channel is accelerated by the increased electric field. On the other hand, electrical repulsion ousts the blocking Mg^{2+} when the cell is depolarized, so that the current–voltage relation becomes linear at membrane potentials more positive than −20 mV.

The non-NMDA receptor-mediated response in the early phase of e.p.s.c. is linear over a wide range of membrane potentials (Fig. 7.7B, filled triangles). In other words, the Mg^{2+}-blocking action occurs only in

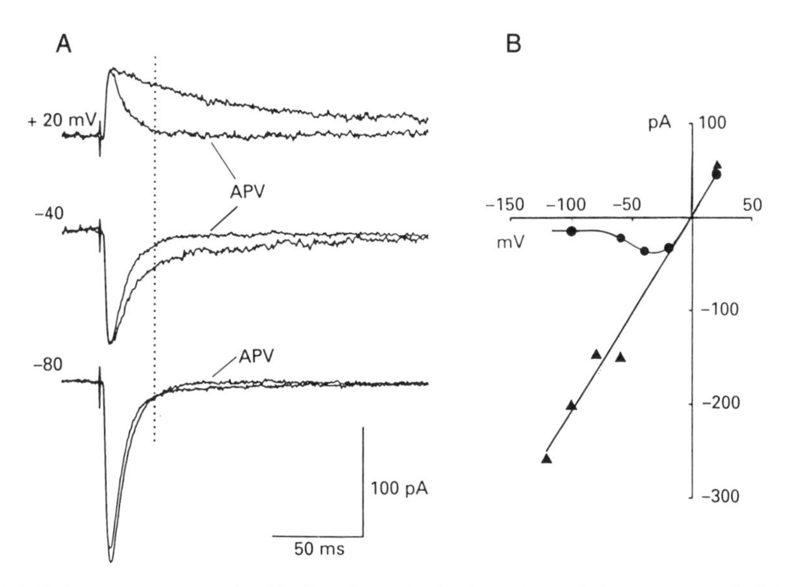

Fig. 7.7. APV, an NMDA receptor antagonist, blocks selectively the late phase of the e.p.s.c. evoked in a CA1 pyramidal neurone by stimulation of the Schaffer collateral and commissural fibres. A, the magnitude of the early phase of the e.p.s.c.s (non-NMDA receptor-mediated component) is measured at the peak of the synaptic response. The magnitude of the late phase measured at 25 ms after the peak (dotted line) reflects the NMDA receptor-mediated synaptic response. The e.p.s.c.s are recorded at three different membrane potentials of the CA1 pyramidal neurone (+20, −40, and −80 mV). B, the relation between the magnitude of e.p.s.c. and the membrane potential for the late component (filled circles) and early component (filled triangles) before the application of APV (50 μM). (Adapted from Hestrin *et al.* 1990.)

the NMDA receptor channel, not in the non-NMDA receptor. The non-NMDA receptor response can be eliminated by its own selective blocker, CNQX (6-cyano-2,3-dihydroxy-7-nitro-quinoxalline; Blake *et al.* 1988; Honore *et al.* 1988). Unlike APV (Fig. 7.7A), CNQX selectively reduces the early phase of the e.p.s.c. (Fig. 7.8). The blocking action by CNQX appears to be stronger at −80 mV than at lower membrane potentials, again due to the voltage-dependent rectification of the NMDA receptor-mediated synaptic response. As shown in Fig. 7.7B, the APV-sensitive NMDA component is very small at a membrane potential of −80 mV. Therefore, CNQX almost eradicates the e.p.s.c. at −80 mV. The implication of these results is important: although central neurones can generate excitatory synaptic reponses by activation of the NMDA and non-NMDA receptors, the contribution of the NMDA receptor to the synaptic response is very small unless the neurones are sufficiently depolarized. As we will see later, this feature of NMDA receptors is linked to the prerequisite of postsynaptic depolarization for the induction of LTP.

Another important difference in the properties of the NMDA and non-NMDA receptors concerns the channel's permeability to Ca^{2+}. The Ca^{2+} permeability in non-NMDA receptors is about 100 times below that in NMDA receptors, except for particular subtypes of the non-NMDA receptor (section 4.1.5). Mayer *et al.* (1987) measured the intracellular Ca^{2+} transient in cultured spinal neurones loaded with a Ca^{2+} indicator, arsenazo III, during the application of NMDA or kainate. If the neurone was depolarized by glutamate agonists, Ca^{2+} influx would be induced through the voltage-gated Ca^{2+} channels. To exclude this possibility, the neurone was voltage-clamped at its resting membrane potential (about -60 mV) in Mg^{2+}-free solution. Under this condition, the application of NMDA induced an inward current (Fig. 7.9Aa). Similarly, kainate produced an inward current by activation of the non-NMDA receptor (Fig. 7.9Ba). Simultaneously recorded intracellular Ca^{2+} transients were clearly larger in response to NMDA (Fig. 7.9Ab) than to kainate (Bb). This Ca^{2+} transient induced by NMDA at the resting potential was blocked by adding Mg^{2+} (Mayer *et al.* 1987). That is, under physiological conditions (in the presence of Mg^{2+}) a sufficient depolarization of the postsynaptic neurone is again a prerequisite in order to elicit a significant Ca^{2+} influx or synaptic response on activation of the NMDA receptor.

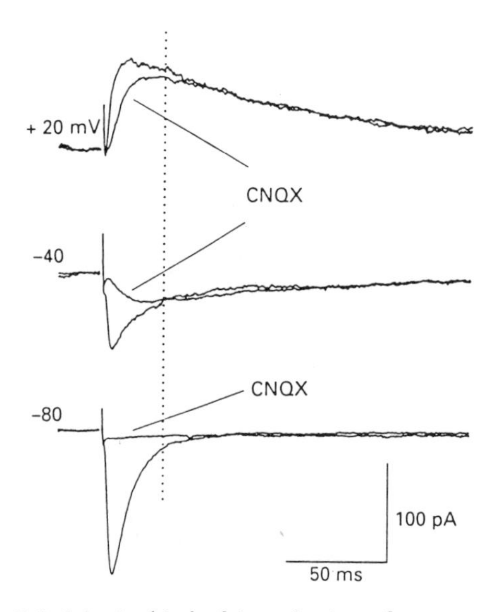

Fig. 7.8. Selective block of the early phase of e.p.s.c.s evoked in a CA1 pyramidal neurone by stimulation of the Schaffer collateral and commissural fibres following the application of CNQX, a non-NMDA receptor antagonist. Dotted line indicates 25 ms after the peak of the e.p.s.c.s. The e.p.s.c.s were recorded at three different membrane potentials of the pyramidal neurone (+20, −40 and −80 mV). (From Hestrin *et al.* 1990.)

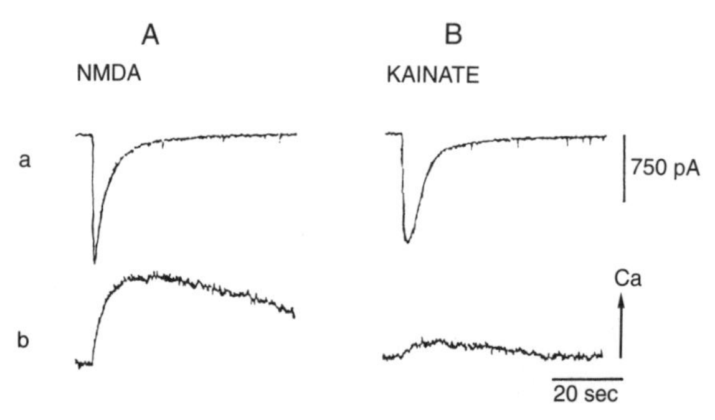

Fig. 7.9. Membrane currents (a) and intracellular Ca^{2+} transients (b) measured in a spinal neurone following a focal application of NMDA (A) or kainate (B). The inward membrane currents were recorded under voltage clamp at a holding potential of about −60 mV. The Ca^{2+} transients were measured with arsenazo III loaded into the neurone through the patch-recording pipette. (From Mayer *et al.* 1987.)

Might metabotropic glutamate receptors play a significant role in the induction or maintenance of long-lasting synaptic plasticity? Clues emerged following the cloning of the first metabotropic glutamate receptor (mGluR1; Houamed *et al.* 1991; Masu *et al.* 1991), which was followed by the clonings of four additional members of this receptor family (mGluR2–mGluR5; Abe *et al.* 1992; Nakanishi 1992; Tanabe *et al.* 1992). Their primary structures again conformed to the motif of seven transmembrane segments, characteristic of G-protein-linked receptors (section 5.1.2). We have seen that some metabotropic glutamate receptors (mGluR1, mGluR5) stimulate the inositol phosphate signal transduction, whereas other receptor subtypes (e.g., mGluR2, mGluR3) inhibit the cyclic AMP production (section 5.3.1). These second messengers regulated by the metabotropic receptor can modulate the sensitivity of ionotropic receptors or even transmitter release (section 5.3.1). Thus, it is quite likely that metabotropic glutamate receptors play an important role in the modulation, induction or maintenance of long-term plasticity at central synapses (Zheng and Gallagher 1992). Some agents proposed as agonists for the metabotropic glutamate receptor (mGluR) such as *trans*-ACPD (Palmer *et al.* 1989; Manzoni *et al.* 1990) and APB (L-2-amino-4-phosphonobutyrate, also known as AP4; Baskeys and Malenka 1991; Nawy and Jahr 1991) are not highly specific, however. In view of the multiple types of mGluR, profiles of their respective actions must await the development of agonists or antagonists highly specific for each individual type. Nevertheless, using MCPG (α-methyl-4-carboxyphenylglycine), a newly developed mGluR blocker, the involvement of mGluRs in the induction of LTP is now strongly supported (Bashir *et al.* 1993; Bartolotto *et al.* 1994).

7.2.2 Long-term potentiation at central synapses

From the results illustrated in Figs 7.7 and 7.8, it is clear that the monosynaptic e.p.s.c. evoked in CA1 pyramidal neurones by stimulation of the Schaffer collaterals and commissural fibres comprises the NMDA and non-NMDA receptor-mediated components. A brief tetanic stimulation (10–100 Hz for 1–20 sec) of the presynaptic fibres in this pathway produces LTP of the synaptic response in the CA1 pyramidal neurones (Schwarzkroin and Wester 1975; Andersen *et al.* 1977; Lynch *et al.* 1977; Nicoll *et al.* 1988). The primary question now is whether the LTP is a receptor-specific phenomenon. If so, which receptor is responsible for the induction of the LTP? And which receptor-mediated synaptic response is being potentiated?

The first provocative finding was that tetanic stimulation of the Schaffer collateral-commissural pathway failed to induce LTP of the monosynaptic response in

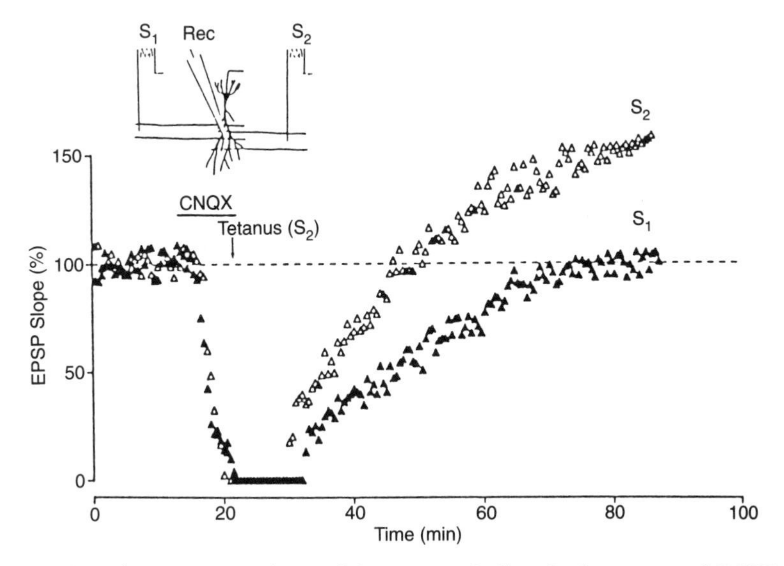

Fig. 7.10. The induction of LTP by tetanic stimulation of the presynaptic fibres in the presence of CNQX. Synaptic responses were recorded with an extracellular electrode placed in the CA1 region in response to stimulation of the Schaffer collateral-commissural pathway at two sites (S_1 and S_2; see inset). Test stimuli were applied at 0.1 Hz. CNQX was applied to the bath 15 min after the observations of the control responses. Two tetanic stimuli (100 Hz for 1 sec, 20 s apart) were applied to S_2, and CNQX was immediately washed away. The synaptic response to stimulation of S_1 recovered to the pre-tetanic, control level, whereas the response to stimulation of S_2 showed potentiation. (From Kauer *et al.* 1988.)

CA1 pyramidal neurones when the NMDA receptor blocker, APV, was applied during the period of tetanic stimulation (Collingridge *et al.* 1983; Wigstrom and Gustaffson 1984; Harris *et al.* 1984). Therefore, activation of NMDA receptors must be essential for the induction of LTP. Can LTP be induced in the presence of the non-NMDA receptor blocker, CNQX? In the experiment illustrated in Fig. 7.10, the synaptic response was recorded from a population of CA1 pyramidal cells with an extracellular electrode to separate stimulation of two sites (S_1 and S_2) in the Schaffer collateral-commissural pathway. One of the two inputs (S_2) was tetanically stimulated in the presence of CNQX. By the time of the tetanic stimulation, the synaptic response was already attenuated by the CNQX. The CNQX was then washed away immediately after the tetanic stimulation. The synaptic response to the non-tetanized control input (S_1) recovered to the original level, whereas the synaptic response to stimulation of the tetanized input (S_2) exceeded the original level, showing LTP (Fig. 7.10). In short, LTP can be induced by tetanic stimulation of the presynaptic fibres even when the non-NMDA receptor is blocked. Moreover, the CNQX-sensitive non-NMDA receptor-mediated synaptic response is enhanced during LTP. These and other additional results implied two important features (Kauer *et al.* 1988; Muller *et al.* 1988): First, it is the NMDA receptor (but not the non-NMDA receptor) that is responsible for the induction of LTP by tetanization; second, it is the non-NMDA receptor that mediates the synaptic response potentiated during the LTP. In other words, the LTP in the CA1 pyramidal cell is induced by the NMDA receptor but expressed by the non-NMDA receptor (but see below).

As we have discussed in section 7.2.1, the NMDA receptor channel has two unique properties: a high Ca^{2+} permeability and rectification by Mg^{2+}-induced block. Might these properties be relevant to the induction of LTP by the NMDA receptor? Intracellular injections of Ca^{2+} chelators, EGTA (Lynch *et al.* 1983) or nitr-5 (Malenka *et al.* 1988), into CA1 pyramidal cells prevented the generation of LTP. These observations point to an increase in the intracellular Ca^{2+} of the postsynaptic neurone as an essential step for LTP. Nitr-5 is a synthetic Ca^{2+} chelator with a photochemically labile domain (Gurney *et al.* 1987). On photolysis by a light flash, the binding affinity of nitr-5 to Ca^{2+} decreases by about 40-fold. Therefore, when a CA1 pyramidal cell is loaded with nitr-5 and Ca^{2+}, the intracellular Ca^{2+} is maintained at a low level, but a flash of light suddenly increases the intracellular Ca^{2+} level, as the Ca^{2+} 'caged'

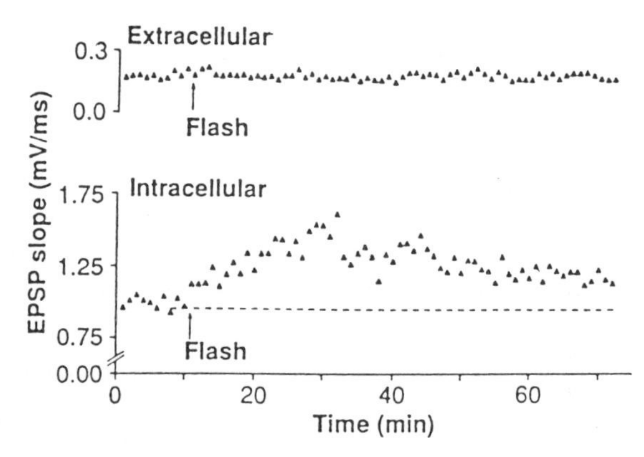

Fig. 7.11. Effects of increased intracellular Ca^{2+} levels in a CA1 pyramidal neurone on its excitatory postsynaptic potentials (e.p.s.p.s or EPSP) evoked by stimulation of the Schaffer collateral-commissural pathway. Upper panel, the monosynaptic e.p.s.p.s recorded from a population of CA1 cells with an extracellular electrode. Lower panel, intracellular e.p.s.p.s were recorded from a CA1 pyramidal cell previously loaded with nitr-5 and Ca^{2+}. A flash of light enhanced the e.p.s.p.s of the loaded pyramidal cell without affecting e.p.s.p.s in other pyramidal cells. (From Malenka *et al.* 1988.)

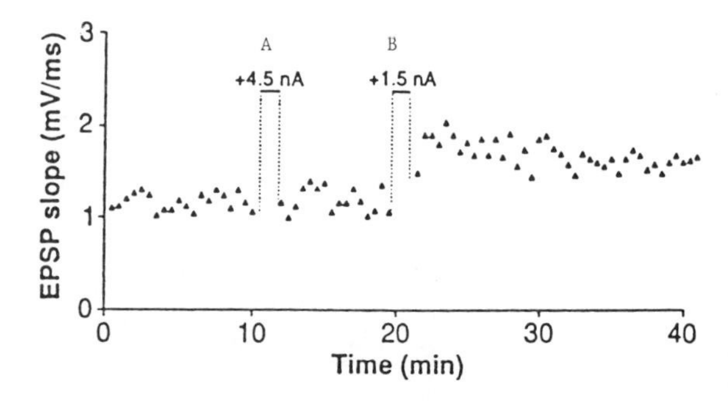

Fig. 7.12. The induction of LTP by postsynaptic depolarization. Monosynaptic e.p.s.p.s (or EPSP) were evoked in a CA1 pyramidal cell by stimulation of the Schaffer collateral-commissural pathway at a frequency of 0.1 Hz. At A, the pyramidal cell was depolarized by a large current beyond the Ca^{2+} equilibrium potential (4.5 nA) for 1 min. At B, the pyramidal cell was similarly depolarized to 0 mV with a small current (1.5 nA). Long-lasting potentiation of the e.p.s.p.s occurred after the appropriate amount of postsynaptic depolarization without presynaptic tetanization. (From Malenka *et al.* 1988.)

in nitr-5 is freed by a decrease in the binding affinity. As shown in Fig. 7.11, the monosynaptic e.p.s.p.s evoked in the CA1 pyramidal cell by stimulation of the presynaptic pathway were potentiated by a flash of light under such conditions (Malenka *et al.* 1988). In other words, the induction of LTP by activation of the NMDA receptor can be mimicked simply by increasing Ca^{2+} levels in the postsynaptic neurone. Thus, the high Ca^{2+} permeability in the NMDA receptor channel seems to be indispensable to its role in the induction of LTP.

How would the rectification by Mg^{2+} block in the NMDA receptor channel affect the induction of LTP? When a pyramidal cell is deliberately hyperpolarized during tetanic stimulation of the presynaptic fibres, it does not develop LTP (Kelso *et al.* 1986; Malinow and Miller 1986; Sastry *et al.* 1986). As shown in Fig. 7.7B, the Mg^{2+} block of the synaptic current through the NMDA receptor channel is enhanced by hyperpolarization, so that its Ca^{2+} influx would be minimized under this condition. Thus, the induction of LTP requires two conditions: activation of the NMDA receptor and a sufficient depolarization of the postsynaptic neurone Recall that this condition, co-activation of a presynaptic fibre and its postsynaptic neurone, is actually equivalent to Hebbian synapse or Hebbian conjunction (Hebb 1949; section 6.3). Tetanic stimulation of the presynaptic fibres would produce a large depolarization in the postsynaptic neurone because of temporal summation of e.p.s.p.s evoked by individual stimuli. In fact, LTP can be induced without tetanic stimulation of the presynaptic fibres if the presynaptic stimuli are applied together with postsynaptic depolarizing pulses (Kelso *et al.* 1986; Sastry *et al.* 1986; Gustafsson *et al.* 1987). In the experiment shown in Fig. 7.12, for example, monosynaptic e.p.s.p.s were recorded from a CA1 pyramidal neurone following stimulation of the Schaffer collateral-commissural pathway at a low frequency (0.1 Hz). At B in Fig. 7.12, the pyramidal cell was depolarized to about 0 mV for 1 min (with a current of 1.5 nA), while identical presynaptic stimuli were applied continuously at the same frequency. This procedure induced LTP, consistent with the assumption that the tetanic stimulation of the presynaptic fibres required for LTP is simply a device to produce postsynaptic depolarization. If the postsynaptic cell were excessively depolarized beyond the equilibrium potential for Ca^{2+}, no Ca^{2+} would enter even when the NMDA receptor-mediated synaptic current was enhanced by the depolarization. Under such conditions, the synaptic response is not potentiated (Fig. 7.12 at A, depolarized with a current of 4.5 nA).

Figure 7.13 summarizes the possible mechanisms underlying LTP in the hippocampal CA1 pyramidal cell (Malenka *et al.* 1989a). Stimulation of the presynaptic fibres causes the release of glutamate (Glu) from their

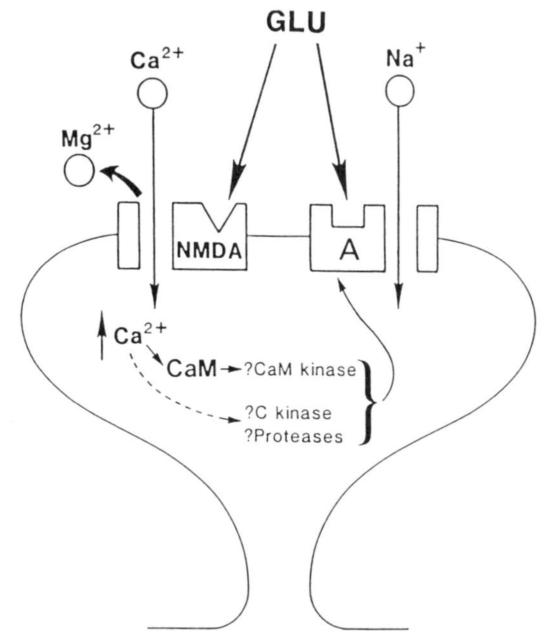

Fig. 7.13. Schematic diagram for a possible mechanism of LTP at the postsynaptic receptor site in a CA1 hippocampal pyramidal neurone. Glutamate (GLU) released from the presynaptic fibres activates the NMDA and non-NMDA (A) receptors. Synaptic depolarization removes Mg^{2+} blocking the NMDA receptor channel, thereby enhancing its Ca^{2+} influx. The Ca^{2+} influx activates some intracellular second messenger system, which in turn potentiates the synaptic response at the non-NMDA receptor-mediated synapse. (Adapted from Malenka *et al.* 1989a.)

terminals, which activates NMDA receptors as well as non-NMDA receptors (indicated by A). When activation of the presynaptic fibres is conjoined with a sufficient depolarization of the postsynaptic neurone, either by summation of e.p.s.p.s due to high-frequency tetanic stimulation of the presynaptic fibres or by depolarizing pulses applied to the postsynaptic cell, the synaptic current flowing through the NMDA receptor channel is enhanced, thereby increasing the intracellular Ca^{2+}. It is this Ca^{2+} that is responsible for the induction of LTP. However, the site at which LTP is developed appears to be the non-NMDA receptors (Fig. 7.10). Then, some intracellular signals must be elicited by Ca^{2+} entering through the NMDA receptor to induce synaptic potentiation at the non-NMDA receptor. In this view, the mechanism of LTP appears to reside in the postsynaptic element. However, as we shall discuss later, such postsynaptic receptor alterations alone cannot account for the mechanism of LTP completely.

We have seen that an increase in intracellular Ca^{2+} levels may activate calmodulin (CaM)-dependent kinase or protein kinase C (Fig. 7.4). Might LTP in CA1 hippocampal neurones be triggered by activation of some protein kinase? Several non-specific protein kinase inhibitors were found to block the generation of LTP (Lovinger *et al.* 1987; Malinow *et al.* 1988, 1989). The involvement of calmodulin (CaM)-dependent kinase was examined by a synthetic calmodulin-binding peptide (CBP) which binds to CaM and inhibits competitively the CaM-dependent kinase. Injections of CBP into CA1 pyramidal cells blocked the generation of LTP without affecting the normal synaptic response (Malenka *et al.* 1989b). This finding was further supported by a newly developed technique, termed the *gene targeting* procedure (Thomas and Capecchi 1987). This procedure is essentially a site-directed mutagenesis whereby the gene or chromosomal locus of interest is inactivated or completely 'knocked out' (for principles of this procedure see *Box E* in Hall 1992). With this procedure Silva *et al.* (1992) produced a mouse strain carrying a mutation (a *null mutation*) in the gene encoding α-CaM kinase II. The mutant mice appeared behaviourally normal, and the e.p.s.p.s recorded from CA1 pyramidal cells remained unaffected. The majority (10 out of 12) of the mutants, however, failed to develop LTP in CA1 pyramidal cells. Undoutedly, mice lacking the expression of α-CaM kinase II are markedly deficient in the generation of LTP. Yet 2 out of the 12 mutant mice exhibited normal LTP, suggesting that CaM kinase II is involved in but not absolutely necessary for the development of LTP. O'Dell *et al.* (1991b) found that tyrosine kinase inhibitors block LTP in CA1 pyramidal neurones. Disruption of one of the tyrosine kinase genes (fyn mutants) reduced the expression of LTP but did not eliminate it completely (Grant *et al.* 1992). Thus it seems that the expression of LTP relies on the sum of several regulatory elements, rather than on activation of a single protein kinase. Indeed, Frey *et al.* (1993) showed that the late long-lasting phase of LTP (>1–3 hours) in the rat CA1 region requires activation of cyclic AMP-dependent protein kinase A. Moreover, the involvement of cyclic GMP-dependent protein kinase (PKG) in the induction of LTP was also suggested (Zhuo *et al.* 1994).

The nature of the intracellular signal for LTP thus remains equivocal. Intriguingly, however, the application of protein kinase inhibitors (Malinow *et al.* 1988) or CBP (Malenka *et al.* 1989b) blocked LTP, but synaptic potentiation still persisted for 30–60 min. Davies *et al.* (1989) suggested that the early phase (30–60 min) of the LTP may be due to an increase in the amount of transmitter released by a presynaptic impulse, whereas the late phase of LTP may be maintained by increased sensitivity of the non-NMDA receptor. If increasing amounts of glutamate were released by the presynaptic impulse, the synaptic responses mediated by both the NMDA receptor and the non-NMDA receptor would be potentiated. In fact, Bashir *et al.* (1991) reported that the NMDA receptor-mediated synaptic response recorded in the presence of CNQX exhibits a potentiation for about 60 min after a brief tetanic stimulation of the presynaptic fibres. Thus, the basic hypothesis that LTP is induced by the NMDA receptor and expressed by the non-NMDA receptor (Fig. 7.13) must be elaborated into a more complicated scheme.

The most reliable test of whether the LTP stems from an increase in transmitter release (presynaptic mechanism) or in the sensitivity of the postsynaptic receptor (postsynaptic mechanism) is the quantal analysis of the synaptic response. We have discussed several uncertainties regarding quantal analysis at central synapses (section 3.2.2; Korn and Faber 1987, 1991). Unfortunately, in fact, some confusing results emerged from such analyses in CA1 neurones. At first, an increase in the mean number of transmitter quanta released by the presynaptic impulse was reported to be responsible for the LTP (Bekkers and Stevens 1990; Malinow and Tsien 1990), but other studies found an increase in quantal size of the synaptic response underlying the LTP (Foster and McNaughton 1991; Manabe *et al.* 1992). Moreover, the coexistence of the pre- and postsynaptic mechanisms in the LTP was also suggested (Kullmann and Nicoll 1992; Larkman *et al.* 1992). It is baffling that the same type of analysis yields contradictory conclusions in different hands. Until the source of these discrepancies is found, misconceptions concerning the valid LTP mechanisms are inevitable.

The consensus is that the induction of LTP is a receptor-specific phenomenon, resulting from activation of the NMDA receptor. In other words, LTP must be triggered in the postsynaptic neurone. If, then, LTP is due in part to increased transmitter release, one must postulate some mechanism that conveys an LTP signal from the site of induction (postsynaptic neuron) to the site of transmitter release (presynaptic terminal). In cultured hippocampal neurones, activation of the postsynaptic NMDA receptors actually increased transmitter release from the presynaptic terminals, as evidenced by an increase in the frequency of spontaneous miniature synaptic currents (Malgaroli and Tsien 1992). Figure

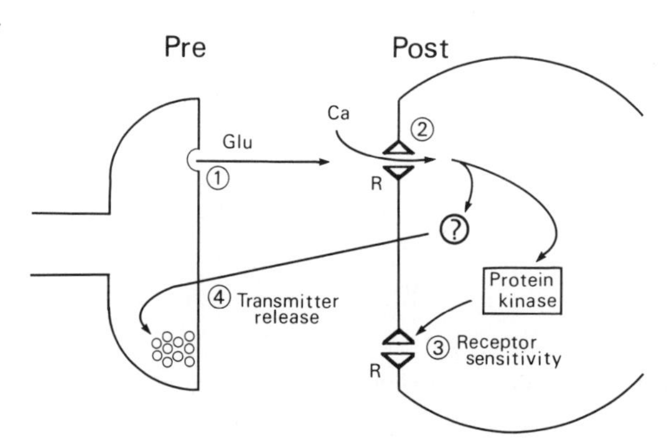

Fig. 7.14. A hypothetical chain of four processes involved in long-term potentiation of synaptic transmission. Step 1, release of glutamate from the presynaptic terminal. Step 2, Ca^{2+} influx into the postsynaptic neurone through activated NMDA receptor channels. Step 3, enhancement of sensitivity of non-NMDA receptors by some protein kinase via Ca^{2+} influx. Step 4, diffusion of a signal molecule produced by increased Ca^{2+} levels from the postsynaptic neurone to the presynaptic terminal across the synaptic cleft. This signal molecule is assumed to enhance transmitter release. (Modified, based on Stevens 1989.)

7.14 is a possible scheme to explain a two-phased LTP (cf. Stevens 1989). Glutamate released from the nerve terminals during presynaptic tetanus (step 1) activates the NMDA receptor, thereby eliciting Ca^{2+} influx (step 2) into the pyramidal neurone. The increased intracellular Ca^{2+} level is assumed to cause activation of protein kinases, which in turn increases sensitivity of the glutamate receptor (step 3). Furthermore, activation of the NMDA receptors may yield some signal molecule (encircled '?'), which diffuses into the presynaptic terminal across the synaptic cleft (step 4), thereby enhancing transmitter release.

The novel concept in this hypothesis is the presence of trans-synaptic signals to mediate the LTP. The postulated signal, if present, must be membrane-permeable. A membrane-permeable molecule, arachidonic acid or its metabolite, has been proposed as a candidate for such trans-synaptic signals. Arachidonic acid is produced by phospholipase A_2, which catalyses the hydrolysis of membrane phospholipids (Fig. 7.4). Stimulation of NMDA receptors in cultured striatal neurones was found to release arachidonic acid metabolites (Dumuis *et al.* 1988). Moreover, Bliss *et al.* (1990) showed that the application of arachidonic acid increases the release of glutamate from the dentate granular cells, and that inhibition of phospholipase A_2 (hence, inhibition of the formation of arachidonic acid) blocks the LTP induced in the dentate granule cells by tetanic stimulation of the perforant path (pathway 1 in Fig. 7.6). These results are consistent with the notion that arachidonic acid or its metabolites can act as a retrograde message from postsynaptic neurones to their presynaptic terminals to enhance transmitter release during the LTP (Bliss *et al.* 1990; Hawkins and Kandel 1990). However, exogenously applied arachidonic acid showed no enhancement of the synaptic response recorded from CA1 hippocampal neurones (O'Dell *et al.* 1991a). Also, arachidonic acid increased the response of NMDA receptors to glutamate in hippocampal and cerebellar neurones (O'Dell *et al.* 1991a; Miller *et al.* 1992). Thus, some inconsistent results still remain unexplained.

Another candidate proposed for the trans-synaptic signal is *nitric oxide* (NO). The free radical gas NO is synthesized from the amino acid arginine by NO synthase. NO is now considered as a possible neuronal messenger because it is released by activation of NMDA receptors in the central nervous system (Box H). The results reported by four groups for the action of NO were consistent (Bohme *et al.* 1991; O'Dell *et al.* 1991a; Schuman and Madison 1991; Haley *et al.* 1992). The LTP induced in CA1 pyramidal cells was blocked by the application of an NO synthase inhibitor, *N*-nitro-L-arginine (Fig. 7.15A, open triangles). LTP was also blocked when haemoglobin was added to the bathing solution (Fig. 7.15B, filled triangles). Haemoglobin is known to have a high affinity of binding to NO (Box H), so that the haemoglobin would restrain the diffusible free NO. The effect of haemoglobin is specific to its binding site to NO, because LTP remains unaffected when the normally ferrous form (Fe^{2+}) of the iron in the heme moiety of haemoglobin is oxidized to the ferric form (Fe^{3+}) to yield methaemoglobin (Fig. 7.15B, open triangles). Haemoglobin cannot cross the cell membranes; hence, these results imply that the NO involved in LTP must travel in the extracellular space. Furthermore, the results illustrated in Fig. 7.16 indicate that the site of NO synthesis is in the postsynaptic CA1 pyramidal cell. In this experiment, the LTP was elicited by pairing postsynaptic depolarization applied through the intracelluar electrode with low-frequency stimula-

Box H Nitric oxide as a neuronal messenger

Vasodilatation or relaxation of vascular muscle has been known to be mediated by *endothelium-derived relaxing factor* (EDRF) released from the endothelium lining the smooth muscle (Vanhoutte *et al.* 1986). Palmer *et al.* (1987) found that the EDRF is nitric oxide (NO), and soon after it was noted with surprise that NMDA receptor activation releases NO from cerebellar cells (Garthwaite *et al.* 1988). NO has now come to be viewed as an important neuronal messenger (Garthwaite 1991; Bredt and Snyder 1992; Fazelli 1992; Mccall and Vallance 1992).

NO, a free radical gas, is freely permeable through the cell membrane. NO is highly reactive but unstable, with a half-life of seconds. It is synthesized from arginine by NO synthase. The primary structure of the cloned brain NO synthase showed a calmodulin-binding site and a phosphorylation site for cyclic AMP-dependent protein kinase (Bredt *et al.* 1991). Presumably, the Ca^{2+} influx induced by activation of NMDA receptors stimulates NO synthase via calmodulin.

Both relaxation of vascular smooth muscle and activation of NMDA receptors in the cerebellum are associated with an increase in cyclic GMP levels, hence one of NO's target molecules must be guanylate cyclase. But whether cyclic GMP elevation is responsible for the long-term synaptic plasticity produced by NO is unclear. Cytoplasmic guanylate cyclase is a haem-containing enzyme, and NO binds to its haem to activate the enzyme (Ignarro 1989). NO has a high affinity for haemoglobin, reacting with its haem group (Gibson and Roughton 1957), and may interact with many chemical substances other than guanylate cyclase (Bredt and Snyder 1992), including enzymes with iron (or haem), to trigger signal transduction.

The distribution of the brain NO synthase in neurones and glia cells of the central nervous system departs from that predicted from the experimental results concerning LTP induction by NO. For example, the brain NO synthase occurs in the dentate granular cells but not in pyramidal cells of the hippocampus (Bredt and Snyder 1992). This discrepancy suggests the presence of other forms of the NO synthase. Alternatively, some other messenger may be involved. Indeed, CO can enhance synaptic responses in hippocampal neurones in a manner similar to that produced by NO (Zhuo *et al.* 1993).

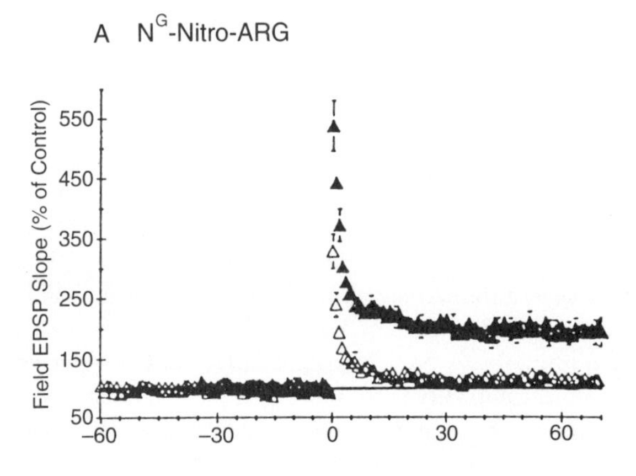

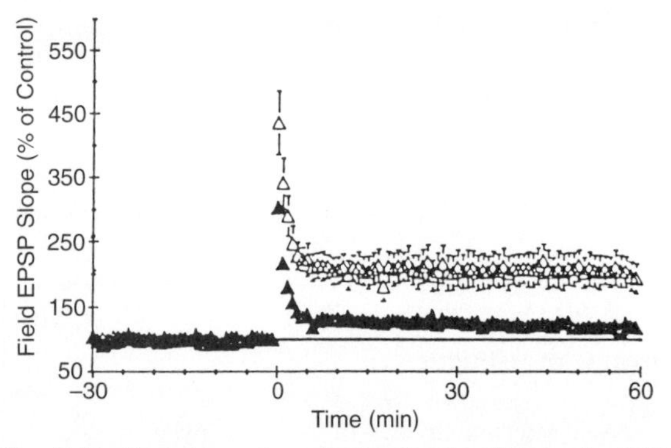

Fig. 7.15. Experimental results implicating NO in the LTP induced in the hippocampal CA1 region. A, tetanic stimulation of the Schaffer collateral-commissural pathway (at time 0) produces LTP of the synaptic response recorded from a population of the CA1 neurones with an extracellular electrode (filled triangles). The LTP was blocked by an NO synthase inhibitor (open triangles). B, the LTP was also blocked by the application of haemoglbin (filled triangles) which binds NO, but not by methaemoglobin (open triangles) which does not bind NO. (From O'Dell *et al.* 1991a.)

tion of the presynaptic fibres (in a manner shown in Fig. 7.12B). LTP observed under this condition (Fig. 7.16A) was blocked when an NO synthase inhibitor, N^G-methyl-L-arginine, was injected into the pyramidal cell (B). Presumably, Ca^{2+} influx in the postsynaptic neurone induced by presynaptic tetanization stimulates NO synthase via calmodulin (Box H). How NO potentiates synaptic responses remains unclear, however. For example, synaptic activation of NMDA receptors prior to presynaptic tetanization is found to inhibit the generation of LTP in CA1 pyramidal cells (Huang *et al.* 1992); this inhibition of LTP appears to be mediated by NO (Izumi *et al.* 1992). Moreover, Zhuo *et al.* (1993) found that NO enhances in hippocampal neurones only when NO is applied at the same time as weak presynaptic tetanic stimulation. It is not clear why a long-term increase in transmitter release induced by NO requires simultaneous presynaptic activity.

The involvement of second messengers in LTP is implied by whole-cell recordings from postsynaptic neurones. In the whole-cell recording configuration, diffusible cytoplasmic constituents will exchange with the solutes in the patch pipette (Marty and Neher 1983), risking removal of some cytoplasmic factors crucial for the induction of LTP. Indeed, in CA1 pyramidal cells the LTP cannot be induced unless the triggering stimuli (e.g., presynaptic tetanization) are applied within 30 min after rupturing the patch membrane of the postsynaptic neurone (Malinow and Tsien 1990; see also Bekkers and Stevens 1990). Interestingly, the LTP so evoked can last an hour or more (Malinow and Tsien 1990), suggesting that the factor necessary to induce LTP may not be required to maintain LTP. Still, the LTP observed under whole-cell recording conditions may not employ the normal physiological mechanism, so cautious generalization only is in order.

LTP is a fascinating phenomenon. Although the current hypotheses for its underlying mechanisms are still unsatisfactory and contradictory in part, how LTP is induced is now steadily unfolding. Undoubtedly, we should be able to delineate the molecular basis for long-term neuronal information storage soon.

7.2.3 Long-term depression at central synapses

The induction of LTP in the hippocampus requires concurrent activation of a postsynaptic neurone and its presynaptic fibres. This accords with Hebb's prediction for the expression of synaptic plasticity (Hebb 1949). Marr (1969) and Albus (1971) postulated a similar mechanism for plastic changes at the synapses formed on cerebellar Purkinje cells by parallel fibres (Fig. 6.1B). Indeed, as noted in section 6.3, the experimental results showed the induction of plastic depression at the synapse formed on the Purkinje cell by the parallel fibres following co-activation of the two elements (Ito *et al.* 1982; Ito, 1989, 1990).

Long-term depression (LTD) of parallel fibre–Purkinje cell synaptic transmission initially observed in the rabbit (Ito *et al.* 1982) was confirmed in cerebellar slices excised from the guinea pig (Sakurai 1987). In this *in vitro* preparation, excitatory synaptic potentials

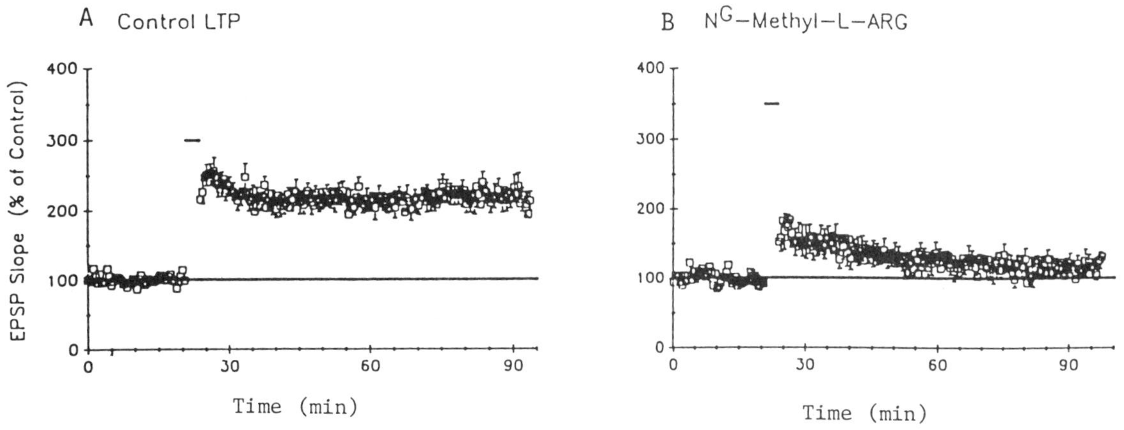

Fig. 7.16. A, LTP induced by pairing (for duration of bar) postsynaptic depolarization of a CA1 hippocampal neurone with low-frequency stimulation of the Schaffer collateral-commissural pathway; compound trace is the average points from 21 cells. B, same as in A, but an NO synthase inhibitor was injected into the CA1 pyramidal cell; the average points from 11 cells. (From O'Dell *et al.* 1991a.)

were recorded intracellularly from the Purkinje cell following stimulation of parallel fibres (Fig. 7.17A). When the climbing and parallel fibres were concurrently stimulated at 4 Hz for 25 sec, the amplitude of e.p.s.p.s evoked in the Purkinje cell by stimulating the parallel fibres was depressed for more than 60 min (Fig. 7.17B). Repetitive stimulation of the climbing fibre alone at the same frequency did not affect the e.p.s.p.s evoked by stimulating the parallel fibres. Repetitive stimulation of the parallel fibres alone did not depress the e.p.s.p.s; instead, it produced a moderate potentiation lasting 10–50 min (Sakurai 1987).

The transmitter released at the parallel fibre–Purkinje synapse is glutamate, and its action is mediated by non-NMDA receptors (Gallo *et al.* 1982; Hirano and Hagiwara 1988; Kano *et al.* 1988). Therefore, iontophoretic application of glutamate to the Purkinje cell may be substituted for stimulation of parallel fibres. When glutamate was iontophoresed and the climbing fibres stimulated, the glutamate sensitivity of Purkinje cells was subsequently depressed for about 60 min (Ito *et al.* 1982). The LTD thus appears to come from reduced sensitivity of the glutamate receptor at the parallel fibre–Purkinje cell synapse. This possibility was further examined in a simplified preparation where the neural circuitry of the cerebellar cortex was replicated in cultures of neurones dissociated from rat embryos. In a culture dish, cerebellar granular cells and inferior olivary neurones formed functional excitatory synapses on the Purkinje cell equivalent to the parallel-and climbing fibre–Purkinje cell synapses, respectively (Hirano and Hagiwara 1988; Hirano 1990a). Hirano (1990b) then recorded synaptic responses from a Purkinje cell with the whole-cell recording technique following stimulation of a granular cell or an inferior olivary neurone (Fig. 7.18A). When both were stimulated concurrently at 2 Hz for 20 sec, the synaptic current induced by stimulating the granule cell was subsequently depressed (Fig. 7.18B), confirming the LTD process observed *in vivo*. Moreover, this prolonged synaptic depression was associated with a decrease in the quantal amplitude of the synaptic response (Hirano 1991), suggesting a decrease in the receptor sensitivity.

How might the receptor sensitivity at the parallel fibre–Purkinje cell synapse be depressed if the synapse is activated at the same time as stimulation of the climbing fibre? Each climbing fibre originating from the inferior olive nucleus delivers powerful excitatory input to a single Purkinje cell, climbing up on its dendrites and thereby forming many synaptic contacts (Eccles *et al.* 1967). The synaptically excited Purkinje cell then generates Na^+ action potentials in its cell body and Ca^{2+} action potentials in its dendrites (Llinás and Sugimori 1980). When a Ca^{2+}-chelating agent, EGTA, was injected into the dendrites through the recording electrode, the joint stimulation of the climbing and parallel fibres failed to induce the LTD (Sakurai 1990). This failure implicates the Ca^{2+} influx associated with dendritic action potentials of the Purkinje cell as one factor responsible for the induction of LTD. In fact, stimulation of the climbing fibre is not essential for inducing LTD. Pairing the depolarization of the Purkinje cell with

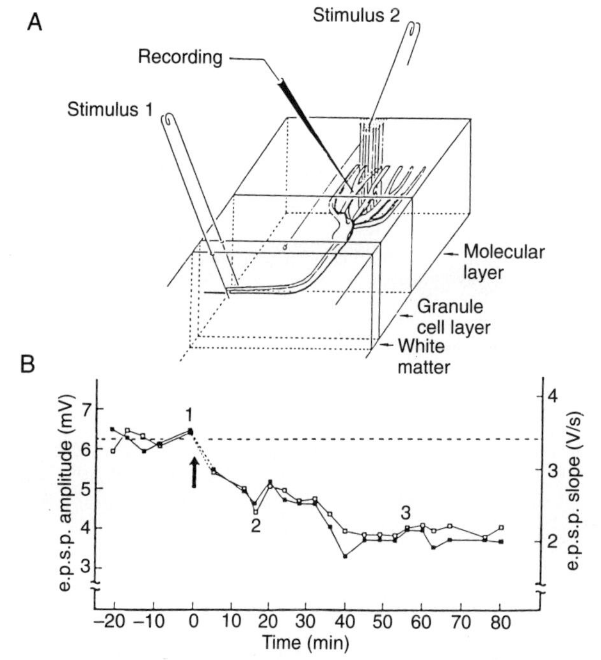

Fig. 7.17. Long-term depression observed in slice preparations of the guinea pig cerebellum. A, the experimental arrangement: intracellular recordings were made from the dendritic region of a Purkinje cell, with presynaptic stimuli applied to the climbing fibres (stimulus 1) or the parallel fibres (stimulus 2). B, depression of the e.p.s.p. amplitude (open squares) and the slope of the e.p.s.p. rising phase (filled squares) obtained by stimulation of the parallel fibres following conjunctive stimulation of the climbing and parallel fibres at 4 Hz for 25 sec. The numbers (1, 2, 3) attached to the curves are irrelevant. (From Sakurai 1987.)

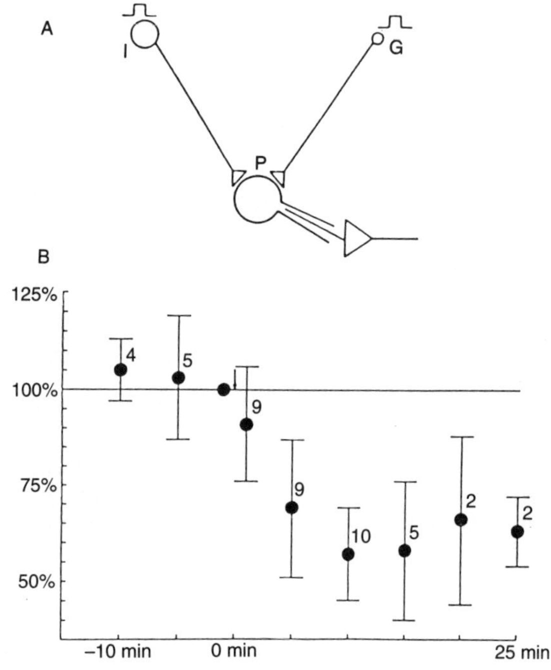

Fig. 7.18. Long-term depression observed in cultured inferior olivary neurones (I), Purkinje cells (P) and cerebellar granular cells (G) dissociated from embryonic rats. A, whole-cell recordings from a Purkinje cell on I or G stimulation. B, depression of monosynaptic excitatory synaptic currents evoked in the Purkinje cell by stimulation of single granular cells following repetitive joint stimulation of I and G at 2 Hz for 20 s applied at time 0. Each point represents the mean change of the synaptic response relative to the response immediately before the joint stimulation. Vertical bars, standard deviations; the number of experiments is indicated near each point. (From Hirano 1990b.)

stimulation of the parallel fibres can produce LTD at the parallel fibre–Purkinje cell synapse (Hirano 1990c; Crepel and Jailard 1991). Hence the co-activation of a postsynaptic neurone (Purkinje cell) and its presynaptic fibres (parallel fibres) is the precondition for LTD, again conforming to Hebb's prediction.

When the parallel fibres alone were repeatedly stimulated without concurrent depolarization of the Purkinje cell, the efficiency at the parallel fibre–Purkinje cell synapse was enhanced (Sakurai 1987; Hirano 1990c; Crepel and Jaillard 1991). In other words, potentiation induced by repetitive stimulation of the parallel fibres at their synapses on a Purkinje cell can be converted into LTD by concurrent Ca^{2+} influx into the Purkinje cell. Because of its powerful synaptic action, stimulation of a climbing fibre invariably generates action potentials in the Purkinje cell, thus effecting co-activation of the presynaptic fibre (climbing fibre) and its postsynaptic neurone (Purkinje cell). Why then does the climbing fibre–Purkinje cell synaptic function remain unchanged following repetitive stimulation of the climbing fibre? An obvious possibility is that the effect of increased cytoplasmic Ca^{2+} levels in the Purkinje cell is specific for the type of transmitter receptor; on co-activation, increased Ca^{2+} levels in the Purkinje cell decrease the sensitivity of the receptor only at the parallel fibre–Purkinje cell synapse, not at the climbing fibre synapse. The transmitter mediated at the climbing fibre–Purkinje cell synapse was believed to be aspartate (Wiklund *et al.* 1982; Kimura *et al.* 1985; but see below). Studies that compared the effects of different glutamate analogues revealed two intriguing results (Kano and

Kato 1987; Linden *et al.* 1991): first, stimulation of climbing fibres or depolarization of Purkinje cells produces LTD of the parallel fibre–Purkinje cell synapse when joined with iontophoresis of glutamate or quisqualate but not of aspartate or kainate. Second, LTD is manifested as a depression in response to iontophoresis of glutamate or non-NMDA receptor agonists but not of aspartate or NMDA. These results imply that the receptor mediated at the parallel fibre–Purkinje cells synapse is a quisqualate- (or AMPA-) preferring non-NMDA receptor and that co-activation of this receptor and the Purkinje cell is essential for the induction of LTD. Also, the sensitivity of the aspartate receptor seems to be unaffected by increased Ca^{2+} levels in the Purkinje cell. The transmitter mediated at the climbing fibre–Purkinje cell synapse may not be aspartate; it is presumably glutamate instead (Hirano 1990a; Zhang *et al.* 1990). Recall that the ionotropic glutamate receptor has several subtypes (Nakanishi 1992), some of which have a higher affinity for quisqualate than for kainate, while others prefer kainate (Egebjerg *et al.* 1991; Sakimura *et al.* 1992). It is possible that the receptor mediated at the climbing fibre–Purkinje cell synapse is a kainate-selective glutamate receptor, whereas that at the parallel fibre–Purkinje cell synapse is a quisqualate-(or AMPA-) selective glutamate receptor. The receptor sensitivity would be then modulated by Ca^{2+} in a manner specific for the receptor subtype, even though the transmitter is glutamate in both cases.

Increased intracellular Ca^{2+} levels in the Purkinje cell alone are not sufficient to induce LTD; for LTD, Ca^{2+} enrichment must occur at about the same time as activation of the parallel fibre–Purkinje cell synapse. Therefore, Ca^{2+}, a co-factor, must be linked to another process triggered by activation of the receptor at the synapse. Linden *et al.* (1991) suggested that LTD requires co-activation of ionotropic and metabotropic receptors by glutamate released from the parallel fibres. But what type of metabotropic receptor is involved in LTD? Shigemoto *et al.* (1994) found that antibodies specific for the $mGluR_1$ metabotropic receptor subunit block the induction of LTD in cultured Purkinje cells. The $mGluR_1$ subunit is known to be coupled to the signal transduction of inositol phosphate pathway, raising intracellular Ca^{2+} levels and probably activating protein kinase C (Masu *et al.* 1991; Abe *et al.* 1992). Shibuki and Okada (1991) reported that the cerebellar LTD can be blocked by haemoglobin or an NO synthase inhibitor, suggesting that NO acts in the LTD. These results were not confirmed in cultured Purkinje cells, however, where LTD was induced by iontophoretic applications of glutamate in conjunction with depolarization of the Purkinje cell (Linden and Connor 1992).

Since the Purkinje cell is an inhibitory neurone, the cerebellar LTD is ultimately manifested as synaptic potentiation at its target neurone (Ito 1990). LTD is not restricted to inhibitory neurones, however, being seen in CA1 hippocampal neurones (Stanton and Sejnowski 1989) and also in neurones of the neocortex (Tsumoto 1990). In the hippocampus, the test and conditioning pulses produce LTD when delivered out of phase, but LTP when given in phase (Stanton and Sejnowski 1989). Furthermore, the LTD in hippocampal neurones is not blocked by an NMDA receptor blocker, APV, but is blocked by a metabotropic glutamate receptor blocker (Sejnowski *et al.* 1990; Bolshakov and Siegelbaum 1994). In fact, Kato (1993) suggested that activation of postsynaptic metabotropic glutamate receptors is sufficient for inducing LTP in visual cortical neurones. In neurones of the visual cortex, the same presynaptic tetanic stimulation induces either LTP or LTD, depending on the level of depolarization of the postsynaptic neurone (Artola *et al.* 1990): a greater depolarization is required for LTP. Very likely, then, the intracellular Ca^{2+} level in the postsynaptic neurone is crucial for the induction of LTP or LTD (Lisman 1989). In fact, an injection of a Ca^{2+} chelator, EGTA, into a visual cortical neurone switched the plastic expression from LTP to LTD in response to the same triggering stimuli (Kimura *et al.* 1990; Tsumoto 1990). Further analysis of this switch mechanism could provide important insights into the mechanisms underlying LTP and LTD. In any event, the mechanisms for the LTD observed in the hippocampus or neocortex seem quite different from those for the cerebellar LTD. Bolshakov and Siegelbaum (1994) suggested that the LTD in the hippocampus is due to a decrease in transmitter release, in contrast to reduced sensitivity in the postsynaptic receptor previously reported for the cerebellar LTD (Ito *et al.* 1982; Hirano 1991).

As with LTP in the hippocampus, attempts have been made to identify specific enzymes or protein kinase involved in the induction or expression of LTD in the mammalian brain (Ito 1990; Tsumoto 1990). Unfortunately, a lack of specificity in the pharmacological agents applied to the preparation, the elusivenss of the precise neuronal network, and the difficulty in isolating a homogenous group of neurones for biochemical analyses have so far made it not feasible to trace the molecular cascade involved in long-term synaptic plasticity at mammalian central synapses. For this approach to succeed, one needs preparations that preserve long-term synaptic plasticity in a simple nervous system. *Aplysia*, for instance, has a much simpler nervous system, yet its synaptic transmission undergoes plastic changes that last for hours and days. Is it possible to make more detailed studies on long-term plasticity at the molecular level in *Aplysia*? If so, what kind of molecular change would we find? This is the subject of the next chapter.

Summary and prospects

It is almost certain that the initial step for the induction of synaptic plasticity is accumulation of intracellular Ca^{2+}, which can occur in the presynaptic terminal or in the postsynaptic neurone. Increased intracellular Ca^{2+} levels may link depolarization or receptor activation to the second intracellular messenger system. When increased Ca^{2+} concentrations in presynaptic terminals subside to normal levels rapidly by Ca^{2+} buffering and transport systems, the resultant synaptic plasticity is relatively short-lasting, as exemplified by post-tetanic potentiation. Long-term alterations in synaptic function require the involvement of the second messenger system.

The elucidation of mechanisms that underly synaptic plasticity has two aspects. One is identification of the second messengers and protein mediators of synaptic plasticity, and the other is identification of biophysical processes involved in the alteration of synaptic function. Biophysically, synaptic plasticity occurs by changes in transmitter release, by changes in the sensitivity of the postsynaptic receptor or by a combination of the two. At peripheral synapses, these alternatives can be distinguished by quantal analysis of synaptic responses. At central synapses, however, quantal analysis cannot be applied reliably at present. Undoubtedly, this is the major obstacle to the study of central synaptic plasticity.

Many examples of long-term plasticity so far reported for peripheral synapses are induced and maintained by activation of the second messenger system and protein kinases in their presynaptic terminals. By contrast, long-term plasticity at central synapses appears to be triggered by cytoplasmic factors (second messengers and protein kinases) in the postsynaptic neurone. Again, the initial step for triggering central synaptic plasticity is an increase in the intracellular Ca^{2+} level. The receptor-mediated increase in intracellular Ca^{2+} levels in central neurones can occur in any of four ways: by activation of NMDA receptors, by activation of Ca^{2+}-permeating non-NMDA receptors, by opening of voltage-gated Ca^{2+} channels due to non-NMDA receptor-induced depolarization, and by activation of metabotropic glutamate receptors which lead to stimulation of phosphatidylinositol hydrolysis.

It is likely that long-term plasticity at central synapses is mediated by more than one type of protein kinase. In addition, there may be trans-synaptic messengers which convey signals from the postsynaptic neurone to its presynaptic terminal. Candidates for such retrograde messengers have been proposed but not proven. Moreover, although LTP and LTD were initially described in different regions of the central nervous system, both are now known to be induced at the same synapse, and plasticity can even be switched from LTP to LTD at a single synapse simply by changing the intracellular Ca^{2+} in the postsynaptic neurone. Thus, the expression of long-term synaptic plasticity appears to be vulnerable to slight alterations in the experimental procedure. Presumably, the sequence of multiple second messengers and protein kinases activated during synaptic plasticity is finely regulated. Some contradictory results reported by different investigators with the seemingly identical analysis may stem from minor differences in the experimental conditions.

Suggested reading

Regulation of intraterminal calcium levels

Augustine, G. J. and Neher, E. (1992). Neuronal Ca^{2+} signalling takes the local route. *Current Opinion in Neurobiology*, 2, 302–7.

Blaustein, M. P. (1988). Calcium transport and buffering in neurons. *Trends in Neurosciences*, 11, 438–43.

Miller, R. J. (1991). The control of neuronal Ca^{2+} homeostasis. *Progress in Neurobiology*, 37, 255–85.

Synaptic plasticity at peripheral synapses

Atwood, H. L. and Wojtowicz, J. M. (1986). Short-term and long-term plasticity and physiological differentiation of crustacean motor synapses. *International Review of Neurobiology*, **28**, 275–362.

Kuba, K. and Kumamoto, E. (1990). Long-term potentiation in vertebrate synapses: a variety of cascades with common subprocesses. *Progress in Neurobiology*, 34, 197–269.

Zucker, R. S. (1989). Short-term synaptic plasticity. *Annual Review of Neuroscience*, 12, 13–31.

Functional properties of NMDA and non-NMDA receptor channels

Alford, S., Frenguelli, B. G., Schofield, J. G., and Collingridge, G. L. (1993). Characterization of Ca^{2+}

signals induced in hippocampal CA1 neurones by the synaptic activation of NMDA receptors. *Journal of Physiology (London)*, **469**, 693–716.

Hestrin, S., Nicoll, R. A., Perkel, D. J., and Sah, P. (1990). Analysis of excitatory action of in pyramidal cells using whole-cell recording from rat hippocampal slices. *Journal of Physiology (London)*, **422**, 203–25.

Jahr, C. E. and Lester, A. J. (1992). Synaptic excitation mediated by glutamate-gated ion channels. *Current Opinion in Neurobiology*, **2**, 270–4.

Mayer, M. L. and Westbrook, G. L. (1987). The physiology of excitatory amino acids in the vertebrate nervous system. *Progress in Neurobiology*, **28**, 197–276.

Traynelis, S. F., Silver, R. A., and Cull-Candy, S. G. (1993). Estimated conductance of glutamate receptor channels activated during EPSCs at the cerebellar mossy fibre-granule cell synapse. *Neuron*, **11**, 279–89.

Long-term potentiation

Bliss, T. V. P. and Collingridge, G. L. (1993). A synaptic model of memory: long-term potentiation in the hippocampus. *Nature*, **361**, 31–9.

Frey, U., Huang, Y. Y., and Kandel, E. R. (1993). Effects of cAMP simulate a late stage of LTP in hippocampal CA1 neurons. *Science*, **260**, 1661–4.

Kauer, J. A., Malenka, R. C., and Nicoll, R. A. (1988). A persistent postsynaptic modification mediates long-term potentiation in the hippocampus. *Neuron*, **1**, 911–17.

Malgaroli, A. and Tsien, R. W. (1992). Glutamate-induced long-term potentiation of the frequency of miniature synaptic currents in cultured hippocampal neurons. *Nature*, **357**, 134–9.

Manabe, T., Renner, P., and Nicoll, R. A. (1992). Postsynaptic contribution to long-term potentiation revealed by the analysis of miniature synaptic currents. *Nature*, **355**, 50–5.

Silva, A. J., Stevens, C. F., Tonegawa, S., and Wang, Y. (1992). Deficient hippocampla long-term potentiation in α-calcium-calmodulin kinase II mutant mice. *Science*, **257**, 201–6.

Zalutskey, R. A. and Nicoll, R. A. (1990). Comparison of two forms of long-term potentiation in single hippocampal neurons. *Science*, **248**, 1619–24.

Long-term depression

Hirano, T. (1991). Synaptic formations and modulations of synaptic transmissions between identified cerebellar neurones in culture. *Journal of Physiology (Paris)*, **85**, 145–53.

Ito, M. (1990). Long-term depression in the cerebellum. *Seminars in Neurosciences*, **2**, 381–90.

Linden, D. J. (1994). Long-term synaptic depression in the mammalian brain. *Neuron*, *12*, 457–72.

Sakurai, M. (1987). Synaptic modification of parallel fibre–Purkinje cell transmission in *in vitro* guinea-pig cerebellar slices. *Journal of Physiology (London)*, **394**, 463–80.

Sakurai, M. (1990). Calcium is an intracellular mediator of the climbing fibre in induction of cerebellar long-term depression. *Proceedings of the National Academy of Sciences of the USA*, **87**, 3383–5.

Tsumoto, T. (1990). Long-term potentiation and depression in the cerebral neocortex. *Japanese Journal of Physiology*, **40**, 573–93.

8

Molecular cascades in learning

Learning is behavioural modification resulting from experience (Thompson 1967). Since all behaviour reflects actions of the nervous system, behavioural modification must result from changes in the behaviour of the nervous system (Thompson 1967). Hence, lasting plasticity in neuronal function is a prerequisite for learning. In the previous chapter, we examined biophysical substrates for neuronal plasticity in the vertebrate. At the synaptic level, plasticity is expressed by changes in the amount of transmitter released or in the sensitivity of postsynaptic receptors for the transmitter. Although each synaptic function that is modulated during plasticity must be described individually, we can begin to discern some features common to several systems at the molecular level. For example, persistent plastic changes appear to require the activation of intracellular second messengers or the phosphorylation of certain proteins (section 7.1.2). Admittedly, in vertebrate synapses this generalization rests on suggestive evidence and many details are lacking. Which intracellular messengers are relevant to the expression and maintenance of plastic change? Which proteins are phosphorylated? And what kind of behavioural modification results from plastic changes at vertebrate synapses? We do not know. In this sense synaptic plasticity in the vertebrate is merely a phenomenon loosely implicated in learning.

Aplysia, a sea slug, has a relatively simple nervous system and shows experience-dependent behavioural modifications (Kandel 1979). The neuronal circuitry mediating the behavioural modification is well delineated, and some of the neurones involved are so large they are readily isolated for biochemical assay. *Aplysia* is thus an ideal preparation for studying molecular changes associated with learning. We will now examine the molecular cascade underlying learning in *Aplysia* and see to what extent molecular mechanisms of learning share common features in different species.

8.1 MOLECULAR CHANGES ASSOCIATED WITH LEARNING

Aplysia's gill, its respiratory organ, is normally covered by the mantle shelf. Seawater entering the mantle cavity supplies oxygen to the gill and is expelled through the siphon (Kandel 1979). Applying a weak tactile stimulus to the siphon or mantle shelf causes the gill to contract and withdraw into the mantle cavity (Kandel 1976). This response, the *gill-withdrawal reflex*, is analogous to defensive withdrawal reflexes common in vertebrates. With repeated stimulation of the siphon, the gill-withdrawal reflex progressively weakens, a phenomenon termed *habituation*. Habituation refers to a progressive decrease in behavioural response when an initially novel stimulus is repeatedly applied (Harris 1943). Once habituation occurs, the response remains depressed for hours and days. Habituation is one of the simplest forms of learning (Thompson and Spencer 1966; Kandel 1976).

There is another simple form of learning, termed *sensitization* (Grether 1938; Spencer *et al*. 1966). Following a strong (noxious) stimulus applied to the tail or head, for example, *Aplysia* shows enhancement of the gill-withdrawal reflex (Kandel 1976). This behavioural modification, sensitization, reflects a priming mechanism that increases arousal and attention (Dudai 1989).

Habituation of the gill-withdrawal reflex stems from depression at the synapses between the sensory neurones arising from the siphon and the motor neurones innervating the gill (see below). The molecular basis for this synaptic depression remains unknown, however. By contrast, sensitization in *Aplysia* has been studied extensively, and at a molecular level, so this form of learning will be our primary focus. We shall trace the neuronal correlates of sensitization in the gill-withdrawal reflex first.

8.1.1 Neuronal correlates of sensitization in *Aplysia*

The gill muscle of *Aplysia* is innervated by several motor neurones located in the abdominal ganglion. The abdominal ganglion also contains clusters of sensory neurones. One group of about 24 cells innervates the siphon, and another cluster of about 24 innervates the mantle shelf. These two clusters of sensory neurones carry sensory signals from the siphon or mantle shelf through their peripheral processes; with their short central processes the sensory neurones form monosynaptic excitatory connections on the motor neurones innervating the gill. Intracellular stimulation of the gill motor neurones in the abdominal ganglion produces contractions of the gill muscle; intracellular recordings from the sensory neurones in the abdominal ganglion show action potentials in response to tactile stimuli applied to the siphon or mantle shelf; and intracellular stimulation of the sensory neurones evokes monosynaptic e.p.s.p.s in the gill motor neurone. In this way, the neuronal pathway responsible for the behavioural gill-withdrawal reflex was delineated (Kandel 1976). This reflex pathway is illustrated in Fig. 8.1 where, for

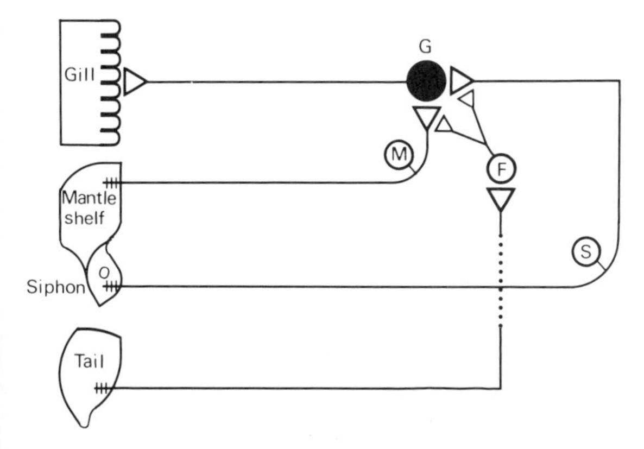

Fig. 8.1. A schematic circuit for the gill-withdrawal reflex in *Aplysia*. The peripheral organs are indicated on the left (Periphery), and the neurone cell bodies in central ganglia on the right (Central ganglion). G represents the motor neurones innervating the gill. M, cell bodies of the sensory neurones arising from the mantle shelf and projecting to the gill motor neurone. S, cell bodies of the sensory neurones arising from the siphon and projecting to the gill motor neurone. F, facilitatory interneurones activated by stimulation of the tail, which facilitate transmission from the sensory neurone to the gill motor neurones. The pathway from the tail to F (dotted line) remains unknown.

simplicity, each group of sensory or motor neurones is represented by a single cell.

Habituation of the gill-withdrawal reflex induced by iterative tactile stimulation of the siphon is associated with a decrement in the monosynaptic e.p.s.p.s evoked in the gill motor neurone by intracellular stimulation of the sensory neurone (Kandel 1976). This synaptic depression is due to a decrease in the number of transmitter quanta released from the sensory terminal (Castellucci and Kandel 1974). Similarly, sensitization of the behavioural gill-withdrawal reflex following a strong stimulus at the tail was associated with enhancement of monosynaptic e.p.s.p.s evoked in the gill motor neurone by stimulation of the sensory neurone (Kandel 1976). This synaptic enhancement is due to an increase in quantal transmitter release from the sensory terminal (Castellucci and Kandel 1976).

Habituation is confined to the synapses whose presynaptic fibres are stimulated repeatedly. Thus, habituation is a *homosynaptic* depression. By contrast, although the strong sensitizing stimulus at the tail does not excite the siphon sensory neurone, it enhances synaptic transmission from the sensory neurone to the gill motor neurone. Thus, sensitization is a *heterosynaptic* event. Even though the sensitizing stimulus does not elicit an action potential in the siphon sensory neurone, the sensitizing pathway must form synaptic connections with the sensory neurone or its terminals to facilitate the transmitter release. Therefore, the mechanism underlying sensitization is assumed to be *presynaptic facilitation* (Fig. 8.1).

The exact neuronal pathway involved in sensitizing the gill-withdrawal reflex is not known. Suggestively, several *facilitatory interneurones* (Fig. 8.1, F) intercalated in the sensitization-triggering pathway from the tail facilitate synaptic transmission from the siphon sensory neurone to the gill motor neurone. One group of facilitatory interneurones (termed *CB1 cells*) located in the cerebral ganglion contains serotonin (Hawkins 1989). Another group of facilitatory interneurones (*L29 cells*) is present in the abdominal ganglion, but its transmitter remains unidentified (Hawkins *et al.* 1981; Hawkins and Schacher 1989). A third group of facilitatory interneurones contains the *small cardioactive peptides* (Abrams *et al.* 1984). Interestingly, although each group uses a different transmitter to induce sensitization of the gill-withdrawal reflex, their underlying mechanisms appear to be identical (Kandel *et al.* 1987; Hawkins and Kandel 1990; see below). The sensitization induced by serotonin (5–hydroxytriptamine; 5HT) is the best understood. We will therefore discuss the neural mechanisms for sensitization of the gill-withdrawal reflex using the 5HT system as our example.

As illustrated by the simplified neuronal network in Fig. 8.1, strong stimuli at the tail excite the facilitatory interneurone (F), which projects to the siphon sensory neurone and enhances its transmitter release. Serotonin immunohistochemistry showed that serotonergic neural processes, presumably arising from a group of the facilitatory interneurones, come in close contact with the siphon sensory cell bodies and their terminals (Kistler *et al.* 1985). How might 5HT released from the facilitatory interneurone affect the siphon sensory neurone? Recall the S-channel, a serotonin-sensitive K^+ channel discussed in section 5.2.2. The S-channel is present in *Aplysia* sensory neurones and closes in response to activation of metabotropic 5HT receptors (Fig. 5.7). In fact, it is the siphon sensory neurone that shows this response to 5HT. Activation of the metabotropic 5HT receptor stimulates adenylate cyclase, which in turn elevates the cyclic AMP level in the sensory neurone. The closure of the S-channel is then induced by the action of cyclic AMP-dependent protein kinase or protein kinase A (Shuster *et al.* 1985; Fig. 5.8).

How would the closure of the S-channel lead to enhancement of transmitter release from the siphon sensory neurone? This process is illustrated in Fig. 8.2. In this experiment (Castellucci *et al.* 1980), the application of 5HT to the siphon sensory neurone was replaced by directly injecting protein kinase A into the sensory neurone (Fig. 8.2A). This procedure clearly increased the monosynaptic e.p.s.p.s evoked in the gill motor neurone by intracellular stimulation of the sensory neurone (Fig. 8.2B). Concomitantly, the action potential of the sensory neurone was prolonged (Fig. 8.2C). What do these results mean? The action potential of a nerve cell is initiated by an increase in the Na^+ conductance and terminated by an increase in the K^+ conductance. Therefore, blockade of the K^+ conductance (closure of the K^+ channel) defers termination of the action potential, thereby prolonging its duration. The action potential broadening (i.e., prolonged depolarization) then enhances the Ca^{2+} influx into the sensory terminal, and more transmitter is released (Castellucci *et al.* 1980; Klein and Kandel 1980; Siegelbaum *et al.* 1982).

In addition to broadening the action potential in sensory neurones, 5HT is now known to facilitate directly the process of transmitter release from the sensory neurone in a Ca^{2+}-independent manner, by as yet unidentified mechanisms which promote exocytosis (Dale and Kandel 1990). Moreover, the sensitizing stimuli activate other excitatory and inhibitory interneurones, which may modulate the sensitization of the gill-withdrawal reflex (Frost *et al.* 1988). Clearly the neural pathway for sensitization illustrated in Fig. 8.1 is over-simplified. But there is little doubt that sensitization of the gill-withdrawal reflex is mediated mainly, if not exclusively, by the cyclic AMP-dependent phosphorylation system in the sensory neurone (Braha *et al.* 1990).

Both habituation and sensitization of the gill-withdrawal reflex have short- and long-term forms. For example, following 10 regularly spaced tactile stimuli to the siphon (e.g., 1 min apart), habituation of the gill-withdrawal reflex lasts only several hours; habitua-

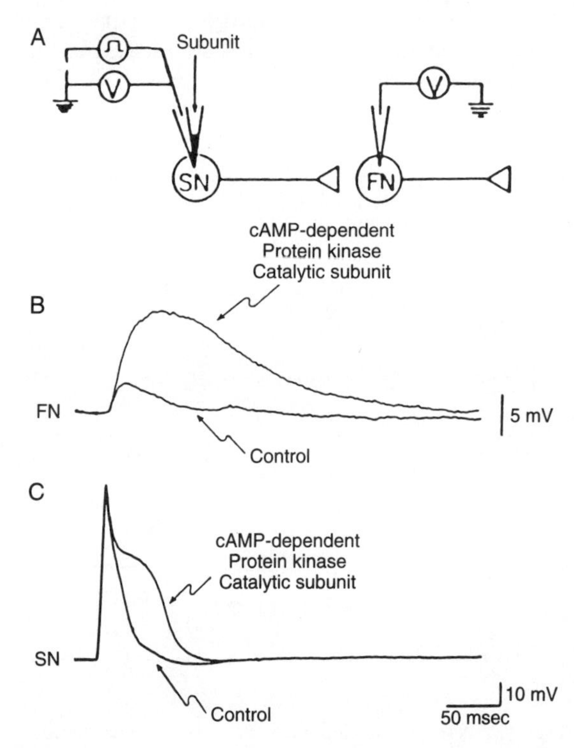

Fig. 8.2. Effects of intracellular injections of the active (catalytic) subunit of cyclic AMP-dependent protein kinase into the *Aplysia* sensory neurone. A, the experimental procedure illustrating the injection of the catalytic subunit into a sensory neurone (SN). After intracellular stimulation of the sensory neurone, monosynaptic e.p.s.p.s were recorded from the motor neurone (follower neurone; FN). B, the FN monosynaptic e.p.s.p.s recorded before (control) and after the injection of the catalytic subunit. C, the SN action potentials recorded before (control) and after the catalytic subunit injection. (Adapted from Castellucci *et al.* 1980.)

tion lasts for weeks, however, if the 10-stimuli train is repeated four times (Carew *et al.* 1972). Similarly, sensitization can be prolonged by iteration of the sensitizing stimulus: One stimulus applied to the tail produces short-term sensitization lasting minutes to hours, whereas repeatedly applied stimuli sensitize for days to weeks (Pinsker *et al.* 1973). The long-term sensitization of the gill-withdrawal reflex was again associated with increased monosynaptic e.p.s.p.s evoked in the gill motor neurone by stimulation of the siphon sensory neurone (Frost *et al.* 1985). That is, both short- and long-term memory employ the same synaptic locus.

Habituation or sensitization can be induced by a single type of stimulus and need not be associated with other stimuli. Thus, both are *non-associative forms of learning*, as distinct from the *associative learning* of classical or Pavlovian conditioning. Recall that Pavlov's dog salivated in response to the sound of a bell alone (*conditioned stimulus*; CS) only after training that paired the CS with feeding (*unconditioned stimulus*; US). Classical conditioning can also be shown in the gill-withdrawal reflex, using a tactile stimulus to the siphon as the CS and a strong stimulus to the tail as the US. Specific timing wherein the CS precedes the US by 0.5–1 s significantly enhances the gill-withdrawal reflex, beyond the level produced by the sensitizing stimulus alone (Carew *et al.* 1981, 1983). The synaptic locus involved in this classical conditioning is again the siphon sensory-gill motor synapse, and synaptic enhancement is associated with the action potential broadening in the siphon sensory neurone (Hawkins *et al.* 1983).

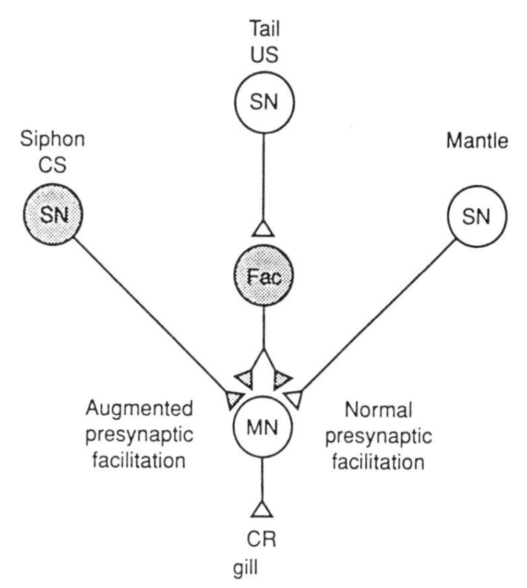

Fig. 8.3. A hypothetical diagram for classical conditioning of the gill-withdrawal reflex in *Aplysia*. SN, sensory neurones. MN, the motor neurones innervating the gill. Fac, facilitatory interneurones involved in sensitization of the reflex. When the CS applied to the siphon precedes the US at the tail, the siphon sensory terminal and the facilitatory interneurone are virtually co-activated (shading). This co-activation is assumed to be responsible for classical conditioning or conditioned response (CR). (Adapted from Hawkins and Kandel 1990.)

Why would transmission at this synapse be enhanced beyond the sensitized level when the CS to the siphon precedes the US at the tail? The simplified diagram shown in Fig. 8.3 illustrates a possible mechanism for classical conditioning. An unconditioned stimulus (US) at the tail activates a group of facilitatory interneurones (Fac); the gill-withdrawal reflex is then sensitized as a result of the presynaptic facilitation exerted by the facilitatory interneurones on the sensory neurones of both the siphon and the mantle shelf. If the sensitizing stimulus is repeatedly applied, long-term sensitization ensues. When the sensitizing stimulus (US) is paired with the CS applied to the siphon, activation of the siphon sensory terminal just precedes the arrival of each facilitatory internuncial impulse at the interneurone terminal. This condition is equivalent to Hebbian conjunction, because the presynaptic terminal (the terminal of the facilitatory interneurone) and its postsynaptic element (the siphon sensory terminal) become virtually coactivated. Hawkins *et al.* (1983) suggested that the Ca^{2+} influx into the siphon sensory terminal by each action potential would reinforce the cyclic AMP signalling triggered by 5HT released from the facilitatory interneurone. This notion is similar to the mechanism postulated for LTP in the hippocampal CA1 region (section 7.2.2). Thus, in this model classical conditioning observed in the gill-withdrawal reflex is actually an elaboration of sensitization based on the activity-dependent amplification of presynaptic facilitation (Hawkins *et al.* 1983). Consistent with this suggestion, the cyclic AMP level elevated in cultured *Aplysia* sensory neurones by a brief application of 5HT is indeed further enhanced immediately after excitation of the sensory neurones, and this activity-dependent amplification effect is abolished in a Ca^{2+}-free solution (Abrams 1985; Hawkins and Kandel 1990).

The conditioning can be converted from short- to long-term by adding more training sessions. In reviewing long-term alterations at vertebrate synapses, three aspects of reaction appear to underly long-term plastic changes: activation of the second messenger system, phosphorylation by protein kinases, and autophosphorylation of the protein kinase (section 7.1.2). So, how might short- and long-term plastic changes be distinguished at the molecular level? What kind of molecular interaction might be added by pairing a CS with a sensitizing stimulus? Would short-term sensitization, long-term sensitization and classical conditioning require different forms of information storage? We examine these questions next.

8.1.2 How neuronal information is stored

Long-term sensitization or classical conditioning of the gill-withdrawal reflex can be considered a modification of short-term sensitization caused by repetition of the triggering stimulus or by pairing the triggering stimulus with another stimulus. Short-term sensitization is thus the basic framework for these plastic changes. We will therefore focus on short-term sensitization first, then trace the molecular modifications that are induced by additional stimuli.

By the early 1970s, information gathered from different sources had implicated cyclic AMP as the second messenger in a variety of physiological responses (Langan 1969; Miyamoto *et al.* 1969; Sutherland 1972). Speculation then arose that cyclic AMP might be the intracellular mediator in transferring and amplifying the sensitizing signal elicited by the tail stimulus in *Aplysia*. From the ensuing investigation has come a wealth of new insights into the molecular mechanisms of learning. First, cyclic AMP levels in the abdominal ganglion indeed increase following application of the sensitizing stimulus (Cedar *et al.* 1972). Second, this action can be reproduced by exposing the abdominal ganglion to serotonin. Third, serotonin not only elevates cyclic AMP in the abdominal ganglion but also enhances synaptic transmission from the sensory neurone to the gill motor neurone (Brunelli *et al.* 1976). As shown in Fig. 8.4A, elevation of cyclic AMP in the abdominal ganglion induced by exposure to serotonin has a time course like that of short-term sensitization produced at the sensory-gill motor neurone synapse by the sensitizing stimuli (Fig. 8.4B; Cedar and Schwartz 1972; Kandel *et al.* 1976). Presumably, the facilitatory interneurone activated by the sensitizing stimulus releases 5HT, which in turn elevates cyclic AMP levels in the sensory neurones. Microbiochemical assay by Bernier *et al.* (1982) indeed proved that it is the sensory neurones whose cyclic AMP rises after 5HT is applied to the abdominal ganglion.

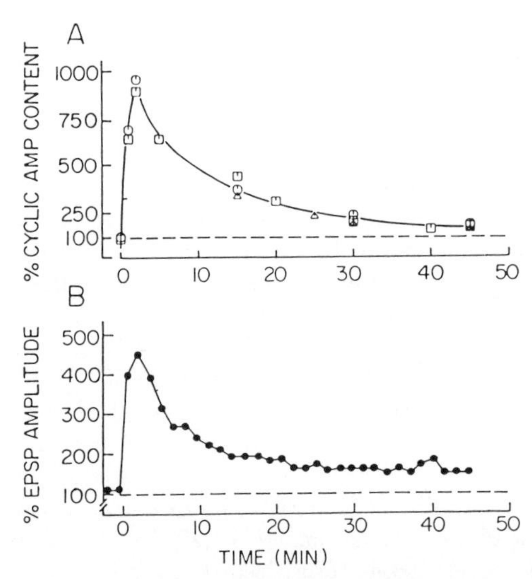

Fig. 8.4. Time courses of increased cyclic AMP levels in the abdominal ganglion (A) and sensitization of monosynaptic excitatory synaptic responses in the gill motor neurone (B). A, cyclic AMP was measured in the abdominal ganglion of *Aplysia* after the exposure of the ganglion to 0.2 mM 5HT. B, monosynaptic e.p.s.p.s were recorded from a gill motor neurone in response to intracellular stimulation of a sensory neurone after the application of sensitizing stimuli. (From Kandel *et al.* 1976.)

Cyclic AMP (3′,5′-AMP) is synthesized from ATP by activation of adenylate cyclase and hydrolysed to 5′-AMP by the action of cyclic nucleotide phosphodiesterase (Fig. 8.6, step 3). Does 5HT elevate cyclic AMP by activating the cyclase or by inhibiting the phosphodiesterase? In the tissue homogenate prepared from *Aplysia* sensory neurones, 5HT increased the activity of the adenylate cyclase (Yovell *et al.* 1987; Abrams *et al.* 1991). In an experiment with labelled ATP, 5HT also clearly enhanced the rate of synthesis of labelled cyclic AMP in the sensory neurones (Bernier *et al.* 1982). Hence 5HT seems to increase cyclic AMP levels by activating adenylate cyclase. The 5HT receptor has several subtypes (Bradley *et al.* 1986; Julius 1991; section 4.1.6), but which subtype is responsible for cyclic AMP elevation in the sensory neurones is not certain. Tellingly, the adenylate cyclase in the homogenate of the sensory neurones is persistently activated when a non-hydrolyzable GTP-analogue, GTP-γS is applied (Yovell *et al.* 1987). Evidently, stimulation of the adenylate cyclase by 5HT is mediated by the G-protein-linked 5HT receptor, that is, metabotropic 5HT receptors (Fig. 8.6, step 1).

Electrophysiological studies show that sensitization is due to an increase in the amount of transmitter released from the sensory neurones; this phenomenon is based on the closure of S-channels as a result of phosphorylation of the channel protein by cyclic AMP-dependent protein kinase (Fig. 8.6, step 5). What sustains a short-term sensitization? For instance, can the S-channel stably maintain its phosphorylated state? In Fig. 8.5A, we see the time course of sensitization following the application of 5HT, measured in terms of the broadened action potential of the sensory neurones (see Fig. 8.2C). As expected, the action potential recorded from a pair of sensory neurones (open and filled circles) was prolonged after 5HT application. The protein kinase inhibitor was then injected into one of the sensory neurones (filled circles), which immediately curtailed the broadening of its spike (Castellucci *et al.* 1982). Clearly, the maintenance of short-term sensitization requires persistently active protein kinase, to keep the S-channel protein phosphorylated. In other words, the S-channel closure induced by phosphorylation itself is too unstable to support sensitization for even a short period.

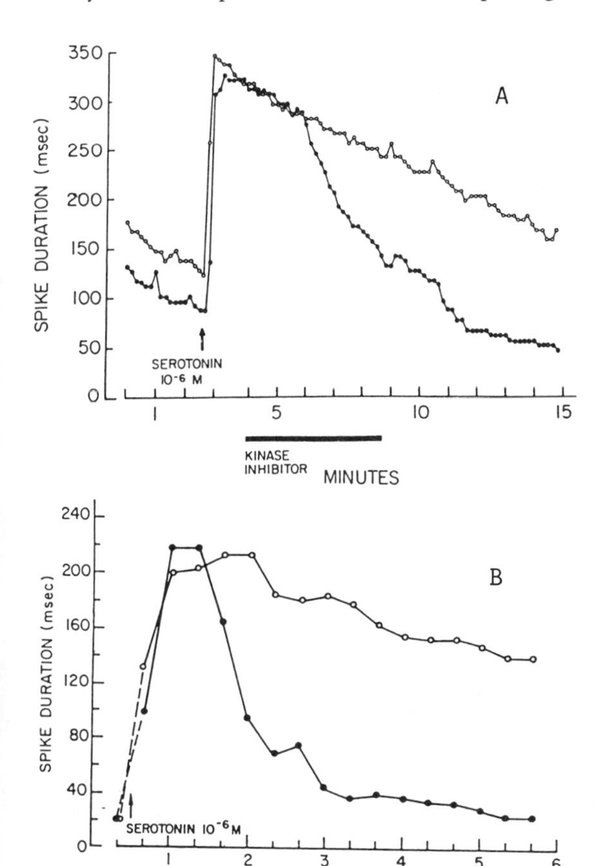

Fig. 8.5. Effects of a protein kinase inhibitor (A) and GDP-βS (B) on the action potential (spike) broadening induced by serotonin. The action potentials were recorded from a pair of sensory neurones (open and filled circles) in *Aplysia*. Applying serotonin prolonged the action potential; injecting either a kinase inhibitor (A) or GDP-βS (B) into one of the sensory neurones (filled circles) curtailed this broadening. (From Schwartz *et al.* 1983.)

We have seen that autophosphorylation of Ca^{2+}-CaM (calmodulin) kinase II can make the kinase active even in the absence of Ca^{2+} (section 7.1.2; Box G). Some cyclic AMP-dependent kinase also shows autophosphorylation in the absence of cyclic AMP (Rangel-Aldao and Rosen 1976). So might short-term sensitization be maintained by autophosphorylation of the protein kinase? No. When a non-hydrolyzable GDP analogue, GDP-βS, is injected into a sensory neurone, the binding of GTP to G-proteins is competitively inhibited (Eckstein *et al.* 1979), so that activation of the 5HT receptor no longer stimulates adenylate cyclase and hence no longer increases cyclic AMP levels in the sensory neurone. The intracellular injection of GDP-βS rapidly terminates the action potential broadening in the sensory neurone (Fig. 8.5B, filled circles; Schwartz *et al.* 1983). Thus the key process in maintaining short-term sensitization appears to be the persistence of cyclic AMP elevation levels in the sensory neurones, which keeps the protein kinase activated. In other words, short-term sensitization lasts as long as cyclic AMP stays high in the sensory neurones, so that the memory must be stored in the form of cyclic AMP elevation. But what upholds the cyclic AMP levels remains unclear.

Figure 8.6 recapitulates the molecular cascade involved in short-term sensitization. Our next question is how sensitization switches from short- to long-term. To make this conversion is simple: one merely repeats the stimulus applied to the tail. Cyclic AMP in the sensory neurones elevated by 5HT is again essential for inducing long-term sensitization, but its maintenance does not require that the cyclic AMP levels stay high (Kandel and Schwartz 1982), as evidenced by a series of experiments described below.

The molecular approach to long-term synaptic plasticity was advanced greatly by reconstruction of the elementary gill-withdrawal circuit in cell culture (Rayport and Schacher 1986). The gill motor neurones and the sensory neurones were dissociated from *Aplysia* and co-cultured in a dish. After 5 days the sensory neurones had formed functional synaptic connections

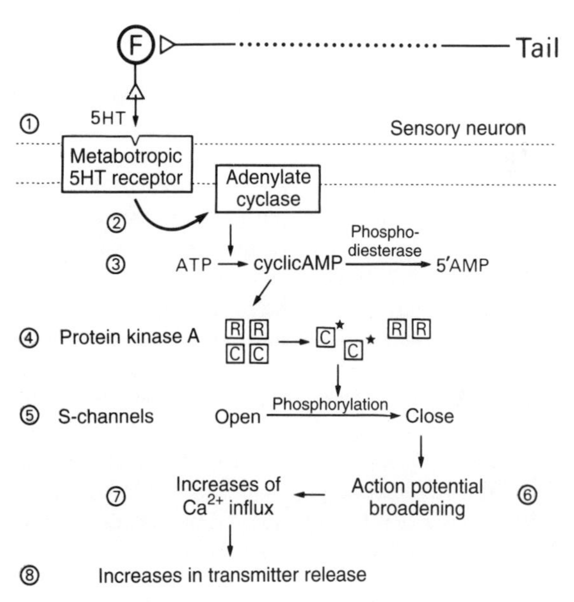

Fig. 8.6. The molecular cascade involved in short-term sensitization. A strong stimulus applied to the *Aplysia* tail activates facilitatory interneurones (F) which release 5HT. The binding of 5HT to its metabotropic receptor in the siphon sensory neurone (step 1) results in activation of adenylate cyclase (step 2), which catalyses the synthesis of cyclic AMP from ATP (step 3). Cyclic AMP binds to regulatory subunits (R) of protein kinase A, thereby releasing active catalytic subunits (C*, step 4). Phosphorylation of the S-channel causes the channel to close (step 5). This broadens the action potential (step 6) and increases Ca^{2+} influx (step 7). Consequently, the amount of transmitter released from the sensory neurone increases (step 8).

with the motor neurones. The synaptic response became habituated when the sensory neurone was stimulated repeatedly and sensitized when 5HT was applied (Rayport and Schacher 1986). A single 5 min application of 5HT produced a few minutes of sensitization, whereas five repetitions at 15 min intervals induced sensitization lasting for more than 24 h (Montarolo *et al.* 1986). Similarly, with a 15 min application of cyclic AMP sensitization lasted just minutes, but it exceeded 24 h after a 2 hr application (Schacher *et al.* 1988; Scholz and Byrne 1988). These parallels suggest that both short-and long-term forms of sensitization share the same intracellular second messenger system. They differ, however, in that long-term sensitization clearly outlasts the application period of cyclic AMP. In other words, although long-term sensitization is triggered by cyclic AMP elevation, it can keep on going without it.

A distinct difference in the molecular mechanisms of short- and long-term forms of sensitization was demonstrated by inhibitors of protein synthesis or RNA synthesis (Montarolo *et al.* 1986). For example, when a reversible inhibitor of protein synthesis, anisomycin, was applied to cultured *Aplysia* neurones, protein synthesis in the neurones was inhibited by over 95 per cent; it recovered to control levels within 3 h after removal of the inhibitor (Montarolo *et al.* 1986). Next, anisomycin was applied to the cultured neurones 1 h before 5HT and maintained for a 3 hr period. The inhibitor did not affect synaptic responses evoked in the gill motor neurone by stimulation of the sensory neurone in culture, but blocked long-term sensitization examined 24 h after multiple applications of 5HT (Montarolo *et al.* 1986). By contrast, short-term sensitization was not affected by the inhibitor (Schwartz *et al.* 1971; Montarolo *et al.* 1986). Long-term sensitization induced by a prolonged application of cyclic AMP to the cultured neurones was also blocked by incubation with anisomycin (Schacher *et al.* 1988). Clearly, then, long-term sensitization requires some proteins newly synthesized by the sensitization-triggering stimuli. The memory for long-term sensitization must be stored in these newly synthesized molecules.

Let us now turn to classical conditioning in *Aplysia*. Classical conditioning differs from non-associative learning in requiring two different stimuli, the conditioned (CS) and unconditioned (US) stimuli. In the gill-withdrawal reflex, activating the sensory neurone immediately before the facilitation interneurone establishes classical conditioning (Fig. 8.3). The rise of cyclic AMP in the sensory neurones can be enhanced significantly when the sensory neurone delivers action potentials just before 5HT is applied (Ocorr *et al.* 1985). In this process, the Ca^{2+} influx during excitation of the sensory neurones is crucial (Abrams 1985). Hence our principal concern is how the CS-evoked Ca^{2+} influx and the US-triggered activation of the 5HT receptor interact to raise cyclic AMP levels in the sensory neurones higher than either stimulus can alone (Fig. 8.7). We have seen that excess free Ca^{2+} in the intracellular milieu binds to a ubiquitous regulatory protein, calmodulin (CaM), which in turn mediates many cellular responses (Fig. 7.4). If Ca^{2+}-CaM (calcium-bound calmodulin) can activate adenylate cyclase, the acquisition of classical conditioning might be accounted for by convergence of the two signals elicited by CS and US on the same enzyme, adenylate cyclase. Eliot *et al.* (1989) actually found that the adenylate cyclase in the homogenate prepared from *Aplysia* ganglia can be stimulated by Ca^{2+}-CaM. Ca^{2+}-

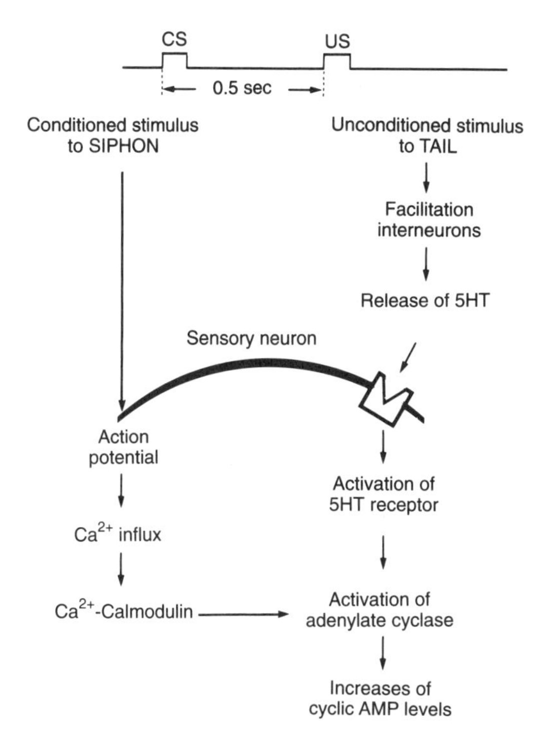

Fig. 8.7. A hypothetical mechanism for classical conditioning of the gill-withdrawal reflex in *Aplysia*. The conditioned stimulus (CS) is applied to the siphon 0.5 s before the unconditioned stimulus (US) to the tail. The US activates metabotropic 5HT receptors in the siphon sensory neurone via facilitation interneurones. This process stimulates adenylate cyclase in the sensory neurone. The CS initiates an action potential in the siphon sensory neurone, during which Ca^{2+} enters the sensory neurone and binds to calmodulin; Ca^{2+}-calmodulin activates adenylate cyclase. The signals evoked by the CS and US thus converge on the same intracellular second messenger.

CaM appears to activate the same cyclase as 5HT activates: the activation of adenylate cyclase by a non-hydrolyzable GTP analogue, GTP-γS (substituting for activation of metabotropic 5HT receptors), was further enhanced by adding Ca^{2+}-CaM (Abrams *et al.* 1991). The normal activity of adenylate cyclase was about doubled by Ca^{2+}-CaM alone, showed a 38-fold increase in response to GTP-γS, and 52–fold with simultaneously applied Ca^{2+}-CaM and GTP-γS. Clearly, adenylate cyclase in *Aplysia* sensory neurones can be dually regulated, and the actions of the two regulators are cooperative, being more than the sum of the two. Abrams *et al.* (1991) also showed that specific temporal pairing of the Ca^{2+} and 5HT is crucial for their cooperative effect. Thus, the two signals (Ca^{2+} and 5HT-receptor activation) elicited by the CS and US converge on the same enzyme (adenylate cyclase) to establish classical conditioning in *Aplysia* (Fig. 8.7).

The studies of learning in *Aplysia* provided three important outcomes. First, the concurrent observations of behavioural modification affirm that learning is indeed based on plastic changes at synapses. Second, the quantitative analysis of the time course of biochemical changes shows that the retention of short-term learning is determined by the life span of a particular intracellular second messenger. Third, the short-and long-term forms of learning previously classified on an empirical basis can now be defined in concrete terms, using the dependence on (or independence of) new protein synthesis as a criterion.

This third outcome clearly distinguishes between short-and long-term forms of learning in their mechanisms for memory storage. We have seen many examples in which second intracellular messengers are activated in association with learning or neuronal plasticity. How might proteins be newly synthesized by activation of the second messenger system? This is the subject of the next section.

8.2 GENE EXPRESSION ASSOCIATED WITH NEURONAL PLASTICITY

The involvement of new protein synthesis in learning was first suggested by behavioural studies. In mice, for example, intracerebral injections of puromycin, an inhibitor of protein synthesis, prevented the retention of avoidance learning acquired by training with electric shocks (Flexner *et al.* 1963; Barondes and Cohen 1966). The process, however—whether the inhibitor indeed blocked the synthesis of proteins induced by the learning process, or whether the inhibitor blocked some constitutive chemical process necessary to maintain learning—remained unclear (Flexner and Goodman 1975). Resolution came with the *in vitro* experiment on long-term sensitization of the gill-withdrawal reflex. The protein inhibitor blocked long-term learning if applied during, but not if applied afterwards, a 2 hour training period (Montarolo *et al.* 1986). Molecules newly synthesized during a critical time window in training therefore must be essential for the long-term retention of learning.

Protein synthesis involves two steps. The first is *transcription*. In this process, the genetic information coded by the nucleotide sequence of DNA is transferred to the RNA synthesized on DNA templates. This RNA is called messenger RNA (mRNA), because after transcription the RNA molecule moves from the nucleus to the cytoplasm to convey its genetic information to the ribosomes. The second step, *translation*, is so named because a protein is now synthesized in the ribosomes by translating the genetic code inherent in the mRNA into amino acids, the constituents of the protein. Long-term sensitization in *Aplysia* is blocked by inhibitors of transcription (actinomycin D, α-amanitin) as well as by inhibitors of translation (anisomycin and emetine; Montarolo *et al.* 1986). Therefore, the retention of long-term sensitization must require the synthesis of new mRNAs—that is, gene expression. For this to occur, some signal induced in the sensory neurones by the sensitizing stimuli must reach the cell nucleus and trigger DNA transcription, besides elevating cyclic AMP in the cytoplasm.

8.2.1 Gene expression by second messengers

Long-term sensitization in *Aplysia* follows repeated applications of 5HT (Montarolo *et al.* 1986) or prolonged applicaton of cyclic AMP to the sensory neurones (Schacher *et al.* 1988; Scholz and Byrne 1988). These results suggest that either cyclic AMP or cyclic AMP-dependent protein kinase (protein kinase A) is involved in the synthesis of the new proteins essential for the long-term maintenance of sensitization. Which of them is involved? We find an important clue in neuropeptide genes whose expression is enhanced by cyclic AMP (Goodman 1990).

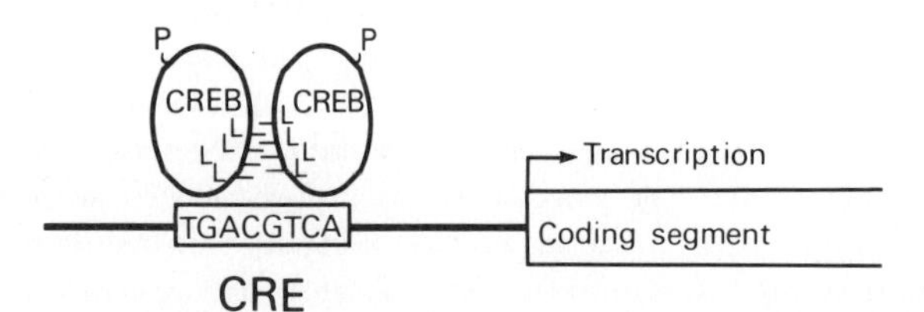

Fig. 8.8. A hypothetical diagram for the regulation of DNA transcription by cyclic AMP. The transcription initiation site of a gene (coding segment) is indicated by the angled arrow. The cyclic AMP-responsive element (CRE) is located upstream from the transcription initiation site as indicated by its consensus nucleotide sequence (TGACGTCA). Its corresponding sequence (ACTGCAGT) on the complementary DNA strand is omitted for simplicity. Specific nuclear CRE-binding (CREB) proteins are phosphorylated by cyclic AMP-dependent protein kinase and form a dimer, binding to the CRE on each DNA strand. Leucine residues (L) form the leucine zipper (short bars) that joins the dimer. The binding of CREB proteins to the CRE induces DNA transcription.

Figure 8.8 illustrates the regulation of the neuropeptide gene expression by cyclic AMP. Some neuropeptide genes, typified by the somatostatin gene, have a conserved nucleotide sequence, comprising eight base pairs (TGACGTCA/ACTGCAGT), upstream from the transcriptional initiation site. This region is termed the *cyclic AMP-responsive element* (CRE), because its presence in the somatostatin gene is essential for transcriptional regulation by cyclic AMP (Montminy *et al.* 1986). Does a particular protein present in the nucleus (nuclear protein) bind to the CRE to trigger DNA transcription of the neuropeptide gene? Yes; Montminy and Bilezikjian (1987) actually purified the protein that binds specifically to the CRE in the somatostatin gene. This *CRE-binding* (CREB) *protein* is phosphorylated when incubated with protein kinase A. The CREB protein cDNA was cloned, and its primary structure near the carboxyl terminus showed four leucines spaced seven amino acid residues apart (Hoeffler *et al.* 1988; Gonzalez *et al.* 1989). This heptad repetition of leucine residues is a molecular motif characteristic of other nuclear proteins involved in gene transcription (section 8.2.2). Since each α-helical turn of a protein is made of 3.6 amino acids (section 2.2.2), the leucine residues located at every seventh position would align on one face at every other turn. Landschultz *et al.* (1988) proposed that this motif, dubbed the *leucine zipper*, interlocks the two molecules in dimerization (Fig. 8.8). In short, the CREB protein is present but inactive in unstimulated nuclei; adding cyclic AMP activates protein kinase A, which phosphorylates CREB proteins; they in turn bind to the CRE of DNA as a dimer and thereby initiate gene transcription (Fig. 8.8; Yamamoto *et al.* 1988; Dwarki *et al.* 1990; Sheng and Greenberg 1990).

This mechanism revealed in the neuropeptide gene was immediately adapted to explain the retention of long-term sensitization in *Aplysia*. Extracts from *Aplysia* sensory neurones proved to contain proteins that bind specifically to the CRE of neuropeptide genes (Dash *et al.* 1990). Might a CREB-like protein in *Aplysia* sensory neurones serve to enhance nuclear transcription? This possibility was tested by injecting double-stranded oligonucleotides (a short stretch of nucleotides) with the CRE sequence into the nucleus of the *Aplysia* sensory neurone. These sensory neurones showed short-term but not long-term sensitization in response to 5HT (Dash *et al.* 1990). By contrast, the sensory neurones showed both forms of sensitization after 5HT when the injected double-stranded oligonucleotides had a sequence similar but not identical to the CRE sequence. These results indicate that the exogenously applied CRE nucleotides compete with the CRE in DNA for the nuclear transcriptional factor, which is identical or very similar to the CREB protein.

In the light of this evidence, the role of protein kinase A activated by cyclic AMP in sensitization of the sensory neurones appears to be two-fold: in the cytoplasm, the protein kinase phosphorylates the S-channel protein, thereby closing the channel; in the nucleus, the protein kinase phosphorylates the nuclear protein, thereby inducing DNA transcription. The former role is involved in the expression of sensitization and is common to both short-and long-term forms, whereas the latter role subserves its long-term retention. Presumably, the switch from short-term to long-term occurs only when the protein kinase A (its catalytic subunit; see below in relation to Fig. 8.6, step 4) is active enough to reach the nucleus (see Emptage and Carew 1993). In other words, gene expression responsible for long-term learning is triggered only when the level of protein kinase activity exceeds a certain threshold.

The general scheme postulated for gene expression induced in long-term sensitization seems reasonable. Yet this scheme omits how gene expression and the resultant new protein synthesis sustain sensitization. What is the action of the newly synthesized molecules responsible for long-term sensitization? Since the protein inhibitors block long-term sensitization induced by cyclic AMP, the molecules in question must interact with some process downstream from the increase in cyclic AMP levels: either activation of protein kinase A or phosphorylation of the S-channel by the kinase. The experimental results described below suggest that the former is the case.

Protein kinase A is composed of two regulatory (R) subunits and two catalytic (C) subunits, the latter's catalytic action being inhibited by the R-subunits in this complex (see Fig. 8.6, step 4). Binding cyclic AMP to the R-subunits releases the C-subunits, an effect referred to as activation of the kinase (Smith *et al.* 1981). If the ratio of the R- to C-subunits somehow decreased, more free C-subunits would be available, hence the kinase would remain active even in the absence of cyclic AMP. To assess this possibility, the R-and C-subunits of protein kinase A isolated from *Aplysia* sensory neurones were quantified separately (Greenberg *et al.* 1987; Bergold *et al.* 1990). Amazingly, a 35 per cent reduction of the R-subunits was observed 24 h after a 2 hr exposure of the sensory neurones to 5HT, whereas the C-subunits remained unchanged. Moreover, this reduction was blocked by anisomycin, although in normal sensory neurones the protein inhibitor had no effect on the amount of R-subunits. Interestingly, 5HT did not decrease the R-subunits of protein kinase A prepared from *Aplysia* muscle, so the regulation of R-subunits by 5HT is specific to neuronal protein kinase A (Bergold *et al.* 1990). Because a 5 min application of 5HT to sensory neurones did not affect the total R-subunits, this mechanism apparently serves long-term rather than short-term sensitization (Greenberg *et al.* 1987; Bergold *et al.* 1990).

These results imply that one type of molecule synthesized in association with long-term sensitization is a specific protease whose activity leads to the degradation of R-subunits, thereby sustaining activation of the protein kinase. On the other hand, as we will discuss in section 9.2.2, long-term sensitization also entails morphological alterations at the sensory-gill motor synapse, implying that additional molecular changes occur. Evidently, the R-subunit depletion is only one of many molecular changes involved in the long-term retention of sensitization.

Barzilai *et al.* (1989) attempted to identify the nature of the proteins newly synthesized in *Aplysia* sensory neurones following prolonged exposure to 5HT. The proteins extracted from the sensory neurones were separated by two-dimensional gel electrophoresis in terms of their size and charge. Newly synthesized proteins were identified by radioactively labelled amino acids which the tissues had incorporated. The changes in protein synthesis induced by 5HT showed three features. First, the changes occurred rapidly, within 15 to 30 min after 5HT application. Second, during this period the rate of synthesis increased for several proteins but decreased for others. Third, some proteins that increased subsided within a few hours, while others kept increasing. Clearly, many proteins are regulated in different ways for their synthesis in association with long-term sensitization. Moreover, a sequential cascade of gene expression must occur, since the nature and number of proteins synthesized change progressively after 5HT stimulation. Might the expression of one gene modify another gene expression? This question leads us to an unexpected linkage between neurobiology and *proto-oncogenes*, the genes related to cancer.

8.2.2 Proto-oncogenes in the nervous system

Until recently, few neurobiologists would have looked to the study of tumour formation for insights into neuronal functions (Hanley 1988). A tumour or cancer characterized by uncontrolled cell proliferation is initiated by changes of the gene or its expression programme. This genetic transformation can be triggered by chemical carcinogens, high-energy radiation or viruses. When the transformation of cultured normal cells by a tumour virus was examined under controlled conditions, specific viral genes were found to induce the cancerous transformation of the host cell (Martin 1970; Bishop 1982; Weinberg 1983). They were termed *oncogenes* (tumour-producing genes) or *viral oncogenes*. Surprisingly, genes homologous to the viral oncogenes were also found in normal host cells (Stehelin *et al.* 1976). It turned out that the oncogenes carried by retroviruses are acquired from the animals in which the viruses replicate. Thus, the cellular counterparts of viral oncogenes are the original form from which the viral copy is made. Hence, their name *proto-oncogenes* (Box I).

The viral oncogene version of the cellular proto-oncogene has a slight alteration that transforms the cell to a cancerous state. Weinberg (1983) remarks: 'Proto-oncogenes would not have been kept almost unchanged over 600 million years unless they were and continue to be indispensable.' What would be the biological functions of the proteins proto-oncogenes encode in normal cells?

So far, about 60 proto-oncogenes have been identified in a variety of species. Although their precise roles are still not clear, their molecular functions mainly appear to involve cell signalling pathways. The protein products encoded by proto-oncogenes can be divided into at least four classes (Bishop 1985): (1) growth factors, (2) the receptors for growth factors and hormones, (3) G-proteins for signal transduction and (4) nuclear proteins for regulation of DNA transcription. Clearly, all these proteins are indispensable for animal life. Among them, it is the fourth class of proto-oncogene products–nuclear transcriptional proteins—that may account for a sequential cascade of gene expression manifested in neuronal long-term plasticity. We now focus on this class of proto-oncogene products.

Two proto-oncogenes, *fos* and *jun*, are the best characterized families that encode nuclear proteins (Hanley 1988; Sheng and Greenberg 1990; Morgan and Curran 1991). These cellular oncogenes are termed c-*fos* and c-*jun*, respectively, to distinguish them from their viral oncogenes, v-*fos* and v-*jun*. The protein Fos encoded by c-*fos* is nearly undetectable in resting cells, but increases rapidly and transiently after stimulation with serum factors, growth factors or other chemical agents. Its prompt induction immediately after stimulation places c-*fos* in a special set of genes termed the *immediate early genes* (Greenberg and Ziff 1984; Sheng and Greenberg 1990; Morgan and Curran 1991).

The concept of immediate early genes emerged from the study of DNA transcription in growth factor-stimulated cells (Cochran *et al.* 1983; Lau and Nathans 1985). When the cell converts from a resting state to a proliferating state in response to growth factors, many proteins are newly synthesized along with or due to cell growth. Are all these proteins synthesized simultaneously in response to the stimulus? It is more likely

Box I **Viral oncogenes and cellular proto-oncogenes**

Infection of chick fibroblasts by the Rous sarcoma (*src*) virus, a chick retrovirus, leads to cancerous transformation of the host cell within 24 hours (Martin 1970). In the host cell the single-stranded RNA of the virus genome is copied into a single strand of DNA by reverse transcriptase and generates a double-stranded DNA by adding a complementary DNA strand. This virus-derived DNA integrates with the host chromosome, transforming the host cell to a cancerous state. The viral genes responsible for cancerous transformation are called viral oncogenes. Thus, *src* is a viral oncogene. Yet Stehelin *et al.* (1976) found that the DNA of normal cells contains nucleotide sequences closely related to the viral *src* gene. To distinguish it from the viral *src* gene (v-*src*), the normal counterpart is called the cellular *src* gene (c-*src*). Clearly, the c-*src* found in normal cells is not of viral origin, since the c-*src* gene comprises non-coding sequences (introns) interspersed in the coding domains (exons; Takeya and Hanafusa 1983). This intron–exon configuration is typical of all animal cell genes. It follows that retrovirus oncogenes are copies of cellular genes, which are thus called proto-oncogenes. In short, retrovirus oncogenes are distorted versions of cellular genes in normal cells. The viral oncogene is not an exact copy of its cellular counterpart, being somewhat crippled. It is from this slight modification that the cancerous transformation erupts.

that certain genes are expressed first and dictate the subsequent gene expression necessary for cell growth. To address this possibility, a growth factor was applied to the cell in the presence of a translational inhibitor of protein synthesis (cycloheximide). Within 20 min several genes were expressed, including c-*fos* (Lau and Nathans 1985). Because the protein synthesis inhibitor was present, the expression of these genes must have been induced directly by the growth factor, not by other newly synthesized proteins. Assuming that cell proliferation is triggered by a sequential cascade of gene expression, the immediate early gene could be its dictator. Perhaps, as Goelet *et al.* (1986) suggested, the learning process induces the expression of immediate early genes that trigger the expression of other genes for retention of long-term memory. If so, precisely how might the expression of immediate early genes regulate the transcription of other genes?

The number of immediate early genes now recognized is close to one hundred (Sheng and Greenberg 1990), but few are also characterized as being c-*fos*. The c-*fos* has at least two regulatory sites upstream from its transcription initiation site (Fig. 8.9A); one is CRE, as seen earlier in neuropeptide genes, and the other is the *serum-response element* (SRE; Treisman 1985). The SRE is so named because this element mediates the gene expression induced by serum factors and other growth factors. Although detailed mechanisms for the gene transcription mediated by these regulatory sites are still unclear (Graham and Gilman 1991), the important point is that the c-*fos* product can act as a transcription factor for other genes. As illustrated in Fig. 8.9B, a heterodimer made of the protein product of c-*fos* (Fos) and the c-*jun* protein product (Jun) specifically binds to a sequence element of DNA, its *activator protein 1* (AP-1) binding site (Sheng and Greenberg 1990; Morgan and Curran 1991). The AP-1 binding site has a motif of seven conserved base pairs (TGACTCA/ACTGAGT), which is similar but not identical to CRE. Fos and Jun resemble the CREB protein in having five leucines spaced seven amino acid residues apart near their carboxyl terminus; this conforms to the leucine zipper and presumably yokes the heterodimer of Fos and Jun. Thus, the mechanism for regulating the expression of other genes by Fos and Jun is quite similar to that for neuropeptide genes (Fig. 8.8). As noted above, the expression of immediate early genes is short-lived. Why might this be? The transcription of c-*fos* induced by serum factors is suppressed by its own protein product, Fos (Sassone-Corsi *et al.* 1988). Such autoregulation may account for

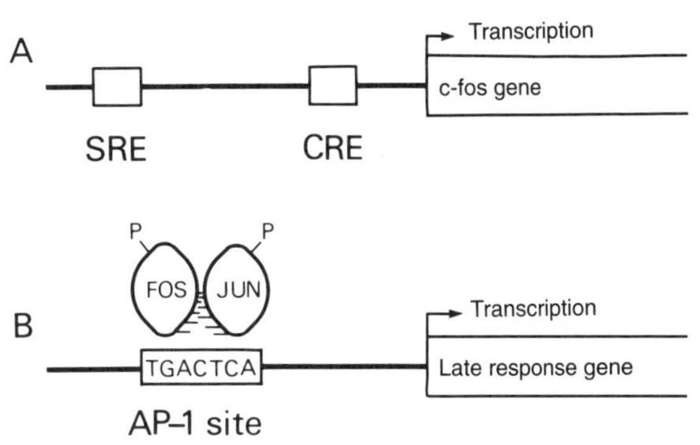

Fig. 8.9. A hypothetical illustration of the expression of a late response gene induced by a protein product (Fos) of an immediate early gene (c-*fos*). A, regulation of the c-*fos* gene expression: gene transcription can be induced via the serum-response element (SRE) or the cyclic AMP-responsive element (CRE). B, transcription of a late response gene by the protein products (Fos and Jun) of the immediate early genes (c-*fos* and c-*jun*). The two proteins form a heterodimer with a leucine zipper (short bars) when they are phosphorylated and bind to the regulatory site (AP-1) upstream from the transcription initiation site. The AP-1 site has a consensus nucleotide sequence (TGACTCA) and induces transcription of the gene when it binds to the Fos-Jun dimer.

the transient expression typical of immediate early genes.

Immunohistochemistry showed that applying noxious stimuli to the rat hindlimb produces Fos proteins in dorsal horn neurones of the lumbar spinal cord within 2 hours, but not in sensory neurones of the dorsal root ganglion (Hunt *et al.* 1987). That is, synaptic transmission seems to induce rapid expression of c-*fos* in certain spinal neurones. High-frequency stimulation of the presynaptic fibres increases mRNA levels of an immediate early gene (*zif*/268) in rat hippocampal granular cells, and this gene expression is blocked by NMDA-receptor antagonists (Cole *et al.* 1989; Wisden *et al.* 1990). In the hippocampus, inhibitors of protein synthesis blocked the late phase of long-term potentiation without affecting its early short-term phase (Stanton and Sarvey 1984; Frey *et al.* 1988, 1993; Otani *et al.* 1989). These results imply that the expression of immediate early genes could link the initiation of some neuronal response to its long-term retention even in the vertebrate.

The sequential induction of 'late genes' by the protein products of immediate early genes is an attractive hypothesis for molecular mechanisms underlying the retention of learning. This hypothesis remains untestable, however, without more information about gene regulation by the immediate early gene products. We need details of at least two aspects. First, how many and which immediate early genes are induced by the learning process? We have seen that CREB proteins, Fos and Jun, act in transcription control. But the leucine zipper motif common to these transcriptional factors is only one of a dozen distinct DNA-binding motifs identified thus far (He and Rosefeld 1991). Until the immediate early genes relevant to long-term plastic changes are identified, the current hypothesis is difficult to test. Second, which 'late genes' are the targets for the protein products of the immediate early genes? Many examples of the expression of c-*fos* or other immediate early genes by neuronal activation have been presented recently (Morgan and Curran 1991). However, the consequence of such gene expression is not yet identified, so the hypothesis remains tenuous, although compelling.

A similar cascade of gene transcription can also be seen in the immune process, which displays both plasticity and memory. Because of these features, the immune process is often considered analogous to a prolonged neuronal response induced by some stimulus. Stimulation of T lymphocytes by a specific antigen, for example, results in the production of interleukin-2 (IL-2), a growth factor for lymphocytes, thereby proliferating the cells involved in the immune response. Specifically, T cell stimulation induces the expression of c-*fos* and c-*jun*, whose protein products (Fos and Jun) in turn bind to the AP-1 binding site of the IL-2 gene and induce IL-2 synthesis (Jain *et al.* 1992a,b). Thus, we know which immediate early genes are induced by antigen stimulation and which gene is targeted by Fos and Jun in T lymphocytes. This gives us a good model for the molecular mechanisms of long-term learning.

So far we have discussed biophysical substrates involved in synaptic plasticity and the molecular basis for the acquisition and retention of learning. Historically, plastic changes of synaptic function were thought to result from morphological alterations at the synapse (section 6.1). This explanation seemed reasonable, and in the absence of biophysical and molecular views of neuronal function it had no competition. But long-standing assumptions provide no reason to rule out the possibility of morphological changes in functional synaptic plasticity. The next chapter takes up the current view of morphological plasticity at synapses.

Summary and prospects

Sensitization of the gill-withdrawal reflex in *Aplysia* is a behaviourally defined non-associative learning. The neuronal circuitry involved in this reflex is delineated at the cellular level, and the synaptic locus at which sensitization takes place is identified. Sensitization has short- and long-term forms. The stimuli that induce the two forms of sensitization are qualitatively identical and different only quantitatively. What causes sensitization to switch from the short- to long-term form? Molecules newly synthesized during the training session are essential for the long-term retention of learning. Apparently, prolonged or stronger sensitizing stimulation can induce a signal which reaches the nucleus and triggers the transcription of new genes encoding the molecules crucial for the long-term storage of neuronal information.

Long-term sensitization is not induced by the synthesis of a single molecule, however. It requires a sequential cascade of gene expression. This cascade of gene expression is dictated by a special set of genes, the immediate early genes. Which immediate early genes are first epxressed by training stimuli? Which 'late genes' are targeted by protein products of the immediate early gene? Might different forms of long-term learning require the expression of their own specific immediate early genes? These questions remain to be explored.

Suggested reading

Plasticity of the gill-withdrawal reflex in Aplysia

Kandel, E. R. (1976). *Cellular basis of behaviour*. Freeman, San Francisco.

Kandel, E. R., Brunelli, M., Byrne, J., and Castellucci, V. (1976). A common presynaptic locus for the synaptic changes underlying short-term habituation and sensitization of the gill-withdrawal reflex in *Aplysia. Cold Spring Harbor Symposia on Quantitative Biology*, **40**, 465–82.

Kandel, E. R. (1979). *Behavioral biology of Aplysia*. Freeman, San Francisco.

The molecular basis of sensitization

Hawkins, R. D. and Kandel, E. R. (1990). Hippocampal LTP and synaptic plasticity in *Aplysia:* possible relationship of associative cellular mechanisms. *Seminars in Neurosciences*, **2**, 391–401.

Kandel, E. R. and Schwartz, J. H. (1982). Molecular biology of learning: Modulation of transmitter release. *Science*, **218**, 433–43.

Kandel, E. R., Klein, M., Hochner, B., Shuster, M., Siegelbaum, S. A., Hawkins, R. D., Glanzman, D. L., Castellucci, V. F., and Abrams, T. (1987). Synaptic modulation and learning: New insights into synaptic transmission from the study of behaviour. In *Synaptic function*, (ed. G. M. Edelman, W. E. Gall, and W. M. Cowan), pp. 471–518. John Wiley, New York.

Proto-oncogenes in long-term neuronal function

Dash, P. K., Hochner, B., and Kandel, E. R. (1990). Injection of the cAMP-responsive element into the nucleus of *Aplysia* sensory neurones blocks long-term facilitation. *Nature*, 345, 718–21.

Goelet, P., Castellucci, V. F., Schacher, S., and Kandel, E. R. (1986). The long and the short of long-term memory—a molecular framework. *Nature*, 332, 419–22.

Sheng, M. and Greenberg, M. E. (1990). The regulation and function of c-*fos* and other immediate early genes in the nervous system. *Neuron*, 4, 477–485.

Wisden, W., Errington, M. L., Williams, S., Dunnett, S. B., Walters, Hitchcock, D., Evan, G., Bliss, T. V. P., and Hunt, S. P. (1990). Differential expression of immediate early genes in the hippocampus and spinal cord. *Neuron*, 4, 603–14.

Morgan, J. I. and Curran, T. (1991). Stimulus-transcription coupling in the nervous system: Involvement of the inducible proto-oncogenes *fos* and *jun*. *Annual Review of Neuroscience*, 14, 421–51.

9

Morphological plasticity at synapses

As early as 1900 Schäfer speculated that the extent of synaptic contacts may vary from time to time as their ramified processes expand or retract. For over 80 years his speculation had remained untested, until the continual 'remodelling' of normal synaptic contacts was proven directly by repeated observation of a synapse over several months in a living animal (Purves *et al.* 1987). Far from being structurally static, the synapse can show dynamic changes under normal and experimental conditions. A plausible sequence, then, is that the structure of a synapse can change over time and that this morphological plasticity in turn contributes to plastic alterations at the synaptic function.

9.1 HOW NERVE FIBRES SPROUT

The morphology of synaptic connections is modified by sprouting: changes of the outgrowth in the most distal end of the nerve axon. Barker and Ip (1966) observed that in normal muscle motor nerve fibres occasionally sprout at three sites, as illustrated in Fig. 9.1: the node of Ranvier (nodal sprout, *ns*), the preterminal region beyond the last myelin segment (preterminal sprout, *ps*), and the nerve terminals (ultraterminal sprout, *us*). As synaptic sites fall vacant due to the degeneration of presynaptic fibres, they may be taken over by axonal sprouts from neighbouring intact nerve fibres. This replenishment occurs both in the central nervous system (Cotman *et al.* 1981) and in the periphery (Brown *et al.* 1981). What signal or stimulus prompts such outgrowth of nerve terminals? What is the functional impact of terminal sprouting? These are the key questions in unlocking the mechanism and significance of morphological plasticity at synapses. We will begin by examining how sprouting is induced at neuromuscular junctions, then go on to the central nervous system and, finally, the functional correlates of sprouting.

9.1.1 Sprouting of motor nerve terminals

It is well known that sprouting can be induced at neuromuscular junctions by sectioning a part of the muscle nerve (partial denervation). Under such conditions, some motor nerve terminals will degenerate, while other neuromuscular junctions remain occupied by the intact motor nerve terminals. Hoffman (1950)

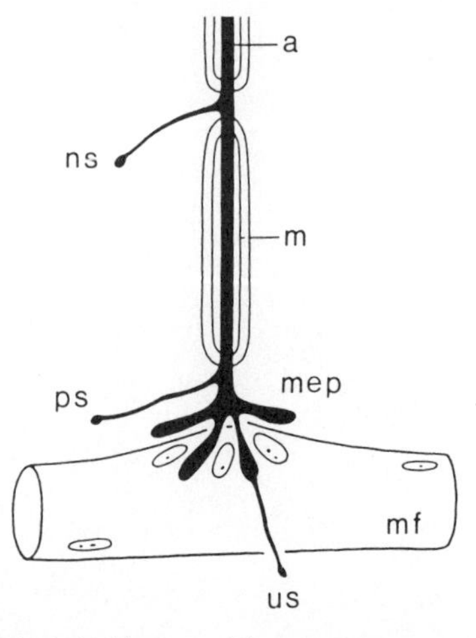

Fig. 9.1. Schmatic diagram of three positions at which sprouts may arise from a motor nerve fibre (a) innervating muscle fibres (mf). Nodal sprout (ns) arises from the node of Ranvier. Preterminal sprout (ps) arises from the motor axon just distal to the last myelin (m) segment. Ultraterminal sprout (us) arises from the motor nerve terminal at the motor end-plate (mep). (From Barker and Ip 1966.)

was the first to describe sprouting from the spared motor nerve terminals of partially denervated muscles in the rat hindlimb. He found that some of the terminal outgrowths formed end-plates on neighbouring denervated muscle fibres (Hoffman 1950; Edds 1953). Later, Brown and Holland (1979) found that chronic stimulation by electrodes implanted in the partially denervated muscle suppresses the formation of terminal sprouts. By implication, the stimulus for sprouting appears to derive from the denervated muscle fibres, as a result of their inactivity. Indeed, terminal sprouting can occur when muscle inactivity is produced by chronic conduction block of the muscle nerve even though the motor nerve fibres are undamaged (Brown *et al.* 1981). Figure 9.2 compares motor nerve terminals in the rat hindlimb muscle on the control, normal side (A) and on the experimental side (B) in which the muscle nerve had been blocked for 10 days with tetrodotoxin (Tsujimoto and Kuno 1988). Each motor nerve terminal normally has a round or oval configuration about 20 μm in diameter (Fig. 9.2A), whereas on the nerve-blocked side most motor nerve terminals (86 per cent) display one or more ultraterminal sprouts that extend beyond the original end-plate zone (Fig. 9.2B; Brown *et al.* 1981; Tsujimoto and Kuno 1988). Again, direct stimulation of the nerve-blocked muscle inhibits the formation of terminal sprouts (Brown *et al.* 1981).

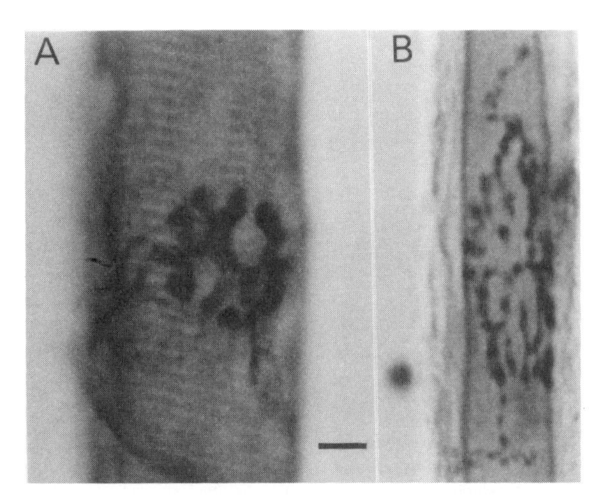

Fig. 9.2. Motor nerve terminals on single muscle fibres dissected from the rat hindlimb on the control side (A) and nerve-blocked side (B). The sciatic nerve on one side had been blocked for 10 days by tetrodotoxin applied locally at the mid-thigh level. A, note the capillary vessel crossing the muscle fibre from the lower right corner and extending toward the terminal. B, one sprout emerges from the upper middle edge of the terminal and another sprout with a bifurcation from the lower middle edge. Bar, 10μm. (From Tsujimoto and Kuno 1988.)

Betz *et al.* (1980) partially paralyzed the rat muscle by applying tetrodotoxin to one of the motor nerve branches. Sprouting ensued both in active (non-paralyzed) motor nerve terminals and in terminals on paralyzed muscle fibres. In other words, eliminating activity in muscle fibres can produce sprouts in nerve terminals on active muscle fibres distant from the paralyzed muscle fibres. How far can the sprouting influence travel from paralyzed muscle fibres? In partially denervated muscles, sprouting was limited to intact nerve terminals located within 200 μm of denervated muscle fibres (Pockett and Slack 1982). Then, it could be anticipated that a substance could emanate from inactive (paralyzed or denervated) muscle fibres. To test this possibility, Henderson *et al.* (1983) applied muscle extracts to spinal neurones in cultures prepared from chick embryos. The ensuing neurite outgrowth of cultured spinal neurones was markedly facilitated. Moreover, this neurite-promoting activity increased 15–fold when the extracts were prepared from chronically denervated muscles (Henderson *et al.* 1983; Rassendren *et al.* 1992). These results imply that skeletal muscle contains a sprouting factor for motor nerve terminals, and that the synthesis or effectiveness of the sprouting factor is enhanced by chronic elimination of muscle activity.

The presence of a sprouting factor in target tissues is not a recent concept. Cajal (1929) noticed two features during the development of epithelial sensory innervation in neonatal animals: first, extensive ramification (sprouting) of nerve fibres occurs only after they reach the epithelium; second, the epithelial target field is innervated almost uniformly, rather than segregated into vast neurite-less spaces and neurite-rich patches. Cajal (1929) suggested that the target epithelium emanates growth-promoting influences (e.g., a sprouting factor), which are neutralized by substances released from the innervating nerve fibres. Thus, virgin epithelium continues to stimulate the neurite growth, which stops when the target field is innervated, resulting in uniform sensory innervation over the area. This notion was elaborated further in an experiment on sensory innervation of salamander skin. Diamond *et al.* (1976) applied colchicine to one of the three nerves in the salamander hindlimb to block its axoplasmic transport. This caused the cutaneous territory of the treated nerve to be innervated by sprouts arising from the two neighbouring intact nerves. Diamond *et al.* (1976) inferred that colchicine had blocked an antisprouting factor nor-

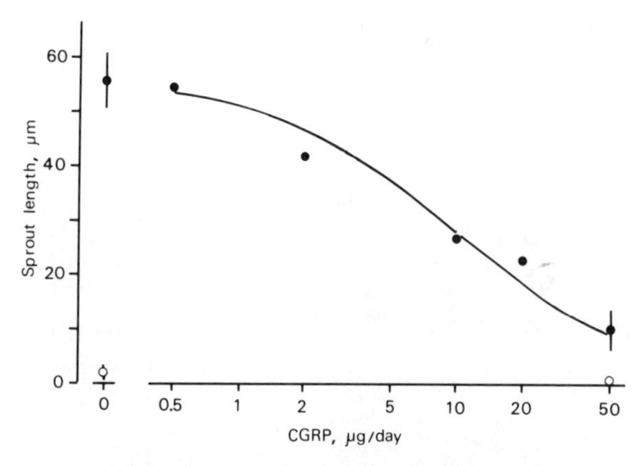

Fig. 9.3. Suppression of disuse-induced sprouting of rat motor nerve terminals by calcitonin gene-related peptide (CGRP). CGRP was injected subcutatneously once a day during a 10 day tetrodotoxin blockade of the sciatic nerve conduction on one side. Ordinate, mean total length of sprouts that emerged from each terminal in disused muscles (filled circles). Abscissa, the amount of CGRP injected daily. Open circles at 0 and 50 μg, sprout length measured on the control side. The points at 0 and 50 μg indicate mean and standard deviation (vertical lines); other points represent the results from one animal. The curve shows the relation expected from a 50 per cent inhibition dose of 10 μg/day. (From Tsujimoto and Kuno 1988.)

mally secreted from the nerve. Specifically, they postulated that nerve terminals normally contain an antisprouting agent, which is transported by axoplasmic flow from their cell bodies to neutralise the effect of a sprouting factor continually secreted from the target tissue.

No attempt was made to define the postulated antisprouting agent until recently, when calcitonin gene-related peptide (CGRP) was proposed to play this role. This candidate boasts two qualifications. First, CGRP is present in both motor and sensory neurones (Rosenfeld *et al.* 1983; Gibson *et al.* 1984) and is transported to their terminals (Takami *et al.* 1985; Matteoli *et al.* 1988; Kashihara *et al.* 1989). Second, as shown in Fig. 9.3, the sprouting of motor nerve terminals observed following chronic block of nerve–muscle activity is suppressed in a dose-dependent manner by daily treatment with CGRP (Tsujimoto and Kuno 1988). CGRP does not affect the morphology of normal motor nerve terminals in active muscles. Whether endogenous CGRP in nerve terminals indeed functions as an antisprouting factor has not been proven yet.

How would muscle inactivity prompt the motor nerve terminal to sprout? What is the endogenous sprouting factor? One participant in this process appears to be the *neural cell adhesion molecule* (*NCAM*). NCAM is a membrane glycoprotein found on the surfaces of neurones and muscle cells, presumably mediating intercellular adhesive interactions. In adult skeletal muscle, NCAM is confined to the neuromuscular junction, but following denervation or paralysis, NCAM synthesis is markedly enhanced, with NCAM being expressed over the entire surface of the muscle fibres (Covault and Sanes 1985; Rieger *et al.* 1985). Booth *et al.* (1990) found that sprouting of motor nerve terminals induced by chronic paralysis of the mouse muscle is suppressed by daily injections of anti-NCAM antibodies. It is likely that the NCAM expressed on the muscle surface by paralysis is essential for the formation of nerve terminal sprouts. The antibodies might have bound to NCAM in the nerve terminals, however. This possibility remains open to examination.

Both *in vivo* and *in vitro*, the growth of neurites is facilitated by adhesion to the surface of the pathway (Letourneau 1983). For example, cultured dissociated neurones grow preferentially along pathways coated with adhesive substances (Letourneau 1975). Intercellular adhesive interaction presumably occurs when two NCAM molecules bind at their extracellular N-termini (*homophilic binding*; Box J). If NCAM on the surface of nerve terminals binds to NCAM newly expressed on the surface of muscle fibres by chronic paralysis, this adhesive interaction between nerve terminals and muscle would then guide the outgrowth of sprouts.

Whereas NCAM is expressed in both nerve and muscle cells, L1, a related glycoprotein (Box J), is localized in neurones and glia (Faissner *et al.* 1984; Keilhauer and Schachner 1985). By applying L1 antibodies, Fischer *et al.* (1986) both reduced axonal *fasciculation* (the formation of bundles) and accelerated the neurite outgrowth of mouse cerebellar neurones in culture. Presumably, the nerve axons exert mutually adhesive forces, forming fasciulation by binding of the cell surface glycoprotein, L1. The presence of L1 antibodies would reduce these forces, producing *defasciculation*. Similarly, L1 antibodies would enhance the neurite growth rate because the reduced mutual adherence among neurites frees them to interact with the pathway substrate.

In chick embryos, L1 antibodies likewise also produce defasciculation of the nerve in hindlimb muscle, multiplying the side branches of the intramuscular nerve (Landmesser *et al.* 1988). It appears that while axon-axon adhesive interactions are mediated by L1 and NCAM, axon-muscle cell interactions are mediated by NCAM but not L1 during the innervation of embryonic

Box J NCAM and cell–cell interactions

During development nerve cells identify and make connections with their own target cells. Specific cell–cell recognition and interaction are thus essential for the developmental patterning of neuronal projections. Edelman (1976) proposed that this cell–cell interaction entails the binding of molecules expressed in cell surfaces. Indeed, immunological assays isolated several cell adhesion molecules (CAM), including NCAM (Edelman 1983), a transmembrane glycoprotein with a long extracellular amino terminus and a short cytosolic carboxyl terminus (Fig. 9.5A). Intercellular adhesive interaction is assumed to be due to *homophilic binding* (binding between identical molecules from apposing cells) of NCAMs at their extracellular regions (Fig. 9.5A). NCAM has many carbohydrate attachment sites, and long chains of polysialic acid bind to NCAM. Polysialic acid is a key modulator of NCAM function, lessening the homophilic binding force as polysialic acid levels rise (Fig. 9.5).

Molecular cloning shows that NCAM has five extracellular domains, each containing a pair of cysteine residues about 60 amino acids apart, which form a disulphide bond (S–S bond; see Figure). These five contiguous segments are homologous to those in members of the immunoglobulin superfamily. Several neural adhesion molecules with the immunoglobulin domain have now been identified (Walsh and Doherty 1991; Rathjen and Jessell 1991). The Figure shows one of them, L1, for comparison with NCAM.

NCAM and L1 represent neural adhesion molecules whose adhesive function does not require Ca^{2+}. A separate class of neural adhesion molecules, termed *cadherins*, functions in a Ca^{2+}-dependent manner (Ranscht 1991; Takeichi 1991).

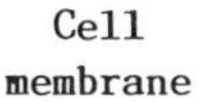

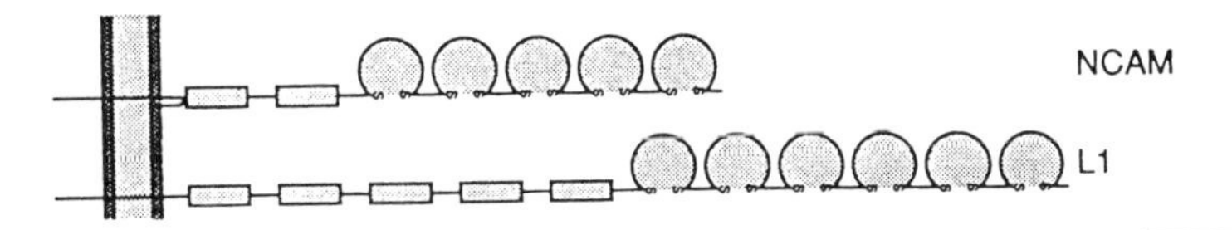

muscle (Landmesser *et al.* 1988). Interestingly, chronic paralysis of chick embryo muscle with curare resulted in defasciculation and side branch proliferation of the intramuscular nerve (Fig. 9.4B), compared with the control muscle (Fig. 9.4A; Dahm and Landmesser 1988; Landmesser *et al.* 1990). Such collateral sprouting to form side branches is analogous to the nodal sprouting illustrated in Fig. 9.1. How might collateral sprouting of intramuscular nerve arise from activity blockade of the embryonic muscle? The answer involves the modulation of surface NCAM by a local carbohydrate, sialic acid, as discussed below.

NCAM has several carbohydrate (sugar) attachment sites, mostly occupied by sialic acid, and removing the sialic acid enhances the homophilic binding of NCAMs. Edelman (1983) attributed this effect to electrostatic repulsion, assuming the negative charge of their sialic acid regions impedes the binding of each pair of NCAMs (Fig. 9.5A). Alternatively, having long chains of polysialic acid packed between the NCAMs may simply prevent close NCAM apposition (Fig. 9.5Ba; Rutishauser 1991; Walsh and Doherty 1991). Thus, polysialic acid may modulate NCAM function. This question was examined in the chick embryo by cleaving polysialic acid with a specific enzyme, endoneuraminidase. Now, blockade of muscle activity failed to induce side branch proliferation of the intramuscular nerve beyond control levels (Fig. 9.4C). According to Landmesser *et al.* (1990), developmental changes in the innervation pattern in embryonic muscle correlate with the amount of polysialic acid, rather than with levels of L1 or NCAM. Indeed, removal of polysialic acid by endoneuraminidase during the period of motor innervation in chick embryos resulted in a reduction of intramuscular nerve branching (Tang and Landmesser 1993). Similarly, defasciculation and increased branching during muscle paralysis are associated with elevated axonal polysialic acid. Clearly, NCAM function can be regulated by polysialic acid levels. We do not yet know how axonal polysialic acid levels depend on muscle activity;

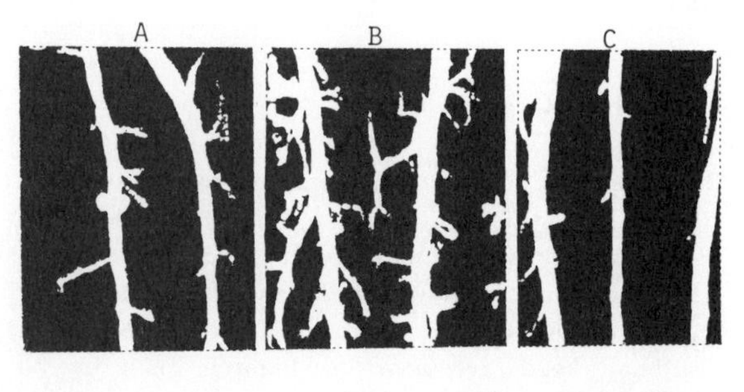

Fig. 9.4. Altered branching patterns of the intramuscular nerve in chick embryos. A, control muscle. B, muscle chronically paralyzed with curare. C, as in B, but also treated with endoneuraminidase N to cleave polysialic acid on NCAMs. (Adapted from Landmesser *et al.* 1990.)

nor whether CGRP, the proposed antisprouting factor, reverses axonal defasciculation induced by curare in embryonic muscle.

The intramuscular nerve branching pattern determines the distribution of ACh receptor clusters in muscles. During activity blockade the additional intramuscular nerve branching should mean more ACh receptor clusters per muscle cell (Dahm and Landmesser 1988). Although this prediction was confirmed with curare in chick embryos, it did not extend to treatment with nicotine or decamethonium, two agents that block muscle activity as much as curare does (Oppenheim *et al.* 1989). This is a puzzling contradiction. A related question is whether ultraterminal sprouting (Fig. 9.2) and axonal defasciculation (Fig. 9.4) result from the mechanical inactivity of the blocked muscle or from its electrical inactivity (the blockade of action potentials). In the cardiac parasympathetic ganglion of the frog, chronic blockade of the presynaptic vagal nerve with tetrodotoxin does not induce sprouting (Roper and Ko 1978). Perhaps, alterations of synaptic morphology induced by blockade of muscle activity result from mechanical or metabolic inactivity—a mechanism for regulating morphology that neuronal synapses may lack. Even in the cardiac parasympathetic ganglion, however, section of one vagal nerve (partial denervation) prompts the spared vagal nerve to sprout (Roper and Ko 1978). We will now turn to sprouting at neuronal synapses, particularly in the central nervous system.

9.1.2 Sprouting in the central nervous system

Although severed axons fail to regenerate in the central nervous system (Cajal 1928), functional loss from

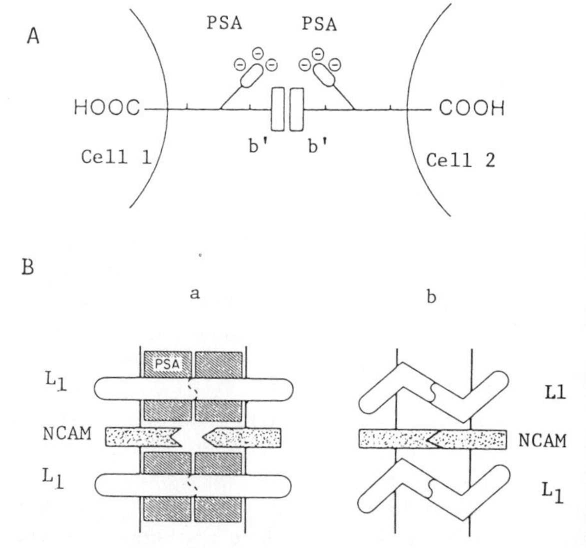

Fig. 9.5. Adhesive interactions between apposing cells by the binding of NCAMs. A, the membranes of cells 1 and 2 incorporate NCAM, an integral membrane glycoprotein with a long extracellular amino terminus and a short cytoplasmic carboxyl (COOH) terminus. Both NCAMs bind together at their extracellular region (b′). The negative charge of polysialic acid (PSA) modulates the homophilic binding of NCAMs, reducing it by repulsion. (From Edelman 1983.) B, changes in the binding force of NCAM and L1 in high (a) and low (b) PSA levels. Vertical lines indicate the cell membranes of two apposing cells. (From Rutishauser 1991.)

lesions of the brain or spinal cord can be considerably compensated for (Kennard 1942). By analogy with sprouting in partially denervated muscle, the functional compensation following injury to the central nervous system was assumed to rely on collateral sprouting. In the first demonstration of sprouting in the central nervous system, Liu and Chambers (1958) severed some dorsal roots on one side in the cat; axonal processes then emerged from the intraspinal branches of the adjacent, spared dorsal root. Liu and Chambers (1958) suggested that spinal synaptic connections lost by *partial deafferentation* (partial elimination of the input pathways) are gradually replaced as sprouts develop from intact intraspinal fibres. Might these sprouts indeed form functional synaptic connections on central neurones?

Morphological evidence of synaptic formation by the presynaptic fibres remaining after partial deafferentation comes from quantitative electron microscopic studies of the rat septal nuclei (Raisman 1977). Some input to the septum originating in the hippocampus traverses the ipsilateral and contralateral fimbriae. After section of

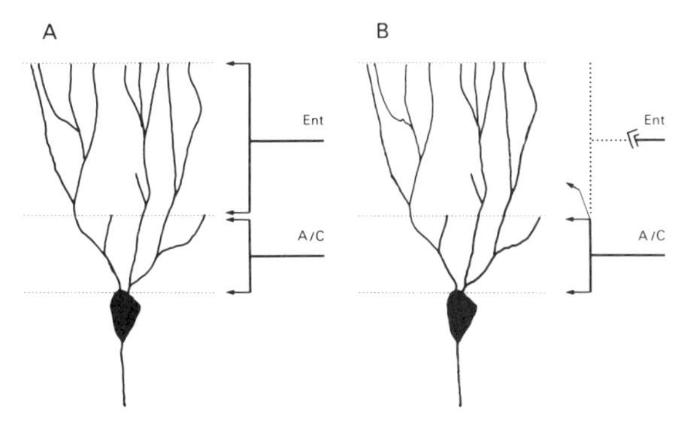

Fig. 9.6. Schematic diagrams of the distribution of synapses on the dendrites of a granule cell in the hippocampal dentate gyrus, (A) normal and (B) after ablation of the ipsilateral entorhinal cortex. A, fibres from the entorhinal (Ent) cortex form synaptic contacts on the distal two-thirds of the granule cell dendrites, whereas the association and commissural (A/C) fibres terminate on the proximal one-third. B, after degeneration of the fibres from the ipsilateral entorhinal cortex (dotted lines), sprouts from the A/C fibres also make synaptic contacts more distally.

one fimbria, many degenerating terminals appear in the dendritic synapses of septal neurones. Gradually, these degenerating terminals are replaced by non-degenerating terminals. It is not that the cut fimbrial fibres regenerate themselves across the lesion site. Rather, the newly recruited terminals arise from the intact presynaptic fibres remaining in the deafferentated region. Virtually every vacated synaptic site is filled by new terminals about 40 days after section of one fimbria (Raisman 1977).

Cotman and Lynch (1976) studied sprouting in granule cells of the hippocampal dentate gyrus following partial deafferentation. The presynaptic input to granule cells has a laminar organization: The perforant pathway from the entorhinal cortex terminates on the outer (distal) two-thirds of the dendrites of granule cells, whereas both commissural and associational fibres from the CA4 pyramidal cells terminate on the inner (proximal) one-third (Fig. 9.6A). Ablation of the ipsilateral entorhinal cortex results in massive degeneration of terminals in the outer dendritic synapses of granule cells, but replacement with new terminals soon follows. The new synapses are formed by sprouts from the contralateral entorhinal cortical fibres and the commissural and associational fibres (CA4 axons). Consequently, the CA4 axons expand their terminations into the distal segment of the granule cell dendrites, well beyond their original synaptic territory (Fig. 9.6B). Consistent with this morphological remapping of synaptic territory is the electrical activity of granule cells evoked by stimulating the commissural and associational fibres: extracellular recording found signals in regions extending more distally along the dendrites than is normal (Cotman and Lynch 1976). These results suggest that the fresh terminals on distal dendrites of granule cells function as excitatory synapses.

The most thoroughly studied on sprouting in the mammalian central nervous system is that of red nucleus neurones in adult cats (Tsukahara 1981, 1987). Red nucleus (RN) neurones receive direct (monosynaptic) excitatory synaptic inputs from the ipsilateral sensorimotor (SM) cortex (corticorubral pathway) and from the contralateral interpositus (IP) nucleus of the cerebellum (cerebellorubral pathway; Fig. 9.7A). Therefore, when intracellular potentials are recorded from an RN neurone, monosynaptic e.p.s.p.s can be produced by stimulating the SM cortex or the IP nucleus. But these e.p.s.p.s differ in shape, with a relatively long time-to-peak (slow rising phase) for those of SM cortex origin versus a shorter time-to-peak for those elicited from the IP nucleus (Fig. 9.7A). These different e.p.s.p. time courses reflect the separate synaptic locus of each input pathway. Because the axons from the IP nucleus terminate predominantly on the RN neurone cell body containing the electrode, these fast rising e.p.s.p.s are recorded faithfully. By contrast, the corticorubral fibres terminate preferentially on the dendrites distant from the RN neurone cell body; hence the e.p.s.p.s elicited at the dendritic synaptic sites are subject to distortion when recorded from the cell body, owing to electrical cable properties (membrane capacitance, membrane resistance, and cytoplasmic resistance) along the dendrites. In particular, the longer the distance between the synaptic site and the cell body (the recording site), the more prolonged is the rising phase of the e.p.s.p. (see Tsukahara *et al.* 1975 for details). Partial deafferentation of RN neurones was induced by electrolytic lesions of the contralateral IP nucleus. Within 10 days, the e.p.s.p.s evoked in RN neurones by SM cortex stimulation showed distinct changes in their shape: an early peak now preceded the original slow peak (Fig. 9.7B). Thus, the time-to-peak of the corticorubral e.p.s.p. is significantly shortened by lesions of the IP nucleus (Fig. 9.7C,D). Tsukahara *et al.* (1975) suggested that sprouts from the corticorubral fibres after partial deafferentation make synaptic contacts mainly on the RN neurone cell body (Fig. 9.7B), as electron microscopic studies indeed confirm (Nakamura *et al.* 1974; Murakami *et al.* 1982).

These studies of lesion-induced sprouting clearly in-

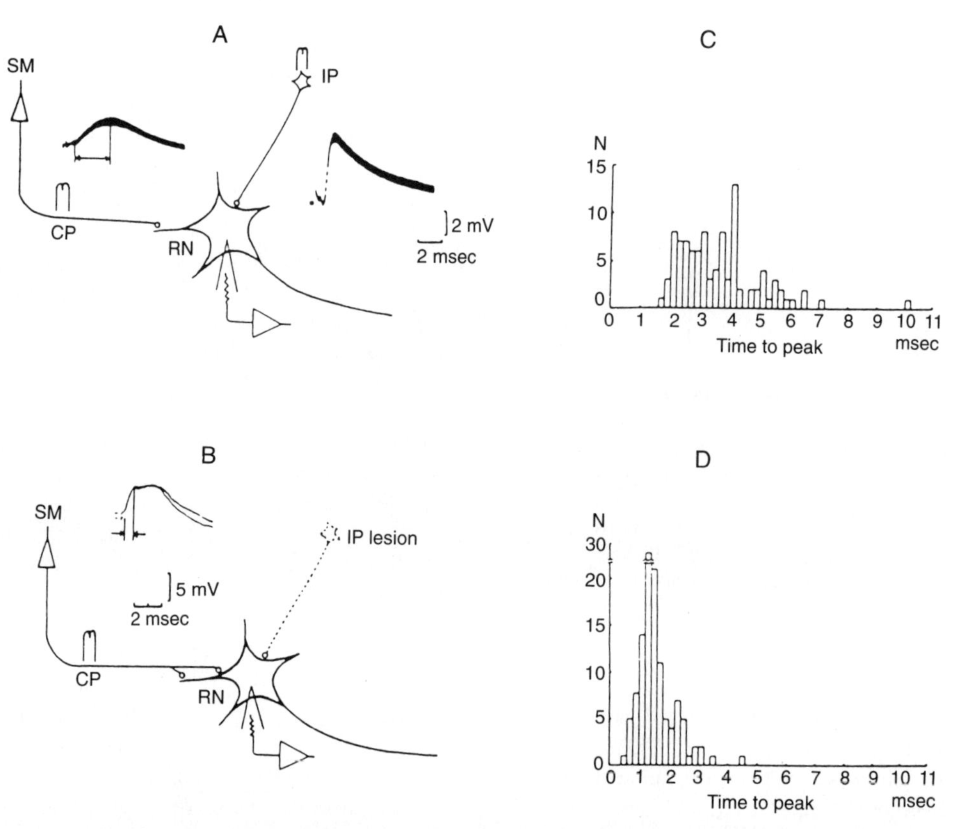

Fig. 9.7. Lesion-induced sprouting in red nucleus (RN) neurones. A, schematic diagram of excitatory inputs from the sensorimotor (SM) cortex and the interpositus (IP) nucleus on RN neurones. Stimulation of the SM cortex or the cerebral peduncle (CP) produces slowly rising e.p.s.p.s (inset), whereas the e.p.s.p.s evoked by stimulation of the IP have a short time-to-peak (inset). B, similar to A, but after ablation of the IP. The e.p.s.p.s evoked by CP stimulation now show an early peak (indicated by arrows) besides the slow peak. C, distribution of the time-to-peak of the e.p.s.p.s evoked by stimulation of the SM or CP in RN neurones in normal cats. The measurements are indicated by the horizontal arrow in A. D, similar to C, but after ablation of the IP. (From Tsukahara 1987.)

dicate that sprouting induced in the central nervous system can form functional synapses. They tell little, however, about the signal or mechanism that initiates sprouting. For example, is it essential that degenerating products be present? Are the vacancies at synaptic sites some kind of stimulus? Does functional loss trigger sprouting from intact fibres? The possibilities are numerous. Let us examine a few.

We have seen that sprouting of motor nerve terminals at neuromuscular junctions can be induced by activity blockade that does not involve damaging the nerve fibres (section 9.1.1). Is it possible to induce axonal sprouting in the central nervous system without degeneration? Yes. This was shown in an experiment on RN neurones to explore the central mechanism in functional compensation of movement disorders (Tsukahara 1987). Movement disorders were induced experimentally by surgical cross-innervation between the two groups of muscles having opposite actions in the cat forelimb: the nerves to the flexor muscle group were cut and sutured to the extensor muscle group for re-innervation, and the nerves to the extensor muscle group were sutured similarly to the flexor muscle group. After one month there was a disproportionately large extension of the elbow during walking, but locomotion was near normal five months later. It was postulated that the control of flexor and extensor spinal motor neurones had been readjusted at the level of RN neurones, which project to them through the rubrospinal pathway. The functional compensation of locomotion was associated with a faster time-to-peak of the monosynaptic e.p.s.p.s evoked in RN neurones by SM cortex stimulation, as observed after lesions of the IP nucleus (see Fig. 9.7D). Again, electron microscopic observations confirmed that the corticorubral fibres terminate on both the RN cell body and its dendrites in the cat (Murakami *et al.* 1984; Tsukahara 1987). The results were taken to imply that the functional compensation of forced movement disorders is achieved by switching the predominant control on RN neurones from the cerebellum (IP nucleus) to the cerebral SM cortex (Tsukahara 1987). Whatever the interpretation, these animals showed no degeneration in the central nervous system; hence, the presence of degenerating products or vacant central synaptic sites can be ruled out as prerequisite for sprouting in the central nervous system.

Tsukahara (1987) and his colleagues went on to investigate the relationship between learning and morphological changes at central synapses. They used a classical conditioning paradigm to evoke the elbow flexion response in the cat: a conditioned stimulus (CS) of electrical shock applied to the SM cortex (corticorubral fibres) and an unconditioned stimulus (US) of electrical shock applied to the forelimb skin. The CS preceded the US by 100 ms, and about 100 trials of the CS–US pairing constituted the daily training. The CS alone did not evoke elbow flexion in the first 4 days of training, but conditioning was acquired by the seventh day (Fig. 9.8). Reversing the CS–US sequence or applying the CS alone extinguished the conditioning within 5 days (Fig. 9.8). Interestingly, this classical conditioning of the elbow flexion response was again accompanied by a faster time-to-peak of the corticorubral e.p.s.p.s in RN neurones and by termination of the corticorubral fibres on the RN neurone cell body under electron microscopy (Tsukahara 1987; Murakami *et al.* 1988).

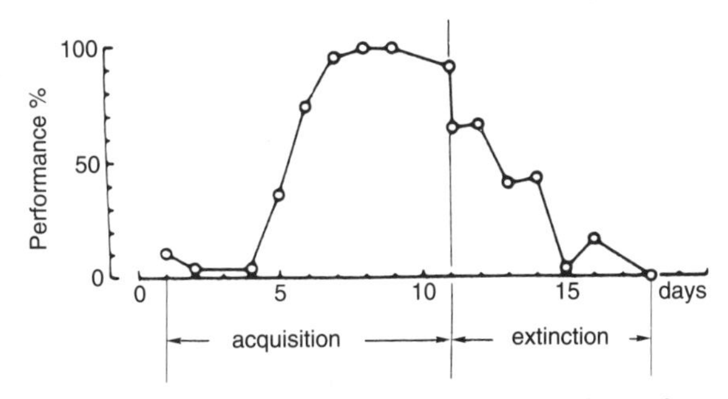

Fig. 9.8. The acquisition and extinction of conditioned responses mediated by the red nucleus. Ordinate, the ratio of the elbow flexion evoked by the conditioned stimulus alone in 24 trials. Abscissa, period since start of training. After day 11, the conditioned-unconditioned stimulus sequence was reversed. (From Tsukahara 1987.)

It is reasonable to assume that the new distribution of the corticorubral fibres on the RN cell bodies is the morphological correlate of the shortened time-to-peak of the corticorubral e.p.s.p.s. Considering the septum and the dentate gyrus sprouting that follow partial deafferentation, the corticorubral fibres almost certainly form new synapses on the RN neurone cell body following lesions of the IP nucleus, which are manifested electrophysiologically by changes in the e.p.s.p. shape. How this bears on the behavioural compensation of movement disorders or behavioural conditioning must be interpreted with caution, however. No causal link between alterations in the corticorubral synapses and the behavioural modification has been substantiated. More quantitative studies are clearly needed to correlate learning with morphological plasticity at synapses. For example, is the rapid extinction of the elbow flexion conditioning associated with the withdrawal of the corticorubral synapses on the RN neurone cell body? Would ablation of the corticorubral pathway prevent the acquisition of the elbow flexion conditioning? Such questions must be answered before sprouting of the corticorubral fibres can be interpreted in relation to learning.

Whether sprouting is responsible for behavioural modification still remains uncertain, despite the compelling observations on long-term morphological and electrophysiological alterations in the corticorubral synapses. In practice, mammalian behavioural modification is too complex for exploring linkages between functional and morphological plastic changes. We shall now return to less complex systems to see whether functional plasticity can indeed be based on morphological changes at the synapse.

9.2 MORPHOLOGICAL CORRELATES OF FUNCTIONAL SYNAPTIC PLASTICITY

At the outset, the search for the underlying mechanisms of synaptic plasticity was intuitively biased toward morphology. Eccles (1953), for instance, postulated that tetanic stimulation of presynaptic fibres causes their terminals to swell, thereby increasing transmitter release. In retrospect, there was no logical reason to think that terminal swelling promotes the quantal release of transmitter. Nevertheless, tetanic stimulation of the dorsal root has indeed been reported to enlarge presynaptic terminals on spinal motor neurones (Darinskii and Korneeva 1969; Illis 1969). These morphological studies received little attention, however, because of unreliable criteria for morphometric measurements and identification of the sensory terminals. Later, however, other attempts with a similar objective have achieved significant success as we shall see.

In principle, there are two approaches to examining the correlation between functional changes and morphological plasticity at synapses. One approach is to investigate morphological synaptic changes following a manipulation known to induce plastic changes in synaptic function. The second, and complementary, approach is to see what functional synaptic alterations accompany particular morphological changes induced by some manipulation. We have seen many examples of long-term synaptic plasticity following repetitive stimulation of presynaptic fibres, from peripheral synapses in the vertebrate sympathetic ganglion and the crayfish neuromuscular junction to central synapses in the mammalian hippocampus (see sections 7.1.2, 7.2.2). The question is, do these synapses show morphological changes in an activity-dependent manner? Also, are morphological changes at synapses associated with their functional alteration? When motor nerve terminals sprout following activity blockade of the neuromuscular junction, for instance, is neuromuscular transmission functionally altered? If so, are the functional and morphological changes related causally?

9.2.1 Activity-dependent morphological changes

The amount of transmitter released at a neuromuscular junction is normally proportional to the size (total length) of its nerve terminal (Kuno *et al.* 1971; Grinnell and Herrera 1981). Therefore, if terminal sprouts induced by disuse (chronic blockade of nerve–muscle activity) form functional connections with the muscle fibres, transmitter release at disused junctions might be enhanced by the added length of the outgrowths. In fact, Snider and Harris (1979) found that both the number of quanta of the transmitter released by a nerve impulse (mean quantum content) and the frequency of spontaneous m.e.p.p.s increase significantly after chronic block of the muscle nerve with tetrodotoxin in the rat. The formation of terminal sprouts at disused neuromuscular junctions might thus be responsible for enhanced transmitter release. But the first puzzle came from the effect of CGRP on disused neuromuscular junctions. As shown in Fig. 9.3, although CGRP suppresses disuse-induced sprouting of motor nerve terminals, transmitter release at the disused junction surprisingly stays enhanced (Tsujimoto and Kuno 1988). That is, terminal sprouting and increased transmitter release can be dissociated. The two processes also differ in their latency of expression: Terminal sprout formation requires at least 3 days of nerve block, whereas transmitter release (measured by the quantal analysis of e.p.p.s) increases within 24 hours of nerve block (Tsujimoto *et al.* 1990). Clearly, terminal sprouting is not a prerequisite to enhancement of transmitter release at disused neuromuscular junctions.

A further test was made to investigate whether sprouting causes additional enhancement of transmitter release. After muscle nerve blockade for 6 days, terminal sprouts were found in about 35 per cent of the disused junctions examined. Transmitter release was then evaluated electrophysiologically (as mean quantum content) at all disused junctions and their total terminal length measured. Although each nerve terminal was longer, on average, at junctions with sprouts than at those without sprouts, the transmitter release was enahanced at both

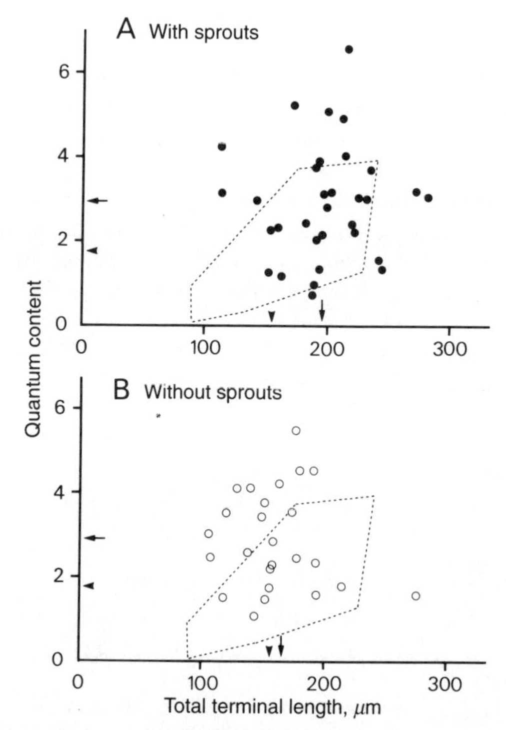

Fig. 9.9 Relationship between total terminal length and mean quantum content at individual neuromuscular junctions observed after 6 days of nerve block. A, terminals with sprouts. B, terminals without sprouts. Arrows, mean values. Arrowheads, mean values at normal junctions. Dotted lines surround the normal range of distribution. (From Tsujimoto *et al.* 1990.)

sites to the same extent (Fig. 9.9). The increased transmitter release and the sprouting at disused junctions thus appear to be two independent processes (Tsujimoto *et al.* 1990). The dotted-line boundaries in Fig. 9.9 show the normal range of quantum content and terminal length at neuromuscular junctions. Both parameters are positively correlated in the rat as in frog neuromuscular junctions (Kuno *et al.* 1971). By contrast, at the disused junctions no positive correlation existed regardless of whether or not the terminals had sprouts (Fig. 9.9). This lack of correlation again suggests that the two parameters are regulated independently during disuse.

Enhanced synaptic efficacy from inactivity of the presynaptic input contradicts current opinion about potentiation, namely that both short-and long-term potentiation require tetanic stimulation of the presynaptic fibres. It remains unclear how chronic blockade of nerve–muscle activity can enhance transmitter release. Interestingly, inactivity-induced synaptic enhancement is not unique to neuromuscular junctions, but also occurs at neuronal synapses. Blocking conduction of sensory nerve fibres with tetrodotoxin enhances the monosynaptic e.p.s.p.s evoked in mammalian spinal motor neurones by stimulating the disused sensory fibres (Gallego *et al.* 1979; Manabe *et al.* 1989; Webb and Cope 1992). Blocking the conduction of presynaptic fibres likewise increases the amplitude of e.p.s.p.s in the mammalian sympathetic ganglion (Gallego and Geijo 1987). Two characteristics of inactivity-induced synaptic enhancement stand out. First, its acquisition is *rapid*; synaptic enhancement occurs within one day of the conduction block of presynaptic fibres (Manabe *et al.* 1989; Tsujimoto *et al.* 1990). Second, it is *long term*; increased synaptic efficacy can last at least 3–4 days after restoration of presynaptic activity (Manabe *et al.* 1989; Tsujimoto *et al.* 1990).

In the sympathetic ganglion, conduction block of presynaptic fibres enhanced transmission even at the synapses formed by presynaptic branches that were central to the blocking site (Gallego and Geijo 1987). This finding is outlined in Fig. 9.10A, where activity blockade of the distal portion of presynaptic fibres (at TTX, between the stellate ganglion, SG, and the superior cervical ganglion, SCG) results in enhancement of transmission both at the synapses formed by the proximal, unblocked branches of the presynaptic fibres (site SG) and at those formed by the distal blocked branches (site SCG). Thus, the local inactivity of presynaptic terminals *per se* (site SCG in Fig. 9.10A) is not the primary factor in synaptic enhancement. Instead, synaptic enhancment must result from changes in the entire presynaptic neurone (Pre in Fig. 9.10A) following the elimination of impulse activity in its distal segment. In other words, the blockade of transmitter release at some of the terminals may initiate a retrograde signal to the presynaptic neurone cell body. It should be noted that in frog muscle, partial denervation increases transmitter release from the remaining motor nerve terminals before any terminal sprouts form (Grinnell 1988).

Figure 9.10B diagrams synaptic enhancement at spinal motor neurones (MN) following conduction block (at TTX) of the sensory fibres arising from the muscle. Each sensory cell body in the dorsal root ganglion (DRG in Fig. 9.10B) has its central process and peripheral process. Presumably, conduction block of the peripheral process causes some changes in the sensory cell body, which in turn result in enhancement at synapses on motor neurones (see also section 11.1.1). By analogy with long-term sensitization in *Aplysia* (section 8.1.2; Montarolo *et al.* 1986), long-term synaptic enhancement induced by inactivity may involve new protein synthesis in the presynaptic neuron. Whether accompanying morphological changes occur at the synapses on sympathetic ganglion cells or spinal motor neurones is not known.

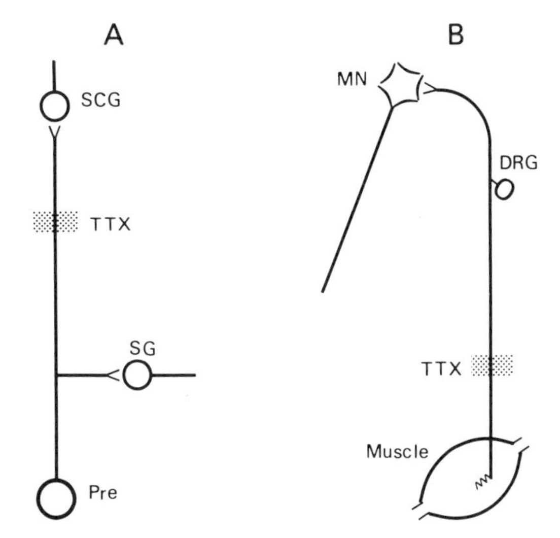

Fig. 9.10. Schematic diagram of synaptic enhancement induced by the blockade of presynaptic impulse. A, activity of preganglionic fibres (Pre) is blocked by the local application of tetrodotoxin (TTX) between the superior cervical ganglion (SCG) and the stellate ganglion (SG). B, activity of sensory fibres arising from muscle spindles is blocked by TTX. This procedure enhances synaptic potentials recorded from spinal motor neurones (MN) in response to stimulation of the sensory fibres. DRG, dorsal root ganglion.

Let us now turn to long-term plasticity at the crayfish neuromuscular junction, where transmitter release is enhanced for several hours following tetanic stimulation of the motor nerve (section 7.1.2; Wojtowicz and Atwood 1988). This plastic change, called long-term facilitation (LTF), involves activation of cyclic AMP-dependent protein kinase in the motor nerve terminal (Dixon and Atwood 1989a). Our primary question is whether tetanic stimulation of the crayfish motor nerve also produces morphological changes at its neuromuscular junction. Transmitter release at this junction increases markedly during tetanic stimulation (tetanic facilitation), prior to the LTF (Fig. 7.5). In electron microscpic studies, serially sectioned motor nerve terminals were examined before (control), during (tetanic facilitation phase) and after (LTF phase) tetanic stimulation (20 Hz for 10 min; Wojtowicz *et al.* 1989). The synaptic contact area remained unchanged, and only slight morphological changes were seen. Fewer synaptic vesicles were near the terminal membrane (within 50 nm, presumably representing releasable vesicles) and the presynaptic *dense bars* (equivalent to active zones in vertebrate synapses, presumably representing release sites) were shorter during the tetanic facilitation phase; both changed in the LTF phase (Wojtowicz *et al.* 1989). Thus, a minor structural modification of motor nerve terminals does occur soon after tetanic nerve stimulation, but its relevance to the LTF remains uncertain. For example, this structural modification could be a consequence of the enhanced transmitter release during tetanic facilitation. When Mearow and Govind (1989) stimulated the crayfish motor nerve at 20 Hz for several days, there were more synaptic vesicles and more dense bars in the motor nerve terminals. Again, is this structural modification responsible for long-term facilitation at this junction? We do not know. Because the LTF can be blocked by a protein kinase A inhibitor or by an adenylate cyclase inhibitor (Dixon and Atwood 1989a), an important test would be whether these inhibitors prevent the morphological modification of motor terminals induced by tetanic stimulation.

Recall the long-term potentiation (LTP) in the hippocampus (section 7.2.2). Are any morphological synaptic changes related to this functional plasticity? Morphometric measurements made on dendritic synapses in the rat CA1 region following tetanic activation of the Schaffer collateral–commissural pathway (see Fig. 7.12) found increased numbers of two particular types: *shaft synapses* and *sessile spine synapses*, classified according to their overall shape (Lee *et al.* 1980; Chang and Greenough 1984). These morphological changes induced by brief tetanus (6 s at 100 Hz) lasted for more than 8 hours, as did functional LTP (Chang and Greenough 1984). When the hippocampal slice was perfused with Ca^{2+}-free solution, the same tetanus failed to induce functional LTP as well as the morphological changes in dendritic synapses (Chang and Greenough 1984). This correlation between morphological and functional plastic changes was also observed when the frequency or the duration of tetanic stimulation was altered: with a long tetanus at low (10 min at 1 Hz) or high (10 min at 100 Hz) frequencies neither functional LTP nor the morphological changes occurred (Chang and Grenough 1984). These results support the assumption that the observed morphological changes in hippocampal synapses are responsible for the functional LTP. However, the general correlation is not sufficient to prove the assumption. To induce LTP in the hippocampal CA1 region requires activation of calmodulin kinase II (CaM kinase II; Malenka *et al.* 1989b; Malinow *et al.* 1989; see 7.2.2). The experimentally induced mutant mice that lack the expression of CaM kinase II show normal synaptic responses in the CA1 pyramidal cell but fail to develop LTP (Silva *et al.* 1992). It would be interesting to examine whether the mutant mice defective in CaM kinase K II also lack the expected morphological changes following tetanic presynaptic stimulation.

Admittedly, the direct evidence linking morphological plasticity to functional plasticity is scanty. Some of the difficulty is inherent in the question itself: it is harder to prove than to disprove a causal relationship between two events, for no matter how high we pile the correlations, a single counter example knocks them all down. Moreover, we simply lack much information about the mechanism underlying activity-dependent morphological changes. Morphological plasticity, such as the multiplication of synapses or the formation of sprouts, must be based on alterations in the gene expression or changes in protein synthesis. If the gene products involved in the morphological changes could be altered selectively, we could resolve the relation between the morphological and functional plastic changes.

Fortunately, we have a model to work with. The analysis of long-term sensitization of the gill-withdrawal reflex in *Aplysia* shows a remarkable correlation between cellular and behavioural modifications and provides molecular insights into its functional expression. This is a long-term plasticity that is associated with morphological changes. We can now ask, what is the molecular basis for such morphological plasticity?

And does the morphological plasticity account for the functional change?

9.2.2 Morphological correlates of sensitization in *Aplysia*

Both habituation and sensitization of the gill-withdrawal reflex in *Aplysia* occur at the same synaptic locus, the synapses formed by the sensory neurones arising from the siphon on the motor neurones innervating the gill (section 8.1.1). Bailey and Chen (1983, 1988a,b) injected horseradish peroxidase (HRP) into a single siphon sensory neurone to label its terminals, followed by electron and light microscopic studies of serial sections of the labelled synaptic terminals. The morphological parameters were defined by the total area of active zones where synaptic vesicles accumulate and by the number of varicosities (swellings with synaptic organelles being equivalent to synaptic terminals). Long-term behavioural change was induced by daily training of *Aplysia*: for habituation, 10 days of repeated applications of seawater jets to the siphon; for sensitization, 4 days of strong electrical stimuli applied to the neck. The total number of active zones, synaptic vesicles or variosities per sensory neurone decreased significantly in habituated animals and increased significantly in sensitized animals, compared with control animals (Bailey and Chen 1988a,b). These morphological changes are outlined in Fig. 9.11. The parallel shifts in the parameters suggest that each varicosity has about as many active zones as usual; it is the number of varicosities that increases or decreases. Since the active zone associated with synaptic vesicles is presumably the site of transmitter release, the observed morphological changes accord with functional plastic changes: more transmitter release in sensitization and less transmitter release in habituation (section 8.1.1). The numerical increase in active zones or varicosities after 4 days of sensitization training persisted for at least one week and had only partly reversed at the end of three weeks. Bailey and Chen (1989) inferred that the extra synaptic terminals (varicosities) contribute to the retention of long-term sensitization—which still leaves the question of what originates the lasting sensitization.

Long-term sensitization occurs rapidly probably before many varicosities have had time to form. Might the acquisition and retention of long-term sensitization therefore be separate processes, employing different mechanisms? Recall that during acquisition of long-term sensitization the regulatory (R) subunit of protein kinase A is reduced, which helps to maintain activation of the kinase (section 8.2.1). Greenberg *et al.* (1987) assumed that the maintenance of the kinase activity by R-subunit reduction is an interim molecular mechanism for enhancing transmitter release until the cell body has forwarded new gene products to its terminals. But the

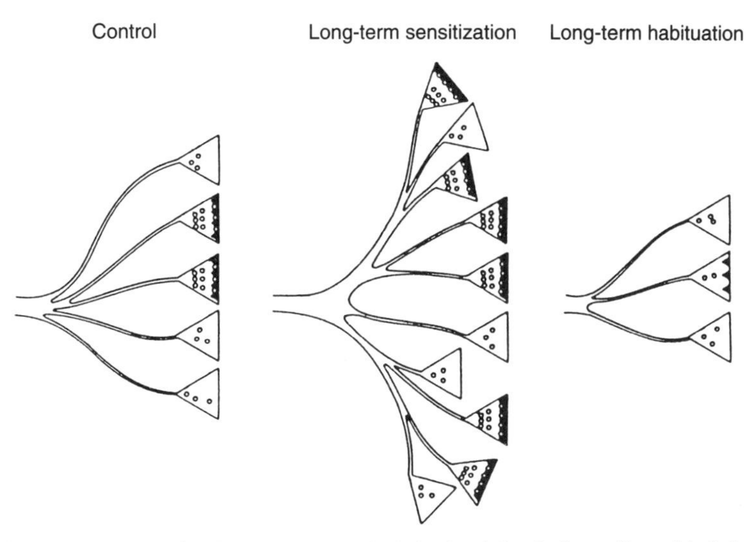

Fig. 9.11. Scheme of the morphology of siphon sensory terminals in the abdominal ganglion of *Aplysia*. Left, control. Middle, sensitized animals. Right, habituated animals. The number of sensory terminals and the number of active zones associated with synaptic vesicles increase in sensitized animals and decrease in habituated animals. (From Bailey and Chen 1988.)

decrease in R-subunits is blocked by anisomycin, an inhibitor of protein synthesis (Bergold *et al.* 1990). Hence the R-subunit reduction cannot be merely an interim mechanism operating only a few minutes or hours; it must ensue from gene expression, induced by some gene product (e.g., a specific protease; section 8.2.1). We have seen that the stimulus for long-term sensitization presumably induces the expression of immediate early genes, whose products then trigger the expression of other genes that support memory retention (section 8.2.2). The reduction of the R-subunits thus would be one among several processes subserving memory retention, and the multiplication of varicosities would be another. Barzilai *et al.* (1989) found that prolonged exposure of *Aplysia* sensory neurones to serotonin increases the rate of synthesis (*up-regulation*) for 10 proteins and decreases it (*down-regulation*) for five proteins. Might some of these changes in protein synthesis be related specifically to the morphological plastic change?

Mayford *et al.* (1992) unveiled an exciting fact: four of the five proteins down-regulated by the sensitizing stimulus are a group of cell surface proteins. These molecules are similar to the cell adhesion molecule (CAM), hence designated *Aplysia* CAMs or apCAMs. The structures of three apCAMs have been predicted from the deduced amino acid sequence from the cDNAs, as shown schematically in Fig. 9.12. Each apCAM contains five immunoglobulin (Ig) domains, characterized by cysteine residues 40–60 amino acids apart, on the extracellular N-terminus. The similarity in the molecular motif between the apCAM (Fig. 9.12) and the NCAM family (Box J) is striking. The apCAM might be involved in intercellular adhesive interactions, thereby regulating axonal fasciculation and the neurite growth rate (comparable to NCAM and L1 in section 7.1.1). The apCAM is expressed in both sensory neurones and motor neurones, but the serotonin-induced down-regulation of apCAMs occurs only in the sensory neurones. In cultured sensory neurones, the amount of the apCAM (measured by a monoclonal antibody) decreased by 20 per cent one hour after the exposure to serotonin (5HT; Mayford *et al.* 1992). Applying the apCAM antibody to cultured sensory neurones induced defasciculation, doubling the number of axonal branches within 3 days. The apCAM reduction occurred in two ways: down-regulation of the pre-existing apCAM on the surface of sensory neurones and down-regulation in the rate of apCAM synthesis. Applying 5HT accelerated the rate of internalization (endocytosis) and degradation of the apCAM on the surface of sensory neurones, cutting the density of the labelled apCAM on the cell surface by 50 per cent within one hour (Bailey *et al.* 1992). Again, the internalization of apCAMs facilitated by 5HT happened only in the sensory neurones, not in the motor neurones. Moreover, both the reduced rate of apCAM synthesis and the internalization of apCAMs by 5HT required new protein synthesis.

This remarkable finding strongly suggests that the down-regulation of apCAM is the basis of the morphological plasticity associated with long-term sensitization. Presumably, the increased number of varicosities in the sensory neurones results from the defasciculation and

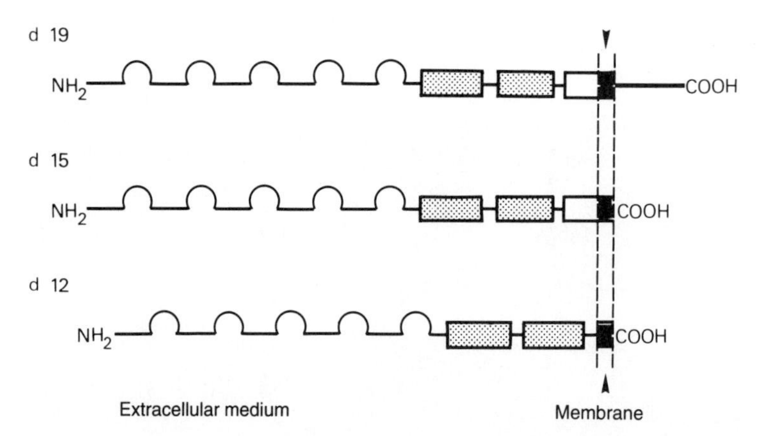

Fig. 9.12. Schematic molecular structures of three *Aplysia* cell adhesion molecules (apCAMs; d 19, d 15, d 12) predicted from their amino acid sequences. Each apCAM has five immunoglobulin domains (semicircles) at its extracellular amino terminus. The d 19 has a transmembrane segment and its carboxyl terminus is located in the cytoplasm, whereas d 15 and d 12 are anchored to the cell membrane by their carboxyl termini without the cytoplasmic domain. Stippled boxes, fibronectin domains. Open boxes, acidic amino acid-rich regions. (Adapted from Mayford *et al.* 1992.)

accelerated neurite growth that follow apCAM reduction. Quite likely, the expression of the apCAM gene in sensory neurones is rapidly and transiently suppressed by the product of some immediate early gene activated by 5HT. If we could block this pathway for apCAM's down-regulation, we could examine whether the 5HT-induced morphological plasticity of sensory neurones is required for functional sensitization to be retained.

An intriguing but puzzling feature of *Aplysia* sensory neurones is that their growth and development depend on the target cell, the motor neurones innervating the gill. Glanzman *et al.* (1990) examined the effect of 5HT on the growth of sensory neurones co-cultured with the gill motor neurone (L_7 cell). After 5 days of co-culture, 5HT was repeatedly applied for about 2 hours, and within 24 hours the sensory neurones had extended their processes significantly. Moreover, this growth was associated with a significant increase in the mean amplitude of e.p.s.p.s evoked in the motor neurone by stimulating individual sensory neurones. In essence, long-term sensitization of the gill-withdrawal reflex was replicated in culture conditions. Glanzman *et al.* (1990) then applied 5HT similarly to cultured sensory neurones in the absence of the target gill motor neurone L_7; surprisingly, the sensory neurones underwent no significant growth. This implies that the morphological change of sensory neurones associated with long-term sensitization requires interaction between the presynaptic and postsynaptic cells for its expression (Glanzman *et al.* 1990).

Does apCAM's down-regulation in sensory neurones by 5HT also require the presence of the target cell? No. The down-regulation or the internalization of apCAM by 5HT occurred equally well in cultured sensory neurones regardless of the presence or absence of the motorneurone (Bailey *et al.* 1992). Furthermore, the application of the apCAM antibody caused cultured sensory neurones to defasciculate and grow out in the absence of co-cultured motor neurones (Mayford *et al.* 1992). Why, then, did cultured sensory neurones fail to respond to 5HT in the absence of the motor neurone? The answer is not yet available. The story of varicosity formation by down-regulation of apCAMs still has a few missing links.

Target dependence also characterizes the morphological differentiation of cultured sensory neurones. This emerged when siphon sensory neurones dissociated from *Aplysia* were co-cultured with either their target gill motor neurone (L_7) or an inappropriate target motor neurone, L_{11} (Glanzman *et al.* 1989). The L_{11} motor neurone is comparable in size to the L_7 and is also located in the abdominal ganglion but never receives synaptic inputs from the siphon sensory neurones *in vivo*. The sensory neurones co-cultured with the appropriate target neurone exhibited elaborate synaptic configurations resembling those found *in vivo*, whereas the sensory neurones co-cultured with L_{11} had less differentiated varicosities (Glanzman *et al.* 1989). Evidently, the presynaptic neurones receive a specific influence from their target cell for morphological differentiation —an example of a neurotrophic effect. Neurotrophic effects can be exerted from a neurone on its target cells or vice versa. The development, differentiation, and maintenance of the normal maturity state in the nervous system all rely on neurotrophic influences. This *neurotrophism* is the subject of Part III.

Summary and prospects

Cajal (1911) postulated that long-term plasticity involves morphological changes in neurones. Even without *a priori* evidence, this proposition held intuitive appeal. Chronic blockade of nerve–muscle activity induces sprouting of motor nerve terminals and enhancement of transmitter release at the disused neuromuscular junctions. At first glance, the morphological plasticity seemed responsible for the functional plasticity, verifying Cajal's hypothesis. Unexpectedly, however, the detailed analyses proved that the two events are expressed independently by activity blockade and have no causal relationship. Tetanic stimulation of the presynaptic pathway induces long-term facilitation (LTF) or potentiation (LTP) at the crayfish neuromuscular junction or in the hippocampal CA1 pyramidal cell; the same stimulus produces morphological changes in these synapses, including an increase in the number of synapses or release sites. In view of the unexpected results in disused neuromuscular junctions, it would be premature to correlate the observed morphological changes with the functional LTF or LTP. In CA1 pyramidal cells, it is not even clear whether the mechanism of LTP is enhanced transmitter release, increased postsynaptic receptor sensitivity or both.

Partial deafferentation causes sprouting from the spared input fibres in both the central and peripheral nervous systems. The synapses newly formed by these sprouts are functional. The aberrant shape of e.p.s.p.s and the abnormally expanded region from which the e.p.s.p.s can be recorded on stimulation of the spared input fibres indicate that the synaptic responses are produced by sprouts from the spared fibres. Thus, the observed functional plasticity must be the consequence of the lesion-induced morphological change. How might partial deafferentation induce sprouting? Could vacated synaptic sites secrete some sprouting factors? The answers remain unknown.

Many guidance cues determine the neurite growth and the patterning of neuronal projections. One class of molecules involved in axon guidance is cell adhesion molecules (CAMs), including NCAMs. A reduction of the NCAMs involved in axon–axon adhesive interactions causes defasciculation. Increasing the NCAMs involved in axon–target adhesive interactions would facilitate the axonal growth. The homophilic binding of molecules of the NCAM family is regulated by sialic acid attached to the molecules. Up-regulation of axonal sialic acid reduces the binding of NCAMs, which would explain the proliferation of side branches that is observed in muscle nerves during activity blockade.

Long-term sensitization of the gill-withdrawal reflex in *Aplysia* is associated with an increase in varicosities formed by the siphon sensory neurones on the gill motor neurones. Interestingly, the stimulus for inducing this long-term sensitization caused down-regulation of *Aplysia* CAMs (apCAMs) in the sensory neurones. How might the down-regulation of apCAMs increase the number of varicosities? Do increased varicosities entirely account for the retention of long-term sensitization? These questions also remain.

Suggested reading

Sprouting at peripheral synapses

Brown, M. C., Holland, R. L., and Hopkins, W. G. (1981). Motor nerve sprouting. *Annual Review of Neuroscience*, 4, 17–42.

Roper, S. and Ko, C. P. (1978). Synaptic remodelling in the partially denervated parasympathetic ganglion in the heart of the frog. In *Neuronal plasticity*, (ed. C. W. Cotman), pp. 1–25. Raven Press, New York.

Tsujimoto, T., Umemiya, M., and Kuno, M. (1990). Terminal sprouting is not responsible for enhanced transmitter release at disused neuromuscular junctions of the rat. *Journal of Neurosicence*, **10**, 2059–65.

Sprouting in the central nervous system

Cotman, C. W. and Nieto-Sampedro, M. (1984). Cell biology of synaptic plasticity. *Science*, **225**, 1287–94.

Raisman, G. (1977). Formation of synapses in the adult rat after injury: similarities and differences between a peripheral and a central nervous site. *Philosophical Transactions of the Royal Society B*, **278**, 349–59.

Tsukahara, N. (1987). Cellular basis of classical conditioning mediated by the red nucleus in the cat. In *Synaptic function* (ed. G. M. Edleman, W. E. Gall, and W. M. Cowan), pp. 447–469. John Wiley, New York.

Neural cell adhesion molecules

Bailey, C. H., Chen, M., Keller, F., and Kandel, E. R. (1992). Serotonin-mediated endocytosis of apCAM: an early step of learning-related synaptic growth in *Aplysia*. *Science*, **256**, 645–9.

Covault, J., Merlie, J. P., Goridis, C., and Sanes, J. R. (1986). Molecular forms of N-CAM and its RNA in developing and denervated skeletal muscle. *Journal of Cell Biology*, **102**, 731–9.

Mayford, M., Brazilai, A., Keller, F., Schacher, S., and Kandel, E. R. (1992). Modulation of an NCAM-related adhesion molecule with long-term synaptic plasticity in *Aplysia*. *Science*, **256**, 638–44.

Rathjen, F. G. and Jessell, T. M. (1991). Glycoproteins that regulate the growth and guidance of vertebrate axons: domains and dynamics of the immunoglobulin/fibronectin type III subfamily. *Seminars in Neurosciences*, 3, 297–307.

Rutishauser, U. (1991). Pleiotropic biological effects of the neural cell adhesion molecule (NCAM). *Seminars in Neurosciences*, 3, 265–70.

Walsh, F. S. and Doherty, P. (1991). Structure and function of the gene for neural cell adhesion molecule. *Seminars in Neurosciences*, 3, 271–84.

Part III Neurotrophism

10

Historical perspective of neurotrophism

10.1 EARLY CONCEPTS OF NEUROTROPHIC FUNCTION

According to Garrison (1929), Magendie was the aversion of the anti-vivisectionists because of his cruel experiments on living animals. In 1824, before the discovery of surgical anaesthesia, Magendie exposed the cranial cavity of the rabbit and cut the trigeminal nerve on one side. Within a few days the cornea on the operated side turned opaque, showing inflammatory ulcerations, and similar changes marred the tongue on the operated side. Magendie attributed these tissue changes to the interruption of some nutritive neural influence. Later, confirming these observations, Samuel (1860) postulated a special group of nerve fibres, *trophischen Nerven* (trophic nerves), whose function is to provide *trophic* (nutritive) support to tissues. Where are these trophic nerve fibres? Heidenhain (1878) proposed the sympathetic nerve.

In Schäfer's *Text-book of physiology* (1900), Sherrington admitted that all the experimental results point to the existence of a nerve-dependent trophic influence. But he noted that every nerve has a trophic influence on its tissue: 'All nerves are trophic nerves.' For Sherrington, normal nerve conduction and the nutritive influence were an inseparable unit of neural function. Hence, the trophic influence would be exerted in the same direction as the nerve conduction, for example, from motor neurone to muscle fibre, but not in reverse: 'Trophic influence in a direction contrary to the ordinary functional relationship and connection, is very obscure.' Yet, even Sherrington recognized an exception in herpes zoster, where inflammation of the dorsal root ganglion exerts a trophic influence in reverse, on the skin.

Another clear example of trophic influence in reverse occurs in the peripheral gustatory receptors, taste buds. Because taste buds disappear after denervation and reappear on nerve regeneration, Olmstead (1920) considered the intact sensory nerve to be a prerequisite for the formation of its peripheral receptor, the taste bud. Next came Torrey's (1934) remarkable finding that denervated taste buds disappear faster when the gustatory nerve is cut near the buds than when sectioned further away. Torrey (1934) envisioned a 'neurosubstance' originating in the sensory neurone cell body and passed along to the tongue, thereby maintaining the formation of the taste buds; the longer the distal nerve stump, the more residual neurosubstance is left to keep the taste buds functioning. This study both pioneered the concept of neurotrophic factors and predicted the existence of axoplasmic transport.

10.2 INDUCTIVE INFLUENCE OF THE MOTOR NEURONE ON MUSCLE

The way in which denervation alters the nutritive condition is illustrated most clearly by atrophy in skeletal muscle. Tower (1937) isolated the lumbar segment of the spinal cord innervating the hindlimb muscles in the dog by transecting the cord rostral and caudal to the lumbar sement and cutting all the lumbar dorsal roots, leaving only motor innervation via the ventral roots intact. Activity of the hindlimb muscles was now virtually absent, and the muscles developed marked atrophy, despite the intact motor innervation. The histological changes in the inactive muscles of the isolated cord preparation were not identical but very similar to those found in denervated muscle (Tower 1937). Trophic influence from motor neurones to muscle thus appears to be due mainly to muscle activity itself and only partly to some contribution by the nerve.

Besides atrophy, many alterations occur in denervated muscle. Among them, two properties have been extensively studied in relation to motor innervation: sensitivity to acetylcholine (ACh) and the speed of muscle contraction. A *law of denervation* was proposed by Cannon and Rosenblueth (1949), stating that following denervation, every innervated tissue becomes supersensitive to chemical agents, including the transmitter. According to Cannon and Rosenblueth (1949), increased sensitivity to nicotine in denervated muscle was first reported by Heidenhain in 1883. In a more quantitative study, Brown (1937) estimated that denervated frog muscle is about ten times more sensitive than normal to ACh. Interestingly, in normal muscle the sensitivity to ACh is restricted to its neuromuscular junction, whereas with denervation the ACh sensitivity spreads along the entire muscle fibre (Kuffler 1943; Axelsson and Thesleff 1959). This development of *extrajunctional ACh sensitivity* is prevented, however, if protein synthesis in the muscle is inhibited at the time of denervation (Fambrough 1970). Clearly, denervation supersensitivity to ACh is due to newly formed extrajunctional ACh receptors. Fambrough (1970) thus inferred that motor neurones must regulate gene activity in muscle fibres. How?

Lomo and Rosenthal (1972) showed that chronic blockade of the peripheral nerve conduction with anaesthetics induces ACh sensitivity along the entire muscle fibre even without denervation and that the ACh supersensitivity in denervated muscle can be prevented by chronic stimulation of the muscle. Undoubtedly, muscle activity (contractions or electrical excitation of muscle fibres) is involved in the regulation of extrajunctional ACh receptors. Oddly, however, the increase in extrajunctional ACh receptors in muscle from chronic nerve block with tetrodotoxin is at most half of that found in denervated muscle (Lavoie *et al.* 1976; Pestronk *et al.* 1976). Therefore, the induction of extracellular ACh receptors in denervated muscle must involve changes in other factors besides loss of muscle activity. Degeneration products following nerve section are one possible factor. Indeed, placing a fragment of degenerating nerve on normal muscle produces local supersensitivity to ACh (Jones and Vrbova 1974). Moreover, some neurotrophic factors emanating from motor neurones also help regulate the ACh sensitivity of muscle (see section 12.1.2).

The discovery that the speed of muscle contraction can be modified by altering motor innervation (Buller *et al.* 1960) prompted analysis of the mechanism of neurotrophic influence. Ranvier (1874) was the first to find that red skeletal muscle contracts more slowly than pale muscle in mammals. The rapid and slowly contracting muscles reflect the properties of their innervating spinal motor neurones (Granit *et al.* 1956; Eccles *et al.* 1958). For instance, the motor neurones innervating the slowly contracting soleus (SOL) muscle discharge (to repeat excitation) at a relatively low frequency (e.g., 10 Hz). By contrast, the motor neurones supplying the fast contracting flexor digitorum longus (FDL) muscle can discharge at high frequencies (e.g., 50 Hz). When Buller *et al.* (1960) re-innervated the fast FDL muscle with the slow SOL nerve, the FDL muscle reduced markedly its speed of contraction. Conversely, the slow SOL muscle reinnervated by the fast FDL nerve accelerated its contraction speed. Buller *et al.* (1960) proposed two possible mechanisms: the contractile properties of a muscle may be determined, first, by the frequency of motor neurone discharge reaching the muscle or, second, by a specific trophic substance emanating from each type of motor neurone. Numerous studies now support the first proposal, that the discharge frequency of motor neurones is the determinant of contractile properties (Eccles *et al.* 1962; Salmons and Vrbova 1969; Buller and Pope 1977).

The experiment summarized in Fig. 10.1 illustrates well this view of the importance of muscle activity pattern (Goldring *et al.* 1981). Here, the nerve to the cat SOL muscle was cut, and the muscle was doubly re-innervated by its original SOL nerve and the foreign FDL nerve (Fig. 10.1A). Two months later, the contraction speed of the SOL muscle became faster when stimulated through the FDL nerve (double arrow) than through the SOL nerve (single arrow, Fig. 10.1B*a*). This difference disappeared, however, when the activity of SOL and FDL motor neurones in the lumbar spinal cord was chronically minimized by transecting the thoracic cord after the dual reinnevation (Fig. 10.1B*b*). Next, both re-innervating nerves were chronically stimulated; the contraction speed of the SOL muscle was now equally reduced when the nerves were stimulated at 10 Hz, but equally accelerated when stimulated at 50 Hz (Fig. 10.1B*c*, B*d*). Evidently, the newly induced contractile properties in a given muscle depend on the pattern of motorneuronal activity, regardless of what type of motor neurone delivers it. The mechanisms underlying all these inductive influences from motoneruon to muscle thus can be accounted for by largely, if not exclusively, in terms of the pattern of muscle activity.

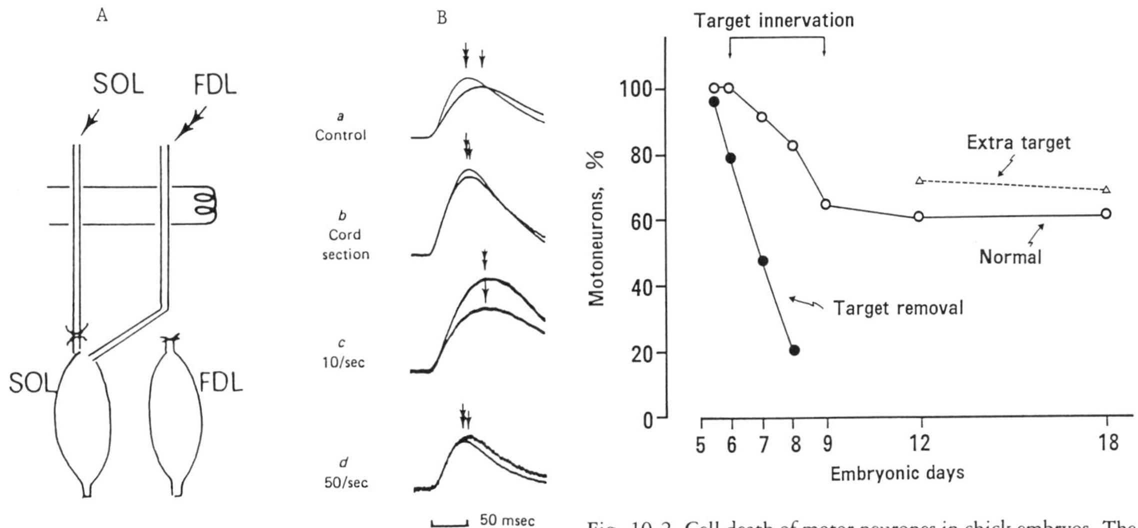

Fig. 10.1. Transformation of contractile properties in the cat soleus (SOL) muscle. A, outline of the experimental procedure. The nerve to the SOL muscle was cut; the SOL muscle was doubly re-innervated by the SOL nerve and the flexor digitorum longus (FDL) nerve. B*a*, contractions of the SOL muscle evoked by stimulation of the SOL nerve (single arrows) or FDL nerve (double arrows) 2 months later. The contractile properties of the SOL muscle were altered by cord transection (*b*), and by chronic nerve stimulation at 10/sec (*c*) or at 50/sec (*d*) during the 2 months. (From Goldring *et al.* 1981.)

Fig. 10.2. Cell death of motor neurones in chick embryos. The lumbar motor neurones form motor innervation (target innervation) between embryonic days 6 and 9. In normal chick embryos, 40 per cent loss of lumbar motor neurones occurs during motor innervation (open circles; natural cell death). Motoneurone death is enhanced if the target is removed before its innervation (filled circles). Natural motor neurone death is reduced if an extra limb bud is grafted (triangles). [The graph is based on data by Hamburger (1958; filled circles), Hamburger (1975; open circles) and Hollyday and Hamburger (1976; triangles).]

10.3 TROPHIC INFLUENCE FROM MUSCLE ON THE MOTOR NEURONE

Despite Sherrington's assertion (section 10.1), some trophic influences undoutedly flow opposite to the direction of the neuronal signal, for example, from muscle fibre to motor neurone. The first clue came from Shorey's (1909) finding that removing the limb buds in chick embryos causes spinal motor neurones to disappear. Hamburger (1934, 1958) confirmed that if the target tissue (wing or limb bud) of chick embryos is excised prior to innervation, motor neurones die (Fig. 10.2, filled circles). This cell death following removal of the target occurs when the motor neurones would normally be making connections with their target (Fig. 10.2, embryonic days 6 to 9). Actually, even with the target intact, a 40 per cent loss of lumbar motor neurones is observed; this process is termed *naturally occurring cell death* (Hamburger 1975; open circles in Fig. 10.2). Again, the timetable of the natural motor neurone death coincides with that of normal neuromuscular connections. Why? It was suggested that an excess of motor neurones compete with one another for the limited amount of trophic factor in the innervated target; any motor neurones that fail to get enough of the factor then die (Prestige 1967; Hamburger 1975). Consistent with this hypothesis, fewer motor neurones die naturally if an additional limb bud is grafted near the original target (Fig. 10.2, triangles; Hollyday and Hamburger 1976). Evidently, motoneuronal survival early in development depends on trophic effects exerted from the target muscle.

Giller *et al.* (1973) dissociated spinal neurones from mouse embryos, culturing them in the presence or absence of muscle cells. The activity of choline acetyltransferase, the enzyme mediating ACh synthesis, in the spinal neurones was at least 10-fold greater when muscle cells were present than in their absence. Obviously not all these cultured spinal neurones were motor neurones, but it seems plain that muscle cells can induce motoneuronal differentiation, evidenced here by increased activity of the enzyme specific for motor neurones. Moreover, increased enzyme activity in spinal neurones by muscle cells can be mimicked simply by applying the *conditioned culture medium*; the medium from which muscle cells are removed after culturing them (Giller *et al.* 1977). This trophic effect on neuronal differentiation therefore must be mediated by chemical factors affiliated with muscle or muscle-derived chemical factors.

A number of examples have since illustrated target-dependent trophic influences on motor neurones. These retrograde influences appear to help maintain motoneuronal survival, induce neuronal differentiation or sustain normal motoneuronal properties (phenotypes) in adulthood. Some of the trophic effects must be exerted by an as yet unidentified retrograde trophic factor.

10.4 THE DISCOVERY OF NEUROTROPHIC FACTORS

Removal of the target tissue in chick embryos (i.e., wing or limb bud extirpation) depletes not only the spinal motor neurones but also the sensory neurones in dorsal root ganglia (Hamburger 1934; Hamburger and Levi-Montalcini 1949). To identify the 'peripheral factors' needed for the innervating neurones to survive, Levi-Montalcini and Hamburger (1951) implanted several tissues, including muscle, brain, liver or skin, in place of the extirpated limb, but the outcome was inconclusive. Bueker (1948) implanted tumour tissue, *mouse sarcoma 180*, after removing part or all of the hindlimb in chick embryos, which produced hyperplasia (enlargement) of the lumbar dorsal root ganglia, while the lumbar motor column became hypoplastic. The sympathetic ganglia also show a marked hyperplasia not only under these conditions (Levi-Montalcini and Hamburger 1951) but even when sarcomas are placed on the extra-embryonic allantoic membrane, a rich vascular bed (Levi-Montalcini and Hamburger 1953). The latter observation clearly indicates that the *growth-promoting agent* is a diffusible substance. To quantify activity of the growth-promoting agent, Levi-Montalcini *et al.* (1954) developed a bioassay system using cultured dorsal root ganglia. Their assay system quickly pinpointed the factor released from sarcomas, which was first termed *nerve growth-promoting factor* and later *nerve growth factor* or NGF (Cohen *et al.* 1954).

How snake venom and mouse salivary gland were found to be a potent source of NGF is well documented (Levi-Montalcini 1975, 1987). It was an entirely unexpected, fortuitous finding. Once a large amount of NGF had been purified from the mouse salivary gland, the amino acid sequence of its principal subunit was determined by Angeletti and Bradshaw (1971), later to be confirmed by the cloning of its cDNA (Scott *et al.* 1983; Ullrich *et al.* 1983). A large supply of NGF from the salivary gland facilitated to prepare its antibodies. Both dorsal root ganglion cells and sympathetic neurones in chick embryos show hyperplasia when treated with NGF and undergo atrophy or cell death when treated with anti-NGF antisera (Levi-Montalcini and Angeletti 1968). The magnitude of natural cell death in dorsal root ganglia of the chick embryo is significantly reduced by applying NGF (Hamburger *et al.* 1981). NGF is transported by axonal flow from the periphery to the neurone cell body (Stockel *et al.* 1976). Moreover, the tissues that receive sympathetic innervation contain high levels of NGF (Korsching and Thoenen 1983a). All these results are consistent with the view that NGF is a retrograde trophic factor derived from the target tissue to affect selectively sensory and sympathetic neurones.

Until the 1980s, NGF was the only neurotrophic factor identified. Another factor, *brain-derived neurotrophic factor* (BDNF), was then purified from pig brain (Barde *et al.* 1982). The molecular cloning of BDNF revealed a primary structure highly homologous to that of NGF (Leibrock *et al.* 1989). A third member of this family, *neurotrophin-3* (NT-3), was also cloned, using oligonucleotides with sequences conserved in NGF and BDNF as probes (Hohn *et al.* 1990; Maisonpierre *et al.* 1990; Rosenthal *et al.* 1990). Clearly, these neurotrophic factors are members of the same gene family, and the gene products are now termed *neurotrophins*. Their biological actions are similar but not identical, and often their effects are additive. Besides the neurotrophin family, *ciliary neurotrophic factor* (CNTF) was cloned (Lin *et al.* 1989; Stockli *et al.* 1989a). CNTF shows no homology with any other trophic factors, and its biological actions clearly differ from those of the neurotrophins.

Gutmann (1976) predicted that 'The analysis of neurotrophic regulations will become clear only after chemical definition of the neurotrophic agents'. Although such analysis has barely begun, the study of neurotrophism is already rapidly accelerating.

11

Neurotrophic factors responsible for neuronal survival

11.1 THE DEFINITION OF NEUROTROPHIC FUNCTION

The meaning of *neurotrophic function* is unfortunately ambiguous. According to Gutmann (1963), we would not even need to postulate neurotrophic effects on muscle if denervation and the absence of motor impulse activity had equivalent effects. Because they differ, neurotrophic function came to be defined as *an influence of the nerve not mediated by impulses* (Gutmann 1963; Miledi 1963; Harris 1974). At first, the increased ACh sensitivity in denervated muscle was listed as a typical phenomenon to exemplify a specific neurotrophic effect, because eliminating impulse activity in the motor nerve was considered to be unlikely to produce it (Gutmann 1963). Yet muscle activity, hence impulse activity in the innervating motor nerve, is now known to be involved in the regulation of extrajunctional ACh receptors (see section 10.2). So might ACh supersensitivity actually be irrelevant to neurotrophic function? Or, alternatively, might some neurotrophic functions be mediated by impulse activity? We must ask once again: what defines neurotrophic function?

According to the definition suggested by Hefti *et al.* (1993), *neurotrophic factors are endogenous soluble proteins regulating survival, growth, morphological plasticity, or synthesis of proteins for differentiated functions of neurones.* This definition neglects trophic functions from the nerve to its target tissues. Historically, the term trophic function was proposed to describe mutual dependencies between neurones and the target cells they innervate. The dependency was deduced from changes in the status or phenotypic expression that arise when one member of the partnership is removed (Purves 1988). The alterations induced in the cell (the innervating neurone or its target cell) by its partner's removal include cell death, failure of normal differentiation, and changes in morphological, biochemical or electrophysiological properties. In this sense, neurotrophic function can be defined as *a neurally mediated influence that maintains the cell's pre-existing or programmed state.* The deprivation of a neurotrophic influence would thus induce anomalous or even deleterious changes in the cell by interfering with its natural state. This definition deliberately does not address any mechanisms of neurotrophic function. The mutual dependencies between neurones and their target cells are presumably mediated by some intercellular signals whose nature is not yet clear—whether a specific chemical substance, impulse activity or both. Since we have no conceptual framework, it would be pointless to define neurotrophic function in mechanistic terms. Whether neurotrophic function occurs independent of neuronal impulse activity is a secondary question, not a defining criterion in itself.

As noted, neurotrophic function is directed from the innervating neurone toward its target cell or is sent the opposite way. Our definition of neurotrophic function described above is applicable to neurotrophic influences exerting in either direction. For instance, if neurotrophic effects exist bidirectionally between a motor neurone and its innervated muscle fibres, both the motor neurone and muscle fibres would fail to show their normally programmed differentiation or they both would alter their pre-existing properties following transection of the motor axon. Despite this definition of neurotrophic function, it is still difficult to visualize what neurotrophic function is in concrete terms. To visualize it, we must now address the mechanisms or at least plausible explanations for the operation of neurotrophic function. Might a given cell receive many different types of trophic influence from its partner? Which phenotypes of the cell are maintained by the trophic influence? What is the nature of the trophic influence? How does it work?

As Gutmann (1976) predicted, the analysis of neurotrophic mechanisms is easier if the nature of the trophic factor is known. Fortunately, several neurotrophic factors have been now identifed. We shall begin with the features of these identifed neurotrophic factors. Although each neurotrophic factor affects many phenotypes, we first focus on the most evident phenotype, cell survival. The purpose of this chapter is to characterize the influence of identified neurotrophic factors, using cell survival as a criterion. As we will see below, these neurotrophic factors can rescue different neurones from cell death. The deprivation of some of these neurotrophic factors causes the neurones to deviate from the normal embryonic programme, resulting in cell death. Chapter 12 deals with neurotrophic influences involved in the early postnatal developmental program. Alterations of the pre-exisitng state of neurones in adulthood by the deprivation of neurotrophic influences are discussed in Chapter 13. From these data, we can at least begin to assemble a conceptual framework for neurotrophic function and speculate on the biological significance of neurotrophism.

As described previously (section 10.4), nerve growth factor (NGF) is the first identified neurotrophic factor that has survival-promoting effects for neurones. Yet, only certain types of neurone display NGF-responsiveness. Clearly, the target-derived trophism throughout the nervous system cannot be accounted for by NGF alone. Indeed, the long-standing presumption that many other neurotrophic factors exist (Barde *et al.* 1983) has been borne out with the discovery of new neurotrophic factors. Studies of these factors and their receptors have substantially refined our insight into neurotrophism (Thoenen 1991).

Several molecules are now recognized as being members of the NGF protein family; they are collectively termed *neurotrophins* and include NGF itself. All neurotrophins share some identical domains while preserving other distinctions within their molecular structure. Consequently, they affect overlapping but distinct neuronal populations. Still, the neurotrophin family alone is insufficient to cover all the target-derived trophic actions observable in the nervous system. Let us now trace the recent progress in studies of neurotrophic factors for different types of neurone. We begin with the neurotrophin family.

11.2 THE NEUROTROPHINS

Barde *et al.* (1982) purified 1 μg of a novel trophic factor, *brain-derived neurotrophic factor* (BDNF), from 1.5 kg of pig brain tissue. BDNF is similar to NGF in its molecular weight and in its ability to support the survival of sensory neurones in chick dorsal root ganglia. Yet, unlike NGF, BDNF neither maintains the survival of sympathetic neurones nor cross-reacts with anti-NGF antibodies. Thus, BDNF is a neurotrophic factor distinct from NGF, even though BDNF shows a high homology with NGF in its molecular structure (Leibrock *et al.* 1989).

Another similar but distinct neurotrophic factor, *neurotrophin-3* (NT-3; Hohn *et al.* 1990; Maisonpierre *et al.* 1990; Rosenthal *et al.* 1990), was so named for being the third member found in the NGF-related neurotrophin family. Two additional neurotrophins, NT-4 and NT-5, have since been discovered. One question is their specificity. Is each neurone responsive to several types of neurotrophin, or will only one type suffice?

11.2.1 Structure and distribution of neurotrophins

The primary structure of NGF was determined by Angeletti and Bradshaw in 1971 and subseqently confirmed by molecular cloning (Scott *et al.* 1983; Ullrich *et al.* 1983). NGF is a dimer of two identical subunits held non-covalently. Figure 11.1 shows this monomer, which comprises 118 amino acids. Characteristically, the NGF monomer has six cysteine (Cys) residues forming three disulphide bridges whose arrangement is presumably a major determinant of NGF's three-dimensional molecular motif. In the primary structure, a high homology (50–60 per cent) with NGF marks all four newly found neurotrophins, BDNF (Leibrock *et al.* 1989), NT-3 (Horn *et al.* 1990; Maisonpierre *et al.* 1990), NT-4 (Hallbook *et al.* 1991), and NT-5 (Berkemeier *et al.* 1991). Each neurotrophin is a protein made of about 120 amino acids and containing the six cysteine residues. So far NT-4 has been detected only in Xenopus ovary (Hallbook *et al.* 1991); NT-5 may be its mammalian homologue (Ip *et al.* 1992a). For lack of detailed information about NT-4 and NT-5, our discussion of functional significance will address mainly the three other neurotrophins, NGF, BDNF, and NT-3.

NGF is the first protein identified as a target-derived trophic factor. Are BDNF and NT-3 also target-derived trophic factors? The high homology in molecular structure strongly implies this, but proof requires that two

Fig. 11.1. The structure of NGF. Six cysteine (Cys) residues form three disulphide bridges (dark rectangles). (From Purves and and Lichtman 1985.)

criteria be met: (1) retrograde transport of the substance in question from the terminal to its neurone cell body, and (2) trophic responsiveness of the neurones to the substance. Let us examine each requirement in turn.

Injection of radioactively labelled NGF, BDNF or NT-3 into the rat sciatic nerve proved their retrograde transport through sensory fibres to dorsal root ganglion (DRG) cells (DiStefano *et al.* 1992). Significantly, more material (about twice as much) was transported to DRG cells with NGF, however, than with BDNF or NT-3. This tendency was more marked in sympathetic neurones, in which radioactive BDNF or NT-3 retrogradely labelled only a small percentage of the neurones labelled with NGF (DiStefano *et al.* 1992). Apparently, the sensory or sympathetic neurones that transport NGF specifically outnumber those transporting BDNF or NT-3. The simplest interpretation of such specificity would be that retrograde transport is receptor-mediated, with different types of neurone expressing different neurotrophin receptors. This notion is supported by the competitive inhibition of retrograde transport between radiolabelled and unlabelled neurotrophins. For instance, adding large quantities of unlabelled BDNF completely blocked the transport of radioactively labelled BDNF in sensory neurones (DiStefano *et al.* 1992). Similarly, NGF and NT-3 show receptor-dependent retrograde transport in sensory neurones.

One drawback of examining retrograde transport by exogenous application is that the true biological relevance is questionable. For example, although labelled NGF injected into the rat spinal cord is retrogradely transported to the DRG cells (Yip and Johnson 1984), implying that the central processes of DRG cells possess NGF receptors, normally no detectable levels of NGF exist in the dorsal root or the spinal cord (Korsching and Thoenen 1985). Thus, the axonal transport of exogenously applied NGF does not prove the transport of endogenous NGF. If evidence for retrograde transport of endogenous neurotrophins is lacking, we need other proof: the presence of the neurotrophin in the target and the presence of the neurotrophin receptor in the neurones. With NGF, three facts strongly endorse its role as a retrograde messenger: first, the tissues in which NGF and its mRNA are expressed are those innervated by sympathetic neurones (Korsching and Thoenen 1983a; Heumann *et al.* 1984); second, although the sympathetic ganglia contain high levels of NGF, they lack the NGF mRNA, ruling out the sympathetic neurones as a site of NGF synthesis (Heumann *et al.* 1984); third, *endogenous* NGF is indeed retrogradely transported along the peripheral nerve (Korsching and Thoenen 1983b).

Unlike NGF, retrograde transport of endogenous BDNF and NT-3 is not yet demonstrated. Also, detailed information is not yet available for the tissue distribution of BDNF and NT-3 and their mRNAs (see review by Yamamori 1992). Moreover, the seeming specificity of receptor-dependent retrograde transport of neurotrophins is betrayed by some promiscuity. For example, unlabelled NT-3 blocks the transport of radiolabelled NT-3 in sensory neurones, but it also impedes the transport of labelled BDNF; similarly, unlabelled BDNF and NGF hinder the transport of labelled NT-3 (DiStefano *et al.* 1992). It seems that a given neurotrophin may crosstalk or cross-react to some extent with multiple receptors. Let us now examine the characteristics of the receptors for these neurotrophins.

11.2.2 Neurotrophin receptors

Basically, the retrograde trophic action of NGF is a three-step process: (1) once released from the target tissue, NGF binds to its receptors on the membrane of nerve terminals; (2) receptor-mediated endocytosis then internalizes the NGF (Schwab *et al.* 1982; Layer and Shooter 1983); (3) the NGF-containing vesicles are transported retrogradely to the neurone cell body (Korsching and Thoenen 1983b), thereby delivering trophic signals to the nucleus (Yankner and Shooter 1979; Schwab *et al.* 1982). Evidently, the neurotrophic action is triggered by the binding of NGF to its receptor. It is well known that NGF has low-and high-affinity classes of receptor (Sutter *et al.* 1979). Do the two types of receptor function independently or cooperatively? Unfortunately, the exact answer is still lacking, despite the recent remarkable progress in studies of NGF receptors. What is still missing?

The molecular cloning revealed that the low-affinity NGF receptor comprises about 400 amino acid residues (D. Johnson *et al.* 1986; Radeke *et al.* 1987). Thus, the estimated molecular weight of the NGF receptor core protein is about 42 500. But because of glycosylation (the attachment of sugars to proteins), which inflates the mature receptor to a 75 000 dalton glycoprotein, this NGF receptor is symbolized by p75. The primary structure of NGF receptor, as deduced from its cDNA, has a single transmembrane segment (Fig. 11.2A). In the extracellular N-terminus domain are four repeating elements, each rich in cysteine residues; these four repeats are required for NGF binding (Welcher *et al.* 1991; Yan and Chao 1991). The relatively short cytoplasmic C-terminus domain suggests that the cytoplasmic region of the p75 NGF receptor may not directly serve signal

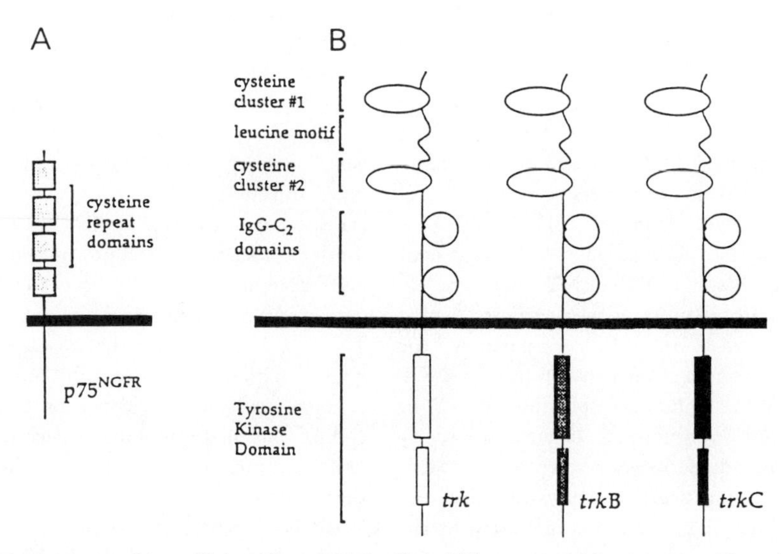

Fig. 11.2. Schematic structure of low-affinity (A) and high-affinity (B) neurotrophin receptors. The cell membrane (heavy horizontal line) separates the extracellular (top) from the cytoplasmic (bottom) region. (Adapted from Chao 1992.)

transduction. Indeed, the internalization of NGF and the resultant signal transduction are known to be mediated only by the high-affinity receptor (Bernd and Greene 1984; Green *et al.* 1986).

Most NGF receptors are those with a low affinity for NGF (dissociation constant K_d being about 10^{-9}M), whereas the rarer, high-affinity NGF receptor has a K_d of about 10^{-11}M (Sutter *et al.* 1979). Surprisingly, the cloned p75 protein has low-affinity for all three neurotrophins. In other words, the low-affinity NGF receptor also serves as a low-affinity BDNF receptor (Rodriguez-Tebar *et al.* 1990) and as a low-affinity NT-3 receptor (Squinto *et al.* 1991). Since all the three neurotrophins share the same low-affinity receptor molecule, this receptor class cannot account for any specificity in the biological action and axonal transport of different neurotrophins. Rather, each neurotrophin must have its own high-affinity receptor, whereby it can exert its specific trophic action.

What do we know about the high-affinity NGF receptors? A clue about their nature came from two related findings. First, the receptors for several other growth factors (e.g., epidermal growth factor and platelet-derived growth factor) are characterized by a tyrosine kinase catalytic site in their cytoplasmic domain (Ullrich and Schlessinger 1990). Second, following NGF application, tyrosine kinase activation indeed occurs in its high-affinity receptor but not in its low-affinity receptor (Meakin and Shooter 1991). Several cell surface proteins with tyrosine kinase activity were found to be encoded by proto-oncogenes, but the nature of their ligands for these putative receptors with tyrosine kinase remained unknown. Eventually, a 140,000 dalton glycoprotein (gp140) with tyrosine kinase activity was first identified as the high-affinity NGF receptor (Kaplan *et al.* 1991; Klein *et al.* 1991). Encoded by the tyrosine kinase proto-oncogene *trk* (also known as *trk*A), this protein was accordingly termed the *trk tyrosine kinase receptor* (see Box K), being symbolized by $gp140^{trk}$. The two other related genes, *trk*B (Klein *et al.* 1989) and *trk*C (Lamballe *et al.* 1991), encode the high-affinity receptors for BDNF (Soppet *et al.* 1991; Squinto *et al.* 1991) and NT-3 (Lamballe *et al.* 1991), respectively. The respective symbols of these gene products are $gp145^{trkB}$ and $gp145^{trkC}$. Figure 11.2B illustrates the schematic structures of these high-affinity neurotrophin receptors. All the *trk* receptor molecules are characterized by a single transmembrane segment motif with a tyrosine kinase domain in their cytoplasmic region.

We remarked on promiscuity in retrograde transport of some neurotrophins (section 11.2.1). A similar promiscuity also occurs between neurotrophins and their receptors. For example, NT-3 can bind to $gp140^{trk}$ and $gp145^{trkB}$, although it reserves a higher affinity for $gp145^{trkC}$. This promiscuity is weaker in neuronal cells than in non-neuronal cells (Ip *et al.* 1993b). Thus NT-3's primary receptor is probably the *trk*C gene product (Chao 1992; Meakin and Shooter 1992; Davies 1994). Undoubtedly, the *trk* tyrosine kinase receptor family is essential for expressing high-affinity neurotrophin receptors, hence for the expression of trophic functions of neurotrophins. What role then remains for the low-affinity neurotrophin receptor?

Box K **The *trk* tyrosine kinase receptor family**

The *trk* proto-oncogene encoding a 140 000 dalton glycoprotein ($gp140^{trk}$ or $gp140^{trkA}$; Martin-Zanca *et al.* 1989), was first isolated from a human colon carcinoma in a chimeric form. Specifically, in the genomically rearranged gene the *trk* gene is fused to a unrelated tropomyosin gene (Martin-Zanca *et al.* 1986). The term *trk* (pronounced 'track') stands for tropomyosin-receptor-kinase (Martin-Zanca *et al.* 1986). By screening with the *trk* as a probe, two additional members of the *trk* gene family, *trk*B (Klein *et al.* 1989) and *trk*C (Lamballe *et al.* 1991), were identified. These *trk* proteins have a long extracellular stretch, a single membrane-spanning region and a cytoplasmic tyrosine kinase domain (Fig. 11.2B). This molecular motif suggests that a ligand binding to the extracellular domain may activate the *trk* protein family, which is translated across the cell membrane into stimulation of its cytoplasmic tyrosine kinase. Several growth factors or hormones are known to exert their actions by activating their receptors with a tyrosine kinase activity (Ullrich and Schlessinger 1990). Which ligands bind to these *trk* proteins? Not knowing the answer at first obscured the physiological relevance of the *trk* proteins, but before long their ligand was unmasked: It was the neurotrophins that act as the ligand for the *trk* proteins.

The low-affinity neurotrophin receptor (p75) may be the foundation on which the high-affinity receptor is constructed, although this notion remains controversial (Bothwell 1991; Chao 1992; Meakin and Shooter 1992). Hempstead *et al.* (1991) reported that expressing either p75 or gp140trk in individual cultured cells yields only NGF binding sites of low affinity, whereas coexpressing them generates high-affinity sites. The hypothetical formation of such high-affinity receptor sites is diagrammed in Fig. 11.3A. It postulates that an NGF monomer or dimer binds to the p75 and gp140^{trk} proteins and that the resultant heterodimerization of the receptor proteins increases receptor affinity to NGF. Since the p75 presumably lacks the signal transduction domain in its cytoplasmic segment, NGF-induced signal transduction would employ the cytoplasmic domain of the gp140 receptor after dimerization. On the other hand, Klein *et al.* (1991) found that even without p75, expressing gp140^{trk} alone develops high-affinity NGF receptors (see also Cordon-Cardo *et al.* 1991). By analogy with receptors of other growth factors (Yarden and Ullrich 1988; Ullrich and Schlessinger 1990), NGF affnity was assumed to increase by dimerization (or oligomerization) of gp140^{trk} proteins (Fig. 11.3B; Klein *et al.* 1991).

These seemingly contradictory findings can be reconsidered with respect to Weskamp and Reichardt's (1991) observation of two distinct classes among the high-affinity NGF receptors. Interestingly, only one class was blocked by the antibodies that recognize the low-affinity neurotrophin receptor. Assuming that the high-affinity receptor can be formed by either the heterodimer of gp140^{trk} and p75 (Fig. 11.3A) or the homodimer of gp140^{trk} (Fig. 11.3B), only the former class should be susceptible to the antibodies against the low-affinity p75 receptor. Unfortunately, the continuing doubts about whether p75 actually subserves the high-affinity NGF receptor clouds our interpretation of NGF's trophic function (see next section). Resolution of this issue is urgently needed.

Admittedly, information about neurotrophin receptors is still fragmentary, yet at least we know that NGF, BDNF, and NT-3 have their own high-affinity receptors. Let us now examine whether each of these different neurotrophins exerts a specific neurotrophic effect.

11.2.3 Trophic actions of neurotrophins

Although NGF, BDNF, and NT-3 share the same low-affinity receptor (p75), the p75 protein does not seem to be responsible for neurotrophin-induced signal transduction. If the low-affinity receptor were entirely irrelevant to NGF's trophic functions, sensory and sympathetic neurones would develop even in the absence of the gene encoding the p75 protein. To explore this possibility, Lee *et al.* (1992) produced targeted mutation of the gene encoding the low-affinity p75 receptor in the mouse. The mutant mice indeed failed to express the p75 mRNA and its protein. Although viable, the mutant mice showed three phenotypical abnormalities in the sensory system: (1) smaller DRGs than those of their normal litter mates; (2) no response to a noxious heat stimulus, suggesting impaired heat sensitivity; (3) no cutaneous innervation by sensory fibres containing substance P or calcitonin gene-related peptide (CGRP). The expression of substance P and CGRP in sensory neurones is known to be up-regulated by NGF (Lindsay and Harmar 1989). Perhaps, the sensory neurones that would normally express these peptides die in the mutant mice, or survive but fail to express the peptides. In any event, the absence of the gene encoding the low-affinity neurotrophin receptor clearly interferes with normal development of sensory neurones.

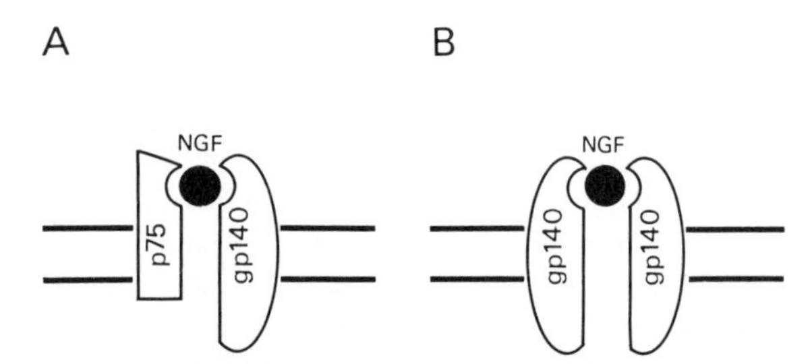

Fig. 11.3. Hypothetical subunit arrangements of the high-affinity NGF receptor. A, a dimer consisting of a low-affinity neurotrophin receptor (gp75) and a gp140^{trk} forms a high-affinity NGF receptor where NGF (filled circle) binds to each subunit. B, a high-affinity NGF receptor like A but formed by dimerization of gp140^{trk}. The cell membrane (two horizontal lines) separates the extracellular (top) from the cytoplasmic (bottom) region.

Surprisingly, the mutant mice were normal in sympathetic innervation and in the size of sympathetic ganglia. Although tyrosine hydroxylase, an enzyme involved in the synthesis of the transmitter, norepinephrine, of sympathetic neurones, is up-regulated by NGF, its levels too were unaffected in sympathetic neurones of the mutant. Thus, deleting the gene encoding the low-affinity neurotrophin receptor causes sensory neurones to develop anomalously without affecting the normal status of sympathetic neurones. How are such disparate effects achieved, despite both types of neurone are NGF-responsive? Davies *et al.* (1993) found that the concentration of NGF required for survival-promoting effects on cutaneous sensory trigeminal neurones is 3- to 4-fold higher in p75-deficient mice than in normal mice. Interestingly, the NGF 'dose–response' relationship for sympathetic neurones was normal in the p75-deficient mice. These results suggest that p75 enhances the sensitivity to NGF in cutaneous sensory neurones and may explain the developmental abnormalities of nociceptive sensory neurones seen in the mutant mice. It is likely that cutaneous sensory neurones are more dependent on NGF than sympathetic neurones during a critical developmental period. Another possibility is that sensory neurones require the low-affinity NGF receptor in Schwann cells for their normal development (Lee *et al.* 1992). In chicks, the low-affinity NGF receptor is transiently expressed in Schwann cells at embryonic stages (Zimmerman and Sutter 1983; Rohrer 1985). If the low-affinity receptor in Schwann cells holds NGF, resultant increase in local NGF concentrations would benefit developing sensory neurones (Johnson *et al.* 1988; see section 11.3.2).

Withholding NGF by injecting anti-NGF antibodies into rat embryos causes massive cell death of DRG sensory neurones (Ruit *et al.* 1992), yet about 30 per cent of the DRG cells survive. Evidently, DRG cells consist of two populations: NGF-dependent and NGF-independent. Might the latter rely on other neurotrophins for survival? If so, each population of DRG cell would express its own high-affinity neurotrophin receptor. For instance, NGF deprivation may kill only those sensory neurones that express the trk mRNA. When Carroll *et al.* (1992) addressed this question, they found that most (about 77 per cent) of the DRG cells express the *trk* mRNA, and indeed about 93 per cent of the trk-expressing neurones die following NGF deprivation. The trkB-and *trk*C-expressing cells comprise about 8 per cent and 15 per cent of the DRG cells, respectively, and these neurones were virtually unaffected by NGF deprivation. These results accord with the notion that NGF-independent DRG cells are supported by other neurotrophins.

DRG cells cultured from chick embryos can be maintained by either NGF or BDNF alone, but jointly adminstering both factors improves the cell survival rate (Lindsay *et al.* 1985). Lindsay *et al.* (1985) suggested that BDNF may be a central target-deprived factor for sensory neurones, whereas NGF is their peripheral target-derived factor, with both factors acting cooperatively to maintain cell survival. If so, each DRG cell would express at least two neurotrophin receptors. Figure 11.4 shows the distribution of rat lumbar DRG neurones expressing different members of the *trk* family. Their frequency histograms as a function of neurone size reveal that the DRG cells expressing trk (*trk*A in Fig. 11.4) are mainly small neurones, whereas those expressing *trk*C are large neurones (Mu *et al.* 1993). DRG cells expressing *trk*B are of intermediate size. Clearly different classes of sensory neurone tend to express distinct members of the *trk* family. Indeed, the mutant mice whose NGF gene

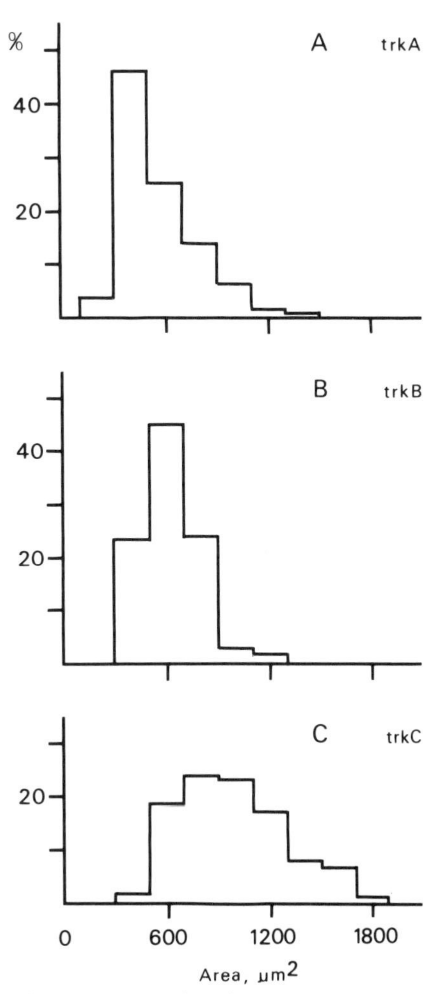

Fig. 11.4. Distribution of lumbar DRG neurones expressing different members of the *trk* family as a function of neuronal size. Ordinate, percentage of trk-expressing neurones having a particular cross-sectional area. Abscissa, cross-sectional area of DRG cell in μm^2. Measurements were made in 21–day-old rats. (Adapted from Mu *et al.* 1993.)

was disrupted show a loss of 70 per cent in the lumbar DRG cells, lacking small sensory neurones (Crowly *et al.* 1994). Evidently, other neurotrophins cannot compensate for the loss of NGF actions on these small sensory neurones or sympathetic neurones. Moreover, essentially the same results were confirmed by disruption of the *trk* gene (Smeyne *et al.* 1994). The mouse with *trk*B gene deletion has a loss of about 30 per cent of DRG cells (Klein *et al.* 1993). The *trk*C-deficient mice exhibited a complete loss of large Ia sensory neurones arising from muscle spindles (Ernfors *et al.* 1994b; Klein *et al.* 1994), supporting the previous suggestion (Hory-Lee *et al.* 1993). These results are consistent with the notion that different types of DRG cell require their own specific neurotrophin for survival. Presumably, BDNF and NT-3 are also a peripheral target-derived trophic factor, and each DRG cell appears to express only one type of high-affinity neurotrophin receptor, being supported by one type of neurotrophic factor: the *one neurone–one neurotrophic factor hypothesis*. The co-localization of different members of the *trk* family in some sensory neurones cannot be ruled out, however (McMahon *et al.* 1994).

So far we have discussed trophic actions in sensory and sympathetic neurones. But originally BDNF was purified from the brain (Barde *et al.* 1982). What is the functional significance of neurotrophins in the central nervous system?

BDNF supports the survival of retinal ganglion cells cultured from rat embryos (J. E. Johnson *et al.* 1986). Since retinal ganglion cells in the rat project to the superior colliculus, Leibrock *et al.* (1989) assumed that BDNF must be synthesized abundantly in this mesencephalic target tissue. In fact, BDNF cDNA was cloned by reverse transcription of mRNAs isolated from the pig superior colliculus (Leibrock *et al.* 1989). Rodrigez-Tebar *et al.* (1989) systematically examined BDNF-dependence of the survival of retinal ganglion cells in chick embryos. At early embryonic stages (day 5) the axons of retinal ganglionic cells do not yet reach their target, and ganglionic cells excised and cultured at this stage can survive independent of BDNF. The dependence on BDNF for survival begins at embryonic day 6, when the axons of retinal ganglionic cells reach the target. These results are reminiscent of those described for NGF responsiveness in mouse trigeminal sensory neurones (Davies *et al.* 1987; section 13.2.1). Both the neurotrophin receptor expression in the neurones and the neurotrophin synthesis in the target occur once the nerve axons arrive at the target. Apparently, BDNF is a retrograde trophic factor for retinal ganglionic cells.

By analogy with its effects on adrenergic sympathetic neurones in the periphery, NGF was first thought to act on catecholaminergic (adrenaline, noradrenaline, and dopamine) neurones in the central nervous system; but not so (Harper and Thoenen 1981). Instead, certain cholinergic central neurones proved responsive to NGF (Korsching 1986). Radioactively labelled NGF injected into the hippocampus and cerebral cortex was transported to the cell bodies of magnocellular cholinergic neurones in the basal forebrain (Seiler and Schwab 1984), neurones that contain endogenous NGF (Korsching *et al.* 1985). Consistent with these results, NGF mRNA was found predominantly in the hippocampus and cortex, the target areas innervated by the magnocellular cholinergic neurones (Korsching *et al.* 1985). Moreover, most cholinergic neurones in the basal forebrain and the striatum of rat express the 140^{trk} mRNA (Holtzman *et al.* 1992). And finally, cell death induced in the basal forebrain cholinergic neurones by axotomy can be prevented by infusing NGF into the brain (Hefti 1986; Williams *et al.* 1986). Thus, NGF appears to be a retrograde trophic factor for the magnocellular cholinergic neurones. It is known that Alzheimer's disease is associated with a drastic depletion of magnocellular cholinergic neurones (Whitehouse *et al.* 1982), so central effects of NGF are clinically relevant (Hefti and Weiner 1986). Recent observations, however, conflict with this notion. Targeted disruption of the NGF gene, for example, did not affect the survival of basal forebrain cholinergic neurones (Crowley *et al.* 1994). Some basal forebrain cholinergic neurones were reported to be responsive to BDNF rather than to NGF (Knusel *et al.* 1992; Morse *et al.* 1993). Yet, mice lacking the BDNF gene did not show any deficiency in cholinergic neurones of the basal forebrain (Jones *et al.* 1994).

A similar clinical relevance was suggested for BDNF in Parkinson's disease, which entails a depletion of mesencephalic (substantia nigra) dopaminergic neurones. Mesencephalic dopaminergic neurones dissociated from rat embryos can be maintained by BDNF, but not by NGF or NT-3 (Hyman *et al.* 1991; Knusel *et al.* 1991). BDNF administration, however, fails to rescue axotomized dopaminergic neurones from cell death (Knusel *et al.* 1992). Furthermore, BDNF mRNA is not expressed in the striatum, which is the target innervated by dopaminergic neurones in substantia nigra (Friedman *et al.* 1991). Thus, no simple scheme can yet be traced for trophic actions of BDNF on central dopaminergic neurones. In fact, central dopaminergic neurones showed no apparent loss in mice lacking the BDNF gene (Ernfors *et* al. 1994; Jones *et al.* 1994).

NT-3 is also called *hippocampus-derived neurotrophic factor* (HDNF; Ernfors *et al.* 1990) because the expression of its mRNA in the central nervous system is confined to the hippocampus and the cerebellum (Ernfors *et al.* 1990; Horn *et al.* 1990). The trkC transcripts are also expressed preferentially in the hippocampus and the cerebellum (Lamballe *et al.* 1991). NT-3 induces the immediate early gene c-fos in hippocampal pyramidal cells (Collazo *et al.* 1992). Clearly, hippocampal pyramidal cells respond to NT-3, but the functional role of NT-3 remains obscure.

The discovery of multiple members in the neurotrophin family has vastly enlarged our knowledge of neurotrophic factors. In the central nervous system, however, except for BDNF's effect on retinal ganglionic cells, BDNF and NT-3 do not yet meet all the criteria defining a target-derived neurotrophic factor. Kokaia *et al.* (1993) and Miranda *et al.* (1993) found that many central neurones coexpress the mRNAs for both neurotrophins and their receptors. It is thus suggested that neurotrophic effects can be exerted by autocrine interactions at least in some types of central neurones (hippocampal and cortical neurones; see also section 11.3.2). Similar autocrine regulation of neuronal survival was also suggested for sensory neurones, sympathetic neurones, and motorneurones at developmental stages (Schecterson and Bothwell 1992). This possibility is reminiscent of autocrine mechanisms seen in proliferation of the helper T cell. The antigen-presenting cell causes the T cell to induce both IL-2 (interleukin-2) and its receptors, thereby stimulating the T cell to proliferate (MacDonald and Nabholz 1986). Although such autocrine mechanisms remain to be explored further, NGF is undoubtedly the best characterized retrograde neurotrophic factor. The basic principles of neurotropism have been derived from NGF's trophic function. Even so, how NGF induces its retrograde trophic effects in the neurone cell body is not clear. We close this section on neurotrophins by summarizing these missing items of information.

11.2.4 Trophic signal of NGF to the cell nucleus

Figure 11.5 illustrates the major gaps in our understanding of the expression of NGF's neurotrophic effects. The current consensus is that NGF is first internalized by endocytosis (Fig. 11.5A, steps 1 and 2), which is mediated exclusively through the high-affinity receptor (Bernd and Greene 1984; Hosang and Shooter 1987), and then transported to the neurone cell body (Fig. 11.5A, step 3; Yanker and Shooter 1979; Schwab *et al.* 1982; Korsching and Thoenen 1983b).

The receptor-mediated internalization of NGF resembles the process for other growth factor receptors with tyrosine kinase activity in non-neuronal cells, typified by the receptor for *epidermal growth factor* (EGF; Ullrich and Schlessinger 1990; Fig. 11.5B). EGF is a mitogen for epidermal cells and for a wide variety of cells in culture (Taylor *et al.* 1972). The binding of EGF to its receptor activates the tyrosine kinase domain in the cytoplasmic segment, which in turn phosphorylates numerous cellular substrates, thereby inducing signal transduction (Carpenter 1992; Fig. 11.5C); the receptor is then internalized and delivered to lysosomes, where it is degraded (Fig. 11.5B).

When EGF's lysosomal degradation is inhibited pharmacologically, EGF can accumulate in the nucleus, but normally the nuclear accumulation is only 1 per cent or less of the total internalized EGF (L. K. Johnson *et al.* 1980; Savion *et al.* 1981). By contrast, about 60 per cent of the NGF bound to the pheochromocytoma cell (PC12 cell) is normally translocated to the nucleus (Yankner and Shooter 1979). These results suggest that NGF both activates its receptor on the cell membrane and reaches the nucleus itself, whereas the signal conveyed to the nucleus through EGF receptors does not rely on EGF directly.

Because the expression of the *trk* protein in cultured cells results in cell proliferation in response to NGF (Cordon-Cardo *et al.* 1991), NGF must induce DNA synthesis via the $gp140^{trk}$ receptor. Presumably, the signal elicited by activating the high-affinity NGF receptor (gp140trk) reaches the nucleus. By what means? In small cultured cells the nucleus is located 5–20 μm from the cell surface, easily accessible to intracellular second messengers activated by the growth factor receptor at the cell membrane. In non-neuronal cells, then, some of the intracellular second messenger signals may reach the nucleus and elicit the nuclear response (e.g., cell proliferation). In neurones (e.g., sensory neurones), on the other hand, the signal elicited in the nerve terminal must travel distances of millimetres, centimetres or even a metre in adult human before reaching the nucleus. Hence, the NGF trophic signal initiated in a nerve terminal must be conveyed to its neurone cell body by retrograde axonal transport.

This unique feature of target-derived trophic signal in neurones raises several questions, but here we need focus on only two. First, does receptor activation by NGF stimulate second messengers, thereby giving rise to a

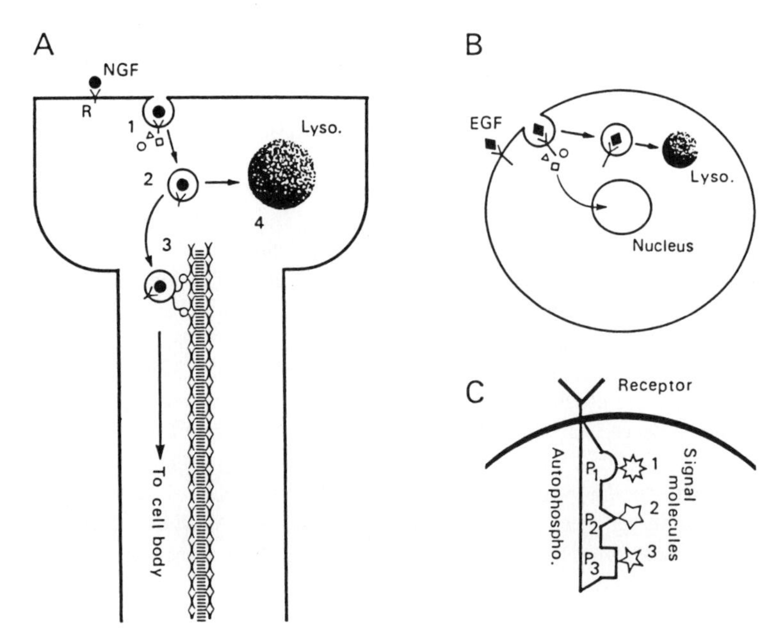

Fig. 11.5. Hypothetical consequences of activation of NGF or epidermal growth factor (EGF) receptor in neuronal and non-neuronal cells. A, a neurone internalizes (1 and 2) NGF and its receptor (R) for retrograde axonal transport (3). Some of the internalized vesicles are degraded (4) in lysosomes (Lyso). Open triangle, circle and square in A and B represent second messengers. B, EGF receptor activation in a non-neuronal cell. Almost all internalized EGF and its receptor-containing vesicles are degraded in lysosomes. Second messengers convey the signal to the nucleus. C, pleiotropic actions of the EGF receptor with tyrosine kinase activity. Its cytoplasmic domain contains multiple active sites that atutophosphorylate on activation (p_1, p_2, p_3) and bind signal molecules (intracellular substrates; 1, 2, 3), thereby inducing multiple signal transduction.

local cellular response at the nerve terminal? If so, is the local response similar to or quite different from the nuclear response elicited by NGF transported to the cell body? NGF's local effect is shown by the experiment illustrated in Fig. 11.6 (Campenot 1982). Sympathetic neurones plated in one compartment (B) of a culture chamber send their neurites into the neighbouring compartments (A and C) if NGF is present there (Fig. 11.6a). Removing NGF from compartment A causes retraction and degeneration of that neurite in the compartment (Fig. 11.6b). Thus, the growth and maintenance of neurites appear to be controlled by NGF locally, without the participation of changes in the neurone cell body or the nucleus (i.e., NGF-induced nuclear response). On the other hand, the sympathetic neurone will survive as long as its cell body (B) or other neurites (C) are exposed to NGF. Hence, neuronal survival must involve the nuclear response—those changes induced in the cell body or the nucleus by NGF. In this example, NGF appears to evoke different local and nuclear responses, but it remains uncertain whether this notion can be generalized to other neurones *in vivo*.

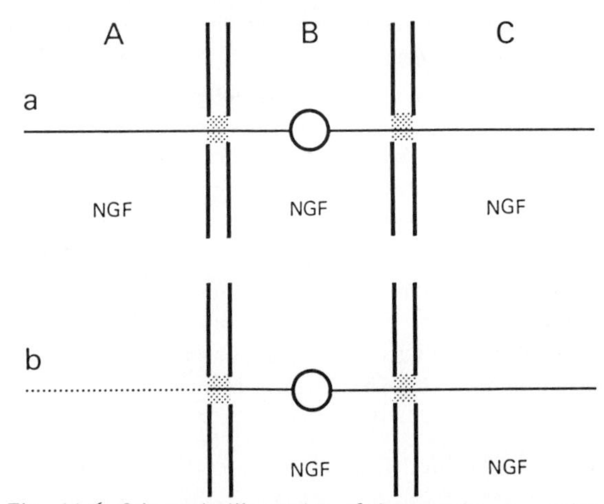

Fig. 11.6. Schematic illustration of the experiment proving the local control of neurite survival by NGF. Dissociated rat sympathetic neurones (circles) plated in the central compartment (B) of a culture dish extend neurites into adjacent comparments (A, C) through grease-filled openings. a, NGF is present in all three compartments. b, removal of NGF from compartment A results in retraction and degeneration of that neurite. (Adapted from Campenot 1982.)

Both NGF and EGF induce a large variety of cellular responses. This behaviour is called the receptor's *pleiotropic* (multiply directed) actions. The EGF receptor has several tyrosine autophosphorylation sites in its cytoplasmic tail (Fig. 11.5C). On receptor activation, these tyrosine kinase sites phosphorylate several cellular substrates (signal proteins), thus opening multiple signalling pathways (Fig. 11.5C; Ullrich and Schlessinger 1990; Carpenter 1992). Because gp140^{trk} receptor is a member of the family of tyrosine kinase receptors (Meakin and Shooter 1992), the local response induced at the nerve terminal by NGF must also be pleiotropic.

The second question raised by NGF's dual sites of action is: how can NGF transported from the nerve terminal induce the nuclear response? Presumably, the retrogradely transported NGF is unloaded in the cell body or the nucleus and then binds to NGF receptors there to launch nuclear signal transduction. Are these NGF receptors newly synthesized and inserted into the nuclear membrane? Or are they cell-surface NGF receptors that travel retrogradely along with NGF? Johnson *et al.* (1987), in fact, found axonal transport of NGF receptors in the rat peripheral nerve—but surprisingly it was the low-affinity p75 molecule (see also Kiss *et al.* 1993). This finding raises further questions. Since NGF internalization at the nerve terminal is mediated exclusively by high-affinity receptors, how does NGF get to the low-affinity receptor for axonal transport? Ehlers *et al.* (1993) found that the high-affinity receptor (*trk*) is also retrogradely transported in the rat sciatic nerve and that this transport can be blocked by the injection of NGF antibodies into the peripheral cutaneous field. It is thus likely that the NGF/*trk* complex is itself transported as a signal. However, how this signal is transmitted to the nucleus remains unknown.

Yanker and Shooter (1979) suggested that the rapid, transient responses evoked by a brief exposure to NGF—such as changes in membrane permeability or neurite extension—result from activation of the cell-surface NGF receptor, whereas long-term changes in RNA and protein synthesis from prolonged exposure to NGF reflect interactions between NGF and the cell nucleus. This notion is reminiscent of short- and long-term neuronal plasticity. Recall that short-term sensitization in *Aplysia* is induced by protein kinase A in the cytoplasm, whereas its long-term form requires phosphorylation of the nuclear protein by the protein kinase (section 8.2.1).

The cells of the immune system communicate through *cytokines*, the factors produced by lymphocytes, macrophages, and epithelial cells (Bigazzi *et al.* 1975; Watanabe *et al.* 1991). As we will see later, some cytokines and particular neurotrophic factors are strikingly similar both in molecular structure of their receptors (section 11.3.1) and in their signal transduction process (section 12.2.3). Yet, unlike neurones, the cells of the immune system lack axonal transport. Neurones must be equipped with a sorting device that routes most of the internalized neurotrophic factor for retrograde transport, rather than to lysosomes for degradation. How the internalized trophic factor is directed to the transport track and how it later exerts its trophic effect in the cell nucleus remain elusive.

11.3 CILIARY NEUROTROPHIC FACTOR

Like motor neurones at early developmental stages, chick ciliary ganglion cells undergo natural cell death; during embryonic days 9–13, about 50 per cent of these cells die (Landmesser and Pilar 1974). Chick ciliary neurones dissociated on embyronic day 7 or 8 die within 24 hrs in culture, but their survival can be maintained with supplemental tissue extracts (Nishi and Berg 1979; Tuttle *et al.* 1980). Best among the tissue extracts tested is the embryonic intraocular musculature, which is the target tissue of ciliary neurones; it contains the highest level of survival factor activity (Adler *et al.* 1979). This result suggested that the surivival factor, termed *ciliary neurotrophic factor* (CNTF: Adler *et al.* 1979), is a target-derived neurotrophic factor for ciliary ganglion cells.

CNTF was then purified from the chick intraocular tissue (Barbin *et al.* 1984) and from the rat sciatic nerve (Manthorpe *et al.* 1986). The molecular cloning of CNTF (Lin *et al.* 1989; Stockli *et al.* 1989) and its receptor (Davis *et al.* 1991) was also achieved. Contrary to the initial assumption, however, the molecular biology of CNTF and its receptor revealed that CNTF lacks any feature characteristic of a target-derived neurotrophic factor. If it is not present in target tissues, what functional significance might CNTF have? Let us examine first the molecular structure and tissue distribution of CNTF and its receptor.

11.3.1 Distribution of CNTF and its receptor

CNTF comprises 200 amino acids. Strikingly, its primary structure shows no homology with any other known proteins, including those with neurotrophic activities (Lin *et al.* 1989; Stockli *et al.* 1989). The N-terminus of CNTF lacks the signal peptide segment of hydrophobic amino acids, which is usually required for secretion from a cell. Moreover, although many cell-surface and secretory proteins are glycosylated, CNTF has no putative glycosylation site. These features suggest that CNTF is a cytosolic protein. Indeed, when expressed in cultured cells, CNTF is detectable in cell extracts but not in culture medium (Lin *et al.* 1989; Stockli *et al.* 1989), indicating that CNTF is confined within the cytoplasm and unable to diffuse across the cell membrane. As noted, CNTF was purified from the sciatic nerve (Manthorpe *et al.* 1986): yet CNTF is not detected in skeletal muscle or skin, again implying that CNTF is not a target-derived factor. In the sciatic nerve, CNTF is confined almost exclusively to Schwann cells (Stockli *et al.* 1991; Friedman *et al.* 1992; Rende *et al.* 1992; Sendtner *et al.* 1992b). Likewise, in the central nervous system astrocytes are the major source of CNTF (Stockli *et al.* 1991). Interestingly, CNTF mRNA is expressed in the rat only after birth (Stockli *et al.* 1989, 1991), precluding CNTF from the regulation of natural cell death at embryonic stages. Nevertheless, because CNTF enables a broad spectrum of neuronal types to survive in culture (see section 11.3.2), CNTF may be a neurotrophic factor even though not derived from target tissues.

Which neurones express the CNTF receptor? Initially, to examine which neurones bind it (Squinto *et al.* 1990), CNTF was linked to a short stretch (just 10 amino acids long) of the protein encoded by a particular proto-oncogene (c-*myc*). The CNTF binding was then assayed immunologically using the monoclonal antibody against this decapeptide as a tag. Besides several neuronal culture cell lines found to bind CNTF, most dorsal root ganglion cells cultured from chick embryos or adult rats had CNTF binding sites. Oddly, however, CNTF did not bind to chick ciliary neurones (Squinto *et al.* 1990), despite the fact that CNTF is purified by survival assays of ciliary neurones (Barbin *et al.* 1984; an explanation of this paradox is proposed in section 11.3.2). This tagging approach eventually led to the molecular cloning of the CNTF receptor (Davis *et al.* 1991). Preliminary studies showed that the CNTF receptor mRNA is widely expressed in the central nervous system; in the periphery, CNTF receptors are expressed in skeletal muscle, the adrenal gland, and the sciatic nerve.

The CNTF receptor comprises 372 amino acids. We have seen that all neurotrophin receptor molecules share the motif of a single transmembrane segment (Fig. 11.2). Similarly, the CNTF receptor molecule has a single hydrophobic domain, which corresponds to the transmembrane segment but lacks an adjoining cytoplasmic hydrophilic segment. Instead, the C-terminus of the mature CNTF receptor is cleaved off, leaving the receptor protein simply anchored to the cell surface (Fig. 11.7A; Davis *et al.* 1991). How can a receptor induce signal transduction without a cytoplasmic domain? The clue to this peculiar phenomenon came from observing signal transduction in a cytokine receptor.

The amino acid sequence of the CNTF receptor resembles that of the *interleukin*-6 (IL-6) receptor. IL-6 is a cytokine produced by lymphocytes, macrophages, and fibroblasts that induces growth and differentiation of many *haematopoietic* (blood-forming) cells (Arai *et al.* 1990); IL-6 also acts on non-haematopoietic cells, including neuronal cells (Hama *et al.* 1989). The extracellular domains of the CNTF receptor and the IL-6 receptor are highly homologous, as indicated in Fig. 11.7A,B (Miyajima *et al.* 1992a). The cytoplasmic domain of the IL-6 receptor is relatively short and lacks tyrosine kinase activity (Yamasaki *et al.* 1988), unlike other growth factor receptors. How, then, does the IL-6 receptor transduce the IL-6 signal? Taga *et al.* (1989) found that the IL-6 receptor works in association with a 130 000 dalton glycoprotein (gp130). The cloning of the human gp130 cDNA revealed that this molecule has a single transmembrane domain (Hibi *et al.* 1990). The gp130 protein has no ligand binding capacity on its own

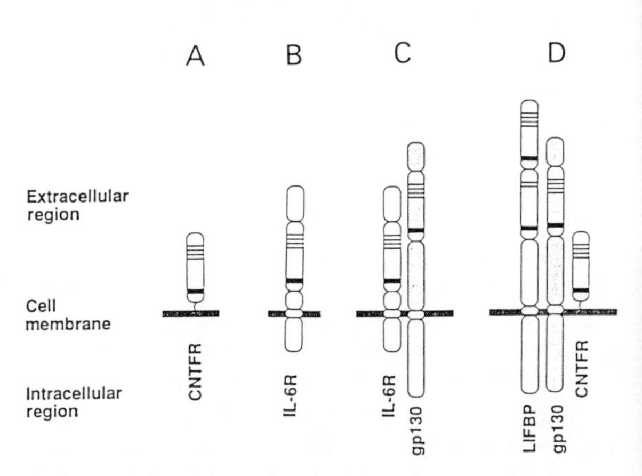

Fig. 11.7. Schematic structure of the receptors for CNTF and IL-6 and their functional form in combination with association proteins. The cell membrane is indicated by horizontal bars. Thin and heavy horizontal lines in the extracellular domain represent conserved amino acid sequences common to different receptor subunits. A, low-affinity CNTF receptor. B, low-affinity IL-6 receptor. C, high-affinity IL-6 receptor formed with gp130. D, high-affinity CNTF receptor formed with LIF binding protein (LIFBP) and gp130. (Adapted from Miyajima *et al.* 1992a.)

but in its presence IL-6 receptors have higher affinity for IL-6. Even an IL-6 receptor mutant that lacks the cytoplasmic domain can still transmit the IL-6 signal if it is associated with gp130 (Taga *et al.* 1989). These observations suggest that the functional high-affinity IL-6 receptor is a dimer composed of the IL-6 receptor and gp130 (Fig. 11.7C; Miyajima *et al.* 1992a). Thus, while IL-6 binds to the extracellular domain of the IL-6 receptor, its signal transduction employs gp130's cytoplasmic domain.

Absence of the cytoplasmic signalling domain marks both the CNTF receptor and the IL-6 receptor. In addition, structural similarities in their extracellular domains imply that the CNTF receptor is related to the haematopoietic growth factor receptor. Davis *et al.* (1991) proposed that as with IL-6, the functional CNTF receptor requires non-ligand-binding gp130 as an association protein. Indeed, CNTF actions on neuronal cells can be blocked by anti-gp130 antibodies (Ip *et al.* 1992b). Moreover, CNTF stimulation induces tyrosine phosphorylation of gp130, directly implicating gp130 in the CNTF signalling process. Finally, CNTF stimulation causes tyrosine phosphorylation of a 190 000 dalton cell-surface protein (gp190) that is physically associated with gp130. Because the gp190 protein matches the *leukaemia inhibitory factor* (LIF) receptor in size (Gearing *et al.* 1991; see below), a tentative theory is that a composite structure forms the functional CNTF receptor. Specifically, once CNTF binds to its receptor, both gp130 and the LIF receptor (or LIF-binding protein, LIFBP) serve as its signal transducing components (Fig. 11.7D; Ip *et al.* 1992b; Miyajima *et al.* 1992a; Taga and Kishimoto 1992).

The proposed involvement of LIF is highly suggestive. Another cytokine, LIF induces a wide variety of activities, many of which resemble those of IL-6. LIF further resembles IL-6 in that they both modulate the growth and differentiation of both haematopoietic cells and neuronal cells. Indeed, some haematopoietic cytokines appear to function as a 'neuropoietic' cytokine (Bazan 1991). It is quite possible that neurotrophic factors and cytokines have basically similar functions; they may have strong evolutional ties. This idea will be addressed in section 12.2.3. For now, let us consider the possible functional role of CNTF.

11.3.2 Lesion-induced neurotrophic actions

Although when CNTF was purified it was expected to maintain the survival of ciliary ganglionic neurones of chick embryos (Barbin *et al.* 1984), as noted earlier CNTF does not bind to these neurones (Squinto *et al.* 1990). Moreover, CNTF fails to prevent natural cell death of chick ciliary ganglionic neurones; instead, CNTF rescues motor neurones from natural cell death (Oppenheim *et al.* 1991). How can these discrepancies be interpreted? We do not know for sure, but suspicions about the criterion used in purifying CNTF may be relevant. Ciliary neurones cultured from chick embryos ordinarily die within 24 hours, but extracts from certain tissues can ensure their survival beyond even 10–20 days (Nishi and Berg 1979; Tuttle *et al.* 1980). The existence of a neurotrophic factor for ciliary neurones was deduced from these protracted survival times. When attempts were made to purify CNTF, however, long-term survival was not the basis of the assay: the ability to support 24 hour survival of ciliary neurones in culture was considered sufficient to characterize the ciliary neurotrophic activity (Barbin *et al.* 1984). This looser standard could well be confounding, for some nutritive molecules might be able to delay neuronal death (e.g., for 24 hours), but not sustain long-term survival (Barde 1988; Eckenstein *et al.* 1990). Thus CNTF purified by short-term survival assays might not be the trophic factor involved in maintaining long-term survival of ciliary neurones.

These doubts have been partly corroborated. When Eckenstein *et al.* (1990) purified a trophic factor from the chick sciatic nerve using long-term survival assays (9 days) of chick embryonic ciliary neurones, a molecule other than CNTF was isolated. This trophic factor, termed *growth-promoting activity* (GPA), was soon cloned (Leung *et al.* 1992). It consists of 195 amino acids, and its sequence shows 50 per cent identity to that of CNTF. The GPA molecule differs from CNTF, however, in three important aspects: first, GPA can maintain long-term survival of ciliary neurones (Eckenstein *et al.* 1990); second its mRNA is expressed in the target tissues of ciliary neurones during the natural cell death period at embryonic stages (Leung *et al.* 1992); third, GPA expressed in cultured cells can be released to the extracellular media (Leung *et al.* 1992). In other words, GPA is not a cytosolic protein and conforms to the general features of a target-derived neurotrophic factor. GPA may be a distinct member of the CNTF family; alternatively, it may be the chicken form of CNTF (Leung *et al.* 1992). Clearly, much more information about GPA is needed to evaluate its physiological significance *in vivo*.

Despite the doubts about short-term survival assays,

CNTF is undeniably present *in vivo*. Responsiveness to CNTF is shown by a large variety of neurones, including sensory, sympathetic, and motor neurones (Thoenen 1991). But recall that CNTF does not appear to be a target-derived neurotrophic factor for neuronal survival at embryonic stages or in maturity. We have seen that this notion is based on three features of CNTF (section 11.3.1): first, CNTF is present exclusively in Schwann cells, never in target tissues (Stockli *et al.* 1989; Rende *et al.* 1992); second, CNTF cannot be secreted into the extracellular medium (Lin *et al.* 1989; Stockli *et al.* 1989); third, CNTF is expressed only after birth, not at embryonic stages (Stockli *et al.* 1989). Given these attributes, what functional role can we ascribe to CNTF?

From the outset Stockli *et al.* (1989) suspected that CNTF exerts its trophic function under pathological conditions, making it a 'lesion factor' (Thoenen 1991): one that functions when tissues are injured. The best illustration of this role for CNTF involves facial nerve transection (axotomy) in the neontal rat. About 80 per cent of the facial motor neurones axotomized at birth die within one week; but when CNTF is applied to the central sump of the cut facial nerve, more than 75 per cent are still alive one week after lesion (Sendtner *et al.* 1990). What does this result imply? Sendtner *et al.* (1990) assumed that CNTF is released from Schwann cells on nerve injury and prevents the cell death of axotomized motor neurones. For instance, CNTF released from Schwann cells may be taken up by the cut nerve end and retrogradely transported to produce a remedial effect on the axotomized motor neurones. But since CNTF expression begins only after birth and remains scarce in neonatal rats, axotomy will cause motoneuronal death unless exogenous CNTF is applied locally. Indeed, as CNTF expression increases motoneuronal survival in rats improves, growing resistant to axotomy in the week or two after birth. Also, CNTF can be transported retrogradely in both motor and sensory neurones when the peripheral nerve is transected (Curtis *et al.* 1993). Thus, this hypothesis explains well how CNTF functions as a lesion factor, but it is not consistent with other experimental results. For instance, axotomy-induced motor neurone death in neonatal rats does not occur if the cut motor fibres are immediately allowed to re-innervate skeletal muscle (Kashihara *et al.* 1987). Since CNTF is not present in skeletal muscle (Stockli *et al.* 1989), axotomy-induced motor neurone death in neonates cannot be accounted for by the absence of CNTF alone.

In the normal sciatic nerve, the CNTF-immunoreactivity is confined to the cytoplasm of intact Schwann cells, whereas after nerve injury it is detectable in the extracellular space (Sendtner *et al.* 1992b). Thus, CNTF seems unavailable for neurones unless nerve lesion occurs. However, this release of CNTF must be transient since the injured cell membrane reseals within about 30 minutes (Yawo and Kuno 1983; Xie and Barrett 1991). It is not clear how long the action of CNTF is maintained after injury.

Going back a step, how is CNTF expression regulated within Schwann cells? If CNTF is a lesion factor, we may expect its expression to be enhanced (up-regulated) following nerve lesion. But on the contrary, the CNTF mRNA levels in the distal region of damaged sciatic nerve decrease within a few days in adult rats (Fig. 11.8), recovering only when the nerve axons regenerate

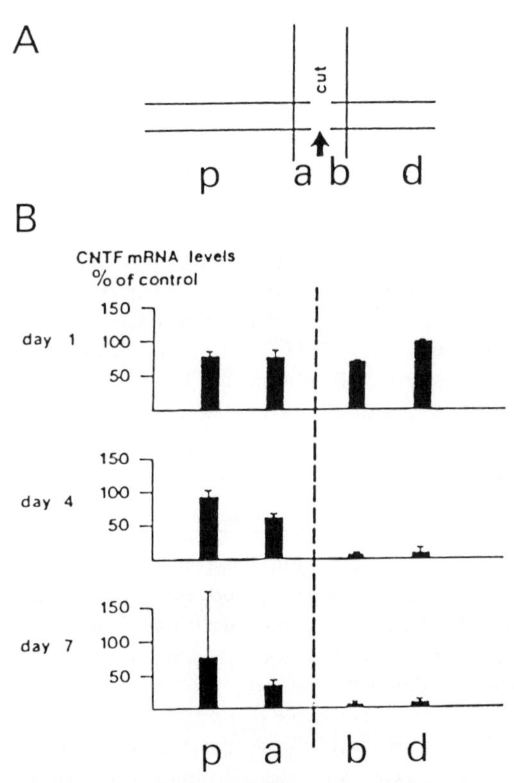

Fig. 11.8. Levels of CNTF mRNA in the rat sciatic nerve after transection. A, diagram of the sciatic nerve assayed: p, proximal to the transection (cut); a and b, 2 mm segments proximal and distal to the lesion, respectively. d, distal to the transection. B, CNTF mRNA levels plotted as a percentage of each region in the normal nerve 1, 4, and 7 days after nerve section. Vertical bars, standard error of the mean. (Adapted from Sendtner *et al.* 1992b.)

(Friedman *et al.* 1992; Sendtner *et al.* 1992b). Thus, CNTF expression in Schwann cells is probably regulated by nerve axon–Schwann cell contacts, but the functional significance of CNTF remains obscure. Masu *et al.* (1993) generated mice with targeted disruption of the CNTF gene. The absence of the CNTF gene did not affect the survival of motor neurones in the mouse up to 4 weeks, and at 28 weeks the animal showed a 25 per cent loss of motor neurones. How these results can be interpreted is not clear. It would be interesting to see if axotomy causes motor neurone death even after maturation in these CNTF null mutant mice.

R. Takahashi *et al.* (1994) found an RNA splicing mutation in the human gene encoding CNTF; the mutant mRNA codes for an anomalous CNTF protein of 62 amino acid residues. Unexpectedly, about 2.3 per cent of 391 Japanese subjects examined showed homozygous mutation in the CNTF gene but display no apparent abnormalities in their neuronal function. In other words, disruption of the CNTF gene causes no obvious abnormality in human.

The down-regulation of CNTF in Schwann cells by nerve lesion is the opposite to the lesion-mediated up-regulation of NGF and its low-affinity receptor (p75). As Fig. 11.9A shows, the NGF mRNA levels in Schwann cells have an initial transient increase and a second prolonged enhancement in the segment just distal to the cut in the sciatic nerve (Heumann *et al.* 1987b; Meyer *et al.* 1992). Similarly, the BDNF mRNA in Schwann cells elevates distal to the lesion in the sciatic nerve, although this up-regulation is markedly different in time course (Fig. 11.9B; Meyer *et al.* 1992). The second phase of the NGF mRNA increase is known to be induced by IL-1 (interleukin-1) released from macrophages which invade the lesion site (Lindholm *et al.* 1987), whereas BDNF expression is not affected by IL-1 (Meyer *et al.* 1992). In addition to NGF and BDNF, Schwann cells in injured nerve newly synthesize the mRNA encoding low-affinity p75 NGF receptors (Fig. 11.9C; Heumann *et al.* 1987b). What would be the functional role of neurotrophins expressed in Schwann cells?

Johnson *et al.* (1988; see also Taniunchi *et al.* 1988) hypothesized that the neurotrophin system in Schwann cells of damaged nerve benefits regenerating nerve fibres. As outlined in Fig. 11.10, after transection of a peripheral nerve, the loss of nerve axon–Schwann cell contacts causes up-regulation of NGF and low-affinity NGF receptors (NGF-R); macrophages invading the lesion site further amplify NGF expression. The newly synthesized NGF molecules are then anchored by the low-affinity NGF receptors (p75) expressed on the cell surface. Because regenerating nerve fibres (e.g., sensory and sympathetic fibres) contain high-affinity NGF receptors, they can pick up the loosely held NGF molecules, internalizing them to facilitate neurite extension. If so, Schwann cells at the lesion site would amount to a 'substitute target' (Heumann *et al.* 1987a) because of locally concentrated neurotrophins. This notion is con-

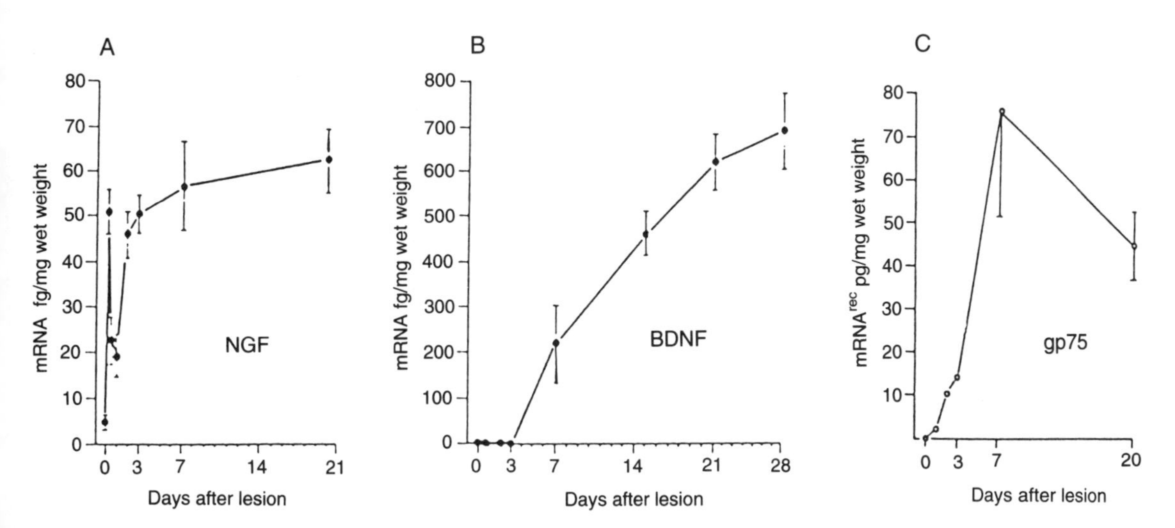

Fig. 11.9. Post-lesion time course of up-regulation for mRNA levels of NGF (A), BDNF (B) and the low-affinity NGF receptor, measured distal to the transection in the rat sciatic nerve. Vertical bars, standard error of the mean. (A and B, adapted from Meyer *et al.* 1992. C, adapted from Heumann *et al.* 1987b.)

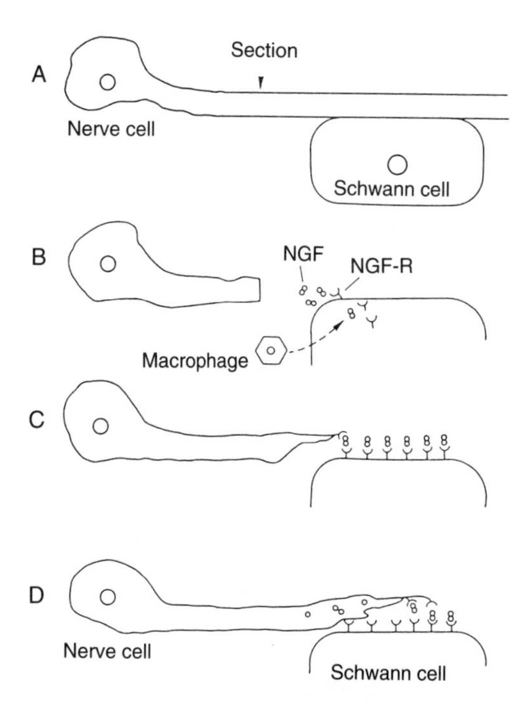

Fig. 11.10. Diagram of the hypothetical role of NGF and low-affinity NGF receptors (NGF-R) synthesized in Schwann cells in a cut peripheral nerve. Following nerve section (A), Schwann cells around degenerating nerve fibres synthesize NGF and NGF-R molecules (B). Macrophages invade the lesion site and release IL-1, which further stimulates NGF synthesis in Schwann cells (B). NGF produced by Schwann cells anchors to the NGF-R expressed on the Schwann cell surface (C), available for uptake by regenerating nerve fibres (D). (Based on Johnson *et al.* 1988.)

sistent with the results observed in a special strain of mouse, C57BL/ola. This strain of animal shows poor regeneration of peripheral sensory fibres after nerve transection, and this feature is associated with low levels of expression of NGF and its p75 receptor mRNAs at the lesion site (Brown *et al.* 1991).

As noted, the CNTF receptor is expressed in skeletal muscle (Davis *et al.* 1991). Helgren *et al.* (1994) found that CNTF prevents significantly denervation-induced atrophy in rat muscle. Thus, CNTF exerts *myotrophic* effects. In other words, upon nerve injury CNTF may promote the survival of axotomized motor neurones (Sendtner *et al.* 1990) and reduce the atrophy of denervated muscle, representing a dual action on neurone and target. This dual action of CNTF was shown in the sexually dimorphic bulbocavernous and its motor neurones (the spinal nucleus of the bulbocavernous; SNB) in rats. The perineal muscles show marked atrophy in neonatal female rats, and correspondingly the number of SNB neurones is three to four times fewer in the female than in the male (Arnold 1984; Breedlove 1986). Both the muscle atrophy and the cell death of SNB neurones in female rats were prevented by daily treatment with CNTF at perinatal stages (Forger *et al.* 1994). CNTF may act directly on SNB neurones as well as on the bulbocavernous. It is also possible that the sparing effect of CNTF on SNB neurones is mediated by its influence on the target bulbocavernous (Forger *et al.* 1994; see also Araki *et al.* 1991).

Exogenously applied CNTF can rescue motor neurones from cell death *in vitro* (Arakawa *et al.* 1990). Yet, CNTF does not fulfil all criteria of a neurotrophic factor for motor neurones *in vivo*. Almost certainly the endogenous neurotrophic factor for motor neurones is distinct from CNTF—but its identity remains ambiguous. Let us see what clues can be found in the area of promoting motor neurone survival.

11.4 MOTONEURONAL TROPHIC FACTORS

By analogy with NGF in relation to sensory and sympathetic neurones, it is reasonable to assume that motor neurones respond to their own survival-promoting factor. What criteria identify a motoneuronal trophic factor? For us to accept a factor, say F, as the neurotrophic factor for motoneuronal survival, three criteria apply: first, F exhibits survival-promoting effects on motor neurones; second, F is available for motor neurones *in vivo*, being present in the target muscle or systemically in blood or interstitial fluid; third, motor neurones express binding sites or receptors for F. In other words, its presence, its action, and the responsiveness to it are the three criteria for identifying a neurotrophic molecule.

Survival-promoting effects are much easier to assay *in vitro* than *in vivo*. But *in vitro* results must be interpreted with caution, because the effects observed in culture conditions do not always reflect those *in vivo* (Barde 1988). Moreover, as noted earlier, trophic effects exerted by an exogenously applied agent do not prove that the agent is an endogenous factor. Ideally, proof should consist of blocking the endogenous factor by its antibodies and subsequently impairing neuronal survival. In reality, however, of all the neurotrophic factor candidates, only NGF meets every defining criterion. It is then necessary to somewhat alleviate the rigour of our definition. In practice, a substance that exhibits survival-promoting effects on motor neurone either *in vivo* or *in* vitro is widely considered to qualify as a potential motoneuronal neurotrophic factor. Let us see which candidates currently meet such modified criteria—and which among them seems likely to be the true motoneuronal trophic factor.

11.4.1 Survival-promoting factors for motor neurones

We have no evidence that NGF has any biological role in motor neurones. The application of NGF prevents natural cell death in the dorsal root ganglia (DRG; Hamburger *et al.* 1981) and sympathetic ganglia (Levi-Montalcini and Angeletti 1968), but not in motor neurones (Oppenheim *et al.* 1982). In neonatal animals, NGF can prevent axotomy-induced cell death in DRG neurones and sympathetic neurones (Hendry 1975; Purves 1975; Miyata *et al.* 1986), but NGF treatment rather enhances cell death in axotomized motor neurones (Miyata *et al.* 1986; Sendtner *et al.* 1992a). Moreover, retrograde axonal transport of NGF occurs in DRG cells and sympathetic neurones, but not in motor neurones (Stockel *et al.* 1975; Brunso-Bechtold and Hamburger 1979). Surprisingly, however, spinal motor neurones in chick embryos (Raivich *et al.* 1985) and in perinatal or neonatal rats transiently express low-affinity NGF (neurotrophin) receptors (p75; Yan *et al.* 1988) and also show transient retrograde axonal transport of NGF (Wayne and Heaton 1988; Yan *et al.* 1988). The expression of the low-affinity NGF receptor mRNA in motor neurones subsides to undetectable levels in adult rats, but revives with axotomy (Ernfors *et al.* 1989; Wood *et al.* 1990). This ability to produce p75 suggests that motor neurones in perinatal rats might be responsive to NGF. But not so: even in neonatal rats, NGF fails to rescue axotomized motor neurones from cell death and does not affect neuronal size (Miyata *et al.* 1986; Yan *et al.* 1988; Sendtner *et al.* 1992a). What functional significance can low-affinity NGF receptors have in motor neurones? By analogy with the role proposed for low-affinity NGF receptors in Schwann cells (Fig. 11.10), Yan *et al.* (1988) suggested that the low-affinity NGF receptors expressed on the motor neurone cell surface anchor neurotrophins, thereby increasing the local neurotrophin concentration to support primary sensory neurones or interneurones as they converge on the motor neurones in early development. Whether or not this speculation is warranted, at least we are sure that no evidence points to NGF as a neurotrophic factor acting on motor neurones themselves.

As described in section 11.3.2, CNTF has survival-promoting effects on cultured motor neurones dissociated from chick embryos (Arakawa *et al.* 1990). Moreover, CNTF significantly reduces the magnitude of natural cell death of chick motor neurones *in vivo* (Oppenheim *et al.* 1991). The features of CNTF, however, contraindicate an endogenous neurotrophic factor on motor neurones (section 11.3.2). How CNTF rescues motor neurones from natural cell death or axotomy-induced cell death is therefore puzzling. Perhaps this survival-promoting effect of CNTF on motor neurones is a secondary consequence of its myotrophic action on their targets (Forger *et al.* 1994; see section 11.3.2). Alternatively, exogenously applied CNTF cross-reacts with the receptor for the structurally homologous GPA (growth-promoting activity) molecule, which is a candidate for the motoneuronal trophic factor (sections 11.3.2 and 11.4.2).

The results from several laboratories now point to the possibility that BDNF is a neurotrophic factor for motor neurones. Exogenously applied BDNF achieves marked survival-promoting effects on motor neurones during the natural cell death period in chick embryos or following axotomy in neonatal rats (Oppenheim *et al.* 1992; Sendtner *et al.* 1992a; Yan *et al.* 1992; Henderson *et al.* 1993; Koliatsos *et al.* 1993); yet BDNF fails to rescue cultured chick motor neurones from cell death (Arakawa *et al.* 1990). Sendtner *et al.* (1992a) suggested that the trophic function of BDNF requires some crucial co-factor absent from cultured conditions. With sensory neurones, however, BDNF supports survival both in culture (Lindsay *et al.* 1985) and *in vivo* (Oppenheim *et al.* 1992). Targeted disruption of the BDNF high-affinity receptor (*trk*B) resulted in the death of the animal on the day after birth because of feeding inability (Klein *et al.* 1993). The *trk*B null mutant mice also showed a significant loss of facial motor neurones (by 70 per cent), lumbar motor neurones (by 35 per cent), trigeminal ganglion cells (by 60 per cent) or DRG cells (by 30 per cent). Presumably, BDNF must be involved in the maintenance of neuronal survival for both motor and sensory neurones. Interestingly, the sacral motor neurones of the mutant mice were not affected, while their lumbar motor neurones were significantly lost. Thus, BDNF appears to be required for survival of a subpopulation of motor neurones. Since NT-3 and NT-5 also have survival-promoting effects on motor neurones (Sendtner *et al.* 1992a; Henderson *et al.* 1993), these neurotrophins or other neurotrophic factors may also be involved in the maintenance of motoneuronal survival.

In fact, unlike targeted disruption of *trk*B (Klein *et al.* 1993a), mice lacking the BDNF gene did not affect survival of facial motor neurones (Ernfors *et al.* 1994; Jones *et al.* 1994). This suggests that NT-5 (also called NT-4/5) supports motoneuronal survival, since *trk*B is the preferred receptor for BDNF and NT-4/5 as well (Davies 1994).

We have seen that CNTF not only supports cultured motor neurones but rescues axotomized motor neurones from cell death *in vivo*. When BDNF similarly reduces the magnitude of axotomy-induced cell death for facial motor neurones in neonatal rats, the surviving motor neurones nonetheless display morphological features of the axon reaction (e.g., cell atrophy and nuclear eccentricity; Sendtner *et al.* 1992a). By contrast, these morphological changes seldom occur in axotomized facial motor neurones treated with CNTF (Sendtner *et al.* 1990). This difference is puzzling but suggests that BDNF and CNTF rescue axotomized motor neurones from cell death, based on different mechanisms.

Arakawa *et al.* (1990) examined the survival-promoting effects of 17 molecules, using motor neurones cultured from chick embryos as a bioassay system. All 17 molecules were known to have survival- or growth-promoting effects on neurones, haematopoietic cells, and other non-neuronal cells. Besides CNTF, *fibroblast growth factor* (FGF) and, to a lesser extent, *insulin-like growth factor* (IGF) showed survival-promoting effects on cultured motor neurones. When CNTF was combined with FGF, CNTF with IGF or FGF with IGF, the survival-promoting effects were additive, suggesting that the three molecules use separate mechanisms.

Two forms of FGF, acidic and basic (also known as FGF-1 and FGF-2), both exert mitogenic effects in a variety of cells in culture (Barde 1988; Yamamori 1992). FGF promotes neurite growth and cell survival in a wide array of neurone types (Yamamori 1992). Tissue distribution of FGF is extensive and abundant. Moreover, acidic or basic FGF is a cytosolic protein and cannot be secreted from cells (Klagsbrun 1989). None of these features supports FGF as a neurotrophic factor specific for motor neurones *in vivo*. Hughes *et al.* (1993) suggested that FGF-5 is a possible target-derived trophic factor for motor neurones. This suggestion is based on three features of FGF-5: first, unlike other forms of FGF, FGF-5 can be released to the extracellular space; second, FGF-5 is expressed in embryonic muscle; third, FGF-5 supports the survival of motor neurones cultured from chick embryos. Whether FGF-5 exhibits survival-promoting effects on motor neurones *in vivo* is not known. Also, whether survival-promoting effects by FGF-5 are specific for motor neurones should be explored.

IGF, like FGF, exists widely in various tissues at relatively high concentrations (Baskin *et al.* 1988). IGF also shows mitogenic effects in a variety of cells in culture. Might FGF and IGF act as co-factors, enhancing the function of other neurotrophic factors? This possibility is suggested by their additive effects on motoneuronal survival when combined with CNTF (Arakawa *et al.* 1990). IGF exerts survival-promoting effects on motor neurones *in vivo* and *in vitro* (Lewis *et al.* 1993; Ness *et al.* 1993). Whether IGF plays an essential role in maintaining motoneuronal survival during development or in maturity remains unknown, however.

In brief, the search for motoneuronal neurotrophic factors is not encouraging. One problem may be that all the molecules discussed here as candidates for the motoneuronal trophic factor were selected for their trophic effects on other cells. While not unlikely, there is no guarantee that the molecules with trophic action on neuronal or non-neuronal cells also act as trophic factors for motor neurones. Motoneuronal trophic factors might well be specific only for motor neurones. If so, a systematic, logical approach would be required for identification of the motoneuronal trophic factor. Such a systematic search for motoneuronal trophic factors has begun, and we examine this next.

11.4.2 The search for motoneuronal trophic factors

Since retrograde neurotrophic factors are derived from target tissues, logically, the search for motoneuronal trophic factors should begin among candidate molecules extracted from skeletal muscle. NGF (or its mRNA) exists in the target tissues innervated by sympathetic neurones (Korsching and Thoenen 1983a; Heumann *et al.* 1984), but not abundantly. For instance, the NGF content in the submandibular gland, atrium or iris of the rat ranges from 0.5 to 2 ng/g wet weight of the target tissue. NGF still manages to exert its neurotrophic action, because the dissociation constant of the high-affinity NGF receptor is about 10^{-11} M (section 11.2.2). If low abundance and high potency are characteristic of all target-derived neurotrophic molecules, the task of purifying the motoneuronal trophic molecule from skeletal muscle is not easy. Nonetheless, some partially purified proteins from chick or rat skeletal muscle showed survival-promoting effects on motor neurones cultured from chick or rat embryos (Dohrmann *et al.* 1986; Smith *et al.* 1986). These findings underlie the continuing search for motoneuronal trophic factors derived from skeletal muscle.

McMananman *et al.* (1988) purified a protein from rat skeletal muscle that stimulates the activity of choline acetyltransferase, the ACh synthetic enzyme, in spinal neurones cultured from rat embryos. This protein is called *choline acetyltransferase development factor* (ChADF). ChADF significantly rescues the natural cell death of lumbar motor neurones of chick embryos *in vivo*, whereas it does not affect the survival of dorsal root ganglion cells or sympathetic preganglionic neurones (McManaman *et al.* 1990). Because ChADF occurs in skeletal muscle and promotes the survival of motor neurones specifically, ChADF may well be a target-derived trophic factor for motor neurones. The molecular weight of ChADF is estimated to be 20 000 to 22 000, but its primary structure remains unknown.

ChADF resembles CNTF in several respects: molecular weight; isoelectric point (4.8 to 5.2), the pH at which the protein has no net charge; ability to rescue motor neurones of chick embryos from natural cell death but not dorsal root ganglion cells *in vivo*. Might they be identical? The estimated amino acid composition of ChADF (McManaman *et al.* 1988) differs slightly but distinctly from that of CNTF (Stockli *et al.* 1989). For example, ChADF is rich in glycine residues, whereas CNTF is rich in leucine residues. Thus, they are distinct molecules but could belong to the same gene product family. GPA, a member of the CNTF family (Leung *et al.* 1992), is also rich in leucine residues, like CNTF rather than ChADF. Unlike CNTF, however, GPA is expressed at early embryonic stages and can be secreted from cells. Since GPA molecule is expressed in the target tissues innervated by the ciliary ganglion cells during the natural cell death period, it may be a target-derived neurotrophic factor for ciliary neurones (Leung *et al.* 1992). But why then is GPA also expressed in the sciatic nerve? Is GPA also expressed in skeletal muscle? Does it promote survival of motor neurones? These issues have yet to be examined.

The CNTF content in the sciatic nerve is relatively high, being about 1 mg/g tissue (Manthorpe *et al.* 1986), whereas the ChADF content in skeletal muscle is abou 40 ng/g tissue (McManaman *et al.* 1988). As noted, if low abundance is indeed a feature of target-derived neurotrophic factors, it will be difficult to purify enough trophic factor to determine its primary structure (amino acid sequence). Fortunately, as we have seen, a method exists for deducing the primary structure of transmitter receptors or ionic channels from their cDNA clones. Lam *et al.* (1992) applied the same strategy to detect cDNA clones encoding neurotrophic factors, using CNTF as an example. Rat C6 glioma cells are known to express CNTF, so mRNAs in these cells must encode CNTF. The mRNAs extracted from rat C6 glioma cells were injected into *Xenopus* oocytes for translation of the exogoneous mRNAs. Extracts from the injected oocytes, but not from uninjected oocytes, contained a factor that stimulates choline acetyltransferase activity in ciliary ganglion cells of chick embryos. Since exogenously applied CNTF has the same effect in ciliary ganglion cells, this bioassay was used to screen the mRNAs encoding CNTF. The mRNAs were then fractionated according to their size. Only a particular fraction directed translation of the products with CNTF-like activity in the injected oocytes. The fractionated mRNA was converted into cDNA, and the cDNA library was found to contain a CNTF clone.

This approach was previously used to clone the receptor for a peptide (Masu *et al.* 1987). It is a useful strategy because, in principle, it lets the primary structure of any protein be determined without information of its partial amino acid sequences. Rassendren *et al.* (1992), with a similar approach, examined the neural control of the expression of a *motor neurone growth-promoting factor* in skeletal muscle, although not its primary structure. The oocyte expression system enabled Kashihara *et al.* (1993) to find that mRNAs purified from the hindlegs of chick embryos encode an intriguing protein: it not only exhibits survival-promoting effects on motor neurones in culture but prevents natural cell death of chick motor neurones as well. Although, unfortunately, the neurotrophic factor for motor neurones still eludes us, the pace of current research promises to identify it soon.

Summary and prospects

Nerve growth factor (NGF) monopolized the 'neurotrophism market' for more than 30 years, until other members of the neurotrophin family were discovered in the late 1980s. Maintenance by BDNF and NT-3 now accounts for the survival of some NGF-unresponsive sensory neurones. Targeted disruption of the NGF, BDNF or NT-3 receptor gene now provides relevant information regarding the role of the neurotrophin family. Different members of the neurotrophin family appear to maintain the survival of different types of sensory neurone, in terms of neurone size or sensory modality.

Targeted disruption of the BDNF receptor (*trk*B) gene results in massive cell death of motor neurones as well as sensory neurones. BDNF is retrogradely transported in motor neurones and promotes survival of motor neurones *in vivo*. Thus, BDNF was suggested to be one of the neurotrophic factors for motor neurones. Surprisingly, however, survival of motor neurones was not affected in mice lacking the BDNF gene. Since *trk*B is the preferred receptor for BDNF and NT-4/5, it may be NT-4/5 that promotes motoneuronal survival *in vivo*.

Another distinct neurotrophic family comprises CNTF and GPA. Although CNTF has survival-promoting effects for a variety of neurones, including motor, sensory, and sympathetic neurones, its physiological role remains elusive. CNTF was suggested to be a 'lesion factor' that functions when tissues are damaged. For now, there is no experimental evidence that directly supports this notion. CNTF gene deletion induces motoneuronal cell death, but it occurs in only 25 per cent of the motor neurones and only when the animal matures. CNTF might protect age-induced neuronal death.

NGF remains a paradigm of the best characterized retrograde neurotrophic factor derived from target tissues. NGF released from a target cell is taken up by the nerve terminal of its innervating sensory or sympathetic neurone, internalized, and transported to the neurone cell body. NGF internalization is mediated exclusively by the high-affinity NGF receptor, *trk*. This receptor has tyrosine kinase activity in its cytoplasmic domain. NGF binding to the receptor induces autophosphorylation of multiple tyrosine residues in the cytoplasmic kinase, which in turn phosphorylates various signal proteins or intracellular substrates. Thus, the signal transduction induced by receptor activation is pleiotropic. Different intracellular second messengers induce local cellular responses, whereas the internalized NGF is transported to the neurone cell body and induces nuclear responses by unknown means. Both the low-affinity neurotrophin receptor (p75) and the high-affinity NGF receptor (*trk*) are transported along with NGF. Two major missing items of information in our understanding of NGF's trophic function are the role of p75 and the relationship between the local and nuclear responses elicited by NGF.

Suggested reading

A general view of neurotrophic factors

Barde, Y. A. (1989). Trophic factors and neuronal survival. *Neuron*, 2, 1525–34.

Ip, N. Y., Maisonpierre, P., Alderson, R., Friedman, B., Furth, M. E., Panayotatos, N., Squinto, S., Yancopoulos, D., and Lindsay, R. M. (1991). The neurotrophins and CNTF: specificity of action towards PNS and CNS neurones. *Journal de Physiologie (Paris)* **85**, 123–30.

Korsching, S. (1993). The neurotrophic factor concept: a reexamination. *Journal of Neuroscience*, 13, 2739–48.

Snider, W. D., Elliott, J. L., and Yan, Q. (1992). Axotomy-induced neuronal death during development. *Journal of Neurobiology*, 23, 1231–46.

Thoenen, H. (1991). The changing scene of neurotrophic factors. *Trends in Neurosciences*, 14, 165–70.

Neurotrophins and their receptors

Carroll, S. L., Silos-Santiago, I., Frese, S. E., Ruit, K. G., Milbrandt, J., and Snider, W. D. (1992). Dorsal root ganglion neurons expressing *trk* are selectively sensitive to NGF deprivation in utero. *Neuron*, 9, 779–88.

Chao, M. V. (1992). Neurotrophin receptors: A window into neuronal differentiation. *Neuron*, 9, 583–93.

Crowley, C., Spencer, S. D., Nishimura, M. C., Chen, K. S., Pitts-Meek, S., Armanin, M. P., Ling, L. H., McMahon, S. B., Shelton, D. L., Levinson, A. D., and Phillips, H. S. (1994). Mice-lacking nerve growth factor display perinatal loss of sensory and sympathetic neurons yet develop basal forebrain cholinergic neurons. *Cell*, 76, 1001–11.

Davies, A. M., Lee, K. F., and Jaenisch, R. (1993). p75–deficient trigeminal sensory neurons have an altered response to NGF but not to other neurotrophins. *Neuron*, 11, 565–74.

DiStefano, P., Friedman, B., Radziejewski, C., Alexander, C., Boland, P., Schick, C. M., Lindsay, R. M., Wiegand, S. J. (1992). The neurotrophins BDNF, NT-3, and NGF display distinct patterns of retrograde axonal transport in peripheral and central neurons. *Neuron*, 8, 983–93.

Ernfors, P., Lee, K. F., Kucera, J., and Jaenisch, R. (1994). Lack of neurotrophin-3 leads to deficiencies in the peripheral nervous system and loss of limb proprioceptive afferents. *Cell*, 77, 503–12.

Jones, K. R., Farinas, I., Backus, C., and Reichardt, L. F. (1994). Targeted disruption of the BDNF gene perturbs brain and sensory neuron development but not motor neuron development. *Cell*, 76, 989–99.

Klein, R., Smeyne, R. J., Wurst, W., Long, L. K., Auerbach, B. A., Joyner, A. L., and Barbacid, M. (1993). Targeted disruption of the *trk*B neurotrophin receptor gene results in nervous system lesions and neonatal death. *Cell*, 75, 113–22.

Lee, K. F., Li, E., Huber, L. J., Landis, S. C., Sharpe, A. H., Chao, M. V., and Jaenisch, R. (1992). Targeted mutation of the gene encoding the low affinity NGF receptor p75 leads to deficits in the peripheral sensory nervous system. *Cell*, 69, 737–49.

Meakin, S. O. and Shooter, E. M. (1992). The nerve growth factor family of receptors. *Trends in Neurosciences*, **15**, 323–31.

Mu, X., Silos-Santiago, I., Carroll, S. L., and Snider, W. D. (1993). Neurotrophin receptor genes are expressed in distinct patterns in developing dorsal root ganglia. *Journal of Neuroscience*, **13**, 4029–41.

Ciliary neurotrophic factor

Barbin, G., Manthorpe, M., and Varon, S. (1984). Purification of the chick eye ciliary neurotrophic factor. *Journal of Neurochemistry*, **43**, 1468–78.

Davis, S., Aldrich, T. H., Valenzuela, D. M., Wong, V., Furth, M. E., Squinto, S. P., and Yancopoulos, G. D. (1991). The receptor for ciliary neurotrophic factor. *Science*, **253**, 59–63.

Davis, S. and Yancopoulos, G. D. (1993). The molecular biology of the CNTF receptor. *Current Opinion in Neurobiology*, **3**, 20–4.

Helgren, M. E., Squinto, S. P., Davis, H. L., Parry, D. J., Boulton, T. G., Heck, C. S., Zhu, Y., Yancopoulos, G. D., Lindsay, R. M., and DiStefano, P. S. (1994). Trophic effect of ciliary neurotrophic factor on denervated skeletal muscle. *Cell*, **76**, 493–504.

Leung, D. W., Parent, A. S., Cachianes, G., Esch, F., Coulombe, J. N., Nikolics, K., Eckenstein, F. P., and Nishi, R. (1992). Cloning, expression during development, and evidence for release of a trophic factor for ciliary ganglion neurons. *Neuron*, **8**, 1045–53.

Masu, Y., Wolf, E., Holtmann, B., Sendtner, M., Brem, G., and Thoenen, H. (1993). Disruption of the CNTF gene results in neuron degeneration. *Nature*, **365**, 27–32.

Stockli, K. A., Lottspeich, F., Sendtner, M., Masiakowski, P., Carroll, P., Gotz, R., Lindholm, D., and Thoenen, H. (1989). Molecular cloning, expression and regional distribution of rat ciliary neurotrophic factor. *Nature*, **342**, 920–3.

Motoneuronal trophic factors

Arakawa, Y., Sendtner, M., and Thoenen, H. (1990). Survival effect of ciliary neurotrophic factor (CNTF) on chick embryonic motoneurons in culture: comparison with other neurotrophic factors and cytokines. *Journal of Neuroscience*, **10**, 3507–15.

Henderson, C. E., Camu, W., Mettling, C., Gouin, A., Poulsen, K., Karihaloo, M., Rullamas, J., Evans, T., McMahon, S. B., Armanini, M. P., Berkemeier, L., Phillips, H. S., and Rosenthal, A. (1993). Neurotrophins promote motor neuron survival and are present in embryonic limb bud. *Nature*, **363**, 266–70.

Koliatsos, V. E., Clatterbuck, R. E., Winslow, J. W., Cayouette, M. H., and Price, D. L. (1993). Evidence that brain-derived neurotrophic factor is a trophic factor for motor neurones *in vivo*. *Neuron*, **10**, 359–367.

Lewis, M. E., Neff, N. T., Contreras, P. C., Strong, D. B., Oppenheim, R. W., Grebow, P. E., and Vaught, J. L. (1993). Insulin-like growth factor-l: potential for treatment of motoneuronal disorders. *Experimental Neurology*, **124**, 73–88.

McManaman, J. L., Oppenheim, R. W., Prevette, D., and Marchetti, D. (1990). Rescue of motorneurons from cell death by a purified skeletal muscle polypeptide: Effects of the ChAT development factor, CDF. *Neuron*, **4**, 891–8.

Sendtner, M., Kreutzber, G. W., and Thoenen, H. (1990). Ciliary neurotrophic factor prevents the degeneration of motor neurons after axotomy. *Nature*, **345**, 440–1.

12

Neurotrophic regulation at developing synapses

We have defined neurotrophic function as a neurally mediated influence that maintains the cell's pre-existing or programmed state. As noted, during embryogenesis neurones are over-produced, so that the final number of neurones is determined by programmed removal—natural cell death—which is regulated by the target-derived neurotrophic factor. The synthesis of target-derived neurotrophic factors presumably is triggered by neuron–target contacts, when the terminals of growing neurones arrive in their target field. Developmental changes at synapses still continue after the connections formed between a neurone and its target cell. For instance, each muscle fibre of the rat has multiple innervation sites at birth, and pruning of redundant connections during the first two weeks after birth eventually leaves only a single neuromuscular junction (Redfern 1970; Brown *et al.* 1976). This change, *synapse elimination*, also occurs at autonomic ganglionic synapses and central synapses at the early postnatal stage (Purves and Lichtman 1985).

At individual synaptic levels, the synaptic specialization develops gradually. In neuromuscular junctions, this specialization begins with the accumulation or clustering of ACh receptors at the junction. Moreover, the kinetics of ACh receptor channels also undergo developmental changes (Schuetze and Role 1987). We saw in section 10.2 that the development of extrajunctional ACh receptors in denervated muscle reflects motoneuronal regulation of gene activity in muscle. This motoneuronal regulation of extrajunctional ACh receptors is mediated primarily, if not exclusively, by electrical or contractile activity of the innervated muscle. By contrast, the developmental changes of ACh receptors appear to involve some motor neurone-derived neurotrophic factors. Evidently, these neurotrophic factors are distinct from those involved in the maintenance of neuronal survival. Several ingenious and meticulous studies have now revealed that some motor neurone-derived neurotrophic factor imprints crucial components required for synaptogenesis and receptor maturation at neuromuscular junctions. Let us trace a series of studies addressed to developmental regulation of ACh receptors by neurotrophic factors.

12.1 DEVELOPMENTAL REGULATION OF TRANSMITTER RECEPTORS

When muscle cells are cultured before motor innervation, many cells show localized patches where ACh receptors densely cluster (Fischbach and Cohen 1973; Anderson *et al.* 1977; Dennis 1981). Initially, these 'hot spots' were thought to be the sites that accept motor innervation to form junctional connections. But the neurones co-cultured with muscle cells, rather than preferentially innervating the pre-existing 'hot spots', formed their junctions in a virgin surface, thereby inducing new clusters of ACh receptors (Anderson and Cohen 1977; Frank and Fischbach 1979). Actually, *in vivo*, there was no evidence for ACh receptor clustering in muscle fibres before innervation, and ACh receptors were found to aggregate only at the site of innervation (Braithwaite and Harris 1979). Thus, nerve contacts appear to trigger the aggregation of ACh receptors.

This phenomenon invites several questions. What induces aggregation of ACh receptors at the neuromuscular junction? Might chemical substances released from the innervating motor neurone be responsible for receptor clustering? Or, alternatively, might muscle activity resulting from motor innervation induce receptor clustering? Where do the aggregated ACh receptor molecules come from? And how are they anchored at the junction?

12.1.1 How transmitter receptors aggregate

Although muscle fibres of rat embryos *in vivo* exhibit no ACh receptor clusters before motor innervation, ACh receptors are already expressed at this stage, being diffusely distributed along the muscle fibres (Bevan and Steinbach 1977; Braithwaite and Harris 1979). Soon after innervation, ACh receptors begin to accumulate at the junction, whereas extrajunctional ACh receptors diminish and become undetectable by one week after birth (Diamond and Miledi 1962; Bevan and Steinbach 1977). In muscle fibres cultured from frog embryos, Anderson *et al.* (1977) labelled the pre-existing ACh receptors with fluorescent α-bungarotoxin (BUTX); when neurones were added to muscle cultures, the labelled ACh receptors redistributed and accumulated at the innervation site within 20 to 40 hours. ACh receptors aggregated at nerve–muscle contacts even when muscle activity in cultures was blocked with high doses of BUTX or curare. ACh receptor clustering at the neuromuscular junction is thus independent of muscle activity. Moreover, when different types of neurone were cocultured with muscle cells, cholinergic neurones were the most effective in inducing ACh receptor aggregation at the nerve contact region (Kidokoro *et al.* 1980). These findings suggest that the receptor clustering at neuromuscular junctions depends on some factor specific for motor neurones, rather than on mechanical neuronal contacts. Might the same factor be required for the maintenance of densely accumulated ACh receptors at mature neuromuscular junctions?

In maturity, the density of ACh receptors at the neuromuscular junction is about 10 000 μm^2, at least 1 000 times greater than in extrajunctional regions (Fambrough and Hartzell 1972; Kuffler and Yoshikami 1975a). After denervation, the density of extrajunctional receptors increases, but the density of the junctional receptors remains unchanged for several weeks (Axelsson and Thesleff 1959; Miledi 1960; Frank *et al.* 1975). Thus, the maintenance of ACh receptor accumulations at mature junctional sites seems to be independent of motor innervation. Or might the junctional ACh receptors simply have a very slow turnover rate? Actually, denervation reduces the half-life of junctional ACh receptors from about 10 days to 1 day, indicating a 10–fold increase in their rate of turnover (Levitt *et al.* 1980; Bevan and Steinbach 1983; Shyng and Salpeter 1990). Therefore, the ACh receptors present at the old junctional region following denervation must be those that are newly synthesized and aggregate at that site. By implication, this accumulation of newly synthesized receptors tells us about the signal-directing receptor aggregation at mature junctions: it is not the motor neurones themselves, but some factor associated with the structure remaining after denervation. Where is the signal located?

Burden *et al.* (1979) examined ACh receptor clustering in regenerating muscle fibres. Figure 12.1a schematically illustrates the cross-section of a normal frog muscle fibre at its neuromuscular junction. When the muscle fibres are damaged together with their nerve, they degenerate and are phagocytized by macrophages; similarly, the nerve terminals degenerate and are phagocytized by Schwann cells. Thus, a week after the nerve and muscle damage, the principal structures remaining are the Schwann cell (S), its basal lamina (SBL), and the basal lamina of the myofibre (MBL; Fig. 12.1b; Sanes *et*

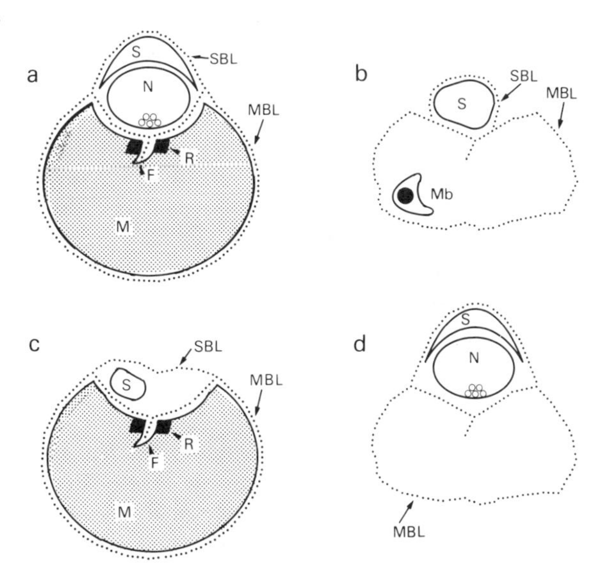

Fig. 12.1. Schematic illustration of regeneration at a neuromuscular junction. a, cross-section of a normal muscle fibre (M) at its neuromuscular junction. S, Schwann cell. SBL, Schwann cell basal lamina. MBL, myofibre basal lamina. N, nerve terminal containing synaptic vesicles (open circles). R, ACh receptor clustering. F, junctional folds. b, degeneration of nerve terminals and muscle fibres one week after damaging nerve and muscle. Mb, myoblasts. c, regeneration of muscle fibres without re-innervation. d, re-innervation of the original junctional myofibre basal lamina without regeneration of muscle fibres. (Adapted from Burden *et al.* 1979, and Sanes and Covault 1985.)

al. 1978). The myofibre basal lamina sheaths now contain only macrophages and myoblasts (Mb), and almost all ACh receptors have disappeared. Myofibres then regenerate within two weeks, and re-innervation begins at about the same time (Sanes *et al.* 1978). The regenerating nerve fibres form new junctions precisely at the original synaptic site (Gutmann and Young 1944; Letinsky *et al.* 1976). How can we tell where the original synaptic site is after denervation? This site can be identified either by staining for acetylcholinesterase remaining in the junctional cleft or by observing the myofibre basal lamina still extending into the original junctional fold (F). Removing a sufficiently long fragment of muscle nerve during muscle damage retards re-innervation, so that the regenerating myofibres are not re-innervated for more than a month. Even without re-innervation, however, ACh receptors still accumulate at the original junctional site (R in Fig. 12.1c; Burden *et al.* 1979). In other words, when the muscle fibres degenerate after their initial innervation, newly synthesized ACh receptors in the regenerating muscle fibres aggregate again at the same site even in the absence of motor re-innervation. This phenomenon suggests that the signal for ACh receptor clustering is imprinted in the neuromuscular junction by the initial motor innervation.

Again, we ask, where—and what—might the signal be? Two important clues exist. First, it is the synaptic portion of the myofibre basal lamina (junctional basal lamina) that directs ACh receptor aggregation at the junctional region (McMahan and Slater 1984). Second, a factor extracted from the basal lamina of the *Torpedo* electric organ induces ACh receptor clustering in cultured muscle cells (Nitkin *et al.* 1987). This signal molecule associated with the junctional basal lamina was named *agrin (Nitkin et al.* 1987), and its cDNA was soon cloned (Rupp *et al.* 1991; Tsim *et al.* 1992). The *agrin hypothesis* (McMahan 1990) now proposes the following scheme for the involvement of agrin in ACh receptor aggregation at the neuromuscular junction.

Agrin is synthesized in motor neurone cell bodies and transported to their peripheral terminals (Magill-Solc and McMahan 1988, 1990). Agrin released from the motor nerve terminal induces ACh receptor aggregation in the muscle surface at the junctional region (Reist *et al.* 1992). Although agrin is present in both motor neurones and muscle (Ferns and Hall 1992), it is the motor neurone-derived form of agrin, not the muscle-derived type, that induces ACh receptor clustering in muscle (Cohen and Godfrey 1992; Reist *et al.* 1992; Ruegg *et al.* 1992). It is still premature to speculate how agrin causes ACh receptors to aggregate. Agrin released from motor nerve teminals presumably binds to its receptors on the muscle surface (Nastuk *et al.* 1991). Recall that the cytoplasmic end of the muscular ACh receptor is attached to the 43 K protein, the protein that helps anchor and cluster ACh receptors at the neuromuscular junction (Froehner *et al.* 1990; see Fig. 4.3B and section 4.1.1). The β-subunit of the ACh receptor appears to be the site of interaction with the 43 K protein (Burden *et al.* 1983). Interestingly, agrin causes the phosphorylation of a tyrosine residue in the ACh receptor β-subunit (Wallace *et al.* 1991). It is possible that activation of the agrin receptor phosphorylates ACh receptors, which in turn induces ACh receptor clustering in association with the 43 K protein. Peng *et al.* (1991) found that basic fibroblast growth factor (bFGF) also induces ACh receptor clustering in muscle cells. It is intriguing to note that this receptor clustering produced by bFGF can be blocked by an inhibitor of tyrosine phosphorylation. Thus, bFGF and agrin may activate a common pathway.

Agrin, as predicted from its cDNA, is a large molecule, comprising about 1950 amino acids (Rupp *et al.* 1991; Tsim *et al.* 1992). This molecule contains several domains, including some like protease inhibitor, laminin, and epidermal growth factor. This multiple domain structure suggests that agrin is not only involved in ACh receptor clustering but also interacts with other proteins, operating as a multifunctional molecule. Agrin is also responsible for the aggregation of acetylcholinesterase in regenerating neuromuscular junctions (Wallace *et al.* 1985; Nitkin *et al.* 1987). It is likely that agrin causes many components related to synaptogenesis to aggregate at the synaptic site. Sanes *et al.* (1978) damaged nerve and muscle, then irradiated the damaged muscle with X-rays to prevent its regeneration. Under this condition, the nerve regenerated into the region of damage within two weeks and again contacted surviving basal laminae (Fig. 12.1d). Moreover, the nerve terminals differentiated normally at the original synaptic site, despite the absence of target myofibres. This phenomenon suggests that some component associated with the junctional basal lamina guides the motor re-innervation and directs the differentiation of nerve terminals. Whether cueing this component is also agrin remains to be examined.

The agrin hypothesis (McMahan 1990) postulates that agrin released from motor nerve terminals during embryogenesis becomes associated with the nascent junctional basal lamina. This link lets it continue interacting with its receptor even after denervation. Such an association with the junctional basal lamina amounts to the imprinting of the signal, although how or in what form agrin is associated with the basal lamina is not clear. We have seen that the ACh receptor density at the mature neuromuscular junction remains unchanged for weeks after denervation. Presumably, agrin in the junctional basal lamina helps maintain junctional ACh receptor clustering after denervation or sustains the aggregation of ACh receptors in regenerating muscle fibres without motor re-innervation. The turnover rate of agrin molecules must be extremely slow.

The induction of ACh receptor clustering by agrin occurs even in the presence of protein synthesis inhibitors (Wallace 1988). Hence, the receptor clustering agrin induces must come at least partly from lateral migration of the pre-existing ACh receptors in the muscle cell membrane (Godfrey *et al.* 1984). As described previously, motor neurones synthesize CGRP (calcitonin gene-related peptide; Rosenfeld et al. 1983), which is transported to their terminals (Kashihara *et al.* 1989). In cultured muscle cells, CGRP increases the synthesis of ACh receptors (Fontaine *et al.* 1986; New and Mudge 1986). Thus, motor neurone-derived CGRP may enhance the synthesis of ACh receptors in muscle, while agrin at the junctional basal lamina helps to aggregate the receptors. Then, the two molecules might act in concert to maintain the high density of junctional ACh receptors.

Denervation induces the expression of extrajunctional ACh receptors (Fambrough 1970; section 8.2), although the junctional ACh receptor density remains unchanged. A similar disparity in expression between the junctional and extrajunctional ACh receptors can be seen during synaptogenesis. After the initial motor innervation, the junctional ACh receptor density increases, whereas the extrajunctional receptors progressively diminish. Evidently, the junctional and extrajunctional ACh receptors in a muscle fibre differ in the way their synthesis is regulated. Furthermore, this reciprocal regulation of the junctional and extrajunctional ACh receptor densitites in the early postnatal period is accompanied by developmental changes in the properties of the junctional ACh receptor channel. As we shall see, neurotrophic factors again play a pivotal role in these developmental programs.

12.1.2 Developmental switch of receptor subunits

In section 4.2.2, we saw that the e.p.p.s recorded from neuromuscular junctions in fetal or neonatal rats are slower in time course than in adult rats (Diamond and Miledi 1962), owing to the developmental change in the mean open time of single-channel currents of ACh receptors (Sakmann and Brenner 1978; Vicini and Schuetze 1985). Moreover, the patch clamp recording technique combined with recombinant DNA technology disclosed that the developmental change of the ACh receptor channel kinetics is induced by a switch in expression from the γ-to the ε-subunit of ACh receptors (Mishina *et al.* 1986). That is, the ACh receptor stoichiometry switches from the fetal form, $\alpha_2\beta\gamma\delta$, to the adult form, $\alpha_2\beta\delta\varepsilon$. We shall now examine how this subunit switch is triggered at early postnatal stages.

Figure 12.2 shows developmental changes in mRNA levels of γ-and ε-subunits of ACh receptors in mouse muscle (Martinou and Merlie 1991; see also Witzemann *et al.* 1991). The γ-subunit mRNA level progressively declines after birth, subsiding to undetectable levels by the end of the second postnatal week. By contrast, the ε-subunit mRNA, barely detectable at birth, increases rapidly during the first two postnatal weeks. Extrajunctional ACh receptors virtually disappear within one week after birth (Diamond and Miledi 1962; Bevan and Steinbach 1977), hence the ε-subunit must be the junctional ACh receptor subunit. Indeed, shortly after birth, the α-subunit mRNA can still be detected in both junctional and extrajunctional regions, while the ε-subunit mRNA is expressed exclusively at the junctional site (Brenner *et al.* 1990). Both the onset and the loca-

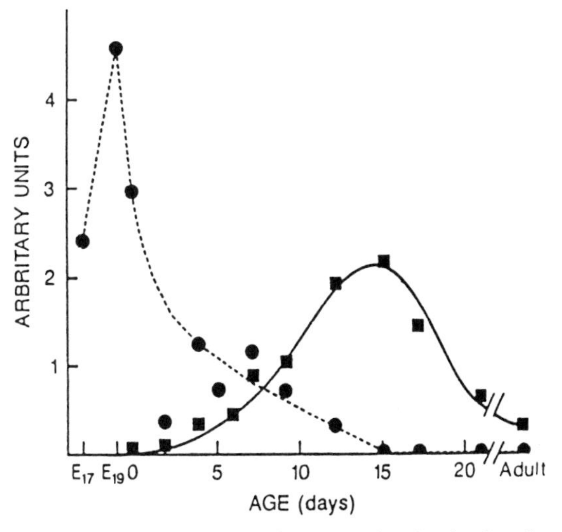

Fig. 12.2. Developmental alterations in the levels of γ-subunit (circles) and ε-subunit (squares) mRNAs in mouse muscle. (From Martinou and Merlie 1991.)

tion of its expression suggest that the synthesis of the ε-subunit in muscle is triggered by motor innervation. Yet, when the muscle is denervated immediately after birth (a few days after the initial innervation), the expression of the ε-subunit mRNA in the muscle continues to increase for at least two weeks, and its expression is still confined to the junctional region (Brenner *et al.* 1990; Witzemann *et al.* 1991). It was thus suggested that the signal directing the ε-subunit expression at the neuromuscular junction is imprinted within a few days of the initial motor innervation, enabling the expression of this receptor subunit to persist even after denervation. This notion is reminiscent of the role of the motor neurone-derived agrin in ACh receptor clustering at the neuromuscular junction (section 12.1.1). Indeed, experiments similar to those illustrated in Fig. 12.1c showed that the ε-subunit mRNA can be expressed in regenerating muscle fibres without motor re-innervation. In these experiments, the hindlimb muscle of adult rats was denervated and treated with marcaine, a myotoxic anaesthetic (Goldman *et al.* 1991), or with the venom of the Australian tiger snake (Brenner *et al.* 1992) to induce complete degeneration of the muscle fibres. The original ε-subunit mRNA was lost during muscle degeneration, but it was expressed again at the old junctional region when muscle fibres regenerated, even in the absence of re-innervation. Evidently, some molecule associated with the junctional basal lamina after the initial motor innervation remains able to induce the expression of the junctional ε-subunit.

A candidate for this molecule is the *ACh receptor-inducing activity* (ARIA). The ARIA is a glycoprotein with a molecular weight of 42 000. This molecule was first purified from chick brain as a candidate for the molecule involved in muscular ACh receptor clustering (Usdin and Fischbach 1986). Subsequently, the ARIA was found to stimulate specifically the ε-subunit gene expression in muscle cells cultured from mouse embryos (Martinou *et al.* 1991). The ARIA cDNA was cloned, and its mRNA was found to be expressed in embryonic motor neurones before their axons reach the target muscle (Falls *et al.* 1993). ARIA seems to have pleiotropic actions by activating several receptor tyrosine kinases in muscle. The physiological function of ARIA remains to be examined.

The sequence of developmental changes in muscular ACh receptors may be summarized as follows (Changeux 1991; Hall and Sanes 1993). First. the fetal form of ACh receptor, $\alpha_2\beta\gamma\delta$, is expressed along the entire muscle fibre prior to innervation. Second, these ACh receptors begin to aggregate at the junctional site soon after innervation—presumably by the action of agrin; meanwhile, the fetal form of ACh receptor is still distributed in both the junctional and extrajunctional regions. Third, the junctional ACh receptors switch from the fetal form to the adult form, $\alpha_2\beta\delta\varepsilon$, at the early postnatal stage, by the action of ARIA; the fetal form of ACh receptor in the extrajunctional region disappears at about the same time. Following denervation of adult muscle, the junctional region preserves the adult form of junctional ACh receptor, whereas the extrajunctional region recapitulates the pre-innervation state, expressing the fetal form of ACh receptor. How might the distributions of the junctional and extrajunctional ACh receptors be regulated independently? Examination of this phenomenon disclosed a surprising mechanism: localized gene expression by distinct types of nucleus in muscle fibres.

A muscle fibre is formed by fusion of a large number of myotubes, so that each muscle fibre contains many nuclei distributed along its entire length. As noted, the adult junctional ACh receptors turn over with a half-life of about 10 days (section 12.1.1). Undoubtedly, the newly synthesized ACh receptors are coded by the mRNAs transcribed from DNA in these nuclei of a muscle fibre. Merlie and Sanes (1985) examined whether ACh receptors are synthesized uniformly along the length of the multinucleated muscle fibre in adult mice. As illustrated in Fig. 12.3*a*,*b*, the phrenic nerve branches run perpendicularly to diaphragm muscle fibres, forming neuromuscular junctions in a narrow band near their midpoint. ACh receptors labelled with radioactive α-bungarotoxin are confined to this 5 mm wide strip of the muscle fibres (Fig. 12.3*c*); thus, this narrow strip is the synapse-rich junctional region, and the rest is the synapse-free extrajunctional region. The remarkable finding of Merlie and Sanes (1985) was that both the α- and δ-subunit mRNA levels are much higher in the synapse-rich muscle tissue than in the synapse-free muscle tissue. Each diaphragm muscle fibre contains about 500 nuclei, but these nuclei are distributed evenly along a muscle fibre, regardless of the synapse-rich or synapse-free regions (Fig. 12.3*d*). Merlie and Sanes (1985) suggested that a few nuclei located just beneath the junctional membrane may be capable of transcribing the ACh receptor mRNA more efficiently than the extrajunctionally located nuclei. In other words, junctional and extrajunctional nuclei in the same muscle fibre may transcribe different sets of genes. This suggestion has won strong support, and it is now known that all the

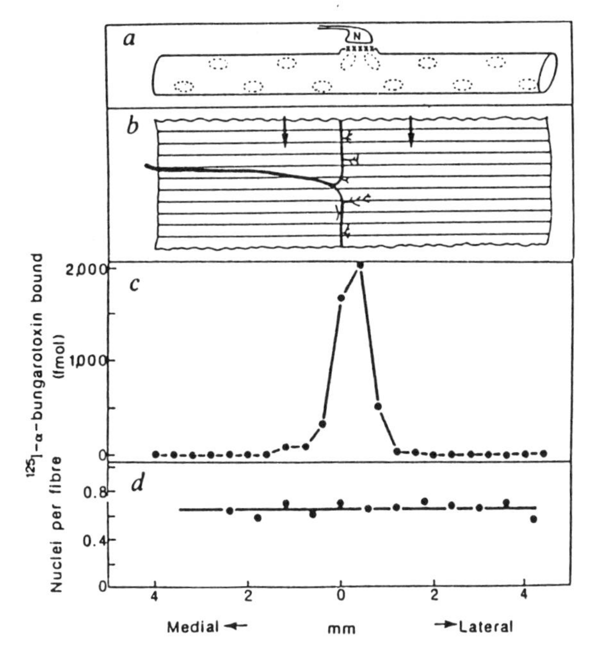

Fig. 12.3. Location of neuromuscular junctions and nuclei in the mouse diaphragm muscle. *a*, a multinucleated muscle fibre with a single neuromuscular junction in the centre. Nuclei are shown by dotted ovals. N, motor nerve terminal. *b*, a strip of diaphragm muscle innervated by the phrenic nerve, which enters from the left side, and its branches form neuromuscular junctions near the midpoint of muscle fibres (between two arrows). *c*, distributions of ACh receptors measured with radioactiven α-bungarotoxin. *d*, distribution of nuclei along diaphragm muscle fibres, determined electron microscopically. (Adapted from Merlie and Sanes 1985.)

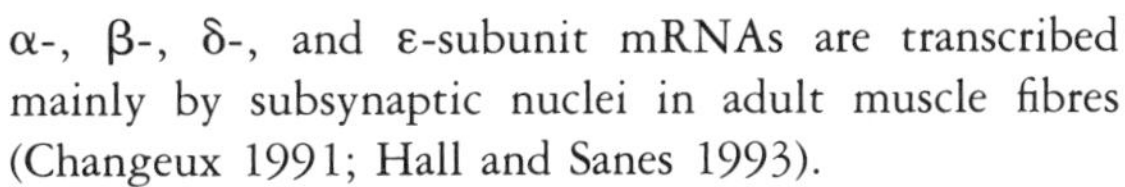

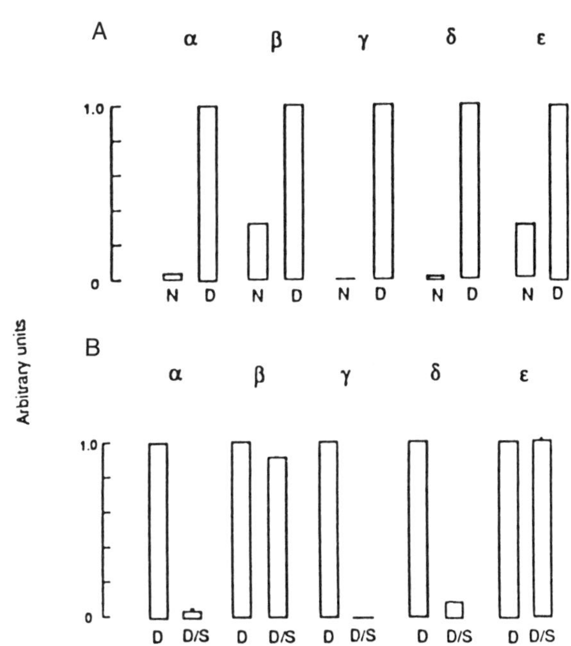

Fig. 12.4. Changes in the mRNA levels of the five ACh receptor subunits in the hindlimb muscle of adult rats. A, comparison between normal (N) and denervated muscles (D). The mRNA level of each subunit is normalized with respect to the value measured 7 days after denervation. B, comparison between denervated muscle (D) and chronically stimulated denervated muscle (D/S). The mRNA level of each subunit is normalized with respect to the value measured 14 days after denervation. Denervated muscle was stimulated chronically for one week after one week of denervation. (Adapted from Witzemann *et al.* 1991.)

α-, β-, δ-, and ε-subunit mRNAs are transcribed mainly by subsynaptic nuclei in adult muscle fibres (Changeux 1991; Hall and Sanes 1993).

This nucleus-specific local expression of the ACh receptor mRNA may account for the high density of ACh receptors at the junctional region in normal muscle. But what accounts for the fetal form of ACh receptor being expressed at extrajunctional regions following denervation? Witzemann *et al.* (1991) examined changes in the expression of mRNAs specific for the five ACh receptor subunits following denervation in adult rat muscles. As shown in Fig. 12.4A, the expression was enhanced in every subunit mRNA within one week after denervation, although the relative increase varied quantitatively in different subunits. For example, the α- and δ-subunit mRNAs showed a 30- to 50-fold increase after denervation, whereas the β- or ε-subunit mRNAs rose only several fold. The γ-subunit mRNA was robustly expressed in denervated muscle, despite being undetectable in normal muscle. When the denervated muscle was stimulated chronically, the expression of the α-, γ-, and δ-subunit mRNAs became markedly suppressed, whereas the ε-subunit mRNA was not affected by electrical stimulation of the denervated muscle (Fig. 12.4B). Importantly, the suppression of the mRNA expression induced by chronic stimulation of denervated muscle occurred only in the extrajunctional region. Although the changes in the β-subunit expression induced by electrical stimulation were relatively small, its expression in the extrajunctional region after denervation was clearly suppressed by muscle activity. These results suggest that neural regulation and muscle activity-dependent regulation of ACh receptors are directed specifically to the receptor mRNA expression in the synaptic and extrajunctional nuclei, respectively.

Figure 12.5 is a hypothetical scheme of the regulation

of ACh receptor subunits in muscle (Hall and Sanes 1993). Although detailed mechanisms still remain uncertain, this diagram emphasizes two points. First, the signals derived from the motor nerve terminal or associated with the junctional extracellular matrix are directed to the synaptic nuclei to up-regulate the junctional ACh receptor subunits. Second, the extrajunctional ACh receptor subunits are synthesized by transcription of DNA in the extrasynaptic nuclei, and their expression is repressed by muscle activity. This concept of regulation by muscle activity requires special comment. Myogenin and MyoD (depicted within an extrasynaptic nucleus in Fig. 12.5) are nuclear proteins that can bind to the regulatory regions of mucle genes to activate their transcription (Pinney *et al.* 1988; Tapscott *et al.* 1988), including binding to the ACh receptor gene (Changeux 1991). Interestingly, myogenin and MyoD expression is repressed by motor innervation. Moreover, their expression increases after denervation, and this increase can be blocked by chronic electrical stimulation of the denervated muscle (Eftimie *et al.* 1991). Thus, it seems reasonable to assume that the expression of extrajunctional ACh receptors in the extrajunctional nuclei is regulated by muscle activity via the up- or down-regulation of these two nuclear proteins (Fig. 12.5). Whether the synaptic nuclei indeed lack myogenin and MyoD needs to be examined, however.

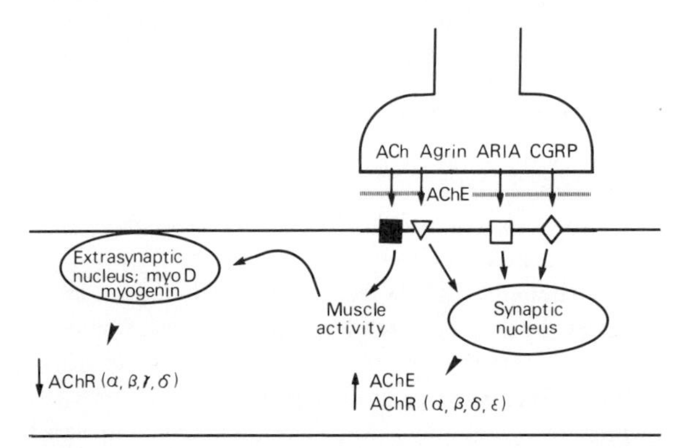

Fig. 12.5. A hypothetical scheme for the regulation of ACh receptors in adult muscle. The junctional ACh receptor (AChR) subunits (α, β, δ, ε) are synthesized by transcription of DNA in synaptic nuclei located just beneath the neuromuscular junction. Extrajunctional ACh receptor subunits (α, β, γ, δ) are synthesized by transcription of DNA in extrajunctional nuclei. Transcription in synaptic nuclei is regulated by the motor neurone-derived factors or the junctional extracellular matrix-associated factors. Transcription in extrajunctional nuclei is regulated by muscle activity via the nuclear proteins myoD and myogenin. (Adapted from Hall and Sanes 1993.)

We have seen that the developmental γ- to ε-subunit switch in the junctional ACh receptor changes the channel properties of the ACh receptor: the single-channel conductance increases, while the channel's mean open time decreases (section 4.2.2). Similar changes in the channel properties of neuronal ACh receptors were observed after presynaptic innervation in chick sympathetic neurones (Moss *et al.* 1989) and ciliary neurones (Margiotta and Gurantz 1989). It is likely that these changes reflect developmental switches of subunits in these neuronal ACh receptors. In frog sympathetic neurones, Marshall (1985) showed that the channel properties of neuronal ACh receptors can be modified in a manner specific to the type of presynaptic fibre. Presumably, each type of presynaptic fibre has its own neurotrophic factor, which can specify either the expression of particular sets of the receptor subunit genes in the postsynaptic neurone or the local aggregation of particular sets of the receptor subunit at the synaptic region.

Developmental subunit switching also occurs in glycine receptors at central inhibitory synapses. In neonatal rats, the glycine-mediated inhibitory postsynaptic currents recorded from spinal motor neurones show a progressive decrease in their time course (Takahashi *et al.* 1992). As noted, the glycine receptor seems to have two types of subunit, α and β (section 4.1.3). In the spinal cord, the α_2-subunit variant predominates in neonatal rats, then is replaced by the α_1 variant (adult form) within three weeks after birth (Akagi *et al.* 1991). The single-channel current of the α_2 homomeric glycine receptor expressed in *Xenopus* oocytes had a significantly longer mean open time than the α_1 homomeric receptor. This behaviour parallels the developmental decrease in the mean open time of glycine receptor single-channel currents recorded from spinal motor neurones (Takahashi *et al.* 1992). Presumably, synaptic contact by an inhibitory neurone on the motor neurone specifies the type of the postsynaptic receptor.

Each central neurone forms thousands of excitatory and inhibitory synapses on its surface. Although every postsynaptic neurone synthesizes many different types of transmitter receptor, each type of receptor molecule would aggregate at an appropriate postsynaptic site if its presynaptic input were to decree what type of postsynaptic receptor should cluster there. It is thus tempting to speculate that each excitatory or inhibitory presynaptic neurone releases a specific agrin-like factor that makes a particular type of receptor molecule aggregate at the synaptic site. The nature of the transmitter in each presynaptic neurone is presumably predetermined according to the developmental programme. Surprisingly, however, the nature of the transmitter synthesized in sympathetic neurones can be altered by a signal from their target cells. The next section illustrates this process, which is mediated by some neurotrophic factor.

12.2 TRANSMITTER CHOICE DURING DEVELOPMENT

As early as 1934, Dale and Feldberg noted that the sympathetic nerve fibres innervating the cat sweat glands are cholinergic, presenting an exception to the general rule that postganglionic sympathetic fibres are adrenergic. This anomalous minority population of sympathetic neurones did not arouse scientific curiosity until the early 1970s, when it was found that the type of transmitter (ACh or norepinephrine) in cultured sympathetic neurones can be controlled experimentally (Patterson 1978). Under certain conditions individual sympathetic neurones can even display both cholinergic and adrenergic transmitter functions. Clearly, the nature of transmitter in sympathetic neurones can be changed when their environment is altered. How does a given sympathetic neurone choose ACh or norepinephrine as its transmitter substance? And how do the sympathetic neurones supplying the sweat gland become cholinergic?

12.2.1 Regulation of transmitter phenotype

Sympathetic ganglionic cells receive cholinergic inputs from the preganglionic neurones, and most of the ganglionic cells release norepinephrine (NE) from their terminals; they are thus cholinoceptive (having ACh receptors) adrenergic (releasing NE) neurones. In the rat, the majority of sympathetic neurones already synthesize NE at birth. When the sympathetic neurones are dissociated from neonatal rats and cultured, these neurones form synapses with one another (Patterson 1978). Since sympathetic neurones are cholinoceptive but lack adrenergic receptors, we would expect the synapses formed between cultured sympathetic neurones to be non-functional. Under some culture conditions, however, stimulation of one sympathetic neurone produces excitatory synaptic potentials in other sympathetic neurones, and the synaptic response is blocked by antagonists of nicotinic ACh receptors (O'Lague *et al.* 1975). These sympathetic neurones can also form functional neuromuscular connections with co-cultured skeletal muscle cells (O'Lague *et al.* 1975). Evidently, sympathetic neurones cultured under certain conditions synthesize ACh and release it on excitation. Let us see what conditions are favourable for developing cholinergic sympathetic neurones.

When sympathetic neurones are cultured in the virtual absence of other cell types, the neurones manifest adrenergic features, whereas the presence of non-neuronal cells causes the sympathetic neurones to become cholinergic (Patterson 1978). The medium that was conditioned by cultures of non-neuronal cells was sufficient to induce cholinergic properties in the cultured sympathetic neurones. In other words, non-neuronal cells appear to secrete factors that turn adrenergic sympathetic neurones into cholinergic neurones. Cells from skeletal muscle or the heart were particularly effective in yielding cholinergic-promoting factors. Under different experimental conditions, adrenergic and cholinergic phenotypes in cultured sympathetic neurones are expressed in a reciprocal manner: in the absence of non-neuronal cells NE synthesis increases, while ACh synthesis is suppressed; in the presence of non-neuronal cells ACh synthesis is promoted at the expense of NE synthesis (Patterson 1978). Thus, sympathetic neurones have plasticity in choice of transmitter. Indeed, such plasticity of transmitter occurs even during normal development; the cholinergic sympathetic neurones innervating the rat sweat glands pass through an adrenergic stage before becoming cholinergic (Landis and Keefe 1983). Hence, initially, all sympathetic neurones must be instructed to become adrenergic neurones. Presumably, those exposed to cholinergic promoting factors at early develpmental stages flip-flop, converting to cholinergic neurones; otherwise, the neurones stay adrenergic.

This notion of conversion raises two questions: what are the cholinergic promoting factors, and how might some sympathetic neurones be exposed to a cholinergic-promoting factor? Weber (1981) partially purified the active cholinergic factor in a conditioned medium from rat heart cultures, and its complete purification as a 45 000 dalton molecule, termed *cholinergic differentiation factor* (CDF), was achieved by Fukada (1985). Subsequently, the cloning of CDF (Yamamori *et al.* 1989) disclosed that CDF is actually identical to *leukaemia inhibitory factor* (LIF), a cytokine. Hence, this factor is now symbolized by CDF/LIF. CDF/LIF exerts a broad range of effects on many cell types; it inhibits proliferation of the leukaemic myeloid cell line and prevents differentiation of embryonic carcinoma cells or embryonic stem cells (Hilton 1992). Moreover, CDF/

LIF promotes the survival of sensory neurones (Murphy *et al.* 1991) and motor neurones (Martinou *et al.* 1992). As noted previously (section 11.3.1), some haematopoietic cytokines appear to function as 'neuropoietic' cytokines. But what is the functional significance of CDF/LIF, in relation to which transmitter the sympathetic neurones choose?

In vitro, the application of CDF/LIF to cultured sympathetic neurones turns them cholineregic. If CDF/LIF is present selectively in some of the target tissues innervated by sympathetic neurones *in vivo*, the cholinergic conversion of sympathetic neurones could occur in a target-dependent manner. Indeed, this was demonstrated by two types of target transplantation experiment. First, the footpad skin containing sweat glands was transplanted to a region prepared by removal of hairy skin in the lateral thorax in neonatal rats (Fig. 12.6; Schotzinger and Landis 1988). The sympathetic neurones innervating piloerector muscles in the hairy skin are normally adrenergic; yet, the grafted footpad enhances choline acetyltransferase (ChAT; ACh synthesis enzyme) activity within one month after transplantation, indicating cholinergic conversion of the innervating sympathetic neurones (Fig. 12.6*c*, filled circles). When sympathetic neurones were destroyed by injecting 6-hydroxydopamine (6-OHDA; adrenergic neurotoxin), the increase of ChAT in the grafted footpad is virtually blocked (Fig. 12.6*c*, open circles). These experiments show that the initially 6-OHDA-sensitive adrenergic sympathetic neurones convert into cholinergic neurones when they innervate sweat glands. In the second type of transplantation experiment, the footpad containing sweat glands was removed and replaced with the parotid gland, which normally receives adrenergic sympathetic innervation, in neonatal rats (Schotzinger and Landis 1990). Under this condition the parotid gland retains catecholamine fluorescence and fails to develop ChAT. Thus, the sympathetic neurones innervating the sweat glands can remain adrenergic, depending on the type of their target tissue. Hence, cholinergic conversion of the sympathetic neurones innervating the sweat glands may be induced by some target-specific factors.

Indeed, Yamamori (1991) found that CDF/LIF mRNA is selectively expressed in rat footpads containing sweat glands. When the gene coding for the CDF/LIF is knocked out in mice by the gene-targeting procedure, however, the sympathetic neurones innervating sweat glands still remain cholinergic as in normal mice (Rao and Landis 1993). It is likely that cholinergic differentiation of the sympathetic neurones is regulated redundantly by multiple cholinergic differentiation molecules, including CDF/LIF, in sweat glands (Landis 1990).

Target-derived cholinergic differentiation factors are clearly distinct in their function from the target-derived NGF. NGF is, of course, essential for survival of sym-

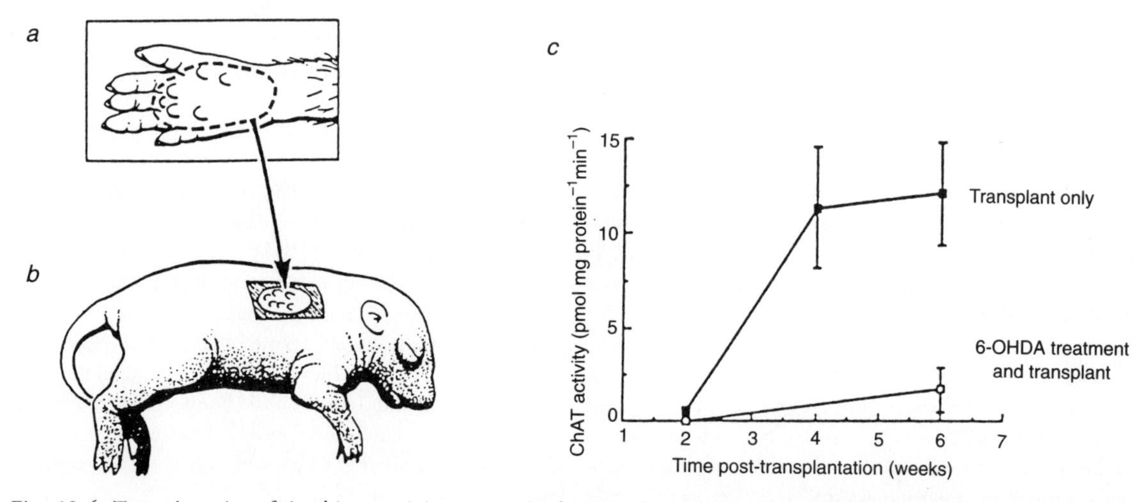

Fig. 12.6. Transplantation of the skin containing sweat glands in the footpad (*a*) to the graft bed prepared by removal of hairy skin of the lateral thorax (*b*) in neonatal rats. *c*, choline acetyltransferase (ChAT) activity in the transplanted footpad increases 4–6 weeks after the operation (filled circles). This increased ChAT reflects the enzyme activity in sympathetic neurones innervating the footpad, since treatment of 6–hydroxydopamine (6–OHDA; adrenergic neurotoxin) blocks the increase in the enzyme activity (open circles). (From Schotzinger and Landis 1988.)

pathetic neurones. NGF also stimulates transmitter differentiation but does not determine which transmitter (NE or ACh) a sympathetic neurone makes. For adrenergic sympathetic neurones, NGF enhances the synthesis of NE, whereas NGF stimulates the synthesis of ACh for cholinergic sympathetic neurones. In this sense, NGF is considered 'permissive' with respect to transmitter production, unlike the 'instructive' signals for cholinergic differentiation such as CDF/LIF (Patterson 1978).

We have seen that CNTF does not have any feature characteristic of a target-derived trophic factor (section 11.3.1). Interestingly, however, CNTF is capable of switching cultured sympathetic neurones from an adrenergic to a cholinergic phenotype (Saadat *et al.* 1989; Rao *et al.* 1992). This action of CNTF may reflect similarities or promiscuity between the CDF/LIF receptor and the CNTF receptor. Recall that the functional CNTF receptor is formed by addition of the CNTF-binding protein to the gp130 plus LIF-binding protein complex (Fig. 11.7D; section 11.3.2). Actually, this gp130-LIF-binding protein complex comprises a functional LIF (or CDF/LIF) receptor (Gearing *et al.* 1992; *Ip et al.* 1992b; see also section 12.2.3). Davis *et al.* (1993) reported that the CNTF-binding protein can exist in a soluble form. Thus, a biological phenomenon that is normally elicited only by CDF/LIF may confer CNTF responsiveness if a released, soluble form of CNTF-binding protein is added to the CDF/LIF receptor (Ip *et al.* 1993a). The cholinergic differentiation action by CNTF may be an example of a widespread role of CNTF in neurotrophic functions, via the CNTF and CDF/LIF receptor complex.

Cholinergic differentiation factors affect sympathetic neurones not only by influencing their choice of transmitter but also by affecting their expression of neuropeptides. In other words, both the transmitter phenotype and the neuropeptide phenotypes of neurones are plastic. Might both phenotypes be regulated by the same instructive signal? Furthermore, what is the biological significance of plastic changes in the neuropeptide phenotype?

12.2.2 Regulation of neuropeptide phenotype

Synaptic vesicles in a nerve terminal can be divided into two classes, small clear vesicles and large dense-core vesicles. The former contains exclusively a classical transmitter (e.g., ACh or NE), whereas the latter contains a classical transmitter plus a neuropeptide (Hökfelt 1991). In sympathetic neurones, neuropeptide Y (NPY) is often coexpressed with NE, and vasoactive intestinal peptide (VIP) with ACh (Lundberg *et al.* 1982). When normally adrenergic sympathetic neurones are converted to cholinergic neurones, by replacing their normal target tissues with sweat glands or by applying CNTF, the neuropeptide expression in the neurones also switches from NPY to VIP (Landis 1990; Patterson and Nawa 1993). This parallel alteration does not mean that a particular classical transmitter is developmentally related to a neuropeptide. In a previous section (12.2.1), we saw that the classical transmitter's phenotype depends on the target tissue. Here again, it is the neuropeptide and the target tissue whose relationship is relevant (Stevens and Landis 1990).

Coulombe and Nishi (1991) observed a clear target-dependent neuropeptide expression in the parasympathetic ciliary ganglion. The chick ciliary ganglion comprises two types of neuron, choroid neurones and ciliary neurones. The former innervates vasculature of the choroid layer, while the latter projects to the ciliary body and iris. Both types of neurone are cholinergic, but only choroid neurones coexpress a neuropeptide, somatostatin (SOM). Dissociated choroid neurones express SOM when co-cultured with choroid layer cells, but not when co-cultured with striated muscle (Coulombe and Nishi 1991). These results suggest that a specific molecule derived from choroid layer cells is responsible for the expression of SOM in choroid neurones. Because ciliary neurones do not express any neuropeptide *in vivo*, the classical transmitter (ACh) and the neuropeptides appear to be regulated independently by the target tissues.

Nawa and Patterson (1990) fractionated heart cell conditioned media and identified three distinct factors that regulate the expression of neuropeptides in cultured sympathetic neurones. One factor (CDF/LIF) increases the expression of ACh, substance P, SOM and VIP, whereas the two other factors specifically promote the expression of SOM and VIP, respectively. The SOM expression in choroid neurones of the chick ciliary ganglion may be accounted for by assuming that one of these factors is present in the vasculature of the choroid layer. Similarly, the presence of CDF/LIF (or related factors) in the sweat glands would explain the concurrent switch of the classical transmitter from NE to ACh and of the neuropeptide from NPY to VIP in the innervating sympathetic neurones. As discussed previously (section 12.2.1), the switch of transmitter phenotype in sympathetic neurones by CNTF is presumably due to activation of the CNTFα-CDF/LIF-gp130 receptor complex. Consistent with this possibility, CNTF increases the expression of the same neuropeptides (substance P, SOM and VIP) in sympathetic neurones as CDF/LIF does (Rao *et al.* 1992). Taken together, these parallels strongly suggest that neuropeptide phenotype in neurones is determined specifically by a factor derived from their target tissues.

Might a given factor enhance the expression of the same peptide in different types of neurone? In contrast with cultured sympathetic neurones, cultured dorsal root ganglion (DRG) cells do not alter levels of substance P, SOM or VIP in response to CNTF. Thus, the effect of peptide-promoting factors can be specific for a neurone type. How can we interpret this neuron-specific effect? The first possibility is that DRG cells lack CNTF receptors, but this possibility can be ruled out because CNTF in fact supports the survival of DRG cells (Thoenen 1991); the CNTF receptor is, in fact, present in DRG cells (Ip *et al.* 1993a). Might DRG cells lack the capacity to up-regulate the neuropeptides? No—DRG cells can increase the neuropeptide levels in response to CDF/LIF, as do sympathetic neurones (Nawa *et al.* 1991). We must then postulate that neuronal survival and neuropeptide expression mediated by CNTF are based on different signal transduction pathways (Rao *et al.* 1992). How a given ligand can induce such distinct signal transduction cascades remains unresolved, however.

Disparity between neuronal survival and peptide expression can also be seen in DRG cells in response to NGF. NGF is not required for the survival of DRG cells cultured from adult rats (Lindsay 1988), but their expression of substance P and CGRP is up-regulated by NGF (Lindsay *et al.* 1989; Inaishi *et al.* 1992). We have seen that NGF is permissive for the expression of transmitter phenotype in sympathetic neurones, no matter whether their transmitter is NE or ACh (section 12.2.1). Unlike substance P or CGRP, however, the expression of VIP is independent of NGF in adult DRG cells (Mulderry and Lindsay 1990). Apparently, NGF acts as an instructive signal for the expression of certain neuropeptides. NGF's selective effect on neuropeptides can be seen following transection of the peripheral sensory fibres (axotomy), which interrupts retrograde trnsport of NGF from the periphery to sensory neurone cell bodies. As expected, following peripheral axotomy the expression of substance P and CGRP in DRG cells is suppressed (Jessel *et al.* 1979; Inaishi *et al.* 1992), whereas the VIP expression is enhanced (Noguchi *et al.* 1989; Villar *et al.* 1989). Although the interpretation of these results is not immediately certain, at least it seems clear that the expression of different neuropeptides in sensory neurones is not affected uniformly by the target-derived NGF.

CGRP is also synthesized in motor neurones, is anterogradely transported and displays trophic functions at neuromuscular junctions, enhancing ACh receptor synthesis (Changeux *et al.* 1992) and inhibiting sprouting of motor nerve terminals (Tsujimoto and Kuno 1988). How are CGRP levels regulated in motor neurones? Unlike the sensory neurones, in motor neurones CGRP levels rise following axotomy (Streit *et al.* 1989; Arvidsson *et al.* 1990). Perhaps, then, the expression of CGRP in motor neurones is down-regulated by the target-derived neurotrophic factor. This possibility was examined in the neurones of the spinal nucleus of the bulbocavernosus (SNB). These SNB motor neurones innervate the perineal striated muscles, and being sexually dimorphic their number is three to four times greater in the male rat than in the female (Arnold 1984). Castration of adult male rats results in atrophy of both the perineal muscles and the SNB motor neurones, and these changes are reversed by treatment with testosterone (Arnold 1984). This hormonal regulation of the size of SNB motor neurones is mediated by their target peripheral muscles (Araki *et al.* 1991). Presumably, the sex hormone acts on the perineal muscles, thereby stimulating the synthesis of retrograde neurotrophic factor supplied to the innervating SNB motor neurones. Thus, the retrograde neurotrophic supply to SNB motor neurones can be interrupted either by axotomy or by castration. As illustrated in Fig. 12.7, the expression of CGRP in SNB motor neurones is enhanced by axotomy of the neurones, by castration or by chronic activity deprivation (Micevych and Kruger 1992; Popper *et al.* 1992). Since axotomy (denervation), castration or chronic inactivity causes muscle atrophy, CGRP levels in SNB motor neurones indeed appear to be down-regulated by some factor associated with metabolic activity of the innervated muscle.

Alternatively, motoneuronal CGRP levels may be up-regulated by some factor associated with muscle inactivity. To address these possibilities, extracts from denervated perineal muscles or the perineal muscles in castrated rats were injected into the intact perineal muscle. This procedure significantly enhanced the expression of CGRP mRNA in SNB motor neurones (Fig. 12.7). In sum, CGRP levels in sensory neurones are up-regulated by peripherally derived NGF, whereas CGRP

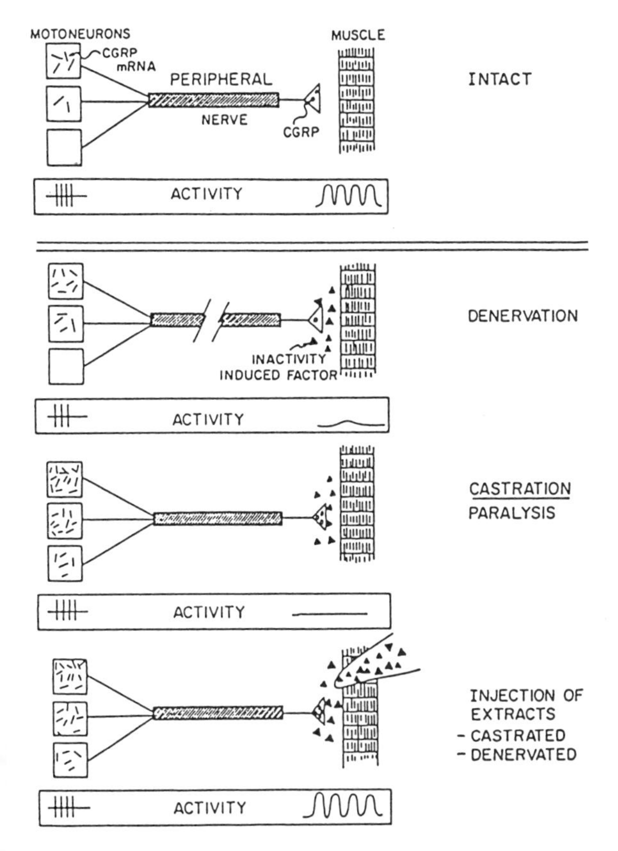

Fig. 12.7. Hypothetical diagram of regulation of CGRP mRNA expression in sexually dimorphic motor neurones innervating the bulbocavernosus. Some factor (filled triangles) released from inactive target muscles or from the target muscles after castration is assumed to be transported retrogradely to the motor neurone cell bodies, thereby enhancing their CGRP mRNA. (From Micevych and Kruger 1992.)

levels in motor neurones are up-regulated by some factor derived from chronically inactive muscle.

Why do neuronal CGRP levels need to be regulated? We do not know. Indeed, the functional significance of CGRP in sensory neurones or motor neurones is not at all clear. CGRP levels in sensory or motor neurones reflect the target-derived trophic signal (NGF or some factor associated with muscle inactivity). Possibly, then, CGRP is a monitor appointed to report the state of the peripheral target tissues. If so, CGRP presumably acts as an intracellular messenger helping the neuronal state to adapt to peripheral environmental changes, besides being as an intercellular messenger. Sakaguchi *et al.* (1992) showed that degeneration of motor fibres induced by section of the rat lumbar ventral roots markedly increases CGRP levels in the lumbar dorsal root ganglia. There must be some 'cross-talk' in neuronal CGRP expression between motor and sensory neurones, but its significance and the underlying mechanisms remain elusive. In the sensory system, CGRP distributes mainly in the neurones with a thin myelinated or unmyelinated fibre that signal nociceptive information (Micevych and Kruger 1992). Kruger *et al.* (1989) suggested that, besides pain-related afferent activation, these sensory neurones mediate the efferent regulation of the peripheral target tissues via CGRP release—the so-called *noceffector* role. CGRP is known to produce vasodilatation, epithepial growth, mineralization of bone, and mobilization of lymphocytes. Thus, CGRP's noceffector role contributes to the trophic maintenance, renewal or repair of the target tissues.

According to Patterson and Nawa (1993), neuropeptides, neurotrophic factors, and cytokines may act in concert as a self-protective device in response to injury and stress. How do they interact? Indeed, we have seen that some haematopoietic cytokines exert neurotrophic effects (section 11.3.1). Could there be any similarity between the regulatory mechanisms of neural development and of haematopoiesis, the blood cell forming process?

12.2.3 Cytokines and neuronal differentiation

Soluble factors generated by antigen-activated lymphocytes mediate the principal features of cellular immunity. These soluble factors were initially referred to as the *lymphokines* (Dumonde *et al.* 1969). But when non-lymphoid cells were also found to produce lymphokine-like substances, the concept of lymphokine signalling in immunology was extended to include many biological mediator substances, collectively termed the *cytokines* (Bigazzi *et al.* 1975). Cytokines are now defined as soluble mediators for cell–cell communication that are produced by T lymphocytes, macrophages and endothelial cells (Watanabe *et al.* 1991). Cytokines include several sets of mediator substances: monokines and *lymphokines* involved in inflammation and immunity, interferons in viral infection, *colony-stimulating factors* in haematopoiesis, *growth factors* in cell proliferation and differentiation, and *interleukins* (ILs) mediating signals between leucocytes (Nathan and Sporn 1991; Watanabe *et al.* 1991). As noted previously (section 12.2.1), both CDF/LIF, a haematopoietic cytokine, and CNTF, a neurotrophic factor, can switch the transmitter phenotype in developing sympathetic neurones from adrenergic to cholinergic. Although CDF/LIF and CNTF share no significant homology in amino acid sequence, CDF/LIF, CNTF, IL-6 and a novel cytokine, *oncostatin M* (OM; Malik *et al.* 1989), reveal a common helical framework in their predicted tertiary folding patterns (Bazan 1991). Interestingly, OM, a cytokine initially identified as a growth inhibitor of melanoma cells, regulates gene expression of sympathetic neurones, thereby stimulating the synthesis of VIP, as CDF/LIF and CNTF also do (Patterson and Nawa 1993).

We have seen that both CNTF and IL-6 receptors share a common feature, requiring non-ligand-binding gp130 as an association protein (Fig. 11.7; section 11.3.1). This receptor molecule motif applies to the CDF/LIF and OM receptors too, as illustrated in Fig. 12.8 (Taga and Kishimoto 1992). Moreover, IL-11, a cytokine with multiple biological functions, is also suggested to mediate its signal transduction via its own IL-11 receptor complex only in association with gp130 (Taga and Kishimoto 1992). Coincidentally IL-11 also regulates gene expression in sympathetic neurones (Patterson and Nawa 1993). Apparently, then, certain cytokines help regulate neural development or neural function and they resemble neurotrophic factors in tertial structure and in the molecular motif of their receptors. This implies that molecules regulating haematopoiesis and neural development have a common evolutionary origin (Bazan 1991; Yamamori and Sarai 1992). In fact, it was suggested that neural development follows a process analogous to blood cell formation (Anderson 1989; Patterson 1990; Nawa *et al.* 1991). Eight main types of mature blood cell originate from a self-renewing population of mutipotential haematopoietic stem cells

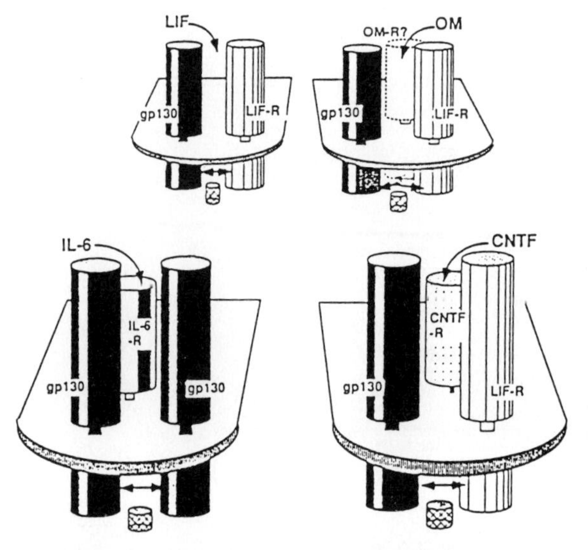

Fig. 12.8. Hypothetical receptor complexes for CDF/LIF, OM, CNTF, and IL-6 (clockwise from upper left). Each receptor complex is associated with one or two gp130, the signal transduction component. In the cytoplasmic side, the gp130 presumably interacts with a protein kinase (e.g., tyrosine kinase; cross-hatched structures). (From Taga and Kishimoto 1992.)

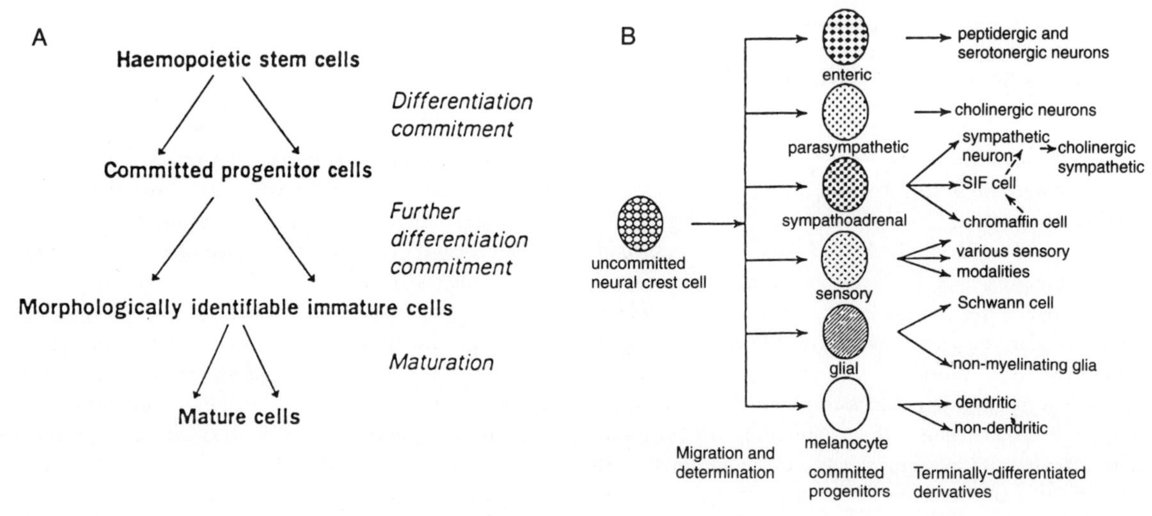

Fig. 12.9. The hierarchy of differentiation and maturation of hematopoietic cells (A) and neural crest-derived neurones (B). In both processes multipotent stem cells are first committed to become progenitor cells, followed by further differentiation and maturation, yielding cells of different types. (A, from Metcalf 1989; B, from Anderson 1989.)

(Fig. 12.9A; Metcalf 1989, 1991). Similarly, several types of neural cell develop from multipotential stem cells in the neural crest (Fig. 12.9B; Anderson 1989; Stemple and Anderson 1992). What commits these multipotential stem cells to one particular cell lineage?

Haematopoietic cells are stimulated to undergo differentiation by a complex network of hormones, growth factors or cytokines. Thus, phenotypic expression (differentiation) of haematopoietic cells is not only determined by inheritence (cell lineage) but can be influenced by environmental signals (see Box L). The haematopoietic cytokines involved in such environmental signals include IL-3, IL-6, IL-11, GM-CSF and CDF/LIF—all of which have trophic influence on neurones. Similar environmental signals are present in the differentiation of phenotypically different cells derived from a common embryonic progenitor in the neural crest (Carnahan and Patterson 1991; Anderson 1993). As shown in Fig. 12.9B, six classes of committed progenitor cell derive from the earliest multipotent neural crest cells.

One of the six, the *sympathoadrenal* (SA) lineage has been investigated in detail. The SA progenitor cell is at least bipotential, being able to differentiate into sympathetic neurones and chromaffin cells, secretory cells. The progenitor cell and its descendents can be identified by specific chemical markers. Treating the progenitor cell with basic FGF (fibroblast growth factor) tends to promote the proliferation and initial neuronal differentiation, whereas glucocoroticoids encourage the chromaffin pathway and suppress the expression of neuron-specific genes (Carnahan and Patterson 1991; Anderson 1993). Developing sympathetic neurones gain responsiveness to NGF for further maturation and survival. The expression of functional NGF receptors ($gp140^{trk}$) appears to be induced by membrane depolarization or electrical activity of committed neuroblasts. Thus, FGF initially promotes neuronal differentiation of embryonic SA progenitors but cannot support their survival; the committed sympathetic neuroblasts then acquire NGF-dependence, and it is NGF that acts as a survival factor for sympathetic neurones.

Again, cytokines can affect neuronal differentiation of the progenitor cells. CNTF or CDF/LIF, for example, inhibits the proliferation of sympathetic neuroblasts (Anderson 1993). Moreover, neuronal differentiation of hippocampal progenitor cells can be induced by IL-5,

Box L Haematopoiesis

Haematopoiesis is the process of blood cell formation. Most mature blood cells are short-lived, hence must be replaced continuously. Blood cells originate from a self-renewing population of multipotential haematopoietic *stem cells* (Fig. 12.9A), located mainly in the bone marrow. Steady levels of blood cells (constitutive haematopoiesis; Miyajima *et al.* 1992b) are assumed to be maintained by the *haematopoietic inductive micro-environment* (HIM). The HIM is formed by stroma cells (macrophages, endothelial cells, fibroblasts, and adipocytes) surrounding the stem cells. The stem cells differentiate into *progenitor cells* committed irreversibly to one or other of the various haematopoietic lineages. Under distressing conditions such as infection or bleeding, a variety of cytokines produced by macrophages and endothelial cells promote the formation of specific types of cell (*inducible haematopoiesis*; Miyajima *et al.* 1992b).

Although neutrophilic granulocytes and macrophages are morphologically distinct, they derive from common progenitor cells (Metcalf 1989). *In vitro*, committed progenitor cells (Fig. 12.9A) proliferate to form clones (colonies) of morphologically identifiable immature cells. This process requires *colony-stimulating factors* (CSF), which are derived from stroma cells *in vivo*. There are at least four CSFs: granulocyte-macrophage (GM-CSF), granulocyte (G-CSF), macrophage (M-CSF), and IL-3 (also known as multispecific CSF). While G-CSF and M-CSF stimulate only granulocyte formation and macrophage formation, respectively, GM-CSF and IL-3 each give rise to different types of blood cell. Initially, however, the progenitor cells can respond to any of the four factors for proliferation. Apparently, the progenitors exhibit receptors for more than one type of CSF. In other words, a single cell responds to more than one cytokine, and many cytokines have overlapping activities. Moreover, cytokines are pleiotropic, exerting multiple biological effects on different types of cell. GM-CSF and IL-3, for example, stimulate choline acetyltransferase activity in mammalian central neurones (Konishi *et al.* 1993). This suggests that signal transduction via the receptors for cytokines does not require any particular factors specific to haematopoietic cells. In fact, reconstitution of the human GM-CSF receptor in mouse fibroblasts resulted in functional transduction of growth-promoting signals in the fibroblasts in response to GM-CSF (Watanabe *et al.* 1993).

-7, -9, and -11 (Mehler *et al.* 1993). These results imply that the regulation of lineage commitment and cellular differentiation requires a hierarchy of growth factors and cytokines in both the neural and haematopoietic systems.

Growth factors and cytokines have broad effects on cell proliferation and gene expression in a wide variety of tissues. Nathan and Sporn (1991) and Patterson and Nawa (1993) refer to all these substances as cytokines, no matter what type of cell they mediate or affect. Thus, CNTF is a cytokine, although it is not produced by lymphocytes, macrophages or epithelial cells, and CDF/LIF is a cytokine, although it exerts neurotrophic effects. This classification is simple, but it redefines the original meaning of *cytokine*. Accepting this classification would mean revising our definition of *neurotrophic factor*, because some neurotrophic factors involved in the maintenance of a cell's programmed state would now be considered cytokines. Admittedly, our existing definition leads to some confusion; 'neuropoietic' factors, for example, are now equivocal, because they are cytokines and neurotrophic factors. The same applies to some neuropeptides. CGRP, for example, can switch the transmitter phenotype in embryonic olfactory neurones from GABAergic to dopaminergic (Denis-Donini 1992). Moreover, CGRP exhibits chemotactic activity for haemapopoietic cells, attracting T lymphocytes (Foster *et al.* 1992). In other words, a neuropeptide can act as a neuropoietic cytokine.

The difficulty of definition and classification of neurotrophic factors and cytokines stems from their pleiotropic (mutiply directed) actions. McDonald and Hendrickson (1993) classified neurotrophins, growth factors, and cytokines based on similarity in their three-dimensional molecular structures (see also Bazan 1991). This approach is not only simpler than a classification based on functional aspects but also may unveil the evolutionary relatedness among these molecules involved in cell proliferation and gene expression. It is almost certain that we will be redefining *neurotrophic factor* in more explicit terms once we understand how neuronal survival is maintained by neurotrophic factors and how neurones differentiate. Meanwhile, instead of formulating a temporary revision, we shall preserve our original definition, recognizing some ambiguity about where neuropoietic cytokines fit in. We defined a neurotrophic factor as a neurally mediated molecule that maintains a cell's pre-existing or programmmed state. Quite likely, however, the physiological function of neurotrophic factors requires the complementary interaction of neurones with the extracellular matrix and various intercellular signals (Barde 1988), including cytokines and neuropeptides.

Summary and prospects

Neurotrophic factors that induce aggregation or developmental switching of transmitter receptors in target tissues are obviously synthesized in the neurone cell bodies and anterogradely transported toward the targets. These neurotrophic factors have two unique features. First, they exert their trophic effects only at a confined locus of the target cell, so that transmitter receptors aggregate or convert to an adult form only at the synaptic site. Second, the trophic effect, once exerted, persists even after elimination of the trophic signal. Agrin responsible for ACh receptor aggregation at neuromuscular junctions appears to linger at the junctional basal lamina, thereby maintaining its trophic effect locally for a long period of time. The ARIA that is responsible for the γ- to ε-subunit switch of ACh receptors at neuromuscular junctions appears to trigger the expression of ε-subunits in a few particular nuclei located immediately beneath the synaptic site. Clearly, this distinct gene transcription by the synaptic nuclei maintains the local, persistent translation of the adult form of ACh receptors at the junctional region.

The mystery of the exceptional cholinergic sympathetic neurones is now accounted for by the presence of cholinergic-promoting trophic factors in the sweat glands. At least one such factor is CDF/LIF, and the sweat glands appear to contain some additional cholinergic promoting factors. Neuropeptides expressed by sympathetic and parasympathetic neurones can also be altered by some trophic factors in the target tissues, independently of the classical transmitter. Thus, the nature of both transmitter substances and neuropeptides in autonomic neurones depends on the target tissues. Neurotrophic factors can be regarded as intercellular signals mediating between neurones and their target tissues. In this respect, they are similar to cytokines, which are soluble mediators for communication in haematopoietic cells. Indeed, some cytokines show neurotrophic effects. Thus, the actions of neurotrophic factors and cytokines partly overlap, which suggests that neuronal development and haematopoiesis have a common evolutionary origin. Similarly, some neuropeptides mediate between neurones and their target tissues, thereby exerting neurotrophic effects. Consequently, it is difficult to distinguish strictly between such cytokines or neuropeptides and neurotrophic factors. Eventually, these molecules may be classified into several distinct superfamilies on the basis of their three-dimensional structures, rather than on their functional basis.

Suggested reading

Developmental changes of transmitter receptors

Changeux, J. P. (1991). Compartmentalized transcription of acetylcholine receptor genes during motor endplate epigenesis. *New Biologist*, 3, 413–29.

Froehner, S. C. (1993). Regulation of ion channel distribution at synapses. *Annual Review of Neuroscience*, 16, 347–68.

Hall, Z. W. and Sanes, J. R. (1993). Synaptic structure and development: The neuromuscular junction. *Neuron*, 10 (Supplement), 99–121.

Martinou, J. C. and Merlie, J. P. (1991). Nerve-dependent modulation of acetylcholine receptor ε-subunit gene expression. *Journal of Neuroscience*, 11, 1291–9.

McMahan, U. J. (1990). The agrin hypothesis. *Cold Spring Harbor Symposia on Quantitative Biology*, 50, 407–18.

Merlie, J. P. and Sanes, J. R. (1985). Concentration of acetylcholine receptor mRNA in synaptic regions of adult muscle fibres. *Nature*, 317, 66–8.

Reist, N. E., Werle, M. J., and McMahan, U. J. (1992). Agrin released by motor neurons induces the aggregation of acetylcholine receptors at neuromuscular junctions. *Neuron*, 8, 865–8.

Simon, A. M., Hoppe, P., and Burden, S. J. (1992). Spatial restriction of AChR gene expression to subsynaptic nuclei. *Development* 114, 545–553.

Witzemann, V., Brenner, H. R., and Sakmann, B. (1991). Neural factors regulate AChR subunit mRNAs at rat neuromuscular synapses. *Journal of Cell Biology*, 114, 125–41.

Transmitter choice by autonomic neurones

Landis, S. C. (1990). Target regulation of neurotransmitter phenotype. *Trends in Neurosciences*, 13, 344–50.

Patterson, P. (1978). Environmental determination of autonomic neurotransmitter functions. *Annual Review of Neuroscience*, 1, 1–17.

Rao, M. S. and Landis, S. C. (1990). Characterization of a target-derived neuronal cholinergic differentiation factor. *Neuron*, 5, 899–910.

Rao, M. S. and Landis, S. C. (1993). Cell interactions that determine sympathetic neuron transmitter phenotype and the neurokines that mediate them. *Journal of Neurobiology*, 24, 215–32.

Schotzinger, R. J. and Landis, S. C. (1988). Cholinergic phenotype developed by noradrenergic sympathetic neurons after innervation of a novel cholinergic target *in vivo*. *Nature*, 335, 637–9.

Yamamori, T., Fukada, K., Aebersold, R., Korsching, S., Fann, M. J., and Patterson, P. H. (1989). The cholinergic neuronal differentiation factor from heart cells is identical to leukemia inhibitory factor. *Science*, 246, 1412–6.

Regulation of neuropeptide levels

Coulombe, J. N. and Nishi, R. (1991). Stimulation of somatostatin expression in developing ciliary ganglion neurones by cells of the choroid layer. *Journal of Neuroscience*, 11, 553–62.

Micevych, P. E. and Kruger, L. (1992). The status of calcitonin gene-related peptide as an effector peptide. *Annals of the New York Academy of Sciences*, 657, 379–96.

Nawa, H. and Patterson, P. H. (1990). Separation and partial characterization of neuropeptide-inducing factors in heart cell conditioned medium. *Neuron*, 4, 269–77.

Rao, M. S., Tyrrell, S., Landis, S. C., Patterson, P. H. (1992). Effects of ciliary neurotrophic factor (CNTF) and depolarization on neuropeptide expression in cultured sympathetic neurons. *Developmental Biology*, 150, 281–93.

Schotzinger, R. J. and Landis, S. C. (1990). Acquisition of cholinergic and peptidergic properties by sympathetic innervation of rat sweat glands requires interaction with normal target. *Neuron*, 5, 91–100.

Neuropoietic cytokines

Anderson, D. J. (1989). The neural crest cell lineage problem: Neuropoiesis? *Neuron*, 3, 1–12.

Anderson, D. J. (1993). Molecular control of cell fate in the neural crest: The sympathoadrenal lineage. *Annual Review of Neurosciences*, 16, 129–58.

Bazan, J. F. (1991). Neuropoietic cytokines in the haematopoietic fold. *Neuron*, 7, 197–208.

Hilton, D. J. (1992). LIF: Lots of interesting functions. *Trends in Biological Sciences*, 17, 72–6.

McDonald, N. Q. and Hendrickson, W. A. (1993). A structural superfamily of growth factors containing a cystine knot motif. *Cell*, 73, 421–4.

Natahn, C. and Sporn, M. (1991). Cytokines in context. *Journal of Cell Biology*, 113, 981–6.

Patterson, P. H. and Nawa, H. (1993). Neuronal differentiation factors/cytokines and synaptic plasticity. *Neuron*, 10 (Supplement), 123–37.

Stemple, D. L. and Anderson, D. J. (1992). Isolation of a stem cell for neurons and glia from the mammalian neural crest. *Cell*, 71, 973–85.

13

Target-dependence of the neuronal state

As discussed in Chapter 11, at least at early developmental stages the survival of neurones is maintained by some specific neurotrophic factors derived from their target tissues. We have also seen that the transmitter phenotype in sympathetic neurones can be modified by alterations of the type of target tissue in developing animals (Section 12.2). Clearly, neuronal survival is not the only phenotype that is regulated by the target-derived neurotrophic factor. Moreover, the target-derived neurotrophic factors responsible for transmitter phenotype are distinct from those involved in the maintenance of neuronal survival. It is not known how many target-derived neurotrophic factors are required for normal neurones to express their different phenotypes. Actually, we do not know how target-derived neurotrophic factors promote neuronal survival or trigger the expression of a phenotype. Might neurones require target-derived trophic factors to maintain the expression of their phenotypes? NGF is required for survival in developing sensory neurones but not in adult sensory neurones. Indeed, as we shall see in section 13.2.2, it is a general rule that the survival of neurones becomes less dependent on their targets in maturity. We defined that neurotrophic factors are required to maintain a cell's pre-existing or programmed state. Might the pre-existing state of neurones in adulthood still depend on target-derived neurotrophic factors? What is the functional significance of target-derived neurotrophic factors in maturity? We now focus on target-dependence of neuronal properties in adulthood.

13.1 REGULATION OF NEURONAL PROPERTIES BY THE TARGET

It is well known that neurone cell bodies undergo several changes when they are separated from their target cells by severing of their axons (*axotomy*). All axotomy-induced changes in the neurone cell body are collectively termed the *axon reaction* (Lieberman 1971). If particular properties in a neurone were maintained by a neurotrophic influence emanating from its target cell, part of the axon reaction would come from the elimination of the trophic influence. Therefore, analysing the axon reaction is one way to identify retrograde neurotrophic actions. For instance, if the neurotrophic effect is mediated by a specific chemical factor, the antibody against this presumptive trophic substance would mimic the axon reaction even without axotomy. Moreover, applying the presumptive trophic substance might reverse the axon reaction.

Spinal motor neurones clearly receive trophic influences from the target muscle, but the nature of their trophic substances remains uncertain (section 11.4). NGF is one retrograde neurotrophic factor known for sympathetic and sensory neurones (section 11.2.3). In the dorsal root ganglion, however, sensory neurones have two axons, a peripheral process projecting to the periphery and a central process projecting to the spinal cord. The two axons thus innervate different targets. Might each axon have a different trophic influence on its shared cell body? If so, the difference would shed light on the probable mechanism of neurotrophic function. We shall now examine the axon reaction in these three cases: spinal motor neurones, sympathetic neurones, and diaxonal sensory cells.

13.1.1 Neuronal alterations induced by axotomy

Since the first description by Downman *et al.* (1953), the changes in electrophysiological properties of spinal motor neurones following axotomy have been studied extensively (Titmus and Faber 1990). Figure 13.1 illustrates action potentials and the after-hyperpolarization (AHP) following each action potential recorded from normal spinal motor neurones of the cat. Recall that mammalian spinal motor neurones innervating the rapid contracting muscle and the slowly contracting muscle differ in their discharge frequency (section 10.2). This difference reflects how long the AHP persists. The generation of action potentials in a motor neurone is suppressed as long as its membrane potential is hyperpolarized by the AHP following each action potential. The longer the AHP, the lower the frequency at which the motor neurone can discharge. Motor neurones innervating the fast contracting medial gastrocnemius (MG) muscle, for example, have a relatively short AHP, compared with the motor neurones innervating the slowly contracting soleus (SOL) muscle (Fig. 13.1B). The MG and SOL motor neurones also differ in their conduction velocity. Thus, because the AHP duration is inversely correlated with the conduction velocity of motor neurones (Fig. 13.2A), the MG (open circles) and SOL (filled circles) motor neurones are segregated into two groups by these two parameters. A few weeks after section of these muscle nerves (axotomy), the conduction velocity is significantly reduced in both types of motor neurone, whereas the AHP duration is significantly reduced only in SOL motor neurones, altering little in the MG motor neurones (Fig. 13.1C,D; Kuno *et al.* 1974a; Foehring and Munson 1990). Thus, axotomy blurs the characteristic differences between MG and SOL motor neurones (Fig. 13.2B). During the development of kittens, both MG and SOL motor neurones progressively increase their conduction velocity. Their AHP durations are uniformly short in neonatal kittens but lengthen with age in SOL motor neurones, whereas MG motor neurones remain fairly steady in AHP over time (Huizar *et al.* 1975). In

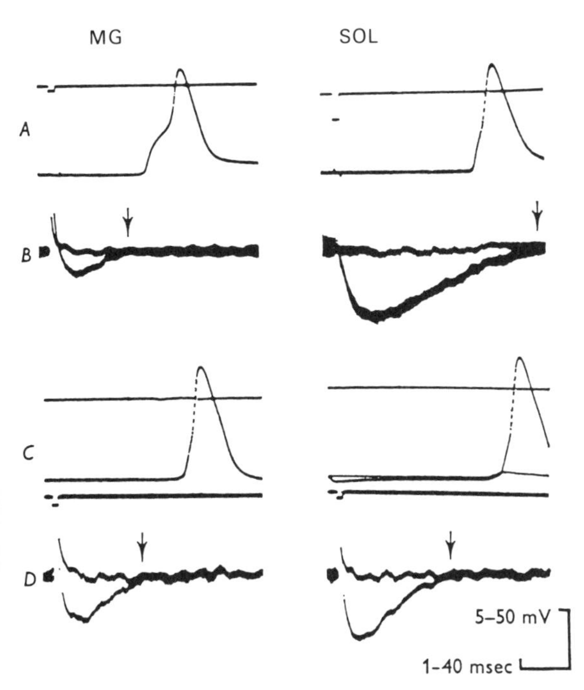

Fig. 13.1. Action potentials recorded from medial gastrocnemius (MG) and soleus (SOL) motor neurones of the cat. A, the action potentials evoked by stimulation of the muscle nerve. B, the after-hyperpolarization following each action potential. Arrows indicate termination of after-hyperpolarization. C and D, similar to A and B, but responses recorded 50 days after section of the MG and SOL muscle nerves. Calibration: A and C, 50 mV and 1 ms; B and D, 5 mV and 40 ms. (From Kuno *et al.* 1974a.)

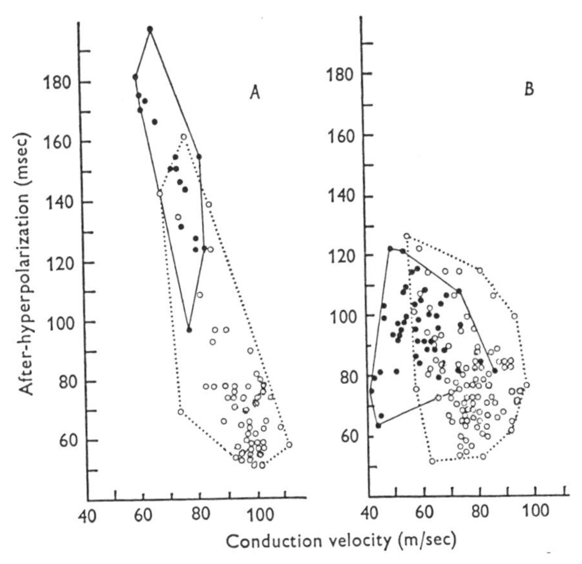

Fig. 13.2. Relation between the duration of after-hyperpolarization and the axonal conduction velocity of motor neurones. Open circles, medial gastrocnemius motor neurones. Filled circles, soleus motor neurones. A, results from normal, unoperated cats. B, results obtained 14–52 days after section of the medial gastrocnemius and soleus muscle nerves. Boundaries indicate the range for soleus (solid line) and medial gastrocnemius (dotted line) motor neurones. (Adapted from Kuno *et al.* 1974a.)

short, axotomy seems to 'de-differentiate' the properties of adult motor neurones, recapitulating their neonatal state. Strikingly, all these motor neurone properties altered by axotomy are able to recover to nearly normal, achieving 're-differentiation' when their regenerated axons resume functional contacts with the target muscle (Kuno *et al.* 1974b; Foehring *et al.* 1986; Titmus and Faber 1990). These results suggest that the properties of adult motor neurones are maintained by some trophic signal from their target muscles. If so, might alterations of muscle activity affect the motor neurone properties? Indeed, the AHP duration in SOL motor neurones was found to shorten when activity of the SOL muscle was deprived for one week by conduction block of the peripheral nerve with locally applied tetrodotoxin (TTX), rather than by peripheral axotomy (Fig. 13.3A; Czeh *et al.* 1978). Moreover, the AHP shortening induced by TTX was reversed by chronic stimulation of the nerve peripheral to the site of nerve block (Fig. 13.3B) but not central to the blocking site (A). What does this result mean? Chronic stimulation of the nerve central to the blocking site does not prevent the TTX-induced atrophy of the SOL muscle, whereas stimulation of the nerve peripheral to the blockade alleviates muscle atrophy. Thus, the AHP duration of SOL motor neurones must depend on the state of the target muscle. In other words, at least some axotomy effects on motor neuronal properties can be mimicked by reducing the activity of the target muscle.

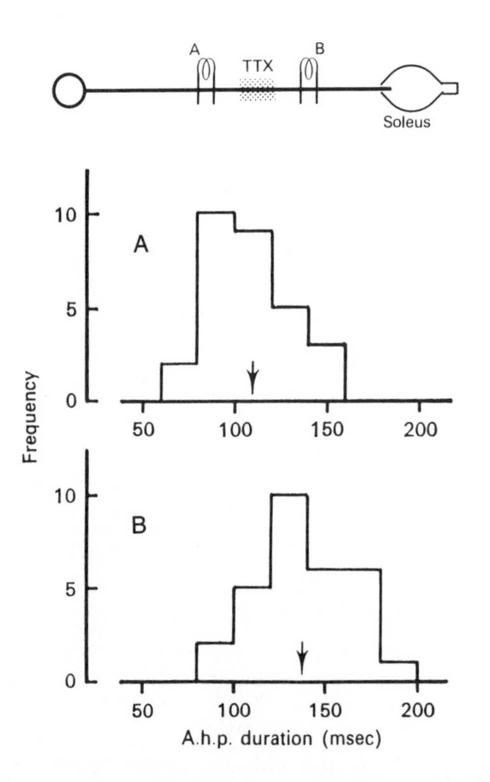

Fig. 13.3. Dependence of the after-hyperpolarization (AHP) duration in soleus motor neurones on activity of the target muscle. Distributions of AHP durations after conduction block of the solues nerve by locally applied tetrodotoxin (TTX) for one week. One week of TTX nerve block was combined with chronic nerve stimulation either central (A) or peripheral (B) to the blockade site (see the top diagram). Arrows indicate the mean durations of AHP. (Adapted from Czeh *et al.* 1978.)

We now turn to axotomy of sympathetic neurones. One major change in mammalian sympathetic neurones produced by axotomy is a reduction in the e.p.s.p. amplitude, as shown in Fig. 13.4*a*,*b* (Purves 1975). The e.p.s.p. amplitude is reduced markedly within a few days of axotomy but recovers when peripheral connections are re-established. This again suggests that axotomy-induced changes in neuronal properties are due to the deprivation of some trophic signal from the target tissues. Purves and Nja (1976) applied a silicone rubber pellet containing NGF to the surface of the sympathetic ganglion at the time of axotomy. This procedure significantly countered the decrease in the e.p.s.p. amplitude induced by axotomy (Fig. 13.4*c*). Axotomy-induced detachment of presynaptic terminals on sympathetic neurones was also prevented by the NGF pellet. Complementary to these findings is the effect of antisera against NGF: in guinea-pigs, systemic treatment for a few days markedly reduced the amplitude of e.p.s.p.s recorded from sympathetic neurones without axotomy (Nja and Purves 1978). In short, at least some axotomy-induced changes in sympathetic neurones can be prevented by NGF, and the axon reaction in sympathetic neurones can be mimicked by anti-NGF antisera without axotomy. Thus, the axon reaction in sympathetic neurones must result from the interruption of the NGF supply from the target tissues. This view is consistent with previous observations that a colchicine-blockade of axonal transport in the peripheral nerve fibres induces changes in their cell bodies similar to the axon reaction (Pilar and Landmesser 1972; Purves 1976).

To examine the axon reaction of a sensory neurone do we sever its peripheral process or its central process? Intriguingly, the sensory neurones show distinct morphological changes, termed *chromatolysis*, following section of their peripheral axons (peripheral axotomy), whereas they are totally unresponsive to section of their

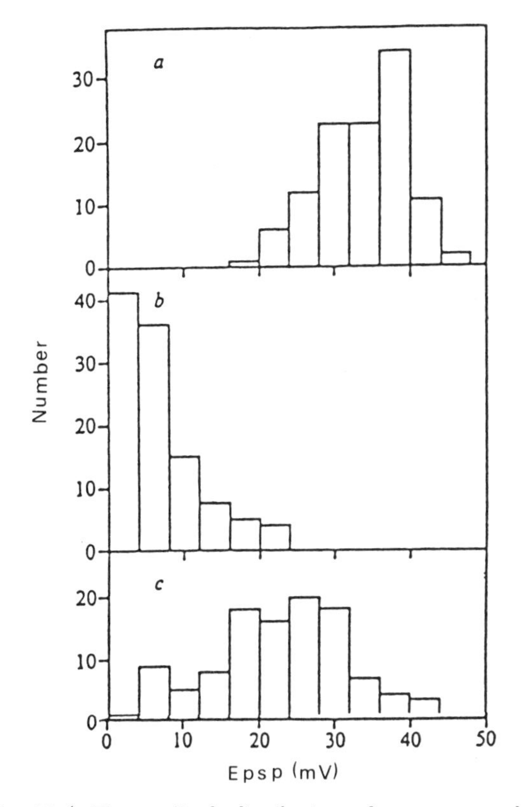

Fig. 13.4. The amplitude distributions of e.p.s.p.s recorded from superior cervical sympathetic ganglion cells of the guinea pig. *a*, results from normal animals. *b*, results obtained 4–7 days after crushing the postganglionic fibres (axotomy). *c*, same as in *b*, but a silicone rubber pellet containing NGF was applied to the ventral surface of the ganglion during axotomy. (Adapted from Purves and Nja 1976.)

central axons (central axotomy; Anderson 1902; Ranson 1914; Lieberman 1971). Consistent with this behaviour, the conduction velocity of sensory neurones decreases 3 weeks after peripheral axotomy but not after central axotomy (Czeh *et al.* 1977; Gallego *et al.* 1987). Clearly, injury *per se* cannot account for the decrease in conduction velocity. In fact, central axotomy does not alter the conduction velocity of the central axon, despite the nerve injury, whereas peripheral axotomy reduces the conduction velocity in the peripheral axon and in the central axon as well, despite intact central connections (Czeh et al. 1977). Moreover, other electrophysiological changes in the sensory neurone cell body follow peripheral axotomy but not central axotomy (Czeh *et al.* 1977; Gallego *et al.* 1987). Thus, changes in the cell bodies of sensory neurones must participate in a reduction of the axonal conduction velocity. That is, a decrease in conduction velocity is one of the functional correlates of the axon reaction.

The conduction velocity is a basic parameter that is characteristic of each neurone. In almost every neurone, axotomy consistently decreases both conduction velocity and fibre diameter (Titmus and Faber 1990). Since fibre diameter is a primary determinant of axonal conduction velocity, and since fibre diameter depends on the number of neurofilaments (Hoffman *et al.* 1987), it has been suggested that axotomy causes down-regulation of neurofilament gene expression in the neurone (Oblinger and Lasek 1988; Titmus and Faber 1990). The expression of neurofilament mRNA in sensory neurones indeed declines within two weeks of peripheral axotomy; yet central axotomy also reduces the neurofilament mRNA level in the sensory neurones with a similar time course (Wong and Oblinger 1990). That central axotomy and peripheral axotomy suppress the neurofilament mRNA similarly yet affect conduction velocity differently is a curious discrepancy. Despite the plausibility of the conjecture, axonal conduction velocity must not be directly related to the expression of neurofilaments of the neurones. On the other hand, the expression of tubulin mRNA in sensory neurones increases after central axotomy but decreases after central axotomy (Wong and Oblinger 1990). A similar reciprocal modulation by central and peripheral axotomy also marks the CGRP content of sensory neurones (Inaishi *et al.* 1992). Clearly, the interactions of a sensory neurone with its central and peripheral targets differ radically.

We have seen that NGF acts as a retrograde neurotrophic factor for sensory neurones as well as for sympathetic neurones. Might NGF likewise reverse the axotomy-induced changes in sensory neurones? Because the dorsal root ganglion (DRG) cells contain substance P and CGRP, their axon reaction can be evaluated quantitatively, using these neuropeptides as a marker. Peripheral axotomy reduces the substance P or CGRP content in the DRG, but both reductions are prevented by infusing NGF into the central stump of the cut peripheral nerve in adult rats (Fitzgerald *et al.* 1985; Wong and Oblinger 1991; Inaishi *et al.* 1992). The same procedure also reverses other axotomy-induced alterations in DRG cells, including some electrophysiological properties (Fitzgerald *et al.* 1985). NGF-infusion fails to counter the decrease in synthesis of neurofilaments displayed by peripheral axotomized sensory neurones, however (Wong and Oblinger 1991). It appears that not all changes in axotomized neurones share the same underlying mechanism or the same neurotrophic effect.

Although eliminating the target activity or the retrograde trophic factor may not account for all the neuronal changes induced by axotomy, there seems little doubt that the neurone properties are regulated by interactions with the innervated targets. We have seen how the contractile properties of skeletal muscle change through re-innervation by a different type of motor neurone—in other words, how motor neurones can specify properties of their target (section 10.2). Might the target tissues also dictate properties of their innervating neurones?

13.1.2 Neuronal changes induced by altered target

As illustrated in Fig. 13.2A, two groups of mammalian spinal motor neurones innervating the slowly contracting soleus (SOL) muscle and the fast contracting medial gastrocnemius (MG) muscle can be distinguished by their electrophysiological properties. We have seen that although axotomy of the motor neurones blurs these differences (Fig. 13.2B), re-innervation of their target muscles restores them. This target-dependent 're-differentiation' of motor neurone properties implies two possible sources. First, even if their characteristic properties are determined by factors intrinsic to each type of motor neurone, the expression of these properties may require axonal contacts with muscle. Second, the characteristic properties may be determined by a target-specific signal from the innervated muscle. If the first case were true, the SOL motor neurones would recover their original properties whether their cut peripheral axons were united to the original SOL muscle or to the foreign fast contracting muscle. In the second case, on the other hand, the SOL motor neurones would adopt the properties of MG motor neurones when united to the fast contracting muscle.

To address these questions, the electrophysiological properties of cat SOL motor neurones were examined after reunion of the cut peripheral SOL nerve to a muscle. The SOL motor neurones recovered their usual long AHPs by about 150 days postoperatively, whether united to the original SOL muscle or to the foreign fast contracting muscle (Kuno *et al.* 1974b). Thus, the motorneurone properties appear to be determined by some intrinsic factor, although their expression requires muscle innervation. Other results were inconsistent with this interpretation, however. For example, Lewis *et al.* (1977) found that the mean conduction velocity of the SOL motor neurone after reunion of the cut SOL nerve to the fast contracting muscle is significantly faster than after reunion to the original SOL muscle. Moreover, MG motor neurones re-innervating the SOL muscle were found to have a significantly longer AHP than those re-innervating the fast contracting muscle (Foehring *et al.* 1987; Foehring and Munson 1990). Thus, the type of target muscle appears to specify the expression of the gene encoding particular properties of the innervating motor neurones.

A remarkable example of target-dependent motoneuronal behaviour involves the rat spinal nucleus of the bulbocavernosus (SNB). SNB motor neurones innervate perineal striated muscles and are more numerous in males than in females, being sexually dimorphic, as briefly described previously (section 12.2.2; Arnold 1984; Breedlove 1986). The size of SNB motor neurones in adult male rats shrinks considerably following castration, but this can be reversed by treatment with testosterone. Clearly, their size is hormonally regulated. Where does testosterone act? Since androgen receptors are present in both the perineal muscles and SNB motor neurones, the neurone size could be regulated either by direct action of the hormone on SNB neurones or by a target-derived factor possibly controlled via hormonal action on the innervated muscle (Breedlove 1986). Araki *et al.* (1991) united the cut SNB axons distally to the grafted SOL muscle, which lacks androgen sensitivity. On the control side, the SNB axons were similarly cut but allowed to re-innervate the original perineal muscle. The SNB motor neurones re-innervating the perineal muscles shrank following castration and were enlarged by testosterone treatment. In contrast, the size of SNB motor neurones innervating the grafted SOL muscle was unaffected by testosterone manipulation. In other words, changes in the size of SNB motorneurones by testosterone manipulation depend on the presence or absence of androgen sensitivity in the re-innervated muscle. Evidently, the target muscle mediates the hormonal regulation of SNB motor neurone size; hence, some retrograde signal from the muscle must control neuronal size. As noted previously (section 11.3.2), prevention of cell deaths of SNB neurones by treating perinatal female rats with CNTF (Forger *et al.* 1994) might result from a secondary influence of CNTF's myotrophic effects on the bulbocavernosus.

A similar regulation of neuronal morphology by interactions with the peripheral target exists in sympathetic neurones. The dendritic length of individual sympathetic neurones shortens after axotomy and increases with target re-innervation (Yawo 1987). When the target size is experimentally altered, the dendritic length of sympathetic neurones correlates well with the target size (Voyvodic 1989). That is, the size of the sympathetic neurone is regulated by its target, presumably by the amount of retrogradely supplied neurotrophic factor, NGF (Purves *et al.* 1988). Administering NGF, in fact, markedly lengthens the dendrites of sympathetic neurones (Snider 1988).

Let us now turn to sensory neurones. Sensory neurones in the dorsal root ganglion (DRG) are not a homogenous population. In terms of sensory modality (or sensory receptors), sensory neurones can be classified into numerous groups by the form of stimulus that best activates them (touch, stretch, pinch, cold, heat, etc.). If a peripheral target influences neuronal properties, sensory neurones with different peripheral receptors may exhibit different properties. Belmonte and Gallego (1983) found that sensory neurones in the petrosal ganglion arising from baroreceptors (mechanoreceptors) in the carotid sinus and those from chemoreceptors in the carotid body show clearly different electrophysiological properties (see also Koerber *et al.* 1988). These results alone do not establish whether the type of peripheral target sensory receptor indeed influences the properties of sensory neurones, however. The experiment of cross-re-innervation of sensory neurones to different peripheral receptive fields is required.

Ritter *et al.* (1991), treating neonatal rats with anti-NGF antisera, observed a loss of small myelinated (Aδ) high-threshold mechanoreceptor sensory fibres. This depletion was not due to cell death of Aδ sensory neurones. Rather, the number of Aδ fibres responsive to hair movement was anomalously high, suggesting that sensory receptors innervated by Aδ fibres had switched from cutaneous high-threshold nociceptors to hair follicles. Ritter *et al.* (1991) suggested that at early postnatal stages NGF stabilizes the epidermal projection of Aδ sensory fibres; a shortage of NGF thus forces the Aδ fibres to withdraw from the epidermis to the dermis, where they erroneously form contacts with hair follicles. Timing is crucial, for a rat treated with anti-NGF antisera two weeks after birth display no abnormality in the sensory modality of Aδ fibres.

These results raise two sets of intriguing questions. First, how might anti-NGF antibodies affect only one type of sensory fibre, the Aδ? Perhaps NGF reserves its trophic effects for only one particular type of sensory neurone. If so, do other types of sensory neurone have customized neurotrophins too? Second, why are anti-NGF antibodies ineffective at late postnatal stages? Perhaps neurones require trophic factors only during earliest development. If so, what induces target-dependent alterations of neurone properties in adults? We have seen various phenotypes of sympathetic and sensory neurones (cell survival, neuronal size, peptide expression, and electrophysiological properties) that are affected by NGF. However, NGF's influence on different phenotypes within a given neurone may differ in quality or quantity. NGF affects the expression of substance P and of neurofilaments in sensory neurones, for example, but not in the same way. Therefore, to address whether a trophic factor acts differently at different developmental stages or among different subtypes of neurone, we must evaluate the trophic effects on a single phenotype. We will now examine developmental changes in the dependence on trophic factors, using neuronal survival as a criterion.

13.2 TARGET-DEPENDENCE OF NEURONAL SURVIVAL

In Chapter 11 we described several identified neurotrophic factors and their effects on neuronal survival. Our principal question in that chapter was which neurotrophic factor rescues which neurone. We have seen many examples that neuronal survival is target-dependent or that the survival of neurones is maintained by a neurotrophic factor derived from their targets. Yet we have not discussed so far how neuronal survival becomes target-dependent or how neurones die when their neurotrophic factor is deprived. As shown in Fig. 10.2 (open circles), the natural cell death of chick lumbar motor neurones commences on embryonic day 6. If the target limb bud is removed on embryonic day 2.5, before its motor innervation, all the lumbar motor neurones eventually die (Fig. 10.2, filled circles)—but they survive at least until day 6. Evidently, the survival of motor neurones is independent of the target until embryonic day 6, but after this critical period they die if not in contact with their target tissue. What makes neuronal survival turn target-dependent at a particular developmental stage? Do the neurones require target-contact for survival throughout life? If not, when does their survival become target-independent? and what makes neuronal survival turn target-independent? Let us examine how developing neurones acquire target-dependence for survival.

13.2.1 The acquisition of target-dependence of neuronal survival

As noted previously, target-dependent neuronal survival

at early developmental stages presumably reflects the presence of a crucial neurotrophic factor in the target (sections 10.3 and 10.4). Thus, the target-dependence of neuronal survival corroborates the cell's dependence on a trophic factor. This notion rests principally on work with NGF. The same notion probably applies to target-dependent survival of motor neurones, but experimental verification is complicated by the lack of identified neurotrophic factors for motor neurones.

NGF is undoutedly a trophic factor essential for the survival of sensory neurones early in development. Yet very young sensory neurones thrive without NGF while their axons are still approaching their targets (see below). This independence raises new questions. How do sensory neurones reach their targets? And what happens when they get there?

Levi-Montalcini (1982) categorized the effects of NGF in the nervous system as the trophic, tropic and transforming. Of this trio, NGF's *tropic* (turning toward) effect requires special comment. When a micropipette containing NGF-rich solution is placed near the growing tip (growth cone) of chick lumbar dorsal root ganglion cells cultured at embryonic day 7 or 12, the growing axon gradually turns toward the NGF source. As shown in Fig. 13.5 (Gundersen and Barrett 1979), the sensory axon will follow the pipette through a series of moves, the growth cone always turning toward the NGF source. Thus, NGF seems able to guide sensory neurones by its tropic effect. Moreover, Menesini-Chen *et al.* (1978) found that concentrated NGF injected daily into the brainstem of neonatal rats causes sympathetic neurones in the paravertebral ganglia to grow anomalously into the brainstem through the dorsal column of the spinal cord. From these results it seemed reasonable to postulate that NGF's tropic effect is responsible for guiding the axonal projections of sensory and sympathetic neurones to their targets, where NGF would be highly concentrated in normal development. Surprisingly, however, Davies *et al.* (1987) proved otherwise.

Davies *et al.* (1987) examined the process by which trigeminal sensory neurones innervate maxillary skin in the mouse embryo. These sensory neurones undergo natural cell death between embyronic days 13 and 19 (E13–19; Davies and Lumsden 1984). The earliest fibres reach the target maxillary epithelium by E11. If the trigeminal sensory axons were guided by NGF in the target tissues, NGF should be present in the maxillary epithelium before E11, and the sensory neurones should possess NGF receptors by then. Neither is the case. NGF is not detected in the maxillary epithelium at E10 or E10.5; it first appears at E11 and reaches a maximal level by E13. The expression of NGF receptors was examined by autoradiography after incubation of trigeminal neurones with readioactive NGF. The fraction of trigeminal neurones labelled with NGF was less than 30 per cent at E10 and increased to 90 per cent on E11.

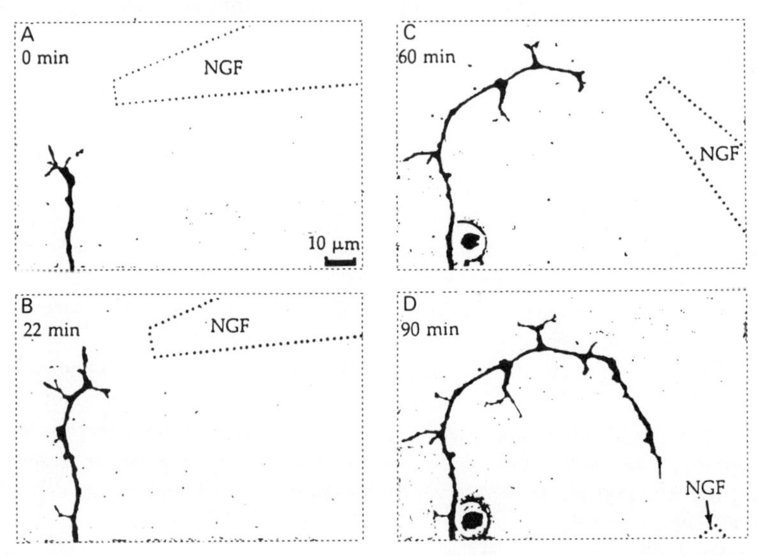

Fig. 13.5. Guidance of axonal growth by NGF in cultured sensory neurones. Dorsal root ganglia excised from chick embryos 7 and 12 days of age were cultured. A micropipette filled with NGF (dotted lines) was placed about 25 μm from the tip of the growth cone and repositioned during the 90 min observation period. Serial photographs (taken at times indicated at upper left) show the growing axon toward the NGF source. (Adapted from Gundersen and Barrett 1979.)

In short, both NGF synthesis and NGF receptor expression coincide with target innervation. By what process? Might the neurone–target contacts trigger NGF receptor expression in the sensory neurones? The trigeminal neurones dissociated at E10 express more NGF receptors when the culture period is prolonged, despite the absence of the target. This result suggests that the neuronal expression of NGF receptors is independent of the target tissue.

Thus, the following scheme is implied for the relationship between developing sensory neurones and NGF at their targets. First, the projection of a sensory neurone to its target develops independently of NGF. Second, the synthesis of NGF in the target tissue is triggered by neuron–target contacts by unknown mechanisms. Third, the expression of NGF receptors is regulated by a programme intrinsic to the neurones. Evidently, the tropic effects of NGF cannot be exerted on the sensory neurones before innervation of their targets, because the sensory neurones lack NGF receptors until they reach their targets. As we discussed previously (section 11.2.3), the survival of chick retinal ganglionic cells also begins to depend on BDNF at embryonic day 6 when the ganglionic axons reach the target (Rodriguez-Tebar *et al.* 1989). How, then, can NGF guide sensory neurones in culture (Fig. 13.5)? It should be noted that the neurones used for the studies of NGF tropic influence described above (Menesini-Chen *et al.* 1978; Gundersen and Barrett 1979) had already innervated their targets, hence, must have had NGF receptors. In other words, NGF exerts tropic effects on neurones once they express NGF receptors.

Without NGF receptors, the sensory neurones cannot experience NGF's trophic effects before target innervation. For example, trigeminal neurones excised from mouse at E9 can grow in culture media containing anti-NGF antisera (Davies and Lumsden 1984). Yet dependence on neurotrophic factor for neuronal survival develops *in vitro* when the neurones are cultured in the absence of their targets (Davies and Lumsden 1984; Davies *et al.* 1987; Ernsberger and Rohrer 1988). Apparently, the onset of NGF dependence is regulated by a 'developmental clock' in the neurones (Davies and Vogel 1991). Thus, the natural cell death is a programmed cell death as a consequence of the expression of NGF receptors in the neurones. Once the sensory neurones become NGF-responsive, their survival becomes NGF-dependent.

As shown in Fig. 10.2 (open circles), the natural mortality of developing motor neurones ceases when the peripheral neuromuscular connections are complete. Two explanations of this behaviour are possible. First, the target-dependence of neuronal survival is limited to a critical period that coincides with the formation of neuromuscular connections, after which the neurones no longer require their targets. Second, the neurones that establish successful neuromuscular connections are supported by the retrograde trophic factor from their targets. The second possibility was favoured by Schmalbruch (1984), who found that transection of the sciatic nerve in neonatal rats causes massive degeneration of lumbar motor neurones (see also Romanes 1946). Earlier work on re-innervation of muscle in neonatal rats found no sign of motor neurone death, however, despite transection of the peripheral motor fibres near their entrance to the muscle (Bennett and Pettigrew 1974; Brown *et al.* 1976). What accounts for this discrepancy? Schmalbruch suggested that in neonatal rats motor neurones will survive if they achieve target contacts soon after section of their motor axons near the muscle. Indeed, most motor neurones (80 per cent) survive when the severed nerve to a hindlimb muscle is allowed to re-innervate the muscle in neonatal rats, whereas most motor neurones (80 per cent) die if re-innervation is prevented for 30 days (Kashihara *et al.* 1987).

How long can axotomized motor neurones survive in neontal rats without their target contacts? Kashihara *et al.* (1987) sectioned a muscle nerve in 4-day-old rats and prevented muscle re-innervation; after various intervals they labelled the surviving motor neurones of the muscle retrogradely with horseradish peroxidase (HRP) applied to the cut peripheral nerve. As shown in Fig. 13.6, about 80 per cent of the motor neurones still survived after two weeks without muscle re-innervation, but a major cell loss then occurred between two and three weeks after the deprivation of the target muscle by nerve section. Thus, motor neurones still depend on the target for survival in the early postnatal period, although they succumb less rapidly to target deprivation than do embryonic motor neurones (Fig. 10.2; see also Snider and Thanedar 1989; Lowrie and Vrbova 1992).

The survival of motor neurones becomes even less dependent on their targets as the animal matures. For example, nine months after amputaton of the hindlimb in kittens (about 50 days of age) only 23 per cent of lumbar motor neurones are dead (Jorgensen and Dyck 1979). The same operation in adult cats (1–3 years old) does not significantly reduce the number of the lumbar motor neurones even 18 months after the operation (Carlson *et al.* 1979). On the other hand, Kawamura and Dyck (1981) examined the lumbar spinal cords of

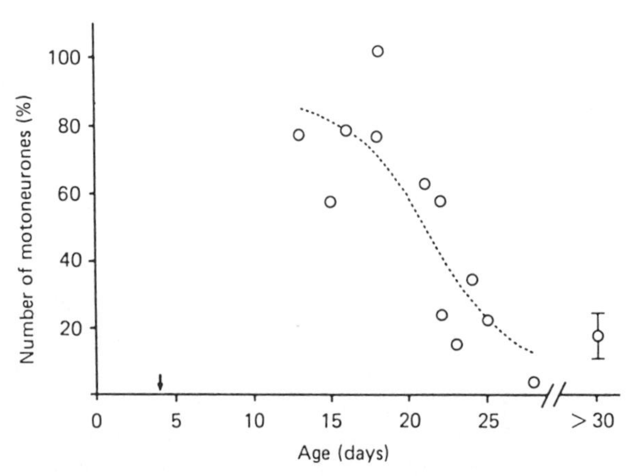

Fig. 13.6. Time course of cell death of neonatal rat motor neurones separated from their target muscle. The nerve to the medial gastrocnemius (MG) muscle was cut on one side 4 days after birth (arrow) and prevented from muscle re-innervation. Between 10 and 40 days postoperatively, the MG motor neurones on both sides were labelled retrogradely with horseradish peroxidase. Ordinate, percentage of the number of MG motor neurones labelled on the operated side relative to that on the contralateral, intact side. (From Kashihara *et al.* 1987.)

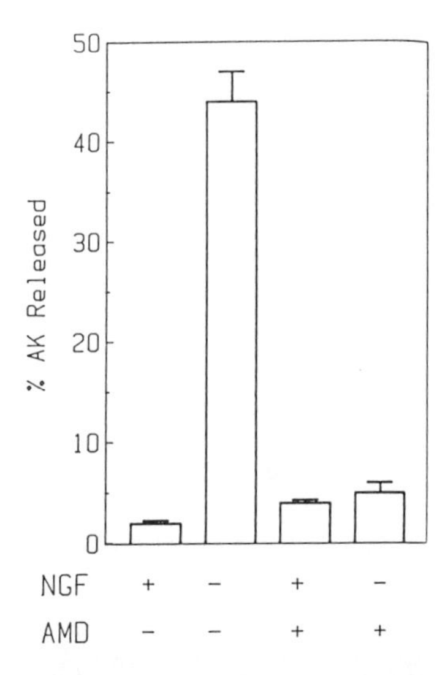

Fig. 13.7. Effect of the RNA synthesis inhibitor, actinomycin-D (AMD), on cell death of cultured sympathetic neurones. Ordinate measures the release into the culture medium of the cytosolic enzyme, adenylate kinase (AK; expressed as a percentage of total releasable AK), an indicator of the fraction of neurones dying in the 48 hour period. +, the presence of NGF or AMD in culture media during the 48 hour period. −, the absence of NGF or AMD. (From Martin *et al.* 1988.

two patients (54 and 60 years old) who died 4.5 and 9 years after unilateral amputation of the lower limb for surgical treatment of gangrene and sarcoma. This study showed unequivocal loss (over 50 per cent) of lumbar motor neurones on the amputated side. In maturity, then motorneuronal survival depends less on the target, but its target-dependence is never completely abolished. Why do neurones die when their targets or target-derived neurotrophic factors are deprived? Why does neuronal survival become less dependent on the target in maturity?

13.2.2 How neuronal death is induced

We have seen that injection of anti-NGF antibodies into rat embryos causes massive cell death of DRG sensory neurones (section 11.2.3). Similarly, the survival of sympathetic neurones is maintained by NGF. In principle, neuronal survival-promoting effects by NGF can be interpreted in two ways: either NGF provides nutritive effects or NGF suppresses the synthesis of deleterious molecules. If the latter is the case, the neurones might be rescued from cell death by adding an inhibitor of protein synthesis even without NGF. To examine these possibilities, sympathetic neurones dissociated from rat embryos (E21) were cultured for one week with NGF, then for two additional days with or without NGF (Martin *et al.* 1988). The ordinate of Fig. 13.7 shows what fraction of the sympathetic neurones died during the final 48 hours; this fraction was estimated by measuring the cytosolic enzyme, adenylate kinase (AK), which is released into the culture medium as membranes rupture in neuronal death. Withholding (−) NGF during the two day period clearly increased neuronal death, and supplying (+) NGF prevented it (Fig. 13.7 left two columns). Inhibiting the RNA synthesis by actinomycin-D (AMD) during the two days, however, markedly suppressed neuronal death, independently of NGF (Fig. 13.7 right two columns). Survival similarly improved with inhibitors of protein synthesis (cycloheximide or puromycine). These results implicate that the cell death of NGF-deprived sympathetic neurones is associated with newly synthesized proteins. Presumably, there are certain proteins deleterious to the neurones whose synthesis NGF inhibits (Martin *et al.* 1988). These presumptive deleterious proteins are called *killer proteins* (Martin *et al.* 1988) or *thanatin* (Martin *et al.* 1992; after the Greek god of death, *Thanatos*). Inhibitors of protein or RNA synthesis also prevent natural and axotomy-induced motor neurone death in chick embryos (Oppenheim *et al.* 1990). As we will see (Fig.

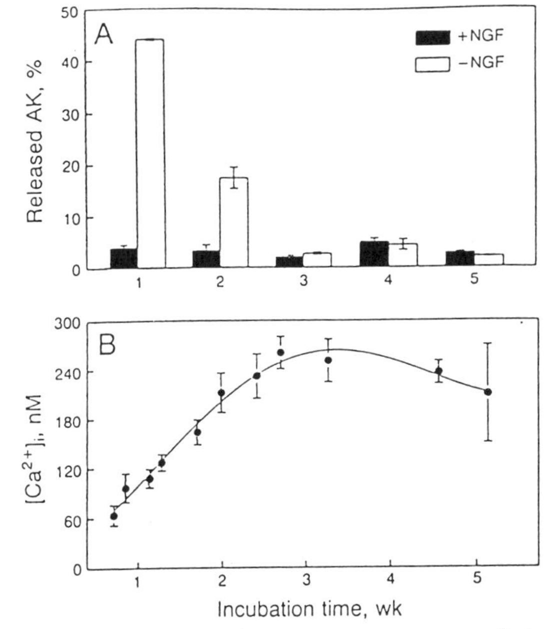

Fig. 13.8. Relationship between the intracellular Ca^{2+} level and NGF-independent survival in cultured sympathetic neurones. A, the fraction of sympathetic neurones dying during 48 hours with (black columns) or without (open columns) NGF in culture, esimated as in Fig. 13.7. Abscissae, weeks of the initial culture, which included NGF. B, the intracellular Ca^{2+} levels of cultured sympathetic neurones, measured with fura-2. (From Koike and Tanaka 1991.)

13.8), the intracellular Ca^{2+} level of cultured sympathetic neurones is very low early in development and increases as the neurones grow. Because Ca^{2+} deficiency impairs cellular protein synthesis (C. O. Brostrom and M. A. Brostrom 1990), the deleterious proteins or thanatin may be scarce in early development, thus enabling neurones to survive even in the absence of the trophic factor or trophic factor receptors. Natural cell death therefore might be programmed by the intracellular Ca^{2+} level (Franklin and Johnson 1992; Johnson *et al.* 1992).

As described in section 13.2.1, motoneuronal survival becomes less dependent on the target in maturity. Sympathetic neurones undergo similar developmental change in their NGF-dependent survival. With mice, the anti-NGF antiserum affects sympathetic neurones far less extensively in adults than in neonates (Angeletti *et al.* 1971; Ruit *et al.* 1990). Furthermore, maturational reduction in NGF-dependence for neuronal survival can be reproduced in cultured sympathetic neurones (Koike *et al.* 1989; Koike and Tanaka 1991; Johnson *et al.* 1992). In the experiment illustrated in Fig. 13.8A, sympathetic neurones dissociated from neonatal rats were cultured for 1–5 weeks in the presence of NGF, then for two more days with (filled columns) or without (open columns) NGF. As we saw earlier (Fig. 13.7, left two columns), more than 40 per cent of the sympathetic neurones die during the two day period if NGF is withheld (Fig. 13.8A, 1). But the longer the initial incubation with NGF, the better the survival rate (Fig. 13.8A). That is, even in culture survival depends less on NGF as sympathetic neurones mature.

The maturational adjustment seems to involve Ca^{2+} levels. Extending the initial incubation allows the intracellular Ca^{2+} level of the sympathetic neurones to rise (Fig. 13.8B), until the 48 hour survival period becomes virtually independent of NGF as the Ca^{2+} level nears 250 nM (Koike and Tanaka 1991). By contrast, sympathetic neurones intially incubated with NGF for only one week are more vulnerable to NGF deprivation. Yet even then neuronal survival of NGF deprivation can be promoted by raising intracellular Ca^{2+} levels; this was done by depolarizing cultured sympathetic neurones with K^+-rich solutions (Koike *et al.* 1989). Franklin and Johnson (1992) and Johnson *et al.* (1992) suggested that at an optimal intracellular Ca^{2+} concentrations (a 'set point'; about 250 nM), developing sympathetic neurones can survive in an NGF-free medium. Reaching the optimal Ca^{2+} level is thus equivalent to having NGF sustain neuronal survival. If the role of NGF is to suppress the synthesis of thanatin, its synthesis may also be suppressed at the optimal intracellular Ca^{2+} level. At this stage, however, no underlying mechanisms are known. Indeed, Edwards *et al.* (1991) showed that NGF can promote the survival of sympathetic neurones independently of protein synthesis (see also Deckwerth and Johnson 1993). Thus, the action of NGF is not mediated solely by suppressing initiation of a cell death programme. How can neuronal death be prevented? To answer this question, we must understand how neuronal death is induced.

From morphological data, Kerr (1969) inferred that rat liver cells can die in two distinct ways, *coagulative necrosis* and *shrinkage necrosis*. The former accompanies pathological cell death resulting from tissue injury, whereas the latter occurs only in scattered individual cells, never in regions of wholesale destruction. Kerr *et al.* (1972) considered shrinkage necrosis as an inherent, programmed phenomenon, part of normal cell turnover of adult tissues or the developmental regulation of cell populations. Based on this hypothesis, Kerr *et al.* (1972) proposed a new term for shrinkage necrosis: *apoptosis*,

Greek for the dropping of petals from flowers or leaves from trees. This term implies that cell loss is actually desirable for the survival of the host (J. J. Cohen *et al.* 1992). Necrosis (deadness) was now reserved for pathological cell death exclusively (Wyllie *et al.* 1980). Figure 13.9 outlines the morphological changes associated with apoptosis (type 1) and necrosis (type 2; Clarke 1990). Apoptosis begins with condensation of both the nucleus and the cytoplasm, which then shrink, fragment, and are engulfed by neighbouring phagocytes (Kerr *et al.* 1972; Wyllie *et al.* 1980; Clarke 1990). By contrast, necrosis is characterized by dilatation of membraneous structures; the cell membrane then ruptures and cellular organelles are dissolved by its own lysosomes (*autophagocytosis*; Wyllie *et al.* 1980; Clarke 1990).

Natural neuronal death is an inherent, programmed biological process, so that it may correspond to apoptosis in terms of its functional implication. Two means of inducing neuronal death—ischaemia and the application of glutamate or its analogues (Choi 1988, 1992; Siesjo 1988; Meldrum and Garthwaite 1990)—are pertinent here. Neuronal death associated with ischaemic damage or glutamate neurotoxicity results from Ca^{2+} influx. By contrast, when sympathetic neurones die from NGF deprivation the presence of extracellular Ca^{2+} is irrelevant (Koike *et al.* 1989). Thus, as in non-neuronal cells, programmed neuronal death and pathologically induced neuronal death may depend on different mechanisms. Do the morphological changes associated with neuronal death accord with those of apoptosis or necrosis? Not exactly. Neuronal death induced by applying glutamate analogues is morphologically akin to necrosis, with only minor divergences (Hajos *et al.* 1986, but see Kure *et al.* 1991). On the other hand, the natural cell death period in chick embryos involves two types of neuronal degeneration (Chu-Wang and Oppenheim 1978; see also O'Connor and Wyttenbach 1974; Pilar and Landmesser 1976).

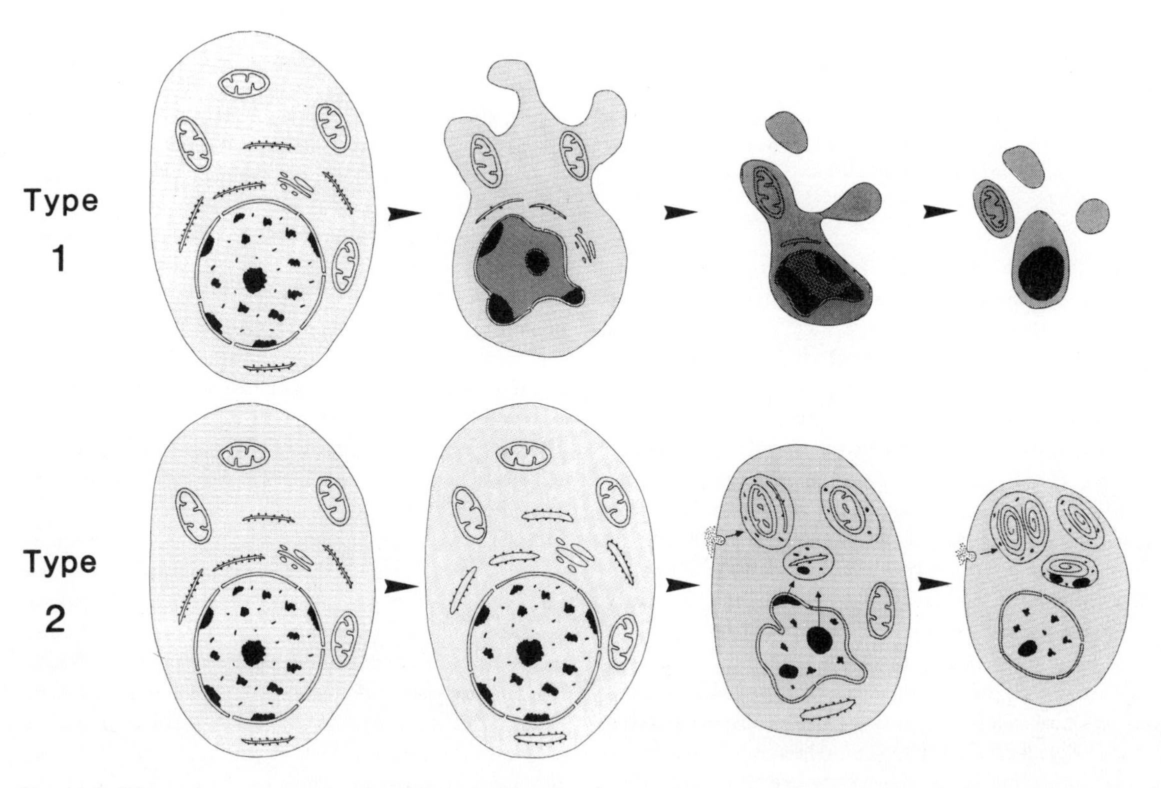

Fig. 13.9. Schematic representations of morphological features of apoptosis (type 1) and necrosis (type 2). Sequential changes of degeneration are shown from left (normal) to right (cell death). Type 1, characterized by condensation, chromatin clumping in the nucleus, convolution in the cell membrane, and loss of ribosomes in the cytoplasm. Type 2, characterized by endocytosis in the cell membrane, dilatation of endoplasmic reticulum, and vacuoles in the cytoplasm. (From Clarke 1990.)

The 'nuclear' type (or type I) exhibits nuclear clumping of chromatin (chromatin condensation) and dissociation of the polyribosomes, while the 'cytoplasmic' type (or type II) features dilatation of membraneous structures. Both types of degeneration occur not only in natural motor neurone death but also in target-deprived motor neurone death in chick embryos (Chu-Wang and Oppenheim 1978; see also Pilar and Landmesser 1976). According to Clarke (1990), the 'nuclear' type of degeneration is similar to apoptosis, whereas the 'cytoplasmic' type may be a variant of necrosis. There is no evidence that two different mechanisms subserve natural neuronal death, and why two distinct types of degeneration appear remains a mystery.

Chromatin condensation seen in apoptosis is associated with DNA fragmentation (Wyllie 1980; Duke *et al.* 1983). Thus, apoptosis appears to involve a process which activates endonuclease or synthesizes new endonuclease; this process in turn cleaves DNA (DNA fragmentation), thereby displaying nuclear condensation (Arends *et al.* 1990). In sum, apopstosis has three features: first, in functional aspects apoptosis is an inherent, programmed biological process; second, morphologically apoptosis begins with nuclear condensation followed by shrinkage, fragmentation, and phagocytosis; third, biochemically DNA is fragmented by endogenous endonuclease. NGF-deprived cell death in cultured sympathetic neurones is also associated with DNA fragmentation (Edwards *et al.* 1991; Deckwerth and Johnson 1993). Biochemical or morphological features of apoptosis were observed when cell death was induced in cultured central neurones by the application of glutamate (Kure *et al.* 1991) or β-amyloid peptide (Loo *et al.* 1993), although these examples do not represent programmed cell death. Actually, the term apoptosis is ambiguous, because it has never been defined in conceptual or mechanistic terms.

Let us now examine the electrophysiological analysis of neuronal death. We have seen that the survival of motor neurones in the early postnatal period is still clearly dependent of their targets (section 13.2.1). For example, as shown in Fig. 13.10, about 80 per cent of facial motor neurones die within 10 days after section of the facial nerve (axotomy) in neonatal rats (Soreide 1981; Sendtner *et al.* 1990; Umemiya *et al.* 1993), with the highest rate of cell death between 4 and 6 days after axotomy. Thus, about half of the motor neurones surviving 4 days after axotomy die during the next two days. The surviving axotomized motor neurones were significantly smaller than normal facial motor neurones in age-matched animals. Consistent with this reduction in neuronal size was a significant decrease of the electrical membrane capacitance, the axotomized motor neurones (Fig. 13.11, filled circles) having, on average, 36 per cent less than normal (open circles). If this reduction in the membrane capacitance were due to decreased cell size, the electrical membrane resistance in the axotomized motor neurones would be 36 per cent higher than normal. In fact, the mean increase of the membrane resistance in the axotomized motor neurones (76 per cent) exceeded this expectation. Moreover, 9 of the 28 axotomized motor neurones showed disproportionately high membrane resistance (>1 GΩ; Fig. 13.11, filled circles). Interestingly, these nine also exhibited a long after-hyperpolarization (AHP), significantly longer in mean duration than that of the other axotomized motor neurones. The axotomized facial motor neurones are thus not homogenous, but contain a distinct subpopulation characterized by an unusually high membrane resistance and long AHP.

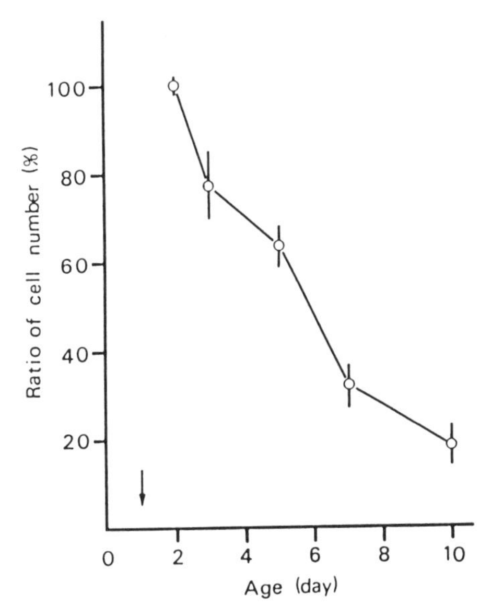

Fig. 13.10. Time course of changes in the number of facial motor neurones following unilateral transection of the facial nerve on the day after birth (arrow) in the rat. Ordinate, the mean number of surviving facial motor neurones on the operated side relative to that on the contralateral, normal side. Vertical bars, standard deviations of mean. (From Umemiya *et al.* 1993.)

The incidence of this subpopulation did not correlate with the period of time after axotomy, but it peaked concurrently with the highest rate of cell death. Could

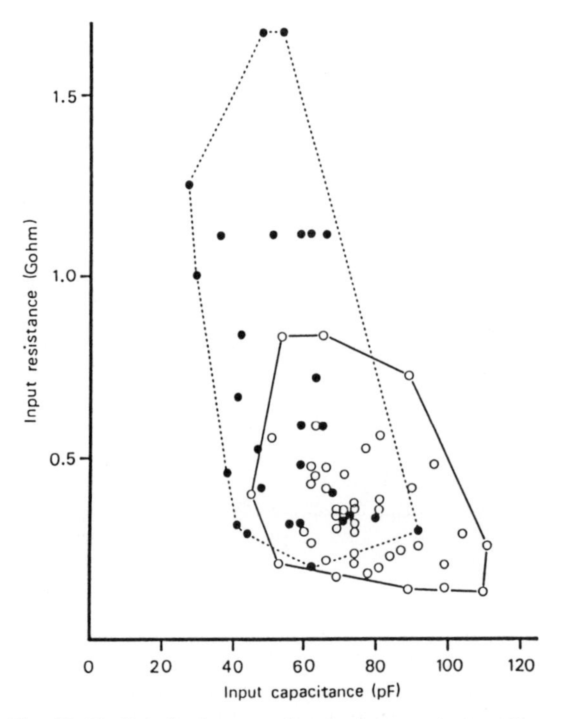

Fig. 13.11. Relation between the membrane resistance (input resistance) and the electrical capacitance (input capacitance) for normal (open circles) and axotomized (filled circles) facial motor neurones observed in 5–day-old rats. Boundaries indicate the range of values for each group. (From Umemiya *et al.* 1993.)

this subpopulation therefore represent axotomized neurones approaching the lethal stage or undergoing cell death (Umemiya *et al.* 1993)? If so, neuronal death induced by axotomy may stem from a loss of the ionic channels that contribute to the resting membrane conductance. Such channel loss must be selective, because these axotomized motor neurones still maintain normal resting and action potentials. How increased membrane resistance (or ionic channel loss) would relate to cell lethality is not clear. We know that lymphocytes and macrophages kill target cells by forming highly permeable channels or pores (Young 1989; Kagan *et al.* 1992). No such process seems indicated in axotomy-induced neuronal death, however.

A. LaVelle and F. W. LaVelle (1958) examined the time course of cell death for facial motor neurones axotomized in neonatal hamsters. Between 24 and 48 hours after axotomy, about 1600 facial motor neurones were lost, but only 70 of the remaining 1400 motor neurones exhibited degenerative signs. This sparsity of degenerating neurones suggests that once an axotomized neurone commences degenerative changes it quickly disappears. In other words, the fatal deterioration is strikingly rapid. This swiftness would account for the difficulty of assessing the morphology and electrophysiological characteristics of neuronal death. On the other hand, from the time course of cell death shown in Fig. 13.10 it seems clear that neuronal death does not occur immediately after axotomy. Therefore, the process of neuronal death triggered by different procedures may be divided into two stages:

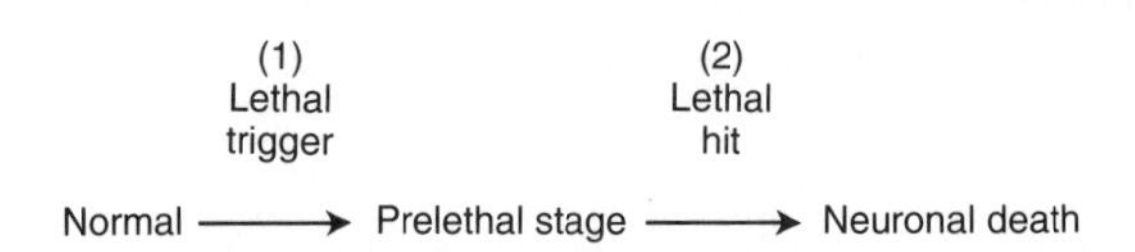

Neuronal death can be triggered by axotomy, neurotrophic factor deprivation, Ca^{2+} loading or cytotoxic metabolic disorder. Under these conditions (prelethal stage) neurones manifest alterations of their properties, which eventually result in cell death. Presumably, it is the transition from prelethal stage to neuronal death that occurs swiftly. The 'lethal hit' responsible for this transition may be rupture of the cell membrane, disintegration of cytoplamic organelles or some as yet unidentified process. Deckwerth and Johnson (1993) examined changes in morphological, biochemical, and molecular events that accompanied cell death of cultured sympathetic neurones following NGF deprivation. Cultured sympathetic neurones can be rescued if NGF is added again within about 20 hours after the onset of NGF deprivation but not beyond this critical period (Edwards *et al.* 1991; Deckwerth and Johnson 1993). In other words, NGF deprivation causes cultured sympathetic neurones to die, but this prelethal stage is reversible up to a certain period of time before the neurone is committed to die. This situation can be shown by the following scheme:

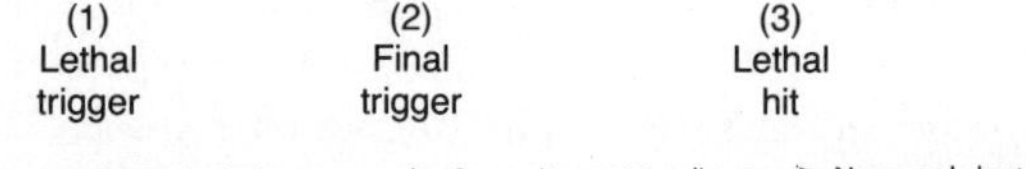

According to Deckwerth and Johnson (1993), the occurrence of the maximal neuronal DNA fragmentation coincides with the time of the commitment of cell death. This suggests that the irreversible neuronal degeneration process is triggered by DNA fragmentation (final trigger in the above scheme), although apoptosis in non-neuronal cell indicates otherwise (Zakeri *et al.* 1993). The crucial questions are how neurones are committed to die and which molecules induce what type of 'lethal hit'. These questions remain unanswered.

Summary and prospects

The size, dendritic geometry, and electrophysiological properties of neurones are regulated by interactions with their targets. In the case of sympathetic neurones, the target-dependent alterations in neuronal morphology and properties can be reproduced by manipulating the target-derived neurotrophic factor, NGF. Thus, the target-dependence of the neuronal state consists in its dependence on the target-derived neurotrophic factor. Besides the cell properties, neuronal survival itself depends on the target or the target-derived neurotrophic factor. The target-dependence of neuronal survival commences approximately when the neurone forms connections with its target, presumably along with the synthesis of the trophic factor in the target tissue and the expression of its receptors in the neurone. Neuronal survival, however, becomes less dependent on the target or the neurotrophic factor as the neurone matures. Yet, even in maturity the target-dependence of neuronal properties is clearly preserved. One possibility is that the target-dependencies of neuronal survival and of neuronal properties are quantitatively different. Thus, in maturity the maintenance of normal neuronal properties may require a relatively large supply of the trophic factor from the target, while a limited amount suffices to maintain neuronal survival. This interpretation probably oversimplifies, however, because replacing the original target with a foreign target can alter certain neuronal properties without affecting others. Hence, the target-dependence of some neuronal properties may be qualitatively different from that of neuronal survival, although no plausible mechanism has yet been proposed.

The natural neuronal death at early developmental stages must entail the same mechanisms as neuronal death from deprivation of the target-derived neurotrophic factor. Since both natural neuronal death and the trophic factor-deprived neuronal death can be prevented by inhibitors of RNA or protein synthesis, the cell death must be triggered by the synthesis of some deleterious proteins, 'thanatin.' How the 'thanatin' kills the neurones is the crucial question. Electrophysiological studies showed a drastic increase in the membrane resistance associated with axotomy-induced neuronal death. Therefore, the rupture of the cell membrane or the insertion of highly permeable channel proteins is unlikely to be primary mechanisms of neuronal death.

The degeneration leading to neuronal death appears to be a very rapid process, taking only a few hours or less. Recording of electrical events from an expiring neurone and identifying crucial molecular events in its final moments would provide important insights into the mechanisms of neuronal death.

Suggested reading

Axotomy-induced alterations in neuronal properties

Kuno, M. (1990). Target dependence of motoneuronal survival: the current status. *Neuroscience Research*, **9**, 155–72.

Snider, W. D., Elliott. J. L., and Yan, Q. (1992). Axotomy-induced neuronal death during development. *Journal of Neurobiology*, **23**, 1231–46.

Titmus, M. J. and Faber, D. S. (1990). Axotomy-induced alterations in the electrophysiological characteristics of neurones. *Progress in Neurobiology*, **35**, 1–51.

Umemiya, M., Araki, I., and Kuno, M. (1993). Electrophysiological properties of axotomized facial motoneurones that are destined to die in neonatal rats. *Journal of Physiology (London)*, 462, 661–78.

How neurotrophic factors function

Davies, A. M., Bandtlow, C., Heumann, R., Korsching, S., Rohrer, H., and Thoenen, H. (1987). Timing and site of nerve growth factor synthesis in developing skin in relation to innervation expression of the receptor. *Nature*, **326**, 353–8.

Davies, A. M. and Vogel, K. S. (1991). Developmental programmes of growth and survival in early sensory neurones. *Philosophical Transactions of the Royal Society B*, **331**, 259–62.

Franklin, J. L., and Johnson, E. M. (1992). Suppression of programmed neuronal death by sustained elevation of cytoplasmic calcium. *Trends in Neurosciences*, **15**, 501–8.

Martin, D. P., Schmidt, R. E., DiStefano, P. S., Lowry, O. H., Carter, J. G., and Johnson, E. M. (1988). Inhibitors of protein synthesis and RNA synthesis prevent neuronal death caused by nerve growth factor deprivation. *Journal of Cell Biology*, **106**, 829–44.

Mechanisms of neuronal death

Chu-Wang, I. W. and Oppenheim, R. W. (1978). Cell death of motoneurons in the chick embryo spinal cord. I. A light and electron microscopic study of naturally occurring and induced cell loss during dvelopment. *Journal of Comparative Neurology*, **177**, 33–58.

Clarke, P. G. H. (1990). Developmental cell death: Morphological diversity and multiple mechanisms. *Anatomy and Embryology*, **181**, 195–213.

Cohen, J. J. and Duke, R. C. (1992). Apoptosis and programmed cell death in immunity. *Annual Review of Immunology*, **10**, 267–93.

Deckwerth, T. L. and Johnson, E. M. (1993). Temporal analysis of events associated with programmed cell death (apoptosis) of sympathetic neurons deprived of nerve growth factor. *Journal of Cell Biology*, **123**, 1207–22.

Edwards, S. N., Buckmaster, A. E., and Tolkovsky, A. M. (1991). The death programme in cultured sympathetic neurones can be suppressed at the post-translational level by nerve growth factor, cyclic AMP, and depolarization. *Journal of Neurochemistry*, **57**, 2140–3.

Ellis, R. E., Yuan, J., and Horvitz, H. R. (1991). Mechanisms and functions of cell death. *Annual Review of Cell Biology*, **7**, 663–98.

Johnson, E. M. and Deckwerth, T. L. (1993). Molecular mechanisms of developmental neuronal death. *Annual Review of Neuroscience*, **16**, 31–46.

Martin, D. P., Ito, A., Horigome, K., Lampe, P. A., and Johnson, E. M. (1992). Biochemical characterization of programmed cell death in NGF-deprived sympathetic neurones. *Journal of Neurobiology*, **23**, 1205–20.

References

Abe, T., Sugihara, H., Nawa, H., Shigemoto, R., Mizuno, N., and Nakanishi, S. (1992). Molecular characterization of a novel metobotropic glutamate receptor mGluR5 coupled to inositiol phosphate/Ca^{2+} signal transduction. *Journal of Biological Chemistry*, **267**, 13361–8.

Abrams, T. W. (1985). Activity-dependent presynaptic facilitation: an associative mechanism in *Aplysia*. *Cellular and Molecular Neurobiology*, **5**, 123–45.

Abrams, T. W., Castellucci, V. F., Camardo, J. S., Kandel, E. R., and Lloyd, P. E. (1984). Two endogenous neuropeptides modulate the gill and siphon withdrawal reflex in *Aplysia* by presynaptic facilitation involving cAMP-dependent closure of a serotonin-sensitive potassium channel. *Proceedings of the National Academy of Sciences of the USA*, **81**, 7956–60.

Abrams, T. W., Karl K. A., and Kandel, E. R. (1991). Biochemical studies of stimulus convergence during classical conditioning in *Aplysia*: dual regulation of adenylate cyclase by Ca^{2+}/calmodulin and transmitter. *Journal of Neuroscience*, **11**, 2655–65.

Adams, D. J., Dwyer, T. M., and Hille, B. (1980). The permeability of endplate channels to monovalent and divalent metal ions. *Journal of General Physiology*, **75**, 493–510.

Adler, E. M., Augustine, G. J., Duffy, S. N., and Charlton, M. P. (1991). Alien intracellular calcium chelators attenuate neurotransmitter release at the squid giant synapse. *Journal of Neuroscience*, **11**, 1496–502

Adler, R., Landa, K. B., Manthorpe, M., and Varon, S. (1979). Cholinergic neurotrophic factors: Intracellular distribution of trophic activity for ciliary neurones. *Science*, **204**, 1434–6.

Akagi, H. and Miledi, R. (1988). Heterogeneity of glycine receptors and their messenger RNAs in rat brain and spinal cord. *Science*, **242**, 270–3.

Akagi, H., Hirai, K., and Hashimura, F. (1991). Cloning of a glycine receptor subtype expressed in rat brain and spinal cord during a specific period of neuronal development. *FEBS Letters*, **281**, 160–6.

Albuquerque, E. X., Deshpande, S. S., Aracava, Y., Alkondon, M., and Daly, J. W. (1986). A possible involvement of cyclic AMP in the expression of desensitization of the nicotinic acetylcholine receptor. A study with forskolin and its analogs. *FEBS Letters*, **199**, 113–20.

Albus, J. S. (1971). A theory of cerebellar function. *Mathematical Biosciences*, **10**, 25–61.

Alder, J., Lu, B, Valtorta, F., Greengard, P., and Poo, M. M. (1992). Calcium-dependent transmitter secretion reconstituted in *Xenopus* oocytes: requirement for synaptophysin. *Science*, **257**, 657–d61.

Alford, S., Frenguelli, B. G., Schofield, J. G., and Collingridge, G. L. (1993). Characterization of Ca^{2+} signals induced in hippocampal CA1 neurones by the synaptic activation of NMDA receptors. *Journal of Physiology (London)*, **469**, 693–716.

Almers, W. (1990). Exocytosis. *Annual Review of Physiology*, **52**, 607–24.

Anad, R., Conray, W. G., Schoepfer, R., Whiting, P., and Lindstrom, J. (1991). Neuronal nicotinic acetylcholine receptors expressed in *Xenopus* oocytes have a pentameric quaternary structure. *Journal of Biological Chemistry*, **266**, 11192–8.

Andersen, P., Sundberg, S. H., Sveen, O., and Wigstrom, H. (1977). Specific long-lasting potentiation of synaptic transmission in hippocampal slices. *Nature*, **266**, 736–7.

Anderson, A. J. and Harvey, A. L. (1987). ω-conotoxin does not block the verapamil-sensitive calcium channels at mouse motor nerve terminals. *Neuroscience Letters*, **82**, 177–80.

Anderson, C. R. and Stevens, C. F. (1973). Voltage clamp analysis of acetylcholine produced end-plate current fluctuations at frog neuromuscular junction. *Journal of Physiology (London)*, **235**, 655–91.

Anderson, D. J. (1989). The neural crest cell lineage problem: Neuropoiesis? *Neuron*, **3**, 1–12.

Anderson, D. J. (1993). Molecular control of cell fate in the neural crest: The sympathoadrenal lineage. *Annual Review of Neuroscience*, **16**, 129–58.

Anderson, H. K. (1902). The nature of the lesions which hinder the development of nerve cells and their processes. *Journal of Physiology (London)*, **28**, 499–513.

Anderson, M. J. and Cohen, M. W. (1977). Nerve-induced and spontaneous redistribution of acetylcholine receptors on cultured muscle cells. *Journal of Physiology (London)*, **268**, 757–73.

Anderson, M. J., Cohen, M. W., and Zorychta, E. (1977). Effects of innervation on the distribution of acetylcholine receptors on cultured muscle cells. *Journal of Physiology (London)*, **268**, 731–56.

Angeletti, P. U., Levi-Montalcini, R., and Caramia, F. (1971). Analysis of the effects of the antiserum to the nerve growth factor in adult mice. *Brain Research*, **27**, 343–55.

Angeletti, R. H. and Bradshaw, R. A. (1971). Nerve growth factor from mouse salivary gland: amino acid sequence. *Proceedings of the National Academy of Sciences of the USA*, **68**, 2417–20.

Aosaki, T. and Kasai, H. (1989). Characterization of two kinds of high-voltage-activated Ca-channel currents in chick sensory neurons: differential sensitivity to dihydropyridines and ω-conotoxin GVIA. *Pflügers Archiv*, **414**, 150–156.

Arai, K., Lee, F., Miyajima, A., Miyatake, S., Arai, N., and Yokota, T. (1990). Cytokines: Coordinators of immune and inflammatory responses. *Annual Review of Biochemistry*, **59**, 783–836.

Arakawa, Y., Sendtner, M., and Thoenen, H. (1990). Survival effect of ciliary neurotrophic factor (CNTF) on chick embryonic motoneurons in culture: comparison with other neurotrophic factors and cytokines. *Journal of Neuroscience*, **10**, 3507–15.

Araki, I., Harada, Y., and Kuno, M. (1991). Target-dependent hormonal control of neuron size in the rat spinal nucleus of the bulbocavernosus. *Journal of Neuroscience*, **11**, 3025–33.

Arends, M. J., Morris, R. G., and Wyllie, A. H. (1990). Aptptosis. The role of the endonuclease. *American Journal of Pathology*, **136**, 593–608.

Arnold, A. P. (1984). Androgen regulation of motor neuron size and number. *Trends in Neurosciences*, **7**, 239–42.

Artalejo, C. R., Ariano, M. A., Perlman, R. L., and Fox, A. P. (1990). Activation of facilitation calcium channels in chromaffin cells by D_1 dopamine receptors through cAMP/protein kinase A-dependent mechanism. *Nature*, **348**, 239–42.

Artalejo, C. R., Dahmer, M. K., Perlman, R. L., and Fox, A. P. (1991a). Two types of Ca^{2+} currents are found in bovine chromaffin cells: facilitation is due to the recruitment of one type. *Journal of Physiology (London)*, **432**, 681–707.

Artalejo, C. R., Mogul, D. J., Perlman, R. L., and Fox, A. P. (1991b). Three types of bovine chromaffin cell Ca^{2+} channels: facilitation increases the opening probability of a 27 pS channel. *Journal of Physiology (London)*, **444**, 213–240.

Artalejo, C. R., Parlman, R. L., and Fox, A. P. (1992a). ω-Conotoxin GVIA blocks a Ca^{2+} current in bovine chromaffin cells that is not the 'classic' N type. *Neuron*, **8**, 85–95.

Artalejo, C. R., Rossie, S., Perlman, R. L., and Fox, A. P. (1992b). Voltage-dependent phosphorylation may recruit Ca^{2+} current facilitation in chromaffin cells. *Nature*, **358**, 63–6.

Artola, A., Brocher, S., and Singer, W. (1990). Different voltage-dependent thresholds for inducing long-term depression and long-term potentiation in slices of rat visual cortex. *Nature*, **347**, 69–72.

Arvidsson, U., Johnson, H., Piehl, F., Cullheim, S., Hökfelt, T., Risling *et al.* (1990). Peripheral nerve section induces increased levels of calcitonin gene-related peptide (CGRP)-like immunoractivity in axotomized motorneurones. *Experimental Brain Research*, **79**, 212–6.

Ascher, P. and Nowak, L. (1988). The role of divalent cations in the N-methyl-D-asparate responses of mouse central neurones in culture. *Journal of Physiology (London)*, **399**, 247–66.

Ashkenazi, A., Peralta, E. G., Winslow, J. W., Ramachandran, J., and Capon, D. J. (1989). Funcitonally distinct G proteins selectively couple different receptors to PI hydrolysis in the same cell. *Cell* **56**, 487–93.

Astrand, P. and Stjarne, L. (1989). On the secretory activity of single varicosities in the sympathetic nerves innervating the rat tail artery. *Journal of Physiology (London)*, **409**, 207–20.

Atchison, W. D. and O'Leary, S. M. (1987). BAY K 8644 increases release of acetylcholine at the murine neuromuscular junction. *Brain Research*, **419**, 315–9.

Atwood, H. L. and Wojtowicz, J. M. (1986). Short-term and long-term plasticity and physiological differentiation of crustacean motor synapses. *International Review of Neurobiology*, **28**, 275–362.

Augustine, G. J. and Charlton, M. P. (1986). Calcium dependence of presynaptic calcium current and post-synaptic response at the squid giant synapse. *Journal of Physiology (London)*, **381**, 619–40.

Augustine, G. J. and Neher, E. (1992a). Neuronal Ca^{2+} signalling takes the local route. *Current Opinion in Neurobiology*, **2**, 302–7.

Augustine, G. J. and Neher, E. (1992b). Calcium requirements for secretion in bovine chromaffin cells. *Journal of Physiology (London)*, **450**, 247–71.

Augustine, G. J., Charlton, M. P., and Smith, S. J. (1985). Calcium entry into voltage-clamped presynaptic terminals of squid. *Journal of Physiology (London)*, **367**, 143–162.

Augustine, G. J., Charlton, M. P., and Horn, R. (1988).

Role of calcium-activated potassium channels in transmitter release at the squid giant synapse. *Journal of Physiology (London)*, **398**, 149–64.

Axelrod, J., Burch, R. M., and Jelsema, C. L. (1988). Receptor-mediated activation of phospholipase A_2 via GTP-binding proteins: arachidonic acid and its metabolites as second messengers. *Trends in Neurosciences*, **11**, 117–23.

Axelsson, J. and Thesleff, S. (1959). A study of supersensitivity in denervated mammalian skeletal muscle. *Journal of Physiology (London)*, **147**, 178–93.

Baehr, W. and Applebury, M. L. (1986). Exploring visual transduction with recombinant DNA techniques. *Trends in Neurosciences*, **9**, 198–203.

Bailey, C. H. and Chen, M. (1983). Morphological basis of long-term habituation and sensitization in *Aplysia*. *Science*, **220**, 91–3.

Bailey, C. H. and Chen, M. (1988a). Long-term memory in *Aplysia* modulates the total number of varicosities of single identified sensory neurons. *Proceedings of the National Academy of Sciences of the USA*, **85**, 2372–7.

Bailey, C. H. and Chen, M. (1988b). Long-term sensitization in *Aplysia* increases the number of presynaptic contacts onto the identified gill motor neuron L_7. *Proceedings of the National Academy of Sciences of the USA*, **85**, 9356–9.

Bailey, C. H. and Chen, M. (1989). Time course of structural changes at identified sensory neuron synapses during long-term sensitization in *Aplysia*. *Journal of Neuroscience*, **9**, 1774–80.

Bailey, C. H., Chen, M., Keller, F., and Kandel, E. R. (1992). Serotonin-mediated endocytosis of apCAM: an early step of learning-related synaptic growth in *Aplysia*. *Science*, **256**, 645–9.

Baker, P. F., Hodgkin, A. L., and Ridgway, E. B. (1971). Depolarization and calcium entry in squid giant axons. *Journal of Physiology (London)*, **218**, 709–755.

Baraban, J. M., Snyder, S. H., and Alger, B. E. (1985). Protein kinase C regulates ionic conductance in hippocampal pyramidal neurons: electrophysiological effects of phorbol esters. *Proceedings of the National Academy of Sciences of the USA*, **82**, 2538–42.

Barbin, G., Manthorpe, M., and Varon, S. (1984). Purification of the chick eye ciliary neurotrophic factor. *Journal of Neurochemistry*, **43**, 1468–78.

Barde, Y. A. (1988). What, if anything, is a neurotrophic factor? *Trends in Neurosciences*, **11**, 343–6.

Barde, Y. A., Edgar, D., and Thoenen, H. (1982). Purification of a new neurotrophic factor from mammalian brain. *EMBO Journal*, **1**, 549–53.

Barde, Y. A., Edgar, D., and Thoenen, H. (1983). New neurotrohic factors. *Annual Review of Physiology*, **45**, 601–12.

Barker, D. and Ip. M. C. (1966). Sprouting and degeneration of mammalian motor axons in normal and deafferented skeletal muscle. *Proceedings of the Royal Society B*, **163**, 538–54.

Barker, J. L. and Nicoll, R. A. (1972). Gamma-aminobutyric acid; role in primary afferent depolarization. *Science*, **176**, 1043–5.

Barnard, E. A., Darlison, M. G., and Seeburg, P. (1987). Molecular biology of the $GABA_A$ receptor: the receptor/channel superfamily. *Trends in Neurosciences*, **10**, 502–9.

Barondes, S. H. and Cohen, H. D. (1966). Puromycin effect on successive phases of memory storage. *Science*, **151**, 594–5.

Barrett, J. N., Magleby, K. L., and Pallotta, B. S. (1982). Properties of single calcium-activated potassium channels in cultured rat muscle. *Journal of Physiology (London)*, **331**, 211–30.

Barzilai, A., Kennedy, T. E., Sweatt, J. D., and Kandel, E. R. (1989). 5–HT modulates protein synthesis and the expression of specific proteins during long-term facilitation in *Aplysia* sensory neurons. *Neuron*, **2**, 1577–86.

Bashir, Z. I., Alford, S., Davies, S. N., Randall, A. D., and Collingridge, G. L. (1991). Long-term potentiation of NMDA receptor-mediated synaptic transmission in the hippocampus. *Nature*, **349**, 156–8.

Bashir, Z. I., Bortolotto, Z. A., Davies, C. H., Berretta, N., Irving, A. J., Seal, A. J., *et al.* (1993). Induction of LTP in the hippocampus needs synaptic activation of glutamate metabotropic receptors. *Nature*, **363**, 347–50.

Baskin, D. G., Wilcox, B. J., Figlewicz, D. P., and Dorsa, D. M. (1988). Insulin and insulin-like growth factors in the CNS. *Trends in Neurosciences*, **11**, 107–11.

Baskys, A. and Malenka, R. C. (1991). Agonists at metabotropic glutamate receptors presynaptically inhibit EPSCs in neonatal rat hippocampus. *Journal of Physiology (London)*, **444**, 687–701.

Bastian, J. and Nakajima, S. (1974). Action potential in the transverse tubules and its role in the activation of skeletal muscle. *Journal of General Physiology*, **63**, 257–78.

Baumgold, J. (1992). Muscarinic receptor-mediated stimulation of adenylyl cyclase. *Trends in Pharmacological Sciences*, **13**, 339–40.

Bazan, J. F. (1991). Neuropoietic cytokines in the hematopoietic fold. *Neuron*, **7**, 197–208.

Bekkers, J. M. and Stevens, C. F. (1989). NMDA and non-NMDA receptors are co-localized at individual excitatory synapses in cultured rat hippocampus. *Nature*, **341**, 230–3.

Bekkers, J. M. and Stevens, C. F. (1990). Presynaptic mechanism for long-term potentiation in the hippocampus. *Nature*, **346**, 724–9.

Bekkers, J. M., Richerson, G. B., and Stevens, C. F. (1990). Origin of variability in quantal size in cultured hippocampal neurones and hippocampal slices. *Proceedings of the National Academy of Sciences of the USA*, **87**, 5359–62.

Belardetti, F. and Siegelbaum, S. A. (1988). Up-and down-modulation of single K channel function by distinct second messengers. *Trends in Neurosciences*, **11**, 232–8.

Belmonte, C. and Gallego, R. (1983). Membrane properties of cat sensory nurones with chemoreceptor and baroreceptor endings. *Journal of Physiology (London)*, **342**, 603–14.

Beltz, B. S. and Kravitz, E. A. (1983). Mapping of serotonin- like immunoreactivity in the lobster nervous system. *Journal of Neuroscience*, **3**, 585–602.

Benne, R., Van Den Burg, J., Brakenhoff, J. P. J., Slool, P., Van Boom, J. H., and Tromp, M. C. (1986). Major trasncript of the frameshifted *coxll* gene from trypanosome mitochondria contains four nucleotides that are not encoded in the DNA. *Cell*, **46**, 819–26.

Bennett, M. K. and Scheller, R. H. (1993). The molecular machinery for secretion is conserved from yeast to neurons. *Proceedings of the National Academy of Sciences of the USA*, **90**, 2559–63.

Bennett, M. R. and Pettigrew, A. G. (1974). The formation of synapses in reinnervated and cross-reinnervated striated muscle during development. *Journal of Physiology (London)*, **241**, 547–73.

Bennett, M. R., Jones, P., and Lavidis, N. A. (1986). The probability of quantal secretion along visualized terminal branches at amphibian (*Bufo marinus*) neuromuscular synapses. *Journal of Physiology (London,)* **379**, 257–74.

Bennett, M. R., Karunanithi, S., and Lavidis, N. A. (1991). Probabilistic scretion of quanta from nerve terminals in toad (*Bufo marinus*) muscle modulated by adenosine. *Journal of Physiology (London)*, **433**, 421–34.

Bennett, M. V. L. (1974). *Synaptic transmission and neuronal interaction*. Raven Press, New York.

Bennett, M. V. L., Aljure, E., Nakajima, Y., and Pappas, G. D. (1963). Electrotonic junctions between teleost spinal neurons: Electrophysiology and ultrastructure. *Science*, **141**, 262–64.

Bennett, M. V. L., Barrio, L. C., Bargiello, T. A., Spray, D. C., Hertzberg, E., and Saez, J. C. (1991). Gap junctions: new tools, new answers, new question. *Neuron*, **6**, 305–320.

Beranek, R. and Hnik, P. (1959). Long-term effects of tenotomy on spinal monosynaptic response in the cat. *Science*, **130**, 981–2.

Bergold, P. J., Sweatt, J. D., Winicov, I., Weiss, K. R., Kandel, E. R., and Schwartz, J. H. (1990). Protein synthesis during acquisition of long-term facilitation is needed for the persistent loss of regulatory subunits of the *Aplysia* cAMP-dependent protein kinase. *Proceedings of the National Academy of Sciences of the USA*, **87**, 3788–91.

Berkemeier, L. R., Winslow, J. W., Kaplan, D. R., Nikolics, K., Goeddel, D. V., and Rosenthal, A. (1991). Neurotrophin-5: a novel neurotrophic factor that activates trk and trkB. *Neuron*, **7**, 857–66.

Bernd, P. and Greene, L. A. (1984). Association of ^{125}I-nerve growth factor with PC12 pheochromocytoma cells. Evidence for internalization via high-affinity only for long-term regulation by nerve growth factor of high- and low-affinity receptors. *Journal of Biological Chemistry*, **259**, 15509–16.

Bernier, L., Castellucci, V. F., Kandel, E. R., and Schwartz, J. H. (1982). Facilitatory transmitter cuases a selective and prolonged increase in adenosin 3′:5′-monophosphate in sensory neurons mediating the gill and siphon withdrawal reflex in *Aplysia*. *Journal of Neuroscience*, **2**, 1682–91.

Berridge, M. (1987). Inositol trisphosphate and diacylglycerol: two interacting second messengers. *Annual Review of Biochemistry*, **56**, 159–93.

Bertrand, D., Devillers-Thiery, A., Revah, F., Galzi, J. L., Hussy, N., Mulle, C., Betrand, S., Ballivet, M., and Changeux, J. P. (1992). Unconventional pharmacology of a neuronal nicotinic receptor mutated in the channel domain. *Proceedings of the National Academy of Sciences of the USA*, **89**, 1261–5.

Betz, H. (1987). Biology and structure of the mammalian glycine receptor. *Trends in Neurosciences*, **10**, 113–7.

Betz, H. (1990). Ligand-gated ion channels in the brain: the amino acid receptor superfamily. *Neuron*, **5**, 383–92.

Betz, W. J., Caldwell, J. H., and Ribchester, R. R. (1980). Sprouting of active nerve terminals in partially inactive muscles of the rat. *Journal of Physiology (London)*, **303**, 281–97.

Bevan, S. and Steinbach, J. H. (1977). The distribution of α-bungarotoxin binding sites on mammalian skeletal muscle develping *in vivo*. *Journal of Physiology (London)*, **267**, 195–213.

Bevan, S. and Steinbach, J. H. (1983). Denervation increases the degradation rate of acetylcholine receptors at end-plates *in vivo* and *in vitro*. *Journal of Physiology (London)*, **336**, 159–77.

Bident, T. J., Peter-Riesch, B., Schlege, W., and Wollheim, C. B. (1987). Ca^{2+}-mediated generation of inositol 1,4,5–trisphosphate and inositol 1,3,4,5–tetrakisphosphate in pancreatic islets. *Journal of Biological Chemistry*, **262**, 3567–71.

Bigazzi, P. E., Yoshida, T., Ward, P. A., and Cohen, S. (1975). Production of lymphokine-like factors (cytokines) by simian virus 40–infected and simian virus 40–transformed cells. *American Journal of Pathology*, **80**, 69–78.

Birnbaumer, L. (1992). Receptor-to-effector signalling through G proteins: Roles for βγ dimers as well as α subunits. *Cell*, **71**, 1069–72.

Bishop, J. M. (1982). Oncogenes. *Scientific American*, **246**(3), 68–78.

Bishop, J. M. (1985). Viral oncogenes. *Cell*, **42**, 23–38.

Black, I. B., Adler, J. E., Dreyfus, C. F., Friedman, W. F., LaGamma, E. F., and Roach, A. H. (1987). Biochemistry of information storage in the nervous system. *Science*, **236**, 1263–8.

Blackman, J. G., Ginsborg, B. L., and Ray, C. (1963). Synaptic transmission in the sympathetic ganglion of the frog. *Journal of Physiology (London)*, **167**, 355–73

Blake, J. F., Brown, M. W., and Collingridge, G. L. (1988). CNQX blocks acidic amino acid induced depolarizations and synaptic components mediated by non-NMDA receptors in rat hippocampal slices. *Neuroscience Letters*, **89**, 182–6.

Blankenship, J. E. and Kuno, M. (1968). Analysis of spontaneous subthreshold activity in spinal motoneurons of the cat. *Journal of Neurophysiology* **31**, 195–209.

Blaustein, M. P. (1988). Calcium transport and buffering in neurons. *Trends in Neurosciences*, **11**, 438–43.

Bliss, T. V. P. and Collingridge, G. L. (1993). A synaptic model of memory: long-term potentiation in the hippocampus. *Nature*, **361**, 31–9.

Bliss, T. V. P. and Gardner-Medwin, A. R. (1973). Long-lasting potentiation of synaptic transmission in the dentate area of the unanaesthetized rabbit following stimulation of the perforant path. *Journal of Physiology (London)*, **232**, 357–74.

Bliss, T. V. P. and Lomo, T. (1973). Long-lasting potentiation of synaptic transmission in the dentate area of the anaesthetized rabbit following stimulation of the perforant path. *Journal of Physiology (London)*, **232**, 331–56.

Bliss, T. V. P., Clements, M. P., Errington, M. L., Lynch, M. A., and Williams, J. H. (1990). Presynaptic changes associated with long-term potentiation in the dentate gyrus. *Seminars in Neurosciences*, **2**, 345–54.

Bloedel, J., Gage, P. W., Llinás, R., and Quastel, D. M. (1966). Transmitter release at the squid giant synapse in the presence of tetrodotoxin. *Nature*, **212**, 49–50.

Bodian, D. (1937). The structure of the vertebrate synapse. A study of the axon endings on Mauthner's cell and neighboring centres in the goldfish. *Journal of Comparative Neurology*, **68**, 117–45.

Bodian, D. (1942). Cytological aspects of synaptic function. *Physiological Reviews*, **22**, 146–69.

Bohme, G. A., Bon, C., Stutzmann, J. M., Doble, A., and Blanchard, J. C. (1991). Possible involvement of nitric oxide in long-term potentiation. *European Journal of Pharmacology*, **199**, 379–81.

Bolshakov, V. Y. and Siegelbaum, S. A. (1994). Post-synaptic induction and presynaptic expression of hippocampal long-term depression. *Science*, **264**, 1148–520.

Bönigk, W., Altenhofen, W., Müller, F., Dose, A., Illing, M., Molday, R. S. *et al.* (1993). Rod and cone photoreceptor cells express distinct genes for cGMP-gated channels. *Neuron*, **10**, 865–77.

Booth, C. M., Kemplay, S. K., and Brown, M. C. (1990). An antibody to neural cell adhesion molecule impairs motor nerve terminal sprouting in a mouse muscle locally paralysed with botulinum toxin. *Neuroscience*, **35**, 85–91.

Bormann, J. (1988). Electrophysiology of GABA and GABA receptor subtypes. *Trends in Neurosciences*, **11**, 112–6.

Bortolotto, Z. A., Bashir, Z. I., Davies, C. H., and Collingridge, G. L. (1994). A molecular switch activated by metebotropic glutamate receptors regulates induction of long-term potentiation. *Nature*, **368**, 740–3.

Bothwell, M. (1991). Keeping track of neurotrophin receptors. *Cell*, **65**, 915–8.

Boulter, J., Connolly, J., Deneris, E., Goldman, D., Heinemann, S., and Patrick, J. (1987). Functional expression of two neuronal nicotinic acetylcholine receptors from cDNA clones identifies a gene family. *Proceedings of the National Academy of Sciences of the USA*, **84**, 7763–7.

Bourne, H. R. and Nicoll, R. (1993). Molecular machines integrate coincident synaptic signals. *Neuron*, **10 (Supplement)**, 65–75.

Boyd, I. A. and Martin, A. R. (1956). The end-plate potential in mammalian muscle. *Journal of Physiology (London)*, **132**, 74–91.

Bradley, P. B., Engel, G., Feniuk, W., Forzand, J. R., Humphrey, P. P. A., Middlemiss, D. N. *et al.* (1986). Proposals for the classification and nomenclature of functional receptors of 5–hydroxytryptamine. *Neuropharmacology*, **25**, 563–76.

Braha, O., Dale, N., Hochner, B., Klein, M., Abrams, T. W., and Kandel, E. R. (1990). Second messengers involved in the two processes of presynaptic facilitation that contribute to sensitization and disinhibition in *Aplysia* sensory neurons. *Proceedings of the National Academy of Sciences of the USA*, **87**, 2040–4.

Braithwaite, A. W. and Harris, A. J. (1979). Neural influence on acetylcholine receptor clusters in embryonic development of skeletal muscles. *Nature*, **279**, 549–51.

Breckenridge, L. J. and Almers, W. (1987a). Final steps in exocytosis observed in a cell with giant secretory granules. *Proceedings of the National Academy of Sciences of the USA*, **84**, 1945–9.

Breckenridge, L. J. and Almers, W. (1987b). Currents through the fusion pore that forms during exocytosis of a secretory vesicle. *Nature*, **328**, 814–7.

Bredt, D. S. and Snyder, S. H. (1992). Nitric oxide, a novel neuronal messenger. *Neuron*, **8**, 3–11.

Bredt, D. S., Hwang, P. M., Glatt, C. E., Lowenstein, C. L., Reed, R. R., and Snyder, S. H. (1991). Cloned and expressed nitric oxide synthase structurally resembles cytochrome P-450 reductase. *Nature*, **351**, 714–8.

Breedlove, S. M. (1986). Cellular analyses of hormone influence on motoneuronal development and function. *Journal of Neurobiology*, 17, 157–76.

Brenner, H. R., Witzemann, V., and Sakmann, B. (1990). Imprinting of acetylcholine receptor messenger RNA accumulation in mammalian neuromuscular synapses. *Nature*, 344, 544–7.

Brenner, H. R., Herczeg, A., and Slater, C. R. (1992). Synapse-specific expression of acetylcholine receptor genes and their products at original synaptic sites in rat soleus muscle fibres regenerating in the absence of innervation. *Development*, **116**, 41–53.

Brightman, M. W. and Reese, T. S. (1969). Junctions between intimately apposed cell membranes in the vertebrate brain. *Journal of Cell Biology*, **40**, 648–77.

Brisson, A. and Unwin, P. N. T. (1985). Quaternary structure of the acetylcholine receptor. *Nature*, **315**, 474–7.

Brock, J. A. and Cunnane, T. C. (1987). Relationship between the nerve action potential and transmitter release from sympathetic postganglionic nerve terminals. *Nature* **326**, 605–7.

Brock, L. G., Coombs, J. S., and Eccles, J. C. (1952). The recording of potentials from motoneurones with an intracellular electrode. *Journal of Physiology (London)*, **117**, 431–60.

Brookes, N. and Werman, R. (1973). The cooperativity of γ-aminobutyric acid action on the membrane of locust muscle fibres. *Molecular Pharmacology*, 9, 571–9.

Brose, N., Petrenko, A. G., Sudhof, T. C., and Jahn, R. (1992). Synaptotagmin: a calcium sensor on the synaptic vesicle surface. *Science*, **256**, 1021–5.

Brostrom, C. O. and Brostrom, M. A. (1990). Calcium-dependent regulation of protein synthesis in intact mammalian cells. *Annual Review of Physiology*, **52**, 577–90.

Brown, A. M. and Birnbaumer, L. (1990). Ionic channels and their regulation by G protein subunits. *Annual Review of Physiology*, **52**, 197–213.

Brown, G. L. (1937). The actions of acetylcholine on denervated mammalian and frog's muscle. *Journal of Physiology (London)*, **89**, 438–61.

Brown, M. C. and Holland, R. L. (1979). A central role for denervated tissues in causing nerve sprouting. *Nature*, **282**, 724–6.

Brown, M. C., Jansen, J. K. S., and Van Essen, D. (1976). Polyneural innervation of skeletal muscle in new-born rats and its elimination during maturation. *Journal of Physiology (London)*, **261**, 387–422.

Brown, M. C., Holland, R. L., and Hopkins, W. G. (1981). Motor nerve sprouting. *Annual Review of Neuroscience*, 4, 17–42.

Brown, M. C., Perry, V. H., Lunn, E. R., Gordon, S., and Heumann, R. (1991). Macrophage dependence of peripheral sensory nerve regeneration: possible involvement of nerve growth factor. *Neuron*, **6**, 359–70.

Brown, T. H. and McAfee, D. A. (1982). Long-term synaptic potentiation in the superior cervical ganglion. *Science*, **215**, 1411–3.

Brown, W. E. L. and Hill, A. V. (1923). The oxygen-dissociation curve of blood, and its thermodynamical basis. *Proceedings of the Royal Society B*, 94, 297–334.

Brunelli, M., Castellucci, V., and Kandel, E. R. (1976). Synaptic facilitation and behavioural sensitization in *Aplysia*: possible role of serotonin and cyclic AMP. *Science*, **194**, 1178–81.

Brunso-Bechtold, J. and Hamburger, V. (1979). Retrograde transport of nerve growth factor in chicken embryo. *Proceedings of the National Academy of Sciences of the USA*, **76**, 1494–6.

Buck, L. and Axel, R. (1991). A novel mutigene family may encode odorant receptors: A molecular basis for odor recognition. *Cell*, **65**, 175–87.

Buckley, K. M., Floor, E., and Kelly, R. B. (1987). Cloning and sequence analysis of cDNA encoding p38, a major synaptic vesicle protein. *Journal of Cell Biology*, **105**, 2447–56.

Bueker, E. D. (1948). Implantation of tumors in the hind limb field of the embryonic chick and the developmental response of the lumbosacral nervous system. *Anatomical Record*, **102**, 369–90.

Buller, A. J. and Pope, R. (1977). Plasticity in mammalian skeletal muscle. *Philosophical Transactions of the Royal Society B*, **278**, 295–305.

Buller, A. J., Eccles, J. C., and Eccles, R. M. (1960). Interactions between motoneurones and muscles in respect of the characteristic speeds of their responses. *Journal of Physiology (London)*, **150**, 417–39.

Burch, R. M., Luini, A., and Axelrod, J. (1986). Phospholipase A_2 and phospholipase C are activated by distinct GTP-binding proteins in response to α_1-adrenergic stimulation in FRTL5 thyroid cells. *Proceedings of the National Academy of Sciences of the USA*, **83**, 7201–5.

Burden, S. J., Sargent, P. B., and McMahan, U. J. (1979). Acetylcholine receptors in regenerating muscle accumulate at original synaptic sites in the absence of the nerve. *Journal of Cell Biology*, **82**, 412–25.

Burden, S. J., DePalma, R. L., and Gottesman, G. S. (1983). Crosslinking of proteins in acetylcholine receptor-rich membranes: association between the beta-subunit and the 43 kD subsynaptic protein. *Cell*, **35**, 687–92.

Burgoyne, R. D. (1990). Secretory vesicle-associated proteins and their role in exocytosis. *Annual Review of Physiology*, **52**, 647–59.

Burnashev, N., Monyer, H., Seeburg, P. H., and Sakmann, B. (1992a). Divalent ion permeability of AMPA receptor channels is dominated by the edited form of a single subunit. *Neuron*, **8**, 189–98.

Burnashev, N., Schoepfer, R., Monyer, H., Ruppersberg, J. P., Gunther, W., Seeburg, P. H. *et al.* (1992b).

Control by asparagine residues of calcium permeability and magnisium blockade in the NMDA receptor. *Science*, 257, 1415–9.

Cajal, S. R. (1911). *Histologie du système nerveux de l'homme et des vertèbres*. Maloine, Paris.

Cajal, S. R. (1928). *Degeneration and regeneration of the nervous system*. (translated and edited by R. M. May). Oxford University Press.

Cajal, S. R. (1929). *Studies on vertebrate neurogenesis*, translated by L. Guth in 1960. Charles C. Thomas, Springfield.

Campenot, R. B. (1982). Development of sympathetic neurons in compartmentalized cultures. II. Local control of neurite survival by nerve growth factor. *Developmental Biology*, **93**, 13–21.

Cannon, W. B. and Rosenblueth, A. (1949). *The supersensitivity of denervated structures*. Macmillan, New York.

Carbone, E. and Lux, H. D. (1984). A low voltage-activated, fully inactivating Ca channel in vertebrate sensory neurones. *Nature*, **310**, 501–502.

Carbone, E. and Lux, H. D. (1987a). Kinetics and selectivity of a low-voltage-activated calcium current in chick and rat sensory neurones. *Journal of Physiology (London)*, **386**, 547–570.

Carbone, E. and Lux, H. D. (1987b). Single low-voltage-activated calcium channels in chick and rat sensory neurones. *Journal of Physiology (London)*, **386**, 571–601.

Carew, T. J., Pinsker, H. M., and Kandel, E. R. (1972). Long-term habituation of a defensive withdrawal reflex in *Aplysia*. *Science*, **175**, 451–4.

Carew, T. J., Walters, E. T., and Kandel, E. R. (1981). Classical conditioning in a simple withdrawal reflex in *Aplysia californica*. *Journal of Neuroscience*, **1**, 1426–37.

Carew, T. J., Hawkins, R. D., and Kandel, E. R. (1983). Differential classical conditioning of a defensive withdrawal reflex in *Aplysia californica*. *Science*, **219**, 397–400.

Carlson, J., Lais, A. C., and Dyck, P. J. (1979). Axonal atrophy from permanent peripheral axotomy in adult cat. *Journal Neuropathology and Experimental Neurology*, **38**, 579–85.

Carnahan, J. F. and Patterson, P. H. (1991). Isolation of the progenitor cells of the sympathoadrenal lineage from embryonic sympathetic ganglia with the SA monoclonal antibodies. *Journal of Neuroscience*, **11**, 3520–30.

Carpenter, G. (1992). Receptor tyrosine kinase substrates: *src* homology domains and signal transduction. *FASEB Journal*, **6**, 3283–9.

Carpenter, R. S., Koenigsberger, R., and Parsons, S. M. (1980). Passive uptake of acetylcholine and other organic cations by synaptic vesicles from *Torpedo* electric organ. *Biochemistry*, **19**, 4373–9

Carpenter, R. W. (1911). The ciliary ganglion of birds. *Folia Neurologica et Biologica*, **5**, 738–54.

Carroll, S. L., Silos-Santiago, I., Frese, S. E., Ruit, K. G., Milbrandt, J., and Snider, W. D. (1992). Dorsal root ganglion neurons expressing *trk* are selectively sensitive to NGF deprivation in utero. *Neuron*, **9**, 779– 88.

Cartaud, J., Benedetti, E. L., Sobel, A., and Changeux, J. P. (1978). A morphological study of the cholinergic receptor protein from *Torpedo marmorata* in its membrane-bound and in its detergent-extracted, purified form. *Journal of Cell Science*, **29**, 313–37.

Castellucci, V. F. and Kandel, E. R. (1974). A quantal analysis of the synaptic depression underlying habituation of the gill-withdrawal reflex in *Aplysia*. *Proceedings of the National Academy of Sciences of the USA*, **71**, 5004–8.

Castellucci, V. F. and Kandel, E. R. (1976). Presynaptic facilitation as a mechanism for behavioural sensitization in *Aplysia*. *Science*, **194**, 1176–8.

Castellucci, V. F., Kandel, E. R., Schwartz, J. H., Wilson, F. D., Nairn, A. C., and Greengard, P. (1980). Intracellular injection of the catalytic subunit of cyclic AMP-dependent protein kinase simulates facilitation of transmitter release underlying behavioural sensitization in *Aplysia*. *Proceedings of the National Academy of Sciences of the USA*, **77**, 7492–6

Castellucci, V. F., Nairn, A., Greengard, P., Schwartz, J. H., and Kandel, E. R. (1982). Inhibition of adenosine 3′:5′-monophosphate-dependent protein kinase blocks presynaptic facilitation in *Aplysia*. *Journal of Neuroscience*, **2**, 1673–81.

Catterall, W. A. (1988). Structure and function of voltage-sensitive ion channels. *Scinece*, **242**, 50–61.

Catterall, W. A. (1991). Functional subunit structure of voltage-gated calcium channels. *Science*, **253**, 1499–500.

Ceccarelli, B., Hurlbut, W. P., and Mauro, A. (1972). Depletion of vesicles from frog neuromuscular junctions by prolonged tetanic stimulation. *Journal of Cell Biology*, **54**, 30–8.

Cedar, H. and Schwartz, J. H. (1972). Cyclic adenosine monophosphate in the nervous system of *Aplysia californica*. II: Effect of serotonin and dopamine. *Journal of General Physiology*, **60**, 570–87.

Cedar, H., Kandel, E. R., and Schwartz, J. H. (1972). Cyclic adenosine monophosphate in the nervous system of *Aplysia californica*. I: Increased synthesis in response to synaptic stimulation. *Journal of General Physiology*, **60**, 558–69.

Chang, F. L. and Greenough, W. T. (1984). Transient and enduring morphological correlates of synaptic activity and efficacy change in the rat hippocampal slice. *Brain Research*, **309**, 35–46.

Changeux, J. P. (1990). The nicotinic acetylcholine receptor: an allosteric protein prototype of ligand-gated ionic channels. *Trends in Pharmacological Sciences*, **11**, 485–92.

Changeux, J. P. (1991). Compartmentalized transcription of acetylcholine receptor genes during motor endplate epigenesis. *New Biologist*, **3**, 413–29.

Changeux, J. P., Duclert, A., and Sekine, S. (1992). Calcitonin gene-related peptides and neuromuscular interactions. *Annals of the New York Academy of Sciences*, **657**, 361–78.

Chao, M. V. (1992). Neurotrophin receptors: a window into neuronal differentiation. *Neuron*, **9**, 583–93.

Charlton, M. P. and Augustine, G. J. (1990). Classification of presynaptic calcium channels at the squid giant synapse: neither T-, L- nor N-type. *Brain Research*, **525**, 133–139.

Charlton, M. P., Smith, S. J., and Zucker, R. S. (1982). Role of presynaptic calcium ions and channels in synaptic facilitation and depression at the squid giant synapse. *Journal of Physiology (London)*, **323**, 173–193.

Charnet, P., Labarca, C., Leonard, R. J., Vogelaar, N. J., Czyzyk, L., Gouin, A. *et al.* (1990). An open-channel blocker interacts with adjacent turns of α-helices in the nicotinic acetylcholine receptor. *Neuron*, **4**, 87–95.

Chen, L. and Huang, L. Y. M. (1991). Sustained potentiation of NMDA receptor-mediated gluatamate responses through activation of protein kinase C by a μ opioid. *Neuron*, **7**, 319–26.

Chen, L. and Huang, L. Y. M. (1992). Protein kinase C reduces Mg^{2+} block of NMDA-receptor channels as a mechanism of modulation. *Nature*, **356**, 521–3.

Choi, D. W. (1988). Glutamate neurotoxicity and diseases of the nervous system. *Neuron*, **1**, 623–34.

Choi, D. W. (1992). Excitotoxic cell death. *Journal of Neurobiology*, **23**, 1261–6.

Chow, R. H., von Ruden, L., and Neher, E. (1992). Delay in vesicle fusion revealed by electrochemical monitoring of single secretory events in adrenal chromaffin cells. *Nature*, **356**, 60–3.

Christie, M., North, R. A., Osborne, P. B., Douglass, J., and Adelman, J. P. (1990). Heteropolymeric potassium channels expressed in *Xenopus* oocytes from cloned subunits. *Neuron*, **2**, 405–11.

Chu-Wang, I. W. and Oppenheim, R. W. (1978). Cell death of motoneurons in the chick embryo spinal cord. I. A light and electron microscopic study of naturally occurring and induced cell loss during development. *Journal of Comparative Neurology*, **177**, 33–58.

Clapham, D. E. and Neer, E. J. (1993). New roles for G-protein βγ-dimers in transmembrane signalling. *Nature*, **365**, 403–6.

Clark, A. J. (1926). The reaction between acetyl choline and muscle cells. *Journal of Physiology (London)*, **61**, 530–46.

Clarke, P. G. H. (1990). Developmental cell death: morphological diversity and multiple mechanisms. *Anatomy and Embryology*, **181**, 195–213.

Claudio, T., Ballivet, M., Patrick, J., and Heinemann, S. (1983). Nucleotide and deduced amino acid sequences of *Torpedo californica* acetylcholine receptor γ subunit. *Proceedings of the National Academy of Sciences of the USA*, **80**, 1111–5.

Clements, J. D., Forsytile, I. D., and Redman, S. J. (1987). Presynaptic inhibition of synaptic potentials evoked in cat spinal motoneurones by impulses in single Ia axons. *Journal of Physiology (London)*, **383**, 153–69.

Cochran, B. H., Reffel, A. C., and Stiles, C. D. (1983). Molecular cloning of gene sequences regulated by platelet-derived growth factor. *Cell*, **33**, 939–47.

Cohen, J. J., Duke, R. C., Fadok, V. A., and Sellins, K. S. (1992). Apoptosis and programmed cell death in immunity. *Annual Review of Immunology*, **10**, 267–93.

Cohen, M. W. and Godfrey, E. W. (1992). Early appearance of and neuronal contribution to agrin-like molecules at embryonic frog nerve-muscle synapses formed in culture. *Journal of Neuroscience*, **12**, 2982–92.

Cohen, M. W., Jones, O. T., and Angelides, K. J. (1991). Distribution of Ca^{2+} channels on frog motor nerve terminals revealed by fluorescent ω-conotoxin. *Journal of Neuroscience*, **11**, 1032–9.

Cohen, S., Levi-Montalcini, R., and Hamburger, V. (1954). A nerve growth-stimulating factor isolated from sarcomas 37 and 180. *Proceedings of the National Academy of Sciences of the USA*, **40**, 1014–8.

Cole, A. J.,Saffen, D. W., Barahan, J. M., and Worley, P. F. (1989). Rapid increase of an immediate early gene messenger RNA in hippocampal neurons by synaptic NMDA receptor activation. *Nature*, **340**, 474–6.

Collazo, D., Takahashi, H., and McKay, R. D. G. (1992). Cellular targets and trophic functions of neurotrophin-3 in the developing rat hippocampus. *Neuron*, **9**, 643–56.

Collier, B. (1969). The preferential release of newly synthesized transmitter by a sympathetic ganglion. *Journal of Physiology (London)*, **205**, 341–52.

Collingridge, G. L., Kehl, S. J., and McLennan, H. (1983). Excitatory amino acids in synaptic transmission in the Schaffer collateral-comissural pathway of the rat hippocampus. *Journal of Physiology (London)*, **334**, 33–46.

Colquhoun, D. and Hawkes, A. G. (1981). On the stochastic properties of single ion channels. *Proceedings of the Royal Society B*, **211**, 205–35.

Conradi, S. (1969). Ultrastructure of dorsal root boutons on lumbosacral motoneurons of the adult cat as revealed by dorsal root section. *Acta Physiologica Scandinavica (Supplement)*, **332**, 85–115.

Conti-Tronconi, B. M. and Raftery, M. A. (1982). The nicotinic cholinergic receptor: correlation of molecular structure with functional properties. *Annual Review of Biochemistry*, **51**, 491–530.

Cooper, E., Couturier, S., and Ballivet, M. (1991). Pentameric structure and subunit stoichiometry of a neuronal nicotinic acetylcholine receptor. *Nature*, **350**, 235–8.

Cope, T. C. and Mendell, L. M. (1982). Parallel fluctuations of EPSP amplitude and rise time with latency at single Ia-fiber-motorneuron connections in the cat. *Journal of Neurophysiology*, 47, 455–68.

Cordon-Cardo, C., Tapley, P., Jing, S., Nanduri, V., O'Rourke, E., Lamballe, F. *et al.* (1991). The *trk* tyrosine protein kinase mediates the mitogenic properties of nerve growth factor and neurotrophin-3. *Cell*, **66**, 173–83.

Corthay, J., Dunant, Y., and Loctin, F. (1982). Acetylcholine changes underlying transmission of a single nerve impulse in the presence of 4–aminopyridine in *Torpedo*. *Journal of Physiology (London)*, **325**, 461–79.

Cotman, C. W. and Lynch, G. S. (1976). Reactive synaptogenesis in the adult nervous system: the effects of partial deafferentation on new synapse formation. In *Neuronal recognition* (ed. S. Barondes), pp. 69–108. Plenum Press, New York.

Cotman, C. W., Nieto-Sampedro, M., and Harris, E. W. (1981). Synapse replacement in the nervous system of adult vertebrates. *Physiological Reviews*, **61**, 684–784.

Coulombe, J. N. and Nishi, R. (1991). Stimulation of somatostatin expression in developing ciliary ganglion neurons by cells of the choroid layer. *Journal of Neuroscience*, **11**, 553–62.

Couteaux, R. and Pecot-Dechavassine, M. (1970). Vesicules synaptiques et poches au niveau des 'zones actives' de la jonction neuromusculaire. *Comptes rendus des seances de l'Academie des Sciences*, **271**, 2346–9.

Covault, J. and Sanes, J. R. (1985). Neural cell adhesion molecule (N-CAM) accumulates in denervated and paralyzed skeletal muscles. *Proceedings of the National Academy of Sciences of the USA*, **82**, 4544–8.

Crepel, F. and Jaillard, D. (1991). Pairing of pre-and postsynaptic activities in cerebellar Purkinje cells induces long-term changes in synaptic efficacy *in vitro*. *Journal of Physiology (London)*, **432**, 123–41.

Crick, F. (1984). Memory and molecular turnover. *Nature*, **312**, 101.

Crowley, C., Spencer, S. D., Nishimura, M. C., Chen, K. S., Pitts-Meek, S., Armanini, M. P. *et al.* (1994). Mice lacking nerve growth factor display perinatal loss of sensory and sympathetic neurons yet develop basal forebrain cholinergic neurons. *Cell*, **76**, 1001–11.

Curtis, R., Adryan, K. M., Zhu, Y., Harkness, P. J., Lindsay, R. M., and DiStefano, P. S. (1993). Retrograde axonal transport of ciliary neurotrophic factor is increased by peripheral nerve injury. *Nature*, **365**, 253–5.

Czeh, G., Kuno, N., and Kuno, M. (1977). Membrane properties and conduction velocity in sensory neurones following central or peripheral axotomy. *Journal of Physiology (London)*, **270**, 165–80.

Czeh, G., Gallego, R., Kudo, N., and Kuno, M. (1978). Evidence for the maintenance of motorneurone properties by muscle activity. *Journal of Physiology (London)*, **281**, 239–52.

Czernik, A. J., Pang, D. T., and Greengard, P. (1987). Amino acid sequences surrounding the cAMP-dependent and calcium/calmodulin-dependent phosphorylation sites in rat and bovine synapsin I. *Proceedings of the National Academy of Sciences of the USA*, **84**, 7518–22.

Dahm, L. M. and Landmesser, L. T. (1988). The regulation of intramuscular nerve branching during normal development and following axtivity blockade. *Developmental Biology*, **130**, 621–44.

Dale, H. H. (1935). Pharmacology and nerve endings. *Proceedings of the Royal Society of Medicine*, **28**, 319–32.

Dale, H. H. and Feldberg, W. (1934). The chemical transmission of secretory impulses to the sweat glands of the cat. *Journal of Physiology (London)*, **82**, 121–8.

Dale, N. and Kandel, E. R. (1990). Facilitatory and inhibitory transmitters modulate spontaneous transmitter release at cultured *Aplysia* sensorymotor synapses. *Journal of Physiology (London)*, **421**, 203–22.

D'Alonzo, A. J. and Grinnell, A. D. (1985). Profiles of evoked release along the length of frog motor nerve terminals. *Journal of Physiology (London)* **359**, 235–58.

Dani, J. A. and Eisenman, G. (1987). Monovalent and divalent cation permeation in acetylcholine receptor channels. *Journal of General Physiology*, **89**, 959–83.

Darinskii, I. and Korneeva, T. E. (1969). Variation in the size and number of synapses on motorneurones of the spinal cord in the frog observed in the case of prolonged stimulation of dorsal roots. *Dokadad Akademii Nauk SSSR*, **188**, 481–4.

Dash, P. K., Hochner, B., and Kandel, E. R. (1990). Injection of the cAMP-responsive element into the nucleus of *Aplysia* sensory neurons blocks long-term facilitation. *Nature*, **345**, 718–21.

Davidoff, R. A. (1972). Gamma-aminobutyric acid antagonism and presynaptic inhibition in the frog spinal cord. *Science*, **175**, 331–3.

Davies, A. and Lumsden, A. (1984). Relation of target encounter and neuronal death to nerve growth factor responsiveness in the developing mouse trigeminal ganglion. *Journal of Comparative Neurology*, **223**, 124–37.

Davies, A. M. (1994). Tracking neurotrophin function. *Nature*, **368**, 193–4.

Davies, A. M. and Vogel, K. S. (1991). Developmental programmes of growth and survival in early sensory neurones. *Philosophical Transactions of the Royal Society B*, **331**, 259–62.

Davies, A. M., Bandtlow, C., Heumann, R., Korsching, S., Rohrer, H., and Thoenen, H. (1987). Timing and site of nerve growth factor synthesis in developing skin in relation to innervation and expression of the receptor. *Nature*, **326**, 353–8.

Davies, A. M., Lee, K. F., and Jaenisch, R. (1993). p75–deficient trigeminal sensory neurons have an altered response to NGF but not to other neurotrophins. *Neuron*, 11, 565–74.

Davies, C. H., Starkey, S. J., Pozza, M. F., and Collingridge, G. L. (1991). $GABA_B$ autoreceptors regulate the induction of LTP. *Nature*, **349**, 609–11.

Davies, J. and Watkins, J. C. (1982). Action of D and L forms of 2–amino-5–phosphonovalerate and 2–amino-4–phosphonobutyrate in the cat spinal cord. *Brain Research*, **235**, 378–86.

Davies, S. N., Lester, R. A. J., Reymann, K. G., and Collingridge, G. L. (1989). Temporally distinct pre-and post-synaptic mechanisms maintain long-term potentiation. *Nature* **338**, 500–3.

Davis, S., Aldrich, T. H., Valenzuela, D. M., Wong, V., Furth, M. E., Squinto, S. P. *et al.* (1991). The receptor for ciliary neurotrophic factor. *Science*, **253**, 59–63.

Davis, S., Aldrich, T. H., Ip, N. Y., Stahl, N., Scherer, S., Farruggella, T. *et al.* (1993). Released form of CNTF receptor α component as a soluble mediator of CNTF responses. *Science*, **259**, 1736–9.

Davis, S. and Yancopoulos, G. D. (1993). The molecular biology of the CNTF receptor. *Current Opinion in Neurobiology*, **3**, 20–4.

De Camilli, P. and Greengard, P. (1986). Synapsin I: A synaptic vesicle-associated neuronal phosphoprotein. *Biochemical Pharmacology*, **35**, 4349–57.

De Camilli, P. and Jahn, R. (1990). Pathways to regulated exocytosis in neurons. *Annual Review of Physiology*, **52**, 625–45.

Deckwerth, T. L. and Johnson, E. M. (1993). Temporal analysis of events associated with programmed cell death (apoptosis) of sympathetic neurons deprived of nerve growth factor. *Journal of Cell Biology*, **123**, 1207–22.

De Jongh, K. S., Warner, C., and Catterall, W. A. (1990). Subunits of purified calcium channels α_2 and δ are encoded by the same gene. *Journal of Biological Chemistry*, **265**, 14738–41.

Delaney, K. R., Zucker, R. S., and Tank, D. W. (1989). Calcium in motor nerve terminals associated with post-tetanic potentiation. *Journal of Neuroscience*, **9**, 3558–67.

Delaney, K. R., Tank, D. W., and Zucker, R. S. (1991). Presynaptic calcium and serotonin-mediated enhancement of transmitter release at crayfish neuromuscular junction. *Journal of Neuroscience*, **11**, 2631–43.

Del Castillo, J. and Katz, B. (1954a). Quantal components of the end-plate potential. *Journal of Physiology (London)*, **124**, 560–73.

Del Castillo, J. and Katz, B. (1954b). Statistical factors involved in neuromuscular facilitation. *Journal of Physiology (London)*, **124**, 574–85.

Del Castillo, J. and Katz, B. (1956). Biophysical aspects of neuromuscular transmission. *Progress in Biophysics and Biophysical Chemistry*, **6**, 121–70.

Del Castillo, J. and Katz, B. (1957). Interaction at end-plate receptors between different choline derivatives. *Proceedings of the Royal Society B*, **146**, 369–81.

Del Castillo, J. and Stark, L. (1952). The effect of calcium ions on the motor end-plate potentials. *Journal of Physiology (London)*, **116**, 507–15.

De Lorenzo, A. J. (1960). The fine structure of synapses in chick ciliary ganglion. *Journal of Biophysical and Biochemical Cytology*, **7**, 31–6.

Deneris, E. S., Connolly, J., Boulter, J., Wada, E., Wada, K., Swanson, L. W. *et al.* (1988). Primary structure and expression of beta2: A novel subunit of neuronal nicotinic acetylcholine receptors. *Neuron*, **1**, 45–54.

Denis-Donini, S. (1992). Calcitonin gene-related peptide influence on central nervous system differentiation. *Annals of the New York Academy of Sciences*, **657**, 344–50.

Dennis, M., Giraudat, J., Kotzyuba-Hibert, F., Goeldner, M., Hirth, C., Chang, J. Y. *et al.* (1988). Amino acids of the *Torpedo mormorata* acetylcholine receptor α subunit labelled by a photoaffinity ligand for the acetylcholine binding site. *Biochemistry*, **27**, 2346–57.

Dennis, M. J. (1981). Development of the neuromuscular junction: inductive interactions between cells. *Annual Review of Neuroscience*, **4**, 43–68.

Derkach, V., Suprenant, A., and North, R. A. (1989). 5–HT_3 receptors are membrane ion channels. *Nature*, **339**, 706–9.

De Robertis, E. D. P. and Bennett, H. S. (1954). Submicroscopic vesicular component in the synapse. *Federation Proceedings*, **13**, 35.

De Robertis, E. D. P. and Bennett, H. S. (1955). Some features of the submicroscopic morphology of synapses in frog and earthworm. *Journal of Biophysics, Biochemistry and Cytology*, **1**, 47–58.

Dhallan, R. S., Yau, K. W., Schrader, K. A., and Reed, R. R. (1990). Primary structure and functional expression of a cyclic nucleotide-activated channel from olfactory neurones. *Nature*, **347**, 184–7.

Diamond, J. and Miledi, R. (1962). A study of foetal and new-born rat muscle fibres. *Journal of Physiology (London)*, **162**, 393–408.

Diamond, J., Cooper, E., Turner, C., and MacIntyre, L. (1976). Trophic regulation of nerve sprouting. *Science*, **193**, 371–7.

Dionne, V. E. and Ruff, R. L. (1977). End-plate current fluctuations reveal only one channel type at frog neuromuscular junction. *Nature*, **266**, 263–5.

DiStefano, P. S., Friedman, B., Radziejewski, C., Alexander, C., Boland, P., Schick, C. M. *et al.* (1992). The neurotrophins BDNF, NT-3 and NGF display distinct patterns of retrograde axonal transport in peripheral and central neurones. *Neuron*, **8**, 983–93.

Dixon, D. and Atwood, H. L. (1989a). Adenylate cyclase system is essential for long-term facilitation at the crayfish neuromuscular junction. *Journal of Neuroscience*, 9, 4246–52.

Dixon, D. and Atwood, H. L. (1989b). Conjoint action of phosphatidylinositol and adenylate cyclase systems in serotonin-induced facilitation at the crayfish neuromuscular junction. *Journal of Neurophysiology*, **62**, 1251–9.

Dixon, R. A. F., Sigal, I. S., Rands, E., Register, R. B., Candelore, M. R., Blake, A. D. *et al.* (1987). Ligand binding to the beta-adrenergic receptor involves its rhodoposin-like core. *Nature*, **326**, 73–7.

Dodge, F. A. and Rahamimoff, R. (1967). Co-operative action of calcium ions in transmitter release at the neuromuscular junction. *Journal of Physiology (London)*, **193**, 419–32.

Dohlman, H. G., Caron, M. G., and Lefkowitz, R. J. (1987). A family of receptors coupled to guanine nucleotide regulatory proteins. *Biochemistry*, **26**, 2657–64.

Dohlman, H. G., Caron, M. G., Strader, C. D., Amlaiky, N., and Lefkowitz, R. J. (1988). Identification and sequence of a binding site peptide of the β-adrenergic receptor. *Biochemistry*, **27**, 1813–7.

Dohrmann, U., Edgar, D., Sendtner, M., and Thoenen, H. (1986). Muscle-derived factors that support survival and promote fibre outgrowth from embryonic chick spinal motor neurones in culture. *Developmental Biology*, **118**, 209–21.

Dolphin, A. C. (1991). Regulation of calcium channel activity by GTP binding proteins and second messengers. *Biochimica et Biophysica Acta* 1091, 68–80.

Dougherty, D. A. and Stauffer, D. A. (1990). Acetylcholine binding by a synthetic receptor: implications for biological recognition. *Science*, **250**, 1558–60.

Douglas, W. W. and Rubin, R. P. (1963). The mechanism of catecholamine release from the adrenal medulla and the role of calcium in stimulus secretion coupling. *Journal of Physiology (London)*, **167**, 288–310.

Downman, C. B. B., Eccles, J. C., and McIntyre, A. K. (1953). Functional changes in chromatolysed motoneurones. *Journal of Comparative Neurology*, **98**, 9–36.

Draguhn, A., Verdorn, T. A., Ewert, M., Seeburg, P. H., and Sakmann, B. (1990). Functional and molecular distinction between recombinant rat $GABA_A$ receptor subtypes by Zn^{2+}. *Neuron*, **5**, 781–8.

Dratz, E. A. and Hargrave, P. A. (1983). The structure of rhodopsin and the rod outer segment disk membrane. *Trends in Biochemical Sciences*, **8**, 128–31.

Dudai, Y. (1989). *The neurobiology of memory*. Oxford University Press.

Dudel, J. (1965). Facilitatory effects of 5–hydroxy-tryptamine on the crayfish neuromuscular junction. *Naunyn-Schmiedebergs Archives of Experimental Pathology and Pharmacology*, **249**, 515–28.

Duke, R. C., Chervenak, R., and Cohen, J. J. (1983). Endogenous endonuclease-induced DNA fragmentation: an early event in cell-mediated cytolysis. *Proceedings of the National Academy of Sciences of the USA*, **80**, 6361–5.

Dumonde, D. C., Wolstencroft, R. A., Panayi, G. S., Matthew, M., Morley, J., and Howson, W. T. (1969). 'Lymphokines': Non-antibody mediators of cellular immunity generated by lymphocyte activation. *Nature*, **224**, 38–42.

Dumuis, A., Sebben, M., Haynes, L., Pin, J. P., and Bockaert, J. (1988). NMDA receptors activate the arachidonic acid cascade system in striatal neurons. *Nature*, **336**, 68–70.

Dunant, Y. (1986). On the mechanism of acetylcholine release. *Progress in Neurobiology*, **26**, 55–92.

Dunant, Y., Jones, G. J., and Loctin, F. (1982). Acetylcholine measured at short time intervals during transmission of nerve impulses in the electric organ of *Torpedo*. *Journal of Physiology (London)*, **325**, 441–60.

Dunlap, K. (1981). Two types of γ-aminobutyric acid receptor on embryonic sensory neurones. *British Journal of Pharmacology*, **74**, 579–85.

Duvoisin, R. M., Deneris, E. S., Patrick, J., and Heinemann, S. (1989). The functional diversity of the neuronal nicotinic acetylcholine receptors is increased by a novel subunit: β_4. *Neuron*, **3**, 487–96.

Dwarki, V. J., Montminy, M., and Verma, I. M. (1990). Both the basic region and the 'leucine zipper' domain of the cyclic AMP response element binding (CREB) protein are essential for transcriptional activation. *EMBO Journal*, **9**, 225–32.

Dwyer, T. M., Adams, D. J., and Hille, B. (1980). The permeability of the endplate channel to organic cations in frog muscle. *Journal of General Physiology*, **75**, 469–92.

Eberhard, D. A. and Holz, R. W. (1987). Cholinergic stimulation of inositol phosphate formation in bovine adrenal chromaffin cells: distinct nicotinic and muscarinic mechanisms. *Journal of Neurochemistry*, **49**, 1634–43.

Eccles, J. C. (1953). *The neurophysiological basis of mind. The principles of neurophysiology*. Clarendon, Oxford.

Eccles, J. C. (1959). The development of ideas on the synapse. In *The Historical development of physiological thought*, (ed. C. McC Brooks and P. F. Cranefield), pp.39–66. Hafner, New York.

Eccles, J. C. (1964). *The physiology of synapses*. Springer-Verlag, Berlin.

Eccles, J. C. (1974). Trophic interactions in the mammalian central nervous system. *Annals of the New York Academy of Sciences*, **228**, 406–23.

Eccles, J. C. (1982). The synapse: From electrical to chemical transmission. *Annual Review of Neuroscience*, **5**, 325–29.

Eccles, J. C. and Krnjević, K. (1959). Presynaptic changes associated with post-tetanic potentiation in the spinal cord. *Journal of Physiology (London)*, **149**, 274–87.

Eccles, J. C. and McGeer, P. L. (1979). Ionotropic and metabotropic neurotransmission. *Trends in Neurosciences*, 2, 39–40.

Eccles, J. C. and McIntyre, A. K. (1953). The effects of disuse and of activity on mammalian spinal reflexes. *Journal of Physiology (London)*, 121, 492–516.

Eccles, J. C. and Rall, W. (1951). Effects induced in a monosynaptic reflex path by its activation. *Journal of Neurophysiology*, 14, 353–76.

Eccles, J. C., Fatt, P., and Koketsu, K. (1954). Cholinergic and inhibitory synapses in a pathway from motor-axon collaterals to motoneurones. *Journal of Physiology (London)*, 126, 524–62.

Eccles, J. C., Eccles, R. M., and Lundberg, A. (1958). The action potential of the alpha motoneurones supplying fast and slow muscles. *Journal of Physiology (London)*, 142, 275–91.

Eccles, J. C., Krnjević, K., and Miledi, R. (1959). Delayed effects of peripheral severance of afferent nerve fibres on the efficacy of their central synapses. *Journal of Physiology (London)*, 145, 204–20.

Eccles, J. C., Eccles, R. M., and Magni, F. (1961). Central inhibitory action attributed to presynaptic depolarization produced by muscle afferent volleys. *Journal of Physiology (London)*, 159, 147–66.

Eccles, J. C., Eccles, R. M., and Kozak, W. (1962). Further investigation on the influence of motoneurones on the speed of muscle contraction. *Journal of Physiology (London)*, 163, 324–39.

Eccles, J. C., Ito, M., and Szentagothai, J. (1967). *The cerebellum as a neuronal machine*. Springer-Verlag, New York.

Eckenstein, F. P., Esch, F., Holbert, T., Blacher, R. W., and Nishi, R. (1990). Purification and characterization of a trophic factor for embryonic peripheral neurones: Comparison with fibroblast growth factor. *Neuron*, 4, 623–31.

Eckstein, F., Cassel, D., Levkowitz, H., Lowe, M., and Selinger, Z. (1979). Guanosine 5′-O-(2–thiophosphate) an inhibitor of adnylate cyclase stimulation by guanine nucleotide and fluoride ions. *Journal of Biological Chemistry*, 254, 9829–34.

Edds, M. V. (1953). Collateral nerve regeneration. *Quarterly Review of Biology*, 28, 260–76.

Edelman, A. M., Blumenthal, D. K., and Krebs, E. G. (1987). Protein serine/threonine kinases. *Annual Review of Biochemistry*, 56, 567–613.

Edelman, G. M. (1976). Surface modulation in cell recognition and cell growth. *Science*, 192, 218–26.

Edelman, G. M. (1983). Cell adhesion molecules. *Science*, 219, 450–7.

Edwards, F. A., Konnerth, A., Sakmann, B., and Takahashi, T. (1989). A thin slice preparation for patch clamp recordings from neurones of the mammalian central nervous system. *Pflügers Archiv*, 414, 600–12.

Edwards, F. A., Konnerth, A., Sakmann, B., and Busch, C. (1990). Quantal analysis of inhibitory synaptic transmission in the dendate gyrus of rat hippocampal slices: a patch-clamp study. *Journal of Physiology (London)*, 430, 213–49.

Edwards, F. A., Gibb, A. J., and Colquhoun, D. (1992). ATP receptor-mediated synaptic currents in the central nervous system. *Nature*, 359, 144–7.

Edwards, F. R., Redman, S. J., and Walmsley, B. (1976). Statistical fluctuations in charge transfer at Ia synapses on spinal motoneurones. *Journal of Physiology (London)*, 259, 665–88.

Edwards, F. R., Harrison, P. J., Jack, J. J. B., and Kullmann, D. M. (1989). Reduction by baclofen of monosynaptic EPSPs in lumbosacral motoneurones of the anaesthetized cat. *Journal of Physiology (London)*, 416, 539–56.

Edwards, S. N., Buckmaster, A. E., and Tolkovsky, A. M. (1991). The death programme in culutured sympathetic neurones can be suppressed at the post-translational level by nerve growth factor, cyclic AMP, and depolarization. *Journal of Neurochemistry*, 57, 2140–3.

Eftimie, R., Brenner, H. R., and Buonanno, A. (1991). Myogenin and MyoD join a family of skeletal muscle genes regulated by electrical activity. *Proceedings of the National Academy of Sciences of the USA*, 88, 1349–53.

Egebjerg, J., Bettler, B., Hermans-Brogmeyer, I., and Heinemann, S. (1991). Cloning of a cDNA for a glutamate receptor subunit activated by kainate but no AMPA. *Nature*, 351, 745–8.

Ehlers, M. D., Kaplan, D. R., and Koliatsos, V. E. (1993). Retrograde transport of neurotrophin receptors. *Society for Neuroscience Abstracts* 19, 538.24.

Ehrenpreis, S. (1959). Interaction of curare and related substances with acetylcholine receptor-like protein. *Science*, 129, 1613–4.

Eliot, L. S., Dudai, Y., Kandel, E. R., and Abrams, T. W. (1989). Ca^{2+}/calmodulin sensitivity may be common to all forms of neural adenylate cyclase. *Proceedings of the National Academy of Sciences of the USA*, 86, 9564–8.

Ellis, R. E., Yuan, J., and Horvitz, H. R. (1991). Mechanisms and functions of cell death. *Annual Review of Cell Biology*, 7, 663–98.

Elmqvist, D. and Quastel, D. M. J. (1965). Presynaptic action of hemicholinium at the neuromuscular junction. *Journal of Physiology (London)*, 177, 463–82.

Emptage, N. J. and Carew, T. J. (1993). Long-term syanptic facilitation in the absence of short-term facilitation in *Aplysia* neurones. *Science*, 262, 253–6.

Engelman, D. M., Goldman, A., and Steitz, T. A. (1982). The identification of helical segments in the polypeptide chain of bacteriorhodopsin. *Methods in Enzymology*, 88, 81–8.

Ernfors, P., Henschen, A., Olson, L., and Persson, H. (1989). Expression of nerve growth factor receptor mRNA is developmentally regulated and increased after axotomy in rat spinal cord motoneurons. *Neuron*, 2, 1605–13.

Ernfors, P., Ibanez, C. F., Ebendal, T., Olson, L., and Persson, H. (1990). Molecular cloning and neurotrophin activities of a protein with structural similarities to nerve growth factor; developmental and topographical expression. *Proceedings of the National Academy of Sciences of the USA* 87, 5454–8.

Ernfors, P., Lee, K. F., and Jaenisch, R. (1994a). Mice lacking brain-derived neurotrophic factor develop with sensory deficits. *Nature*, **368**, 147–50.

Ernfors, P., Lee, K. F., Kucera, J., and Jaenisch, R. (1994b). Lack of neurotrophin-3 leads to deficiencies in the peripheral nervous system and loss of limb proprioceptive afferents. *Cell*, 77, 503–12.

Ernsberger, U. and Rohrer, H. (1988). Neuronal precursor cells in chick dorsal root ganglia: differentiation and survival *in vitro*. *Developmental Biology*, **126**, 420–32.

Eusebi, F., Molinaro, M., and Zani, B. M. (1985). Agents that activate protein kinase C reduce acetylcholine sensitivity in cultured myotubes. *Journal of Cell Biology*, **100**, 1339–42.

Evans, R. J., Derkach, V., and Suprenant, A. (1992). ATP mediates fast synaptic transmission in mammalian neurons. *Nature*, **357**, 503–5.

Fagg, G. E. (1985). L-glutamate, excitatory amino acid receptors and brain function. *Trends in Neurosciences*, **8**, 207–10.

Faissner, A., Kruse, J., Nieke, J., and Schachner, M. (1984). Expression of neural cell adhesion molecule L1 during development, in neurological mutants and in the peripheral nervous system. *Developmental Brain Ressearch*, 15, 69–82.

Falls, D. L., Rosen, K. M., Corfas, G., Lane, W. S., and Fischbach, G. D. (1993). ARIA, a protein that stimulates acetylcholine receptor synthesis, is a member of the Neu ligand family. *Cell*, 72, 801–15.

Fambrough, D. M. (1970). Acetylcholine sensitivity of muscle fibre membranes: mechanism of regulation by motorneurones. *Science*, **168**, 372–3.

Fambrough, D. M. and Hartzell, H. C. (1972). Acetylcholine receptors: number and distribution at neuromuscular junctions in rat diaphragm. *Science*, 176, 189–91.

Fatt, P. (1950). The electromotive action of acetylcholine at the motor end-plate. *Journal of Physiology (London)*, **111**, 408–22.

Fatt, P. and Katz, B. (1951). An analysis of the end-plate potential recorded with an intra-cellular electrode. *Journal of Physiology (London)*, **115**, 320–70.

Fazelli, M. S. (1992). Synaptic plasticity: on the trail of the retrograde messenger. *Trends in Neurosciences*, **15**, 115–7.

Feldberg, W. (1977). The early history of synaptic and neuromuscular transmission by acetylcholine: Reminiscences of an eye witness. In *The pursuit of nature*, pp. 65–81. Cambridge University Press.

Feltz, A. and Trautmann, A. (1982). Desensitization at the frog neuromuscular junction: a biphasic process. *Journal of Physiology (London)*, **322**, 257–72.

Feng, T. P. (1936). Studies on the neuromuscular junction. II. The universal antagonism between calcium and curarizing agencies. *Chinese Journal of Physiology*, **10**, 513–528.

Feng, T. P. (1940). Studies on the neuromuscular junction. XVIII. The local potentials around N-M junctions induced by single and multiple volleys. *Chinese Journal of Physiology*, **15**, 367–404.

Fenwick, E., Marty, M., and Neher, E. (1982). Sodium and calcium channels in bovine chromaffin cells. *Journal of Physiology (London)*, **331**, 599–635.

Ferns, M. J. and Hall, Z. W. (1992). How many agrins does it take to make a synapse? *Cell*, **70**, 1–3.

Fertuck, H. C. and Salpeter, M. M. (1974). Localization of acetylcholine receptor by ^{125}I-labelled α-bungartoxin binding at mouse motor end-plate. *Proceedings of the National Academy of Sciences of the USA*, 71, 1376–8.

Finer-Moore, J. and Stroud, R. M. (1984). Amphipathic analysis and possible formation of the ion channel in an acetylcholine receptor. *Proceedings of the National Academy of Sciences of the USA*, **81**, 155–9.

Fischbach, G. D. and Cohen, S. A. (1973). The distribution of acetylcholine sensitivity over uninnervated and innervated muscle fibres grown in cell culture. *Developmental Biology*, **31**, 147–62.

Fischbach, G. D. and Schuetze, S. M. (1980). A postnatal decrease in acetylcholine channel open time at rat endplates. *Journal of Physiology (London)*, **303**, 125–37.

Fischer, G., Kunemund, V., and Schachner, M. (1986). Neurite outgrowth patterns in cerebellar microexplant cultures are affected by antibodies to the cell surface glycoprotein L1. *Journal of Neuroscience*, **6**, 605–12.

Fitzgerald, M., Wall, P. D., Goedert, M., and Emson, P. C. (1985). Nerve growth factor counteracts the neurophysiological and neurochemical effects of chronic sciatic nerve section. *Brain Research*, **332**, 131–41.

Fleckenstein, A. (1983). History of calcium antagonists. *Circulation Research*, **52**, Supplement 1, 3–16.

Fletcher, P. and Forrester, T. (1975). The effect of curare on the release of acetylcholine from mammalian motor nerve terminals and an estimate of quantum content. *Journal of Physiology (London)*, **251**, 131–44.

Flexner, J. B., Flexner, L. B., and Stellar, E. (1963). Memory in mice as affected by intracerebral puromycin. *Science*, **141**, 57–9.

Flexner, L. B. and Goodman, R. H. (1975). Studies on memory: inhibitors of protein synthesis also inhibit catecholamine synthesis. *Proceedings of the National Academy of Sciences of the USA*, 72, 4660–3.

Flockerzi, V., Oeken, H. J., Hofmann, F., Pelzer, D.,

Cavalie, A., and Trautwein, W. (1986). The purified dihydropyridine- binding site from skeletal muscle T-tubules is a functional Ca channel. *Nature*, **323**, 66–8.

Foehring, R. C. and Munson, J. B. (1990). Motoneuron and muscle-unit properties after long-term direct innervation of soleus muscle by medial gastrocnemius nerve in cat. *Journal of Neurophysiology*, **64**, 847–61.

Foehring, R. C., Sypert, G. W., and Munson, J. B. (1986). Properties of self-innervated motor units in medial gastrocnemius of the cat: II. Axotomized motoneurons and the time course of recovery. *Journal of Neurophysiology*, **55**, 947–65.

Foehring, R. C., Sypert, G. W., and Munson, J. B. (1987). Motor-unit properties following cross-reinnervaion of cat lateral gastrocnemius and soleus muscles with medial gastrocnemius nerve. II. Influence of muscle on motoneurons. *Journal of Neurophysiology*, **57**, 1227–45.

Fonnum, F. (1967). The 'compartation' of choline acetyltransferase within the synaptosomes. *Biochemical Journal*, **103**, 262–70.

Fontaine, B., Klarsfeld, A, Hökfelt, T., and Changeux, J. P. (1986). Calcitonin gene-related peptide, a peptide present in spinal cord motoneurons, increases the number of acetylcholine receptors in primary cultures of chick embryo myotubes. *Neuroscience Letters*, **71**, 59–65.

Forbes, A. (1922). The interpretation of spinal reflexes in terms of present knowledge of nerve conduction. *Physiological Reviews*, **2**, 361–414.

Forbes, A. (1939). Problems of synaptic function. *Journal of Neurophysiology*, **2**, 391–7.

Forger, N. G., Roberts, S. L., Wong, V., and Breedlove, S. M. (1994). Ciliary neurotrophic factor maintains motoneurons and their target muscles in developing rats. *Journal of Neuroscience*, **13**, 4720–6.

Forsythe, I. D. and Clements, J. D. (1990). Presynaptic glutamate receptors depress excitatory monosynaptic transmission between mouse hippocampal neurones. *Journal of Physiology (London)*, **429**, 1–16.

Fosset, M., Jaimovich, E., Delpont, E., and Lazdunski, M. (1983). [^{3}]Nitrendipine receptors in skeletal muscle. Properties and preferential localization in transverse tubules. *Journal of Biological Chemistry*, **258**, 6086–6092.

Foster, C. A., Mandak, B., Kromer, E., and Rot, A. (1992). Calcitonin gene-related peptide is chemotactic for human T lymphocytes. *Annals of the New York Academy of Sciences*, **657**, 397–404.

Foster, M. (1897). *A text book of physiology*, (7th edn). Macmillan, New York.

Foster, T. C. and McNaughton, B. L. (1991). Long-term enhancement of CA1 synaptic transmission is due to increased quantal size, not quantum content. *Hippocampus*, **1**, 79–91.

Fox, A. P., Nowycky, M. C., and Tsien, R. W. (1987a). Kinetic and pharmacological properties distinguishing three types of calcium currents in chick sensory neurones. *Journal of Physiology (London)*, **394**, 149–172.

Fox, A. P., Nowycky, M. C., and Tsien, R. W. (1987b). Single-channel recordings of three types of calcium channels in chick sensory neurones. *Journal of Physiology (London)*, **394**, 173–200.

Frank, E. and Fischbach, G. D. (1979). Early events in neuromuscular junction formation *in vitro*. *Journal of Cell Biology*, **83**, 143–58.

Frank, E., Gautvik, K., and Sommerschild, H. (1975). Persistence of junctional acetylcholine receptors following denervation. *Cold Spring Harbor Symposia on Quantitative Biology*, **40**, 275–81.

Frank, K. and Fuortes, M. G. F. (1957). Presynaptic and post-synaptic inhibition of monosynaptic reflexes. *Federation Proceedings*, **16**, 39–40.

Franklin, J. L. and Johnson, E. M. (1992). Suppression of programmed neuronal death by sustained elevation of cytoplasmic calcium. *Trends in Neurosciences*, **15**, 501–8.

Frey, U., Krug, M., Reymann, K. G., and Matthies, H. (1988). Anisomycin, an inhibitor of protein synthesis, blocks late phases of LTP phenomena in the hippocampal CA1 region *in vitro*. *Brain Research*, **452**, 57–65.

Frey, U., Huang, E. R., and Kandel, E. R. (1993). Effects of cAMP simulate a late stage of LTP in hippocampal CA1 neurons. *Science*, **260**, 1661–4.

Fried, G., Terenius, L., Hökfelt, T., and Goldstein, M. (1985). Evidence for differential localization of noradrenaline and neuropeptide Y in neuronal storage vesicles isolated from rat vas deferens. *Journal of Neuroscience*, **5**, 450–8.

Fried, G., Franck, J., Brodin, E., Born, W., Fischer, J. A., Hiort, W. *et al.* (1989). Evidence for differential storage calcitonin gene-related peptide, substance P and serotonin in synaptosomal vesicles of rat spinal cord. *Brain Research*, **499**, 315–24.

Friedman, B., Scherer, S. S., Rudge, J. S., Helgren, M., Morrisey, D., McClain, J. *et al.* (1992). Regulation of ciliary neurotrophic factor expression in myelin-related Schwann cells *in vivo*. *Neuron*, **9**, 295–305.

Friedman, W. J., Olson, L., and Persson, H. (1991). Cells that express brain-derived neurotrophic factor mRNA in the developing postnatal rat brain. *European Journal of Neuroscience*, **3**, 688–97.

Froehner, S. C. (1986). The role of the postsynaptic cytoskeleton in AChR organization. *Trends in Neurosciences*, **9**, 37–41.

Froehner, S. C. (1993). Regulation of ion channel distribution at synapses. *Annual Review of Neuroscience*, **16**, 347–68.

Froehner, S. C., Luetje, C. W., Scotland, P. B., and Patrick, J. (1990). The postsynaptic 43K protein clusters muscle nicotinic acetylcholine receptors in *Xenopus* oocytes. *Neuron* 5, 403–10.

Frost, W. N., Castellucci, V. F., Hawkins, R. D., and Kandel, E. R. (1985). Monosynaptic connections made by the sensory neurons of the gill-and siphon-withdrawal reflex in *Aplysia* participate in the storage of long-term memory for sensitization. *Proceedings of the National Academy of Sciences of the USA*, **82**, 8266–9.

Frost, W. N., Clark, G. A., and Kandel, E. R. (1988). Parallel processing of short-term memory for sensitization in *Aplysia*. *Journal of Neurobiology*, **19**, 297–334.

Fukada, K. (1985). Purification and partial characterization of a cholinergic neuronal differentiation factor. *Proceedings of the National Academy of Sciences of the USA*, **82**, 8795–9.

Fukuoka, T., Engel, A. G., Lang, B., Newson-Davis, J., Prior, C., and Wray, D. W. (1987a). Lambert-Eaton myasthenic syndrome: I. Early morphological effects of IgG on the presynaptic membrane active zones. *Annals of Neurology*, **22**, 193–9.

Fukuoka, T., Engel, A. G., Lang, B., Newson-Davis, J., and Vincent, A. (1987b). Lamber-Eaton myasthenia syndrome: II. Immunoelectron microscopy localization of IgG at the mouse motor end-plate. *Annals of Neurology*, **22**, 200–11.

Furshpan, E. J. and Potter, D. D. (1959). Transmission at the giant motor synapses of the crayfish. *Journal of Physiology (London)*, **145**, 289–325.

Furuichi, T., Yoshikawa, S., Miyawaki, A., Wada, K., Maeda, N., and Mikoshiba, K. (1989). Primary structure and functional expression of the inositol 1,4,5–trisphosphate-binding protein P_{400}. *Nature*, **342**, 32–8.

Furukawa, T. and Furshpan, E. J. (1963). Two inhibitory mechanisms in the Mauthner neurons of goldfish. *Journal of Neurophysiology*, **26**, 140–76.

Gage, P. and McBurney, R. N. (1972). Miniature end-plate currents and potentials generated by quanta of acetylcholine in glycerol-treated toad sartorius fibres. *Journal of Physiology (London)*, **226**, 79–94.

Gallego, R. and Geijo, E. (1987). Chronic block of the cervical trunk increases synaptic efficacy in the superior and stellate ganglia of the guinea-pig. *Journal of Physiology (London)*, **382**, 449–62.

Gallego, R., Kuno, M., Nunez, R., and Snider, W. D. (1979). Disuse enhances synaptic efficacy in spinal motoneurones. *Journal of Physiology (London)*, **291**, 191–205.

Gallego, R., Ivorra, I., and Morales, A. (1987). Effects of central or peripheral axotomy on membrane properties of sensory neurones in the petrosal ganglion of the cat. *Journal of Physiology (London)*, **391**, 39–56.

Gallo, V., Ciotti, M. T., Coletti, A., Aloisi, F., and Levi, G. (1982). Selective release of glutamate from cerebellar granule cells differentiating in culture. *Proceedings of the National Academy of Sciences of the USA*, **79**, 7919–23.

Galzi, J. L., Devillers-Thiery, A., Hussy, N., Bertrand, S., Changeux, J. P., and Bertrand, D. (1992). Mutations in the channel domain of a neuronal nicotinic receptor convert ion selectivity from cationic to anionic. *Nature*, **359**, 500–5.

Garrison, F. H. (1929). *An introduction to the history of medicine*. Saunders, Philadelphia.

Garthwaite, J. (1991). Glutamate, nitric oxide and cell-cell signalling in the nervous sytem. *Trends in Neurosciences*, **14**, 60–7.

Garthwaite, J., Charles, S. L., and Chess-Williams, R. (1988). Endothelium-derived relaxing factor release on activation of NMDA receptors suggests role as intercellular messenger in the brain. *Nature*, **336**, 385–8.

Gearing, D. P., Thut, C. J., VandenBos, T., Gimpel, S. D., Delaney, P. B., King, J., Price, V., Cosman, D., and Beckmann, M. P. (1991). Leukemia inhibitory factor receptor is structurally related to the IL-6 signal transducer, gp 130. *EMBO Journal*, **10**, 2839–48.

Gearing, D. P., Comeau, M. R., Friend, D. J., Gimpel, S. D., Thut, C. J., McGourty, J., Brasher, K. K., King, J. A., Gillis, S., Mosley, B., Ziegler, S. F., and Cosman, D. (1992). The IL-6 signal transducer, gp130: an oncostatin M receptor and affinity coverter for the LIF receptor. *Science*, **255**, 1434–7.

Gibson, Q. H. and Roughton, F. J. W. (1957). The kinetics and equilibria of the reactions of nitric oxide with sheep haemoglobin. *Journal of Physiology (London)*, **136**, 507–26.

Gibson, S. J., Polak, J. M., Bloom, S. R., Sabate, I. M., Mulderry, P. M., Ghatei, M. A. *et al.* (1984). Calcitonin gene-related peptide immunoreactivity in the spinal cord of man and of eight other species. *Journal of Neuroscience*, **4**, 3101–11.

Giller, E. L., Schrier, B. K., Shainberg, A., Fisk, H. R., and Nelson, P. G. (1973). Choline acetyltransferase activity is increased in combined cultures of spinal cord and muscle cells from mice. *Science*, **182**, 588–9.

Giller, E. L., Neale, J. H., Bullock, P. N., Schrier, B. K., and Nelson, P. G. (1977). Choline acetyltransferase activity of spinal cord cell cultures increased by co-culture with muscle and by muscle-conditioned medium. *Journal of Cell Biology*, **74**, 16–29.

Gilman, A. G. (1984). G proteins and dual control of adenylate cyclase. *Cell*, **36**, 577–9.

Gilman, A. G. (1987). G proteins: tranducers of receptor-generated signals. *Annual Review of Biochemistry*, **56**, 615–49.

Ginsborg, B. L. and Hirst, G. D. S. (1972). The effect of adenosine on the release of the transmitter from the phrenic nerve of the rat. *Journal of Physiology (London)* **224**, 629–45.

Giraudat, J., Dennis, M., Heidmann, T., Haumont, P.

Y., Lederer, F., and Changeux, J. P. (1987). Structure of the high-affinity binding site for noncompetitive blockers of the acetylcholine receptor: ^{3}H]chloropromazine labels homologous residues in the β and α chains. *Biochemistry*, 26, 2410–8.

Glanzman, D. L., Kandel, E. R., and Schacher, S. (1989). Identified target motor neuron regulates neurite outgrowth and synaptic formation of *Aplysia* sensory neurons *in vitro*. *Neuron*, 3, 441–50.

Glanzman, D. L., Kandel, E. R., and Schacher, S. (1990). Target-dependent structural changes accompanying long-term synaptic facilitation in *Aplysia* neurones. *Science*, 249, 799–802.

Godfrey, E. W., Nitkin, R. M., Wallace, B. G., Rubin, L. L., and McMahan, U. J. (1984). Components of *Torpedo* electric organ and muscle that cause aggregation of acetylcholine receptors on cultured muscle cells. *Journal of Cell Biology*, 99, 615–27.

Goelet, P., Castellucci, V. F., Schacher, S., and Kandel, E. R. (1986). The long and the short of long-term memory—a molecular framework. *Nature*, 322, 419–22.

Gold, M. R. and Martin, A. R. (1983). Characteristics of inhibitory post-synaptic currents in brain-stem neurones of the lamprey. *Journal of Physiology (London)* 342, 85–98.

Goldman, D., Deneris, E., Luyten, W., Kochhar, A., Patrick, J., and Heinemann, S. (1987). Members of a nicotinic acetylcholine receptor gene family are expressed in different regions of the mammalian central nervous sytem. *Cell*, 48, 965–73.

Goldman, D., Carlson, B. M., and Staple, J. (1991). Induction of adult-type nicotinic acetylcholine receptor gene expression in noninnervated regenerating muscle. *Neuron*, 7, 649–58.

Goldring, J. M., Kuno, M., Nunez, R., and Weakly, J. N. (1981). Do identical activity patterns in fast and slow motor axons exert the same influence on the twitch time of cat skeletal muscle? *Journal of Physiology (London)*, 321, 211–23.

Gonzalez, G. A., Yamamoto. K. K., Fischer, W. H., Kart, D., Menzel, P., Biggs, W. *et al.* (1989). A cluster of phosphorylation sites on the cyclic AMP-regulated nuclear factor CREB predicted by its sequence. *Nature*, 337, 749–52.

Goodman, R. H. (1990). Regulation of neuropeptide gene expression. *Annual Review of Neuroscience*, 13, 111–27.

Goy, M. F. and Kravitz, E. A. (1989). Cyclic AMP only partially mediates the actions of serotonin at lobster neuromuscular junctions. *Journal of Neuroscience*, 9, 369–79.

Goy, M. F., Schwarz, L., and Kravitz, E. A. (1984). Serotonin- induced protein phosphorylation in a lobster nerve-muscle preparation. *Journal of Neuroscience*, 4, 611–26.

Graham, D., Pfeiffer, F., Simler, R., and Betz, H. (1985). Purification and characterization of the glycine receptor of pig spinal cord. *Biochemistry*, 24, 990–4.

Graham, R. and Gilman, M. (1991). Distinct protein targets for signals acting at the c-*fos* serum response element. *Science*, 251, 189–192.

Granit, R., Henatsch, H. D., and Steg, G. (1956). Tonic and phasic ventral horn cells differentiated by post-tetanic potentiation in cat extensors. *Acta physiologica scandinavica*, 37, 114–26.

Grant, S. G. N., O'Dell, T. J., Karl, K. A., Stein, P. L., Soriano, P., and Kandel, E. R. (1992). Impaired long-term potentiation, spatial learning, and hippocampal development in *fyn* mutant mice. *Science*, 258, 1903–10.

Gray, E. G. (1962). A morphological basis for presynaptic inhibition? *Nature*, 193, 82–3.

Green, S. H., Rydel, R. E., Connolly, J. L., and Greene, L. A. (1986). PC12 cell mutants that possess low- but not high-affinity nerve growth factor receptors neither respond to nor internalize nerve growth factor. *Journal of Cell Biology*, 102, 830–42.

Greenberg, M. E. and Ziff, E. B. (1984). Stimulation of 3T3 cells induces transcription of the c-*fos* protoncogene. *Nature*, 311, 433–8.

Greenberg, S. M., Castellucci, V. F., Bayley, H., and Schwartz, J. H. (1987). A molecular mechanism for long-term sensitization in *Aplysia*. *Nature*, 329, 62–5.

Greengard, P., Jen, J., Nairn, A. C., and Stevens, C. F. (1991). Enhancement of the glutamate response by cAMP-dependent protein kinase in hippocampal neurons. *Science*, 253, 1135–8.

Grenningloh, G., Rienitz, A., Schmitt, B., Methfessel, C., Zensen, M., Beyreuther, K. *et al.* (1987). The strychnine-binding subunit of the glycine receptor shows homology with nicotinic acetylcholine receptors. *Nature*, 328, 215–20.

Grenningloh, G., Pribilla, I., Prior, P., Multhaup, G., Beyreuther, K. Taleb, O. *et al.* (1990). Cloning and expression of the 58 kd β subunit of the inhibitory glycine receptor. *Neuron*, 4, 963–70.

Grether, W. F. (1938). Psudo-conditioning without paired stimulation encountered in attempted backward conditioning. *Journal of Comparative Psychology*, 25, 91–6.

Grinnell, A. D. (1988). Synpaptic plasticity following motor nerve injury in frogs. In *The current status of peripheral nerve regeneration* (ed. T. Gordon, R. B. Stein, and P. A. Smith), pp. 223–34. Liss, New York.

Grinnell, A. D. and Herrera, A. A. (1981). Specificity and plasticity of neuromuscular junctionbs: long-term regulation of motoneuron function. *Progress in Neurobiology*, 17, 203–82.

Gu, Y. and Hall, Z. W. (1988). Immunological evidence for a change in subunits of the acetylcholine receptor in developing and denervated rat muscle. *Neuron*, 1, 117–25.

Gundersen, C. B., Katz, B., and Miledi, R. (1982). The antagonism between botulinum toxin and calcium in motor nerve terminals. *Proceedings of the Royal Society B*, **216**, 369–76.

Gundersen, R. W. and Barrett, J. N. (1979). Neuronal chemotaxis: chick dorsal-root axons turn toward high concentrations of nerve growth factor. *Science*, **206**, 1079–80.

Gurdon, J. B., Lane, C. D., Woodland, H. R., and Marbaix, G. (1971). Use of frog eggs and oocytes for the study of messenger RNA and its translation in living cells. *Nature*, **233**, 177–182.

Gurney, A. M., Tsien, R. Y., and Lester, H. A. (1987). Activation of a potassium current by rapid photochemically generated step increases of intracellular calcium in rat sympathetic neurons. *Proceedings of the National Academy of Sciences of the USA*, **84**, 3496–500.

Gustafsson, B. and Wigstrom, H. (1988). Physiological mechanisms underlying long-term potentiation. *Trends in Neurosciences*, **11**, 156–62.

Gustafsson, B., Wigstrom, H., Abraham, W. C., and Huang, Y. Y. (1987). Long-term potentiation in the hippocampus using depolarizing current pulses as the conditioning stimulus to single volley. *Journal of Neuroscience*, **7**, 774–80.

Gutmann, E. (1963). Evidence for the trophic function of the nerve cell in neuromuscular relations. In *The effect of use and disuse on neuromuscular functions* (ed. E. Gutmann and P. Hnik), pp. 29–34. Elsevier, Amsterdam.

Gutmann, E. (1976). Neurotrophic relations. *Annual Review of Physiology*, **38**, 177–216.

Gutmann, E. and Young, J. Z. (1944). The reinnervation of muscle after various periods of atrophy. *Journal of Anatomy*, **78**, 15–43.

Guy, H. R. (1984). A structural model of the acetylcholine receptor channel based on partition energy and helix packing calculations. *Biophysical Journal*, **45**, 249–61.

Guy, H. R. and Conti, F. (1990). Pursuing the structure and function of voltage-gated channels. *Trends in Neurosciences*, **13**, 201–6.

Guy, H. R. and Seetharmaulu, P. (1986). Molecular model of the action potential sodium channel. *Proceedings of the National Academy of Sciences of the USA*, **83**, 508–12.

Haga, T. (1989). The structure and function of muscarinic acetylcholine receptors. In *Biological transduction mechanisms* (ed. M. Kasai, T. Yoshioka, and H. Suzuki), pp. 115–138. Japan Science Society Press, Tokyo.

Hagiwara, S. and Byerly, L. (1981). Calcium channel. *Annual Review of Neuroscience*, **4**, 69–125.

Hagiwara, S. and Tasaki, I. (1958). A study on the mechanism of impulse transmission across the giant synapse of the squid. *Journal of Physiology (London)*, **143**, 114–137.

Hajos, F., Garthwaite, G., and Gathwaite, J. (1986). Reversible and irreversible neuronal damage caused by excitatory amino acid analogues in rat cerebellar slices. *Neuroscience*, **18**, 417–36.

Haley, J. E., Wilcox, G. L., and Chapman, P. F. (1992). The role of nitric oxide in hippocampal long-term potentiation. *Neuron*, **8**, 211–6.

Hall, Z. W. (1992). *An introduction to molecular neurobiology*. Sinauer Associates, Inc., Sunderland.

Hall, Z. W. and Sanes, J. R. (1993). Synaptic structure and development: The neuromuscular junction. *Neuron*, **10** (Supplement), 99–121.

Hall, Z. W., Gorin, P. D., Silberstein, L., and Bennett, C. (1985). A postnatal change in the immunological properties of the acetylcholine receptor at rat muscle endplates. *Journal of Neuroscience*, **5**, 730–4.

Hallbook, F., Ibanez, C. F., and Persson, H. (1991). Evolutionary studies of the nerve growth factor family reveal a novel member abundantly expressed in *Xenopus* ovary. *Neuron*, **6**, 845–58.

Hama, T., Miyamoto, M., Tsukui, H., Nishino, C., and Hatanaka, H. (1989). Interleukin-6 as a neurotrophic factor for promoting the survival of cultured basal forebrain cholinergic neurones from postnatal rats. *Neuroscience Letters*, **104**, 340–4.

Hamburger, V. (1934). The effects of wing bud extirpation on the development of the central nervous system. *Journal of Experimental Zoology*, **68**, 449–94.

Hamburger, V. (1958). Regression versus peripheral control of differentiaiton in motor hypoplasia. *Journal of Anatomy*, **102**, 365–410.

Hamburger, V. (1975). Cell death in the development of the lateral motor column of the chick embryo. *Journal of Comparative Neurology*, **160**, 535–46.

Hamburger, V. and Levi-Montalcini, R. (1949). Proliferation, differentiation and degeneration in the spinal ganglia of the chick embryo under normal and experimental conditions. *Journal of Experimental Zoology*, **111**, 457–502.

Hamburger, V., Bruso-Bechtold, J. K., and Yip, J. W. (1981). Neuronal death in the spinal ganglia of the chick embryo and its reduction by nerve growth factor. *Journal of Neuroscience*, **1**, 60–71.

Hamill, O. P., Marty, A., Neher, E., Sakmann, B., and Sigworth, F. J. (1981). Improved patch-clamp techniques for high-resolution current recording from cells and cell-free membrane patches. *Pflügers Archiv*, **391**, 85–100.

Hamilton, B. R. and Smith, D. O. (1991). Autoreceptor-mediated purinergic and cholinergic inhibition of motor nerve terminal calcium currents in the rat. *Journal of Physiology (London)*, **432**, 327–41.

Hanley, M. R. (1988). Proto-oncogenes in the nervous system. *Neuron*, **1**, 175–82.

Hanley, M. R. and Jackson, T. (1987). Return of the magnificent seven. *Nature*, **329**, 766–7.

Harper, G. P. and Thoenen, H. (1981). Target cells, biological effects, and mechanism of action of nerve growth factor and its antibodies. *Annual Review of Pharmacology and Toxicology*, **21**, 205–29.

Harris, A. J. (1974). Inductive functions of the nervous system. *Annual Review of Physiology*, **36**, 251–305.

Harris, E. W., Ganong, A. H., and Cotman, C. W. (1984). Long-term potentitaion in the hippocampus involves activation of N-methyl-D-aspartate receptors. *Brain Research*, **323**, 132–7.

Harris, J. D. (1943). Habituatory response decrement in the intact organism. *Psychological Bulletin*, **40**, 385–422.

Hata, Y., Davletov, B., Petrenko, A. G., Jahn, R., and Südhof, T. C. (1993). Interaction of synaptotagmin with the cytoplasmic domains of neurexins. *Neuron*, **10**, 307–15.

Hawkins, R. D. (1989). Localization of potential serotonergic facilitator neurons in *Aplysia* by glyoxylic acid histofluorescence combined with retrograde fluorescent labeling. *Journal of Neuroscience*, **9**, 4214–26.

Hawkins, R. D. and Kandel, E. R. (1990). Hippocampal LTP and synaptic plasticity in *Aplysia*: possible relationship of associative cellular mechanisms. *Seminars in Neurosciences*, **2**, 391–401.

Hawkins, R. D. and Schacher, S. (1989). Identified facilitator neurones L29 and L28 are excited by cutaneous stimuli used in dishabituation, sensitization, and classical conditioning of *Aplysia*. *Journal of Neuroscience*, **9**, 4236–45.

Hawkins, R. D., Castellucci, V. F., and Kandel, E. R. (1981). Interneurones involved in mediation and modulation of gill-withdrawal reflex in *Aplysia*. II. Identified neurones produce heterosynaptic facilitation contributing to behavioural sensitization. *Journal of Neurophysiology*, **45**, 315–26.

Hawkins, R. D., Abrams, T. W., Carew, T. J., and Kandel, E. R. (1983). A cellular mechanism of classical conditioning in *Aplysia*: activity-dependent amplification of presynaptic facilitation. *Science*, **219**, 400–5.

He, X. and Rosenfeld, M. G. (1991). Mechanisms of complex transcriptional regulation: implications for brain development. *Neuron*, **7**, 183–96.

Hebb, D. O. (1949). *The organization of behaviour. A neuropsychological theory*. John Wiley and Sons, London.

Hefti, F. (1986). Nerve growth factor (NGF) promotes survival of septal cholinergic neurones after fimbrial transection. *Journal of Neuroscience*, **6**, 2155–62.

Hefti, F. and Weiner, W. J. (1986). Nerve growth factor and Alzheimer's disease. *Annals of Neurology*, **20**, 275–81.

Hefti, F., Denton, T. L., Knusel, B., and Lapechak, P. A. (1993). Neurotrophic factors: what are they and what are they doing? In *Neurotrophic factors* (ed. S. E. Loughlin and J. H. Fallon), pp. 25–49. Academic Press, San Diego.

Heidenhain, R. P. H. (1878). Uber sekretorische und trophische Drusen-nerven. *Pflügers Archiv*, **17**, 1–67.

Heidmann, T. and Changeux, J. P. (1980). Interaction of a fluorescent agonist with the membrane-bound acetylcholine receptor from *Torpedo marmorata* in the millisecond time range: resolution of an 'intermediate' conformational transition and evidence for positive cooperative effects. *Biochemical and Biophysical Research Communications*, **97**, 889–96.

Heinemann, S. H., Terlau, H., Stühmer, W., Imoto, K., and Numa, S. (1992). Calcium channel characteristics conferred on the sodium channel by single mutations. *Nature*, **356**, 441–3.

Held, H. (1897). Beiträge zur Struktur der Nervenzellen und ihren Fortsatze. *Archiv für Anatomie und Physiolgie*, Leipzig, pp. 204–294.

Helgren, M. E., Squinto, S. P., Davis, H. L., Parry, D. J., Boulton, T. G., Heck, C. S. *et al.* (1994). Trophic effects of ciliary neurotrophic factor on denervated skeletal muscle. *Cell*, **76**, 493–504.

Hempstead, B. L., Martin-Zanca, D., Kaplan, D. R., Parada, L. F., and Chao, M. V. (1991). High-affinity NGF binding requires coexpresssion of the *trk* protooncogene and the low-affinity NGF receptor. *Nature*, **350**, 678–83.

Henderson, C. E., Huchet, M., and Changeux, J. P. (1983). Denervation increases a neurite-promoting activity in extracts of skeletal muscle. *Nature*, **302**, 609–11.

Henderson, C. E., Camu, W., Mettling, C., Gouin, A., Poulsen, K., Karihaloo, M. *et al.* (1993). Neurotrophins promote motor neuron survival and are present in embryonic limb bud. *Nature*, **363**, 266–70.

Henderson, R. and Unwin, P. N. T. (1975). Three-dimensional model of purple membrane obtained by electron microscopy. *Nature*, 257, 28–32.

Hendry, I. A. (1975). The response of adrenergic neurones to axotomy and nerve growth factor. *Brain Research*, **94**, 87–97.

Hepler, J. R. and Gilman, A. G. (1992). G proteins. *Trends in Biochemical Sciences*, 17, 383–7.

Hestrin, S. (1992). Activation and desensitization of glutamate-activated channels mediating fast excitatory synaptic currents in the visual cortex. *Neuron*, **9**, 991–9.

Hestrin, S., Nicoll, R. A., Perkel, D. J., and Sah, P. (1990). Analysis of excitatory action in pyramidal cells using whole-cell recording from rat hippocampal clices. *Journal of Physiology (London)*, 422, 203–25.

Heumann, R., Korsching, S., Scott, J., and Thoenen, H. (1984). Relationship between levels of nerve growth factor (NGF) and its messenger RNA in sympathetic ganglia and peripheral target tissues. *EMBO Journal*. 3, 3183–9.

Heumann, R., Korsching, S., Bandtlow, C., and

Thoenen, H. (1987a). Changes of nerve growth factor synthesis in non-neuronal cells in response to sciatic nerve transection. *Journal of Cell Biology*, **104**, 1623–31.

Heumann, R., Lindholm, D., Bandtlow, C., Meyer, M., Radeke, M. J., Misko, T. P. *et al.* (1987b). Differential regulation of mRNA encoding nerve growth factor and its receptor in rat sciatic nerve during development, degeneration, and regeneration: Role of macrophages. *Proceedings of the National Academy of Sciences of the USA*, **84**, 8735–9.

Heuser, J. E. (1989). Review of electron microscopic evidence favouring vesicle exocytosis as the structural basis for quantal release during synaptic transmission. *Quarterly Journal of Experimental Physiology*, 74, 1051–69.

Heuser, J. E. and Reese, T. S. (1973). Evidence for recycling of synaptic vesicle membrane during transmitter release at the frog neuromuscular junction. *Journal of Cell Biology*, **57**, 315–44.

Heuser, J. E. and Reese, T. S. (1981). Structural changes after transmitter release at the frog neuromuscular junction. *Journal of Cell Biology*, **88**, 564–80.

Heuser, J. E., Reese, T. S., Dennis, M. J., Jan, Y., Jan, L. and Evans, L. (1979). Synaptic vesicle exocytosis captured by quick freezing and correlated with quantal transmitter release. *Journal of Cell Biology*, **81**, 275–300.

Hibi, M., Murakami, M., Saito, M., Hirano, T., Taga, T., Kishimoto, T. (1990). Molecular cloning and expression of an IL-6 signal transducer, gp130. *Cell*, **63**, 1149–57.

Hill, A. V. (1910). The possible effects of the aggregation of the molecules of haemoglobin on its dissociation curves. *Journal of Physiology (London)*, **40**, 4–7*P*.

Hill, D. R. and Bowery, N. G. (1981). ^{3}H-baclofen and ^{3}H-GABA bind to bicuculline-insensitive $GABA_B$ sites in rat brain. *Nature*, **290**, 149–52.

Hille, B. (1992a). *Ionic channels of excitable membranes*, (2nd edn). Sinauer Assoc. Inc., Sunderland.

Hille, B. (1992b). G protein-coupled mechanisms and nervous signaling. *Neuron*, **9**, 187–95.

Hilton, D. J. (1992). LIF: lots of interesting functions. *Trends in Neurosciences*, **17**, 72–6.

Hirano, T. (1990a). Synaptic transmission between rat inferior olivary neurones and cerebellar Purkinje cells in culture. *Journal of Neurophysiology*, **63**, 181–9.

Hirano, T. (1990b). Depression and potentiation of the synaptic transmission between a granule cell and a Purkinje cell in rat cerebellar culture. *Neuroscience Letters*, **119**, 141–4.

Hirano, T. (1990c). Effects of postsynaptic depolarization in the induction of synaptic depression between a granule cell and a Purkinje cell in rat cerebellar culture. *Neuroscience Letters*, **119**, 145–7.

Hirano, T. (1991). Differential pre-and postsynaptic mechanisms for synaptic potentiation and depression between a granule cell and a Purkinje cell in rat cerebellar culture. *Synapse*, 7, 321–3.

Hirano, T. and Hagiwara, S. (1988). Synaptic transmission between rat cerebellar granule and Purkinje cells in dissociated cell culture: effects of excitatory-amino acid transmitter antagonists. *Proceedings of the National Academy of Sciences of the USA*, **85**, 934–8.

Hirning, L. D., Fox, A. P., McCleskey, E. W., Olivera, B. M., Thayer, S. A., Miller, R. J., and Tsien, R. W. (1988). Dominant role of N-type Ca channels in evoked release of norepinephrine from sympathetic neurons. *Science*, **239**, 57–61.

Hirokawa, N., Sobue, K., Kanda, K., Harada, A., and Yorifuji, H. (1989). The cytoskeletal architecture of the presynaptic terminal and molecular structure of synapsin 1. *Journal of Cell Biology*, **108**, 111–26.

Hirst, G. D. S., Redman, S. J., and Wong, K. (1981). Post-tetanic potentiation and facilitation of synaptic potentials evoked in cat spinal motoneurones. *Journal of Physiology (London)*, **321**, 97–109.

Ho, K., Nichols, C. G., Lederer, W. J., Lytton, J., Vassilev, P. M., Kanazirska, M. V. *et al.* (1993). Cloning and expression of an inwardly rectifying ATP-regulated potassium channel. *Nature*, **362**, 31–38.

Hodgkin, A. L. and Huxley, A. F. (1952). A quantitative description of membrane current and its application to conduction and excitation in nerve. *Journal of Physiology (London)*, **117**, 500–44.

Hoeffler, J. P., Meyer, T. E., Yun, Y., Jameson, J. L., and Habener, J. F. (1988). Cyclic AMP-responsive element DNA-binding protein: structure based on a cloned placental cDNA. *Science*, **242**, 1430–3.

Hoffman, H. (1950). Local reinnervation in partially denervated muscle: a histophysiological study. *Australian Journal of Experimental Biology and Medical Science*. **28**, 383–97.

Hoffman, P. N., Cleveland, D. W., Griffin, J. W., Landes, P. W., Cowan, N. J., and Price, D. L. (1987). Neurofilament gene expression: A major determinant of axonal caliber. *Proceedings of the National Academy of Sciences of the USA*, **84**, 3472–6.

Hohn, A., Leibrock, J., Bailey, K., and Barde, Y. A. (1990). Identification and characterization of a novel member of the nerve growth factor/brain-derived neurotrophic factor family. *Nature*, **344**, 339–41.

Hökfelt, T. (1991). Neuropeptides in perspective: the last ten years. *Neuron*, 7, 867–79.

Hökfelt, T., Holets, V., Staines, W., Meister, B., Melander, T., Schalling, M. *et al.* (1986). Coexistence of neuronal messengers—an overview. *Progress in Brain Research* **68**, 33–70.

Hollmann, M., O'Shea-Greefield, A., Rogers, S. W., and Heinemann, S. (1989). Cloning by functional expression of a member of the glutamate receptor family. *Nature*, **342**, 643–8.

Hollmann, M., Hartley, M., and Heinemann, S. (1991). Ca^{2+} permeability of KA-AMPA-gated glutamate receptor channels depends on subunit composition. *Science*, **252**, 851–3.

Hollyday, M. and Hamburger, V. (1976). Reduction of the naturally occurring motor neuron loss by enlargement of the periphery. *Journal of Comparative Neurology*, **170**, 311–20.

Holzman, D. M., Li, Y., Parada, L. F., Kinsman, S., Chen, C. K., Valletta, J. S. *et al.* (1992). $P140^{trk}$ mRNA marks NGF-responsive forebrain neurones: evidence that *trk* gene expression is induced by NGF. *Neuron*, **9**, 465–78.

Honore, T., Davies, S. N., Drejer, J., Fletcher, E. J., Jacobsen, P., Lodge, D. *et al.* (1988). Quinoxalinediones: Potent competitive non-NMDA glutamate receptor antagnoists. *Science*, **241**, 701–3.

Hopp, T. P. and Woods, K. R. (1981). Prediction of protein antigenic determinants from amino acid sequences. *Proceedings of the National Academy of Sciences of the USA*, **78**, 3824–8.

Horn, A., Leibrock, J., Bailey, K., and Barde, Y. A. (1990). Identification and characterization of a novel member of the nerve growth factor/brain-derived neurotrophic factor family. *Nature*, **344**, 339–41.

Hory-Lee, F., Russell, M., Lindsay, R. M., and Frank, E. (1993). Neurotrophin 3 supports the sruvival of developing muscle sensory neurones in culture. *Proceedings of the National Academy of Science of the USA*, **90**, 2613–7.

Hosang, M. and Shooter, E. M. (1987). The internalization of nerve growth factor by high-affinity receptors of pheochromocytoma PC12 cells. *EMBO Journal*, **6**, 1197–202.

Hoshi, T., Rothlein, J., and Smith, S. J. (1984). Facilitation of Ca^{2+}-channel currents in bovine adrenal chromaffin cells. *Proceedings of the National Academy of Sciences of the USA*, **81**, 5871–5.

Houamed, K. M., Kuijper, J. L., Gilbert, T. L., Haldeman, B. A., O'Hara, P. J., Mulvihill, E. R. *et al.* (1991). Cloning, expression, and gene structure of a G protein-coupled glutamate receptor from rat brain. *Science*, **252**, 1318–21.

Huang, L. Y., Catterall, W. A., and Ehrenstein, G. (1978). Selectivity of cations and nonelectrolytes for acetylcholine-activated channels in cultured muscle cells. *Journal of General Physiology*, **71**, 397–410.

Huang, Y. Y., Colindo, A., Selig, D. K., and Malenka, R. C. (1992). The influence of prior synaptic activity on the induction of long-term potentiation. *Science*, **255**, 730–3.

Hubbard, J. I. and Schmidt, R. F. (1963). An electrophysiological investigation of mammalian motor nerve terminals. *Journal of Physiology (London)*, **181**, 810–29.

Hubel, D. H. and Wiesel, T. N. (1965). Binocular interaction in striate cortex of kittens reared with artificial squint. *Journal of Neurophysiology*, **28**: 1041–59.

Hucho, F., Layer, P., Kiefer, H., and Bandini, G. (1976). Photoaffinity labeling and quaternary structure of the acetylcholine receptor from *Torpedo californica*. *Proceedings of the Nattional Academy of Sciences of the USA*, **73**, 2624–8.

Hucho, F., Oberthur, W., and Lottspeich, F. (1986). The ion channel of the nicotinic acetylcholine receptor is formed by the homologous helices M II of the receptor subunits. *FEBS Letters*, **205**, 137–42.

Huganir, R. L. and Greengard, P. (1983). Cyclic AMP-dependent protein kinase phosphorylates the nicotinic acetylcholine receptor. *Proceedings of the National Academy of Sciences of the USA* **80**, 1130–4.

Huganir, R. L. and Greengard, P. (1990). Regulation of neurotransmitter receptor desensitization by protein phosphorylation. *Neuron*, **5**, 555–67.

Hughes, R. A., Sendtner, M., Goldfarb, M., Lindholm, D., and Thoenene, H. (1993). Evidence that fibroblast growth factor 5 is a major muscle-derived survival factor for cultured spinal motoneurons. *Neuron*, **10**, 369–77.

Huizar, P., Kuno, M., and Miyata, Y. (1975). Differentiation of motoneurones and skeletal muscles in kittens. *Journal of Physiology (London)*, **252**, 465–79.

Hume, R. I., Dingledine, R., and Heinemann, S. F. (1991). Identification of a site in glutamate receptor subunits that controls calcium permeability. *Science*, **253**, 1028–31.

Hunt, S. P., Pini, A., and Evan, G. (1987). Induction of c-*fos*-like protein in spinal cord neurons following sensory stimulation. *Nature*, **328**, 632–4.

Hyman, C., Hofer, M., Barde, Y. A., Juhasz, M., Yancopoulos, G. D., Squinto, S. P. *et al.* (1991). BDNF is a neurotrophic factor dopaminergic neurons of the substantia nigra. *Nature*, **350**, 230–232.

Ignarro, L. J. (1989). Heme-dependent activation of soluble guanylate cyclase by nitric oxide: regulation of enzyme activity by prophyrins and metalloprophyrins. *Seminars in Hematology*, **26**, 63–76.

Iino, M., Ozawa, S., and Tsuzuki, K. (1990). Permeation of calcium through excitatory amino acid receptor channels in cultured rat hippocampal neurones. *Journal of Physiology (London)*, 424, 151–65.

Illis, L. S. (1969). Enlargement of spinal cord synapses after repetitive stimulation of a single posterior root. *Nature*, **223**, 76–7.

Imoto, K., Methfessel, C., Sakmann, B., Mishina, M., Mori, Y., Konno, T. *et al.* (1986). Location of a delta-subunit region determining ion transport through the acetylcholine channel. *Nature*, **324**, 670–4.

Imoto, K., Busch, C., Sakmann, B., Mishina, M., Konno, T., Nakai, J. *et al.* (1988). Rings of negatively charged amino acids determine the acetylcholine receptor channel conducatnce. *Nature*, **335**, 645–8.

Imoto, K., Konno, T., Nakai, J., Wang, F., Mishina,

M., and Numa, S. (1991). A ring of uncharged polar amino acids as a component of channel constriction in the nicotinic acetylcholine receptor. *FEBS Letters*, **289**, 193–200.

Inaishi, Y., Kashihara, Y., Sakaguchi, M., Nawa, H., and Kuno, M. (1992). Cooperative regulation of calcitonin gene-related peptide levels in rat sensory neurons via their central and peripheral processes. *Journal of Neuroscience*, **12**, 518–24.

Inui, M., Saito, A., and Fleischer, S. (1987). Purification of the ryanodine receptor and identity with feet structure of junctional terminal cisternae of sarcoplasmic reticulum from fast skeletal muscle. *Journal of Biological Chemistry*, **262**, 1740–7.

Ip, N. Y., Ibanez, C. F., Nye, S. H., McClain, J., Jones, P. F., Gies, D. R. *et al.* (1992a). Mammalian neurotrophic-4: Structure, chromosomal localization, tissue distruction, and receptor specificity. *Proceedings of the National Academy of Sciences of the USA*, **89**, 3060–4.

Ip, N. Y., Nye, S. H., Boulton, T. G., Davis, S., Taga, T., Li, Y. *et al.* (1992b). CNTF and LIF on neuronal cells via shared signalling pathways that involve the IL-6 signal transducing receptor component gp 130. *Cell*, **69**, 1121–32.

Ip, N. Y., McClain, J., Barrezueta, N. X., Aldrich, T. H., Pan, L., Li, Y. *et al.* (1993a). The α-component of the CNTF receptor is required for signaling and defines potential CNTF targets in the adult and during development. *Neuron*, **10**, 89–102.

Ip, N. Y., Stitt, T. N., Tapley, P., Klein, R., Glass, D. J., Fandl, J. *et al.* (1993b). Similarities and differences in the way neurotrophins interact with the trk receptors in neuronal and nonneuronal cells. *Neuron*, **10**, 137–49.

Isacoff, E. Y., Jan, Y. N., and Jan, L. Y. (1990). Evidence for the formation of heteromultimeric potassium channels in *Xenopus* oocytes. *Nature*, **345**, 530–534.

Israel, M., Dunant, Y. and Manaranche, R. (1979). The present status of the vesicular hypothesis. *Progress in Neurobiology*, **13**, 237–75.

Ito, M. (1989). Long-term depression. *Annual Review of Neuroscience*, **12**, 85–102.

Ito, M. (1990). Long-term depression in the cerebellum. *Seminars in Neurosciences*, **2**, 381–90.

Ito, M., Sakurai, M., and Tongroach, P. (1982). Climing fibre induced depression of both mossy fibre responsiveness and glutamate sensitivity of cerebellar Purkinje cells. *Journal of Physiology (London)*, **324**: 113–34.

Izumi, Y., Clifford, D. B., and Zorumski, C. F. (1992). Inhibition of long-term potentiation by NMDA-mediated nitric oxide release. *Science*, **257**, 1273–6.

Jack, J. J. B., Redman, S. J., and Wong, K. (1981). The components of synaptic potentials evoked in cat spinal motoneurones by impulses in single group Ia afferents. *Journal of Physiology (London)*, **321**, 65–96.

Jahn, R., Schiebler, W., Ouimet, C., and Greengard, P. (1985). A 38,000–daltion membrane protein (p38) present in synaptic vesicles. *Proceedings of the National Academy of Sciences of the USA*, **82**, 4137–41.

Jahr, C. E. and Lester, A. J. (1992). Synaptic excitation mediated by glutamate-gated ion channels. *Current Opinion in Neurobiology*, **2**, 270–4.

Jain, J., McCaffrey, P. G., Valge-Archer, V. E., and Rao, A. (1992a). Nuclear factor of activated T cells contains Fos and Jun. *Nature*, **356**, 801–4.

Jain, J., Valge-Archer, V. E., and Rao, A. (1992b). Analysis of the AP-1 sites in the IL-2 promotor. *Journal of Immunology*, **148**, 1240–50.

James, W. (1890). *The principles of psychology*. Henry Holt, New York.

Jessell, T., Tsunoo, A., Kanazawa, Z., and Otsuka, M. (1979). Substance P: Depletion in the dorsal horn of rat spinal cord after section the peripheral processes of primary sensory neurons. *Brain Research*, **168**, 247–59.

Johnson, D., Lanahan, A., Buck, C. R., Sehgal, A., Morgan, C., Mercer, E. *et al.* (1986). Expression and structure of the human NGF receptor. *Cell*, **47**, 545–54.

Johnson, E. M. and Deckwerth, T. L. (1993). Molecular mechanisms of developmental neuronal death. *Annual Review of Neuroscience*, **16**, 31–46.

Johnson, E. M., Taniuchi, M., Clark, H. B., Springer, J. E., Koh, S., Tayrien, M. W. *et al.* (1987). Demonstration of the retrograde transport of nerve growth factor (NGF) receptor in the peripheral and central nervous system. *Journal of Neuroscience*, **7**, 923–9.

Johnson, E. M., Taniuchi, M., and DiStefano, P. S. (1988). Expression and possible function of nerve growth factor receptors on Schwann cells. *Trends in Neurosciences*, **11**, 299–304.

Johnson, E. M., Koike, T., and Franklin, J. (1992). A 'calcium set-point hypothesis' of neuronal dependence on neurotrophic factor. *Experimental Neurology*, **115**, 163–6.

Johnson, J. E., Barde, Y. A., Schwab, M., and Thoenen, H. (1986). Brain-derived neurotrophic factor supports the survival of cultured rat retinal ganglion cells. *Journal of Neuroscience*, **6**, 3031–8.

Johnson, L. K., Vlodavsky, I., Baxter, J. D., and Gospodarowicz, D. (1980). Nuclear accumulation of epidermal growth factor in cultured rat pituitary cells. *Nature*, **287**, 340–3.

Jones, K. A. and Baughman, R. W. (1991). Both NMDA and non-NMDA subtypes of glutamate receptors are concentrated at synapses on cerebral cortical neurons in culture. *Neuron*, **7**, 593–603.

Jones, K. R., Farinas, I., Backus, C., and Reichardt, L. F. (1994). Targeted disruption of the BDNF gene perturbes brain and sensory neuron development but not motor neuron development. *Cell*, **76**, 989–99.

Jones, R. and Vrbova, G. (1974). Two factors responsible

for the development of denervation hypersensitivity. *Journal of Physiology* (*London*), **236**, 517–38.

Jones, S. W. and Salpeter, M. M. (1983). Absence of [^{125}I]α-bungarotoxin binding to motor nerve terminals of frog, lizard and mouse. *Journal of Neuroscience*, **3**, 326–31.

Jorgensen, D. and Dyck, P. J. (1979). Axonal underdevelopment from axotomy in kittens. *Journal of Neuropathology and Experimental Neurology*, **38**, 571–8.

Julius, D. (1991). Molecular biology of serotonin receptors. *Annual Review of Neuroscience*, **14**, 335–60.

Kagan, B. L., Baldwin, R. L., Munoz, D., and Wisnieski, B. J. (1992). Formation of ion-permeable channels by tumor necrosis factor-α. *Science*, **255**, 1427–30.

Kanamori, M., Naka, M., Asano, M., and Hidaka, H. (1981). Effect of N-(6–aminohexyl)-5–chloro-l-naphtalenesulfonamide and other calmodulin antagonists (calmodulin-interacting agents) on calcium-induced contraction of rabbit aortic strips. *Journal of Pharmacology and Experimental Therapeutics*, **217**, 494–9.

Kandel, E. R. (1976). *Cellular basis of behaviour*. Freeman, San Francisco.

Kandel, E. R. (1979). *Behavioral biology of Aplysia*. Freeman, San Francisco.

Kandel, E. R. and Schwartz, J. H. (1982). Molecular biology of learning: modulation of transmitter release. *Science*, **218**, 433–43.

Kandel, E. R., Brunelli, M., Byrne, J., and Castellucci, V. (1976). A common presynaptic locus for the synaptic changes underlying short-term habituation and sensitization of the gill-withdrawal reflex in *Aplysia*. *Cold Spring Harbor Symposia on Quantitative Biology*, **40**, 465–82.

Kandel, E. R., Klein, M., Hochner, B., Shuster, M., Siegelbaum, S. A., Hawkins, R. D. *et al.* (1987). Synaptic modulation and learning: New insights into synaptic transmission from the study of behaviour. In *Synaptic function*, (ed. G. M. Edelman, W. E. Gall, and W. M. Cowan), pp.471–518. John Wiley, New York.

Kano, M. and Kato, M. (1987). Quisqualate receptors are specifically involved in cerebellar synaptic plasticity. *Nature*, **325**, 276–9.

Kano, M., Kato, M., and Chang, H. S. (1988). The glutamate receptor subtype mediating parallel fibre-Purkinje cell transmission in rabbit cerebellar cortex. *Neuroscience Research*, **5**, 325–37.

Kao, P. N., Dwork, A. J., Kaldany, R. R. J., Silver, M. L., Wideman, J., Stein, S. *et al.* (1984). Identification of the α-subunit half-cysteine specifically labelled by an affinity reagent for the acetylcholine receptor binding site. *Journal of Biological Chemistry*, **259**, 11662–5.

Kaplan, D. R., Hempstead, B. L., Martin-Zanca, D., Chao, M. V., and Parada, L. F. (1991). The *trk* proto-oncogene product: a signal transducing receptor for nerve growth factor. *Science*, **252**, 554–8.

Karlin, A. (1969). Chemical modification of the active site of the acetylcholine receptor. *Journal of General Physiology* **54**, 245s–64s.

Karlin, A. (1974). The acetylcholine receptor: Progress report. *Life Science*, **14**, 1385–415.

Karlin, A. (1980). Molecular properties of nicotinic acetylcholine receptors. In *The cell surface and neuronal functions* (ed. C. W. Cotman, G. Poste, and G. L. Nicolson) pp. 191–260, North-Holland, Amsterdam.

Kasai, H. and Aosaki, T. (1989). Modulation of Ca-channel current by an adenosine analog mediated by a GTP-binding protein in chick sensory neurons. *Pflügers Archiv*, **414**, 145–9.

Kashihara, Y., Kuno, M., and Miyata, Y. (1987). Cell death of axotomized motoneurones in neonatal rats, and its prevention by peripheral reinnervation. *Journal of Physiology* (*London*), **386**, 135–48.

Kashihara, Y., Sakaguchi, M., and Kuno, M. (1989). Axonal transport and distribution of endogenous calcitonin gene-related peptide in rat peripheral nerve. *Journal of Neuroscience*, **9**, 3796–802.

Kashihara, Y., Takasu, C., and Kuno, M. (1993). Messenger RNAs from chick muscle encode a motoneuronal survival-promoting factor. *Neuroscience Letters* **163**, 208–10.

Kato, N. (1993). Dependence of long-term depression on postsynaptic metabotropic glutamate receptors in visual cortex. *Proceedings of the National Academy of Sciences of the USA*, **90**, 3650–4.

Katz, B. (1962). The transmission of impulses from nerve to muscle and the subcellular unit of synaptic action. *Proceedings of the Royal Society B*, **155**, 455–77.

Katz, B. (1969). *The release of neural transmitter substances*. Charles C. Thomas, Springfield.

Katz, B. (1971). Quantal mechanism of neural transmitter release. *Science*, **173**, 123–126.

Katz, B. and Miledi, R. (1965a). The measurement of synaptic delay, and the time course of acetylcholine release at the neuromuscular junction. *Proceedings of the Royal Society B*, **161**, 483–95.

Katz, B. and Miledi, R. (1965b). The effect of calcium on acetylcholine release from motor terminals. *Proceedings of the Royal Society B*, **161**, 496–503.

Katz, B. and Miledi, R. (1965c). The quantal release of transmitter substance. In *Studies in physiology*, (ed. D. R. Curtis and A. K. McIntyre), pp. 118–25. Springer-Verlag, New York.

Katz, B. and Miledi, R. (1966). Input-output relation of a single synapse. *Nature*, **212**, 1242–1245.

Katz, B. and Miledi, R. (1967). A study of synaptic transmission in the absence of nerve impulses. *Journal of Physiology* (*London*), **192**, 407–36.

Katz, B. and Miledi, R. (1968). The role of calcium in neuromuscular facilitation. *Journal of Physiology* (*London*), **195**, 481–92.

Katz, B. and Miledi, R. (1969). Tetrodotoxin-resistant electrical activity in presynaptic terminals. *Journal of Physiology (London)*, 203, 459–87.

Katz, B. and Miledi, R. (1971). The effect of prolonged depolarization on synaptic transfer in the stellate ganglion of the squid. *Journal of Physiology (London)*, **216**, 503–12.

Katz, B. and Miledi, R. (1972). The statistical nature of the acetylcholine potential and its molecular components. *Journal of Physiology (London)*, **224**, 665–99.

Katz, B. and Miledi, R. (1977). Transmitter leakage from motor nerve endings. *Proceedings of the Royal Society B*, **196**, 59–72.

Katz, B. and Miledi, R. (1979). Estimates of quantal content during 'chemical potentiation' of transmitter release. *Proceedings of the Royal Society B*, **205**, 369–78.

Katz, B. and Miledi, R. (1981). Does the motor nerve impulse evoke 'non-quantal' transmitter release? *Proceedings of the Royal Society B*, **212**, 131–7.

Katz, B. and Thesleff, S. (1957). A study of the 'desensitization' produeced by acetylcholine at the motor end-plate. *Journal of Physiology (London)*, **138**, 63–80.

Kauer, J. A., Malenka, R. C., and Nicoll, R. A. (1988). A persistent postsynaptic modification mediates long-term potentiation in the hippocampus. *Neuron* 1, 911–7.

Kaupp, U. B., Niidome, T., Tanabe, T., Terada, S., Bonigk, W., Stühmer, W. *et al.* (1989). Primary structure and functional expression from complementary DNA of the rod photoreceptor cyclic GMP-gated channel. *Nature*, **342**, 762–6.

Kawamura, Y. and Dyck, P. J. (1981). Permanent axotomy by amputation results in loss of motor neurons in man. *Journal of Neuropathology and Experimetnal Neurology*, **40**, 658–66.

Keilhauer, G. and Schachner, M. (1985). Differential inhibition of neuron-neuron, neuron-astrocyte and astrocyte-astrocyte adhesion by L1, L2 and N-CAM antibodies. *Nature*, **316**, 728–30.

Keller, B. U., Hollmann, M., Heinemann, S., and Konnerth, A. (1992). Calcium influx through subunits GluR1/GluR3 of kaniante/AMPA receptor channels is regulated by cAMP dependent protein kinase. *EMBO Journal*, 11, 891–6.

Kelso, S. R., Ganong, A. H., and Brown, T. H. (1986). Hebbian synapses in hippocampus. *Proceedings of the National Academy of Sciences of the USA*, **83**, 5326–30.

Kelso, S. R., Nelson, T. E., and Leonard, J. P. (1992). Protein kinase C-mediated enhancement of NMDA currents by metabotropic glutamate receptors in *Xenopus* oocytes. *Journal of Physiology (London)*, **449**, 705–18.

Kennard, M. A. (1942). Cortical reorganization of motor function. Studies on series of monkeys of various ages from infancy to maturation. *Archives of Neurology and Psychology*, **48**, 227–40.

Kennedy, M. B. (1989). Regulation of neuroneal function by calcium. *Trends in Neurosciences*, **12**, 417–20.

Kerr, J. F. R. (1969). An electron-microscope study of liver cell necrosis due to heliotrine. *Journal of Pathology*, **97**, 557–62.

Kerr, J. F. R., Wyllie, A. H., and Currie, A. R. (1972). Apoptosis: a basic biological phenomenon with wide-ranging implications in tissue kinetics. *British Journal of Cancer*, **26**, 239–57.

Kerr, L. M. and Yoshikami, D. (1984). A venom peptide with a novel presynaptic blocking action. *Nature*, **308**, 282–4.

Kidokoro, Y., Anderson, M. J., and Gruener, R. (1980). Changes in synaptic potential properties during acetylcholine receptor accumulation and neurospecific interaction in *Xenopus* nerve-muscle cell culture. *Developmental Biology*, **78**, 464–83.

Kim, Y. I. and Neher, E. (1988). IgG from patients with Lambert-Eaton syndrome blocks voltage-dependent calcium channels. *Science*, **239**, 405–8.

Kimura, F., Tsumoto, T., Nishigori, A., and Yoshimura, Y. (1990). Long-term depression but not potentiation is induced in Ca^{2+}-chelated visual cortex neurons. *Neuroreport*, 1, 65–8.

Kimura, H., Okamoto, K., and Sakai, Y. (1985). Pharmacological evidence for L-asparatate as the neurotransmitter of cerebellar climbing fibres in the guinea pig. *Journal of Physiology (London)*, **365**, 103–19.

Kiss, J., Shooter, E. M., and Patel, A. J. (1993). A low-affinity nerve growth factor receptor antibody is internalized and retrogradely transported selectivley into cholinergic neurones of the rat basal forebrain. *Neuroscience*, **57**, 297–305.

Kistler, H. S., Hawkins, R. D., Koester, J., Steinbusch, H. W. M., Kandel, E. R., and Schwartz, J. H. (1985). Distribution of serotonin-immunoreactive cell bodies and processes in the abdominal ganglion of mature *Aplysia*. *Journal of Neuroscience*, 5, 72–80.

Kistler, J., Stroud, R. M., Klymkowsky, M. W., Lalancette, R. A., and Fairclough, R. H. (1982). Structure and function of an acetylcholine receptor. *Biophysical Journal*, 37, 371–83.

Klagsbrun, M. (1989). The fibroblast growth factor family: structural and biological properties. *Progress in Growth Factor Research*, 1, 207–35.

Klein, M. and Kandel, E. R. (1980). Mechanism of calcium current modulation underlying presynaptic facilitation and behavioural sensitization in *Aplysia*. *Proceedings of the National Academy of Sciences of the USA*, 77, 6912–6.

Klein, M., Camardo, J., and Kandel, E. R. (1982). Serotonin modulates a specific potassium current in the sensory neurons that show presynaptic facilitation in *Aplysia*. *Proceedings of the National Academy of Sciences of the USA*, 79, 5713–7.

Klein, R., Parada, L. F., Coulier, F., and Barbacid, M. (1989). *trk*B, a novel tyrosine protein kinase receptor expressed during mouse neural development. *EMBO Journal*, **8**, 8060–4.

Klein, R., Jing, S., Nanduri, V., O'Rourke, E., and Barbacid, M. (1991). The *trk* proto-oncogene encodes a receptor for nerve growth factor. *Cell*, **65**, 189–97.

Klein, R., Smeyne, R. J., Wurst, W., Long, L. K., Auerbach, B. A., Joyner, A. L., and Barbacid, M. (1993). Targeted disruption of the *trk*B neurotrophin receptor gene results in nervous system lesions and neonatal death. *Cell*, **75**, 113–22.

Klein, R., Silos-Santiago, I., Smeyne, R. J., Lira, A. S., Brambrilla, R., Bryant, S. *et al.* (1994). Disruption of the neurotrophin-3 receptor gene *trk*C eliminates Ia muscle afferents and results in abnormal movements. *Nature*, **368**, 249–51.

Knusel, B., Winslow, J. W., Rosenthal, A., Burton, L. E., Seid, D. P., Nikolics, K. *et al.* (1991). Promotion of central cholinergic and dopaminergic neurone differentiation by brain-derived neurotrophic factor but not neurotrophin-3. *Proceedings of the National Academy of Sciences of the USA*, **88**, 961–5.

Knusel, B., Klaus, D. B., Winslow, J. W., Rosenthal, A., Burton, L. E., Widmer, H. R. *et al.* (1992). Brain-derived neurotrophic factor administration protects basal forebrain cholinergic but not nigral dopaminergic neurons from degenerative changes after axotgomy in the adult rat brain. *Journal of Neuroscience*, **12**, 4391–402.

Koerber, H. R., Druzinsky, R. E., and Mendell, L. M. (1988). Properties of somata of spinal dorsal root ganglion cells differ according to peripheral receptor innervated. *Journal of Neurophysiology*, **60**, 1584–96.

Koike, T. and Tanaka, S. (1991). Evidence that nerve growth factor dependence of sympathetic neurons for survival *in vitro* may be determined by levels of cytoplasmic free Ca^{2+}. *Proceedings of the National Academy of Sciences of the USA*, **88**, 3892–6.

Koike, T., Martin, D. P., and Johnson, E. M. (1989). Role of Ca^{2+} channels in the ability of membrane depolarization to prevent neuronal death induced by trophic factor-deprivation: evidence that levels of internal Ca^{2+} determine nerve growth factor dependence of sympathetic ganglion cells. *Proceedings of the National Academy of Sciences of the USA*, **86**, 6421–5.

Kokaia, Z., Bengzon, J., Metsis, M., Kokaia, M., Persson, H., and Lindvall, O. (1993). Coexpression of neurotrophins and their receptors in neurons of the central nervous system. *Proceedings of the National Academy of Sciences of the USA*, **90**, 6711–5.

Koliatsos, V. E., Clatterbuck, R. E., Winslow, J. W., Cayouette, M. H., and Price, D. L. (1993). Evidence that brain-derived neurotrophic factor is a trophic factor for motor neuron *in vivo*. *Neuron*, **10**, 359–67.

Konishi, Y., Chui, D. H., Hirose, H., Kunishita, T., and Tabira, T. (1993). Trophic effect of erythropoietin and other hematopoietic factors on central cholinergic neurons in vitro and in vivo. *Brain Research*, **609**, 29–35.

Konorski, J. (1948). *Conditioned reflexes and neurone organization*. University Press, Cambridge.

Korn, H. and Faber, D. S. (1975). An electrically mediated inhibition in goldfish medulla. *Journal of Neurophysiology*, **38**, 452–71.

Korn, H. and Faber, D. S. (1987). Regulation and significance of probabilistic release mechanisms at central synapses. In *Synaptic function*, (ed. G. M. Edelman, W. E. Gall, and W. M. Cowan), pp. 57–108, John Wiley, New York.

Korn, H. and Faber, D. S. (1991). Quantal analysis and synaptic efficacy in the CNS. *Trends in Neurosciences*, **14**, 439–45.

Korn, H., Triller, A., Mallet, A., and Faber, D. S. (1981). Fluctuating responses at a central synapse: *n* of binomial fit predicts number of stained presynaptic boutons. *Science*, **213**, 898–901.

Korn, H., Mallet, A., Triller, A., and Faber, D. S. (1982). Transmission at a central synapse. II. Quantal description of release with a physical correlate for binomial *n*. *Journal of Neurophysiology*, **48**, 679–707.

Korn, H., Burnod, Y., and Faber, D. S. (1987). Spontaneous quantal currents in central neuron match predictions from binomial analysis of evoked responses. *Proceedings of the National Academy of Sciences in the USA*, **84**, 5981–5.

Korsching, S. (1986). The role of nerve growth factor in the CNS. *Trends in Neurosciences*, **9**, 570–3.

Korsching, S. (1993). The neurotrophic factor concept: reexamination. *Journal of Neuroscience*, **13**, 2739–48.

Korsching, S. and Thoenen, H. (1983a). Nerve growth factor in sympathetic ganglia and corresponding target organs of the rat: correlation with density of sympathetic innervation. *Proceedings of the National Academy of Sciences of the USA*, **80**, 3513–6.

Korsching, S. and Thoenen, H. (1983b). Quantitative demonstration of the retrograde axonal transport of endogenous nerve growth factor. *Neuroscience Letters*, **39**, 1–4.

Korsching, S. and Thoenen, H. (1985). Nerve growth factor supply for sensory neurons: site of origin and competition with the sympathetic nervous system. *Neuroscience Letters*, **54**, 201–5.

Korsching, S., Auburger G., Heumann, R., Scott, J., and Thoenen, H. (1985). Levels of nerve growth factor and its mRNA in the central nervous system of the rat correlate with cholinergic innervation. *EMBO Journal*, **4**, 1389–93.

Koshland, D. E., Nemethy, G., and Filmer, D. (1966). Comparison of experimental binding data and theoretical models in proteins containing subunits. *Biochemistry*, 5, 365–85.

Koyano, K., Kuba, K., and Minota, S. (1985). Long-term potentiation of transmitter release induced by repetitive presynaptic activities in bull-frog sympathetic ganglia. *Journal of Physiology (London)*, 359, 219–33.

Kozak, W. and Westerman, R. A. (1961). Plastic changes of monosynaptic responses from tenotomized muscles in cats. *Nature*, **189**, 753–5.

Krnjević, K. (1974). Chemical nature of synaptic transmission in vertebrates. *Physiological Reviews*, 54, 418–540.

Kruger, L., Silverman, J. D., Mantyh, P. W., Sternini, C., and Brecha, N. C. (1989). Peripheral patterns of calcitonin gene-related peptide general somatic sensory innervation: cutaneous and deep terminations. *Journal of Comparative Neurology*, **280**, 291–302.

Krupinski, J., Coussen, F., Bakalyar, H. A., Tang, W. J., Feinstein, P. G., Orth, K. *et al.* (1989). Adenylyl cyclase amino acid sequence: possible channel- or transporter-like structure. *Science*, 244, 1558–64.

Kuba, K. and Koketsu, K. (1978). Synaptic events in sympathetic ganglia. *Progress in Neurobiology*, 11, 77–169.

Kuba, K. and Kumamoto, E. (1990). Long-term potentiation in vertebrate synapses: A variety of cascades with common subprocesses. *Progress in Neurobiology*, 34, 197–269.

Kubo, T., Bujo, H., Akida, I., Nakai, J., Mishina, M., and Numa, S. (1988). Location of a region of the muscarinic acetylcholine receptor involved in selective effector coupling. *FEBS Letters*, 241, 119–25.

Kubo, Y., Baldwin, T. J., Jan, Y. N., and Jan, L. Y. (1993a). Primary structure and functional expression of a mouse inward rectifier potassium channel. *Nature*, 362, 127–33.

Kubo, Y., Reuveny, E., Slesinger, P. A., Jan, Y. N., and Jan, L. Y. (1993b). Primary structure and functional expression of a rat G-protein-coupled mucarinic potasium channel. *Nature*, **364**, 802–6.

Kuffler, S. W. (1943). Specific excitability of the endplate region in normal and denervated muscle. *Journal of Neurophysiology*, **6**, 99–110.

Kuffler, S. W. (1980). Slow synaptic responses in autonomic ganglia and the pursuit of a peptide transmitter. *Journal of Experimental Biology*, **89**, 257–86.

Kuffler, S. W. and Yoshikami, D. (1975a). The distribution of acetylcholine sensitivity at the post-synaptic membrane of vertebrate skeletal twitch muscles: ionophoretic mapping in the micron range. *Journal of Physiology (London)*, 244, 703–30.

Kuffler, S. W. and Yoshikami, D. (1975b). The number of transmitter molecules in a quantum: an estimate from iontophoretic application of acetylcholine at the neuromuscular synapse. *Journal of Physiology (London)*, 251, 465–82.

Kullmann, D. M. and Nicoll, R. A. (1992). Long-term potentiation is associated with increases in quantal content and quantal amplitude. *Nature*, 357, 240–4.

Kuno, M. (1964a). Quantal components of excitatory synaptic potentials in spinal motoneurones. *Journal of Physiology (London)*, 175, 81–99.

Kuno, M. (1964b). Mechanisms of facilitation and depression of the excitatory synaptic potential in spinal motoneurones. *Journal of Physiology (London)*, 175, 100–12.

Kuno, M. (1971). Quantum aspects of central and ganglionic synaptic transmission in vertebrates. *Physiological Reviews*, 51, 647–78.

Kuno, M. (1990). Target dependence of motoneuronal survival: the current status. *Neuroscience Research*, 9, 155–72.

Kuno, M. and Miyahara, J. T. (1969). Non-linear summation of unit synaptic potentials in spinal motoneurones. *Journal of Physiology (London)*, 201, 465–77.

Kuno, M. and Weakly, J. N. (1972). Quantal components of the inhibitory synaptic potential in spinal motoneurones of the cat. *Journal of Physiology (London)*, 224, 287–303.

Kuno, M., Turkanis, S. A., and Weakly, J. N. (1971). Correlation between nerve terminal size and transmitter release at the neuromuscular junction of the frog. *Journal of Physiology (London)*, 213, 545–56.

Kuno, M., Miyata, Y., and Munoz-Martinez, E. J. (1974a). Differential reaction of fast and slow α-motoneurones to axotomy. *Journal of Physiology (London)*, 240, 725–39.

Kuno, M., Miyata, Y., and Munoz-Martinez, E. J. (1974b). Properties of fast and slow α-motoneurones following motor reinnervation. *Journal of Physiology (London)*, 242, 273–88.

Kurahashi, T. (1989). Activation by odorants of cation-selective conductance in the olfactory receptor cell isolated from the newt. *Journal of Physiology (London)*, 419, 177–92.

Kure, S., Tominaga, T., Yoshimoto, T., Tada, K., and Narisawa, K. (1991). Glutamate triggers internucleosomal DNA cleavage in neuronal cells. *Biochemical and Biophysical Research Communications*, 179, 39–45.

Kutsuwada, T., Kashiwabuchi, N., Mori, H., Sakimura, K., Kushiya, E., Araki, K. *et al.* (1992). Molecular diversity of the NMDA receptor channel. *Nature*, 358, 36–41.

Kyte, J. and Doolittle, R. F. (1982). A simple method for displaying the hydropathic character of a protein. *Journal of Molecular Biology*, 157, 105–32.

Lai, F. A., Erickson, H. P., Rousseau, E., Liu, Q. Y., and Meissner, G. (1988). Purification and reconstitution of the calcium release channel from skeletal muscle. *Nature*, 331, 315–9.

Lam, A., Kloss, J., Fuller, F., Cordell, B., and Ponte, P. A. (1992). Expression cloning of neurotrophic factors using *Xenopus* oocytes. *Journal of Neuroscience Research*, 32, 43–50.

Lamballe, F., Klein, R., and Barbacid, M. (1991). *trk*C, a new member of the *trk* family of tyrosine protein kinases, is a receptor for neurotrophin-3. *Cell*, **66**, 967–79.

Lambert, E. H. and Elmqvist, D. (1971). Quantal components of endplate potentials in the myasthenic syndrome. *Annals of the New York Academy of Sciences*, **183**, 183–99.

Lancet, D. (1986). Vertebrate olfactory reception. *Annual Review of Neuroscience*, **9**, 329–55.

Landis, D. M. D., Hall, A. K., Weinstein, L. A., and Reese, T. S. (1988). The organization of cytoplasm at the presynaptic active zone of a central nervous system synapse. *Neuron*, **1**, 201–9.

Landis, S. C. (1990). Target regulation of neurotransmitter phenotype. *Trends in Neurosciences*, **13**, 344–50.

Landis, S. C. and Keefe, D. (1983). Evidence for neurotransmitter plasticity *in vivo*: developmental changes in properties of cholinergic sympathetic neurons. *Developmental Biology*, **98**, 349–72.

Landmesser, L. and Pilar, G. (1974). Synapse formation during embryogenesis on ganglion cells lacking a periphery. *Journal of Physiology (London)*, **241**, 715–36.

Landmesser, L., Dahm, L., Schultz, K., and Rutishauser, U. (1988). Distinct roles for adhesion molecules during innervation of embryonic chick muscle. *Developmental Biology*, **130**, 645–70.

Landmesser, L., Dahm, L., Tang, J., and Rutishauser, U. (1990). Polysialic acid as a regulator of intramuscular nerve branching during embryonic development. *Neuron*, **4**, 655–67.

Landschulz, W. H., Johnson, P. F., and McKnight, S. L. (1988). The leucine zipper: a hypothetical structure common to a new class of DNA binding proteins. *Science*, **240**, 1759–64.

Lang, B., Newsom-Davis, J., Prior, C., and Wray, D. (1983). Antibodies to motor nerve terminals: an electrophysiological study of a human myasthenic syndrome transferred to mouse. *Journal of Physiology (London)*, **344**, 335–45.

Langan, T. A. (1969). Histone phosphorylation: stimulation by adenosine 3′,5′-monophosphate. *Science*, **162**, 579–80.

Langley, J. N. (1896). On the nerve cell connection of the splanchnic nerve fibres. *Journal of Physiology (London)*, **20**, 223–46.

Langley, J. N. (1909). On the contraction of muscle, chiefly in relation to the presence of 'receptive' substances. Part IV. The effect of curari and some other substances on the nicotine response of the sartorius and gastrocenmius muscles of the frog. *Journal of Physiology (London)*, **39**, 235–95.

Langosch, D., Thomas, L., and Betz, H. (1988). Conserved quaternary structure of ligand-gated ion channels: The postsynaptic glycine receptor is a pentamar. *Proceedings of the National Academy of Sciences of the USA*, **85**, 7394–8.

Langosch, D., Becker, C. M., and Betz, H. (1990). The inhibitory glycine receptor: a ligand-gated chloride channel of the central nervous system. *European Journal of Biochemistry*, **194**, 1–8.

Larkman, A., Hannay, T., Stratford, K., and Jack, J. (1992). Presynaptic release probability influences the locus of long-term potentiation. *Nature*, **360**, 70–3.

Larrabee, M. G. and Bronk, D. W. (1947). Prolonged facilitation of synaptic excitation in sympathetic ganglia. *Journal of Neurophysiology*, **10**, 139–54.

Lau, L. F. and Nathans, D. (1985). Identification of a set of genes expressed during the Go/G1 transition of cultured mouse cells. *EMBO Journal*, **4**, 3145–51.

Laufer, R. and Changeux, J. P. (1987). Calcitonin gene-related peptide elevates cyclic AMP levels in chick skeletal muscle: possible neurotrophic role for a coexisting neuronal messenger. *EMBO Journal*, **6**, 901–6.

LaVelle, A. and LaVelle, F. W. (1958). The nucleolar apparatus and neuronal reactivity to injury during development. *Journal of Experimental Zoology*, **137**, 285–316.

Lavoie, P. A., Collier, B., and Tenenhouse, A. (1976). Comparison of α-bungarotoxin binding to skeletal muscles after inactivity or denervation. *Nature*, **260**, 349–50.

Layer, P. G. and Shooter, E. M. (1983). Binding and degradation of nerve growth factor by PC12 pheochormocytoma cells. *Journal of Biological Chemistry*, **258**, 3012–8.

Lear, J. D., Wasserman, Z. R., and DeGrado, W. F. (1988). Synthetic amphiphilic peptide models for protein ion channels. *Science*, **240**, 1177–81.

Lee, C. Y. (1973). Chemistry and pharmacology of purified toxins from elapid and sea snake venoms. *Proceedings of 5th International Congress of Pharmacolgoy, San Francisco* pp. 210–32.

Lee, K. F., Li, E., Huber, L. J., Landis, S. C., Sharpe, A. H., Chao, M. V., and Jaenisch, R. (1992). Targeted mutation of the gene encoding the low affinity NGF receptor p75 leads to deficits in the peripheral sensory nervous system. *Cell*, **69**, 737–49.

Lee, K. S., Schottler, F., Oliver, M., and Lynch, G. (1980). Brief bursts of high-frequency stimulation produce two types of structural change in rat hippocampus. *Journal of Neurophysiology*, **44**, 247–58.

Lefkowitz, R. J. and Caron, M. G. (1988). Adrenergic receptors. *Journal of Biological Chemistry*, **263**, 4993–6.

Leibrock, J., Lottspeich, F., Hohn, A., Hofer, M., Hengerer, B., Masiakowski, P. *et al.* (1989). Molecular cloning and expression of brain-derived neurotrophic factor. *Nature*, **341**, 149–52.

Leonard, J. P., Nargeot, J., Snutch, T. P., Davidson, N., and Lester, H. A. (1987). Ca channels induced in *Xenopus* oocytes by rat brain mRNA. *Journal of Neuroscience*, **7**, 875–81.

Leonard, R. J., Labarca, C. G., Charnet, P., Davidson,

N., and Lester, H. A. (1988). Evidence that the M_2 membrane-spanning region lines the ion channel pore of the nicotinic receptor. *Science*, 242, 1578–81.

Lester, R. and Jahr, C. (1990). Quisqualate receptor-mediated depression of calcium currents in hippocampal neurons. *Neuron*, 4, 741–9.

Letinsky, M. K., Fischbeck, K. H., and McMahan, U. J. (1976). Precision of reinnervation of original postsynaptic sites in muscle after a nerve crush. *Journal of Neurocytology*, 5, 691–718.

Letourneau, P. C. (1975). Cell-to-substratum adhesion and guidance of axonal elongation. *Developmental Biology*, 44, 92–101.

Letourneau, P. C. (1983). Axonal growth and guidance. *Trends in Neurosciences*, 6, 451–5.

Leube, R. E., Kaiser, P., Seiter, A., Zimbelmann, R., Franke, W. W., Rehm, H. *et al.* (1987). Synaptophysin: molecular organization and mRNA expression as determined from cloned cDNA. *EMBO Journal*, 6, 3261–8.

Leung, D. W., Parent, A. S., Cachianes, G., Esch, F., Coulombe, J. N., Nikolics, K. *et al.* (1992). Cloning, expression during development, and evidence for release of a trophic factor for ciliary ganglion neurons. *Neuron*, 8, 1045–53.

Leveque, C., Hoshino, T., David, P., Shoji-Kasai, Y., Leys, K., Omori, A. *et al.* (1992). The synaptic vesicle protein synaptotagmin associates with calcium channels and is a putative Lambert-Eaton myasthenic syndrome antigen. *Proceedings of the National Academy of Sciences of the USA*, 89, 3625–9.

Levi-Montalcini, R. (1975). NGF: an uncharted route. In *The neurosciences, paths of discovery*, (ed. F. G. Worden, J. P. Swazey, and G. Adelman), pp. 244–265. Massachusetts Institute of Technology Press, Cambridge, Mass.

Levi-Montalcini, R. (1982). Developmental neurobiology and the natural history of nerve growth factor. *Annual Review of Neuroscience*, 5, 341–62.

Levi-Montalcini, R. (1987). The nerve growth factor 35 years later. *Science*, 237, 1154–62.

Levi-Montalcini, R. and Angeletti, P. U. (1968). Nerve growth factor. *Physiological Reviews*, 48, 534–65.

Levi-Montalcini, R. and Hamburger, V. (1951). Selective growth stimulating effects of mouse sarcoma on the sensory and sympathetic nervous system of the chick embryo. *Journal of Experimental Zoology*, 116, 321–61.

Levi-Montalcini, R. and Hamburger, V. (1953). A diffusible agent of mouse sarcoma, producing hyperplasia of sympathetic ganglia and hyperneurotization of viscera in the chick embryo. *Journal of Experimental Zoology*, 123, 233–89.

Levi-Montalcini, R., Meyer, H., and Hamburger, V. (1954). *In vitro* experiments on the effects of mouse sarcomas 180 and 37 on the spinal and sympathetic ganglia of the chick embryo. *Cancer Research*, 14, 49–57.

Levitan, E. S., Blair, L. A. C., Dionne, V. E., and Barnard, E. A. (1988a). Biophysical and pharmacological properties of cloned GABA-A receptor subunits expressed in *Xenopus* oocytes. *Neuron*, 1, 773–81.

Levitan, E. S., Schofield, P. R., Burt, D. R., Rhee, L. M., Wisden, W., Kohler, M. *et al.* (1988b). Structural and functional basis for GABA-A receptor heterogeneity. *Nature*, 335, 76–9.

Levitt, T. A., Loring, R. H., and Salpeter, M. M. (1980). Neuronal control of acetylcholine receptor turnover rate at a vertebrate neuromuscular junction. *Science*, 210, 550–1.

Lewis, D. M., Bagust, J., Webb, S. N., Westerman, R. A., and Finol, H. J. (1977). Axon conduction velocity modified by reinnervation of mammalian muscle. *Nature*, 270, 745–6.

Lewis, M. E., Neff, N. T., Contreras, P. C., Strong, D. B., Oppenheim, R. W., Grelow, P. E. *et al.* (1993). Insulin-like growth factor-l: potential for treatment of motor neuronal disorders. *Experimental Neurology*, 124, 73–88.

Lieberman, A. R. (1971). The axon reaction: a review of the principal features of perikaryal responses to axon injury. *International Review of Neurobiology*, 14, 49–124.

Liley, A. W. (1956). The quantal components of the mammalian end-plate potential. *Journal of Physiology (London)*, 133, 571–87.

Lim, N. F., Nowycky, M. C., and Bookman, R. J. (1990). Direct measurement of exocytosis and calcium currents in single vertebrate nerve terminals. *Nature*, 344, 449–51.

Lin, J. W., Rudy, B., and Llinás, R. (1990a). Funnel-web spider venom and a toxin fraction block calcium current expressed from rat brain mRNA in *Xenopus* oocytes. *Proceedings of the National Academy of Sciences of the USA*, 87, 4538–4542.

Lin, J. W., Sugimori, M., Llinás, R., McGuinness, T. L., and Greengard, P. (1990b). Effects of synapsin I and calcium/calmodulin-dependent protein kinase II on spontaneous neurotransmitter release in the squid giant synapse. *Proceedings of the National Academy of Sciences of the USA*, 87, 8257–61.

Lin, L. F. H., Mismer, D., Lile, J. D., Armes, L. G., Butler, E. T., Vannice, J. L. *et al.* (1989). Purification, cloning, and expression of ciliary neurotrophic factor (CNTF). *Science*, 246, 1023–5.

Linden, D. J. (1994). Long-term synaptic depression in the mammalian brain. *Neuron*, 12, 457–72.

Linden, D. J. and Connor, J. A. (1992). Long-term depression of glutamate currents in cultured cerebellar Purkinje neurons does not require nitric oxide signalling. *European Journal of Neuroscience*, 4, 10–5.

Linden, D. J., Dickinson, M. H., Smeyne, M., and Connor, J. A. (1991). A long-term depression of AMPA currents in cultured cerebellar Purkinje neurons. *Neuron*, 7, 81–9.

Lindholm, D., Heumann, R., Meyer, M., and Thoenen, H. (1987). Interleukin-1 regulates synthesis of nerve growth factor in non-neuronal cells of rat sciatic nerve. *Nature*, **330**, 658–9.

Lindsay, R. M. (1988). Nerve growth factors (NGF, BDNF) enhance axonal regeneration but are not required for survival of adult sensory neurons. *Journal of Neuroscience*, **8**, 2394–405.

Lindsay, R. M. and Harmer, A. J. (1989). Nerve growth factor regulates expression of neuropeptides in adult sensory neurons. *Nature*, **337**, 362–4.

Lindsay, R. M., Thoenen, H., and Barde, Y. A. (1985). Placode and neural crest-derived sensory neurons are responsive at early developmental stages to brain-derived neurotrophic factor. *Developmental Biology*, **112**, 319–28.

Lindsay, R. M., Lockett, C., Sternberg, J., and Winter, J. (1989). Neuropeptide expression in cultures of adult sensory neurons: modulation of substance P and calcitonin gene-related peptide levels by nerve growth factor. *Neuroscience*, **33**, 53–65.

Lisman, J. E. (1985). A mechanism for memory storage insensitive to molecular turnover: a bistable autphosphorylating kinase. *Proceedings of the National Academy of Sciences of the USA*, **82**, 3055–7.

Lisman, J. E. (1989). A mechanism for the Hebb and anti-Hebb processes underlying learning and memory. *Proceedings of the National Academy of Sciences of the USA*, **86**, 9574–8.

Lisman, J. E. and Goldring, M. A. (1988). Feasibility of long-term storage of graded information by the Ca^{2+}/calmodulin-dependent protein kinase molecules of the postsynaptic density. *Proceedings of the National Academy of Sciences of the USA*, **85**, 5320–4.

Lisman, J. E. and Goldring, M. A. (1989). Evaluation of a model of long-erm memory based on the properties of the Ca^{2+}/calmodulin-dependent protein kinase. *Journal de Physiologie (Paris)*, **83**, 187–97.

Liu, C. N. and Chambers, W. W. (1958). Intraspinal sprouting of dorsal root axons. *Archives of Neurology and Psychiatry*. **79**, 46–61.

Livingstone, M. S., Schaeffer, S. F., and Kravitz, E. A. (1981). Biochemistry and ultrastructure of serotonergic nerve endings in the lobster; serotonin and octopamine are contained in different nerve endings. *Journal of Neurobiology*, **12**, 27–54.

Llinás, R. and Sugimori, M. (1980). Electrophysiological properties of *in vitro* Purkinje cell dendrites in mammalian cerebellar slices. *Journal of Physiology* (*London*), **305**, 197–213.

Llinás, R. and Yarom, Y. (1981). Properties of distribution of ionic conductances generating electroresponsiveness of mammalian inferior olivary neurones *in vitro*. *Journal of Physiology* (*London*), **315**, 569–84.

Llinás, R., Blinks, J. R., and Nicholson, C. (1972). Calcium transient in presynaptic terminals of squid giant synapse: detection with aequorin. *Science*, **176**, 1127–9.

Llinás, R., Steinberg, I. Z., and Walton, K. (1981). Presynaptic calcium currents in squid giant synapse. *Biophysical Journal*, **33**, 289–322.

Llinás, R., Sugimori, M., and Simon, S. M. (1982). Transmission by presynaptic spike-like depolarization in the squid giant synapse. *Proceedings of the National Academy of Sciences of the USA*, **79**, 2415–9.

Llinás, R., McGuinness, T. L., Leonard, C. S., Sugimori, M., and Greengard, P. (1985). Intraterminal injection of synapsin I or calcium/calmodulin-dependent protein kinase II alters neurotransmitter release at the squid giant synapse. *Proceedings of the National Academy of Sciences of the USA*, **82**, 3035–9.

Llinás, R., Sugimori, M., Lin, J. W., and Cherksey, B. (1989). Blocking and isolation of a calcium channel from neurons in mammalian and cephalopods utilizing a toxin fraction (FTX) from funnel-web spider poison. *Proceedings of the National Academy of Sciences of the USA*, **86**, 1689–93.

Llinás, R., Sugimori, M., and Silver, R. B. (1992). Microdomains of high calcium concentration in a presynaptic terminal. *Science*, **256**, 677–9.

Lloyd, D. P. C. (1949). Post-tetanic potentiation of response in monosynaptic reflex pathways of the spinal cord. *Journal of General Physiology*, **33**, 147–70.

Lomo T. (1966). Frequency potentiation of excitatory synaptic activity in the dentate area of the hippocampal formation. *Acta Physiologica Scandinavica*, **68** (Supplement 27), 128.

Lomo, T. and Rosenthal, J. (1972). Control of ACh sensitivity by muscle activity in the rat. *Journal of Physiology* (*London*), **221**, 493–513.

Loo, D. T., Copani, A., Pike, C. J., Whittemore, E. R., Walencewicz, A. J., and Cotman, C. W. (1993). Apoptosis is induced by β-amyloid in cultured central nervous system neurons. *Proceedings of the National Academy of Sciences of the USA*, **90**, 7951–5.

Lorente de No, R. (1934). Studies on the structure of the cerebral cortex. II. Continuation of the study of the ammonic system. *Journal of Psychololog and Neurology*, **46**, 113–77.

Lovinger, D. M., Wong, K. L., Murakami, K., and Routtenberg, A. (1987). Protein kinase C inhibitors eliminate hippocampal long-term potentiation. *Brain Research*, **436**, 177–83.

Lowrie, M. B. and Vrbova, G. (1992). Dependence of postnatal motoneurones on their targets: review and hypothesis. *Trends in Neurosciences*, **15**, 80–4.

Lundberg, J. and Hökfelt, T. (1983). Coexistence of peptides and classical neurotransmitters. *Trends in Neurosciences*, **6**, 325–33.

Lundberg, J. M., Hökfelt, T., Anggard, A., Terenius, L., Elder, R., Marker, K. *et al.* (1982). Organizational principles in the peripheral sympathetic nervous system: Subdivision by coexisting peptides (somatostatin-, avian pancreatic polypeptide-, and vasoactive polypeptide-like immunoreactive materials). *Proceedings of the National Academy of Sciences of the USA*, **79**, 1303–7.

Luther, P. W., Yip, R. K., Bloch, R. J., Ambes, A., Lindenmayer, G. E., and Blaustein, M. P. (1992). Presynaptic localization of sodium/calcium exchangers in neuromuscular preparations. *Journal of Neuroscience*, **12**, 4898–904.

Lynch, G. S., Dunwiddie, T., and Gribkoff, V. (1977). Heterosynaptic depression: a postsynaptic correlate of long-term potentiation. *Nature*, **266**, 736–7.

Lynch, G., Larson, J., Kelso, S., Barrionuevo, G., and Schottler, F. (1983). Intracellular injections of EGTA block induction of hippocampal long-term potentiation. *Nature*, **305**, 719–21.

MacDonald, H. R. and Nabholz, M. (1986). T-cell activation. *Annual Review of Cell Biology*, **2**, 231–53.

Maeno, T., Edwards, C., and Anraku, M. (1977). Permeability of the endplate membrane activated by acetylcholine to some organic cations. *Journal of Neurobiology*, **8**, 173–84.

Magendie, M. (1824). De l'influence de la cinquième paire de nerfs sur la nutrition et les fonctions de l'oeil. *Journal de Phsyiologie Experimentale et Pathologique*, **4**, 176–82.

Magill-Solc, C. and McMahan, U. J. (1988). Motor neurons contain agrin-like molecules. *Journal of Cell Biology*, **107**, 1825–33.

Magill-Solc, C. and McMahan, U. J. (1990). Synthesis and transport of agrin-like molecules in motor neurons. *Journal of Experimental Biology*, **153**, 1–10.

Magleby, K. L. (1973). The effect of tetanic and post-tetanic potentiation on facilitation of transmitter release at the frog neuromuscular junction. *Journal of Physiology (London)*, **234**, 353–71.

Magleby, K. L. (1987). Short-term changes in synaptic efficacy. In *Synaptic function*, (ed. G. M. Edelman, W. E. Gall, and W. M. Cowan), pp. 21–56. John Wiley, New York.

Magleby, K. L. and Stevens, C. F. (1972). A quantitative description of end-plate currents. *Journal of Physiology (London)*, **223**, 173–97.

Maisonpierre, P. C., Belluscio, L., Squinto, S., Ip, N. Y., Furth, M. E., Lindsay, R. M. *et al.* (1990). Neurotrophin-3: a neurotrophic factor related to NGF and BDNF. *Science*, **247**, 1446–51.

Makowski, L., Casper, D. L. D., Phillips, W. C., and Goodenough, D. A. (1977). Gap junction structure. II. Analysis of the x-ray diffraction data. *Journal of Cell Biology*, **74**, 629–45.

Malenka, R. C. and Nicoll, R. A. (1990). Intracellular signals and LTP. *Seminars in Neurosciences*, **2**, 335–43.

Malenka, R. C., Kauer, J. A., Zucker, R. S., and Nicoll, R. A. (1988). Postsynaptic calcium is sufficient for potentiation of hippocampal synaptic transmission. *Science*, **242**, 81–4.

Malenka, R. C., Kauer, J. A., Perkel, D. J., and Nicoll, R. A. (1989a). The impact of postsynaptic calcium on synaptic transmission—its role in long-term potentiation. *Trends in Neurosciences*, **12**, 444–50.

Malenka, R. C., Kauer, J. A., Perkel, D. J., Mauk, M. D., Kelly, P. T., Nicoll, R. A. *et al.* (1989b). An essential role for postsynaptic calmodulin and protein kinase activity in long-term potentiation. *Nature*, **340**, 554–7.

Malgaroli, A. and Tsien, R. W. (1992). Glutamate-induced long-term potentiation of the frequency of miniature synaptic currents in cultured hippocampal neurons. *Nature*, **357**, 134–9.

Malik, N., Kallestad, J. C., Gunderson, N. L., Austin, S. D., Neubauer, M. G., Ochs, V. *et al.* (1989). Molecular cloning, sequence analysis, and functional expression of a novel growth regulator, oncostatin M. *Molecular Cell Biology*, **9**, 2847–53.

Malinow, R. and Miller, J. P. (1986). Postsynaptic hyperpolarization during conditioning reversibly blocks induction of long-term potentiation. *Nature*, **320**, 529–30.

Malinow, R. and Tsien, R. W. (1990). Presynaptic enhancement shown by whole-cell recordings of long-term potentiation in hippocampal slices. *Nature*, **346**, 177–80.

Malinow, R., Madison, D. V., and Tsien, R. W. (1988). Persistent protein kinase activity underlying long-term potentiation. *Nature*, **335**, 820–4.

Malinow, R., Schulman, H., and Tsien, R. W. (1989). Inhibition of postsynaptic PKC or CaMKII blocks induction but not expression of LTP. *Science*, **245**, 862–6.

Mallart, A. (1985). Electrical current flow inside perineural sheaths of mouse motor nerves. *Journal of Physiology (London)*, **368**, 565–75.

Mamalaki, C., Stephenson, F. A., and Barnard, E. A. (1987). The GABA-A/benzodiazepine receptor is a heterotetramer of homologous alpha and beta subunits. *EMBO Journal*, **6**, 561–5.

Manabe, T., Kaneko, S., and Kuno, M. (1989). Disuse-induced enhancement of Ia synaptic transmission in spinal motoneurons of the rat. *Journal of Neuroscience*, **9**, 2455–61.

Manabe, T., Renner, P., and Nicoll, R. A. (1992). Postsynaptic contribution to long-term potentiation revealed by the analysis of miniature synaptic currents. *Nature*, **355**, 50–5.

Manthorpe, M., Skaper, S., Williams, L., and Varon, S. (1986). Purification of adult sciatic nerve ciliary neurotrophic factor. *Brain Research*, **367**, 282–6.

Manzoni, O., Fagni, L., Pin, J., Rassendren, F., Poulatt, F., Sladeczek, F. *et al.* (1990). (Trans)-1-amino-cyclopentyl-1,3-dicarboxylate stimulates quisqualate phosphoinositide-coupled receptors but not ionotropic glutamate receptors in striatal neurones and *Xenopus* oocytes. *Molecular Pharmacology*, **38**, 1–6.

Marchbanks, R. M. and Israel, M. (1972). The heterogenecity of bound acetylcholine and synaptic vesicles. *Biochemical Journal*, **129**, 1049–61.

Margiotta, J. F. and Gurantz, D. (1989). Changes in the number, function, and regulation of nicotinic acetylcholine receptors during neuronal development. *Developmental Biology*, **135**, 326–39.

Maricq, A. V., Peterson, A. S., Brake, A. J., Myers, R. M., and Julius, D. (1991). Primary structure and functional expression of the $5HT_3$ receptor, a serotonin-gated ion channel. *Science*, **254**, 432–7.

Marr, D. (1969). A theory of cerebellar cortex. *Journal of Physiology (London)*, **202**, 437–70.

Marshall, L. M. (1985). Presynaptic control of synaptic channel kinetics in sympathetic neurones. *Nature*, **317**, 621–3.

Martin, A. R. (1955). A further study of the statistical composition of the end-plate potential. *Journal of Physiology (London)*, **130**, 114–22.

Martin, A. R. (1965). Quantal nature of synaptic transmission. *Physiological Reviews*, **46**, 51–66.

Martin, A. R. (1977). Junctional transmission II. Presynaptic mechanism. In *The handbook of physiology, section I. The nervous system*, Vol 1, Part 1, (ed. E. R. Kandel), pp. 329–55. American Physiological Society, Bethesda.

Martin, A. R. and Pilar, G. (1963). Dual mode of synaptic transmission in the avian ciliary ganglion. *Journal of Physiology (London)*, **168**, 443–63.

Martin, A. R. and Pilar, G. (1964a). Quantal components of the synaptic potential in the ciliary ganglion of the chick. *Journal of Physiology (London)*, **175**, 1–16.

Martin, A. R. and Pilar, G. (1964b). Presynaptic and post-synaptic events during post-tetanic potentiation and facilitation in the avian ciliary ganglion. *Journal of Physiology (London)*, **175**, 17–30.

Martin, A. R., Patel, V., Faille, L., and Mallart, A. (1989). Presynaptic calcium currents recorded from calyciform *Neuroscience Letters*, **105**, 14–8.

Martin, D. P., Schmidt, R. E., DiStefano, P. S., Lowry, O. H., Carter, J. G., and Johnson, E. M. (1988). Inhibitors of protein synthesis and RNA synthesis prevent neuronal death caused by nerve growth factor deprivation. *Journal of Cell Biology*, **106**, 829–44.

Martin, D. P., Ito, A., Horigome, K., Lampe, P. A., and Johnson, E. M. (1992). Biochemical characterization of programmed cell death in NGF-deprived sympathetic neurones. *Journal of Neurobiology*, **23**, 1205–20.

Martin, G. S. (1970). Rous sarcoma virus: a function required for the maintenance of the transformed state. *Nature*, **227**, 1021–3.

Martin, R. and Miledi, R. (1978). A structural study of the squid synapse after intraaxonal injection of calcium. *Proceedings of the Royal Society B*, **201**, 317–33.

Martinou, J. C. and Merlie, J. P. (1991). Nerve-dependent modulation of acetylcholine receptor ε-subunit gene expression. *Journal of Neuroscience*, **11**, 1291–9.

Martinou, J. C., Falls, D. L., Fischbach, G. D., and Merlie, J. P. (1991). Acetylcholine receptor-inducing activity stimulates expression of the ε-subunit gene of the muscle acetyulcholine receptor. *Proceedings of the National Academy of Sciences of the USA*, **88**, 7669–73.

Martinou, J. C., Martinou, I., and Kato, A. C. (1992). Cholinergic differentiation factor (CDF/LIF) promotes survival of isolated rat embryonic motoneurons *in vitro. Neuron*, **8**, 737–44.

Martin-Zanca, D., Hughes, S. H., and Barbacid, M. (1986). A human oncogene formed by the fusion of truncated tropomyosin and protein tyrosine kinase sequences. *Nature*, **319**, 743–8.

Martin-Zanca, D., Oskam, R., Mitra, G., Copeland, T., and Barbacid, M. (1989). Molecular and biochemical characterization of the human *trk* proto-oncogene. *Molecular and Cellular Biology*, **9**, 24–33.

Marty, A. and Neher, E. (1983). Tight-seal whole-cell recording. In *Single-channel recording* (ed. B. Sakmann and E. Neher), pp. 107–22. Plenum Press, New York.

Marty, A., Tan, Y. P., and Trautmann, A. (1984). Three types of calcium-dependent channel in rat lacrimal glands. *Journal of Physiology (London)*,**357**, 293–325.

Masu, M., Tanabe, Y., Tsuchida, K., Shigemoto, R., and Nakanishi, S. (1991). Sequence and expression of a metabotropic glutamate receptor. *Nature*, **349**, 760–5.

Masu, Y., Nakayama, K., Takami, H., Harada, Y., Kuno, M., and Nakanishi, S. (1987). cDNA cloning of bovine substance-K receptor through oocyte expression system. *Nature*, **329**, 836–8.

Masu, Y., Wolf, E., Holtmann, B., Sendtner, M., Brem, G., and Thoenen, H. (1993). Disruption of the CNTF gene results in motor neuron degeneration. *Nature*, **365**, 27–32.

Matteoli, M., Haimann, C., Torri-Tarelli, F., Polak, J. M., Ceccarelli, B., and De Camilli, P. (1988). Differential effect of α-latrotoxin on exocytosis from small synaptic vesicles and from large dense-core vesicles containing calcitonin gene-related peptide at the frog neuromuscular junction. *Proceedings of the National Academy of Sciences of the USA*, **85**, 7366–70.

Matthews, G. and Watanabe, S. (1988). Activation of single ion channels from toad retinal rod inner segments by cyclic GMP: concentration dependence. *Journal of Physiology (London)*, **403**, 389–405.

Maxwell, D. J., Christie, W. M., Short, A. D., and Brown, A. G. (1990). Direct observations of synapses between GABA-immunoreactive boutons and muscle afferent terminals in lamina VI of the cat's spinal cord. *Brain Research*, **530**, 215–22.

Mayer, L. M. and Westbrook, G. L. (1987a). Permeation and block of N-methyl-D-aspartic acid receptor channels by divalent cations in mouse cultured central neurones. *Journal of Physiology (London)*, **394**, 501–27.

Mayer, M. L. and Westbrook, G. L. (1987b). The physiology of excitatory amino acids in the vertebrate nervous system. *Progress in Neurobiology*, **28**, 197–276.

Mayer, M. L., Westbrook, G. L., and Guthrie, P. B. (1984). Voltage-dependent block by Mg^{2+} of NMDA responses in spinal cord neurones. *Nature*, **309**, 263.

Mayer, M. L., MacDermott, A. B., Westbrook, G. L., Smith, S. J., and Barker, J. L. (1987). Agonist-and voltage-gated calcium entry in cultured mouse spinal cord neurons under voltage clamp measured using arsenazo III. *Journal of Neuroscience*, **7**, 3230–44.

Mayford, M., Barzilai, A., Keller, F., Schacher, S., and Kandel, E. R. (1992). Modulation of an NCAM-related adhesion molecule with long-term synaptic plasticity in *Aplysia*. *Science*, **256**, 638–44.

Mccall, T. and Vallance, P. (1992). Nitric oxide takes centre-stage with newly defined roles. *Trends in Pharmacological Sciences*, **13**, 1–6.

McDonald, N. Q. and Hendrickson, W. A. (1993). A structural superfamily of growth factors containing a cystine knot motif. *Cell*, **73**, 421–4.

McLachlan, E. M. (1975). An analysis of the release of acetylcholine from preganglionic nerve terminals. *Journal of Physiology (London)*, **245**, 447–66.

McMahan, U. J. (1990). The agrin hypothesis. *Cold Spring Harbor Symposia on Quantitative Biology*, **50**, 407–18.

McMahan, U. J. and Slater, C. R. (1984). The influence of basal lamina on the accumulation of acetylcholine receptors at synaptic sites in regenerating muscle. *Journal of Cell Biology*, **98**, 1453–73.

McMahon, S. B., Armanini, M. P., Ling, L. H., and Phillips, H. S. (1994). Expression and coexpression of *trk* receptors in subpopulations of adult primary sensory neurons projecting to identified peripheral targets. *Neuron*, **12**, 1161–71.

McManaman, J. L., Crawford, F. G., Stewart, S. S., and Appel, S. H. (1988). Purification of a skeletal muscle polypepide which stimulates choline acetyltransferase activity in cultured spinal cord neurons. *Journal of Biological Chemistry*, **263**, 5890–7.

McManaman, J. L., Oppenheim, R. W., Prevette, D., and Marchetti, D. (1990). Rescue of motoneurons from cell death by a purified skeletal muscle polypeptide: Effects of the ChAT development factor, CDF. *Neuron*, **4**, 891–8.

Meakin, S. O. and Shooter, E. M. (1991). Molecular investigations on the high-affinity nerve growth factor receptor. *Neuron*, **6**, 153–63.

Meakin, S. O. and Shooter, E. M. (1992). The nerve growth factor family of receptors. *Trends in Neurosciences*, **15**, 323–31.

Mearow, K. M. and Govind, C. K. (1989). Stimulation-induced changes at crayfish (*Procambarus clarkii*) neuromuscular terminals. *Cell Tissue Research*, **256**, 119–23.

Meech, R. W. (1978). Calcium-dependent potassium activation in nervous tissue. *Annual Review of Biophysics and Bioengineering*, **7**, 1–18.

Meech, R. W. and Strumwasser, F. (1970). Intracellular calcium injection activates potassium conductance in *Aplysia* nerve cells. *Federation Proceedings*, **29**, 834.

Meguro, H., Mori, H., Araki, K., Kushiya, E., Kutsuwada, T., Yamazaki, M. *et al.* (1992). Functional characterization of a heteromeric NMDA receptor channel expressed from cloned cDNAs. *Nature*, **357**, 70–4.

Mehler, M. F., Rozental, R., Dougherty, M., Spray, D. C., and Kessler, J. A. (1993). Cytokine regulation of neuronal differentiation of hippocampal progenitor cells. *Nature*, **362**, 62–5.

Meldrum, B. and Garthwaite, J. (1990). Excitatory amino acid neurotoxicity and neurodegenerative disease. *Trends in Pharmacological Sciences*, **11**, 379–87.

Melloni, E. and Pontremoli, S. (1989). The calpains. *Trends in Neurosciences*, **12**, 438–44.

Mendell, L. M. and Weiner, R. (1976). Analysis of pairs of individual Ia-e.p.s.p.s in single motoneurons. *Journal of Neurophysiology*, **49**, 269–89.

Menesini-Chen, M. G., Chen, J. S., and Levi-Montalcini, R. (1978). Sympathetic nerve fibres ingrowth in the central nervous system of neonatal rodent on intracerebral NGF injections. *Archives of Italian Biology*, **116**, 53–84.

Merlie, J. P. and Sanes, J. R. (1985). Concentration of acetylcholine receptor mRNA in synaptic regions of adult muscle fibres. *Nature*, **317**, 66–8.

Metcalf, D. (1989). The molecular control of cell division, differentiation commitment and maturation in haemopoietic cells. *Nature*, **339**, 27–30.

Metcalf, D. (1991). Control of granulocytes and macrophages: molecular, cellular, and clinical aspects. *Nature*, **254**, 529–33.

Methfessel, C., Witzemann, V., Takahashi, T., Mishina, M., Numa, S., and Sakmann, B. (1986). Patch clamp measurements on *Xenopus* laevis oocytes: currents through endogenous channels and implanted acetylcholine receptor and sodium channels. *Pflügers Archiv*, **407**, 577–88.

Meyer, M., Matsuoka, I., Wetmore, C., Olson, L., and Thoenen, H. (1992). Enhanced synthesis of brain-derived neurotrophic factor in the lesioned peripheral nerve: Different mechanisms are responsible for the regulation of BDNF and NGF mRNA. *Journal of Cell Biology*, **119**, 45–54.

Meyer, T., Holowka, D., and Stryer, L. (1988). Highly cooperative opening of calcium channels by inositol 1,4,5–trisphophate. *Science*, **240**, 653–6.

Micevych, P. E. and Kruger, L. (1992). The status of calcitonin gene-related peptide as an effector peptide. *Annals of the New York Academy of Sciences*, **657**, 379–96.

Middleton, P., Jaramillo, F., and Schuetze, S. M. (1986). Forskolin increases the rate of acetylcholine receptor desensitization at rat soleus endplates. *Proceedings of the National Academy of Sciences of the USA*, **83**, 4967–71.

Mikami, A., Imoto, K., Tanabe, T., Niidome, T., Mori, Y., Takeshima, H. *et al.* (1989). Primary structure and functional expression of the cardiac dihydropyridine-sensitive calcium channel. *Nature*, **340**, 230–233.

Miledi, R. (1960). The acetylcholine sensitivity of frog muscle fibres after complete or partial denervation. *Journal of Physiology (London)*, **151**, 1–23.

Miledi, R. (1963). An influence of nerve not mediated by impulses. In *The effect of use and disuse on neuromuscular functions*, (ed. E. Gutmann and P. Hnik), pp. 35–40. Elsevier, Amsterdam.

Miledi, R. (1967). Spontaneous synaptic potentials and quantal release of transmitter in the stellate ganglion of the squid. *Journal of Physiology (London)*, **192**, 379–406.

Miledi, R. (1973). Transmitter release induced by injection of calcium ions into nerve terminals. *Proceedings of the Royal Society B*, **183**, 421–5.

Miledi, R. (1980). Intracellular calcium and desensitization of acetylcholine receptors. *Proceedings of the Royal Society B*, **209**, 447–52.

Miledi, R. (1982). A calcium-dependent transient outward current in *Xenopus laevis* oocytes. *Proceedings of the Royal Society B*, **299**, 401–11.

Miledi, R. and Slater, C. R. (1966). The action of calcium on neuronal synapses in the squid. *Journal of Physiology (London)*, **184**, 473–98.

Miledi, R., Molenaar, P. C., and Polak, R. L. (1978). α-Bungartoxin enhances transmitter 'released' at the neuromuscular junction. *Nature*, **272**, 641–3.

Miles, K., Greengard, P., and Huganir, R. L. (1989). Calcitonin gene-related peptide regulates phosphorylation of the nicotinic acetylcholine receptor in rat myotubes. *Neuron*, **2**, 1517–24.

Miller, B., Sarantis, M., Traynelis, S. F., and Attwell, D. (1992). Potentiation of NMDA receptor currents by arachidonic acid. *Nature*, **355**, 722–5.

Miller, C. (1989). Genetic manipulation of ion channels: A new approach to structure and mechanism. *Neuron*, **2**, 1195–205.

Miller, R. J. (1987). Multiple calcium channels and neuronal function. *Science*, **235**, 46–52.

Miller, R. J. (1991). The control of neuronal Ca^{2+} homeostasis. *Progress in Neurobiology*, **37**, 255–85.

Miller, S. G. and Kennedy, M. B. (1986). Regulation of brain type II Ca^{2+}/calmodulin-dependent protein kinase by autophosphorylation: A Ca^{2+}-triggered molecular switch. *Cell*, **44**, 861–70.

Milner, P. M. and Penfield, W. (1955). The effect of hippocampal lesions on recent memory. *Transactions of the American Neurological Association*, **80**, 42–8.

Minota, S., Kumamoto, E., Kitakoga, O., and Kuba, K. (1991). Long-term potentiation induced by a sustained rise in the intraterminal Ca^{2+} in bull-frog sympathetic ganglia. *Journal of Physiology (London)*, **435**, 421–38.

Mintz, I. M., Venema, V. J., Swiderek, K. M., Lee, T. D., Bean, B. P., and Adams, M. E. (1992). P-type calcium channels blocked by the spider toxin ω-Aga-IVA. *Nature*, **355**, 827–829.

Miranda, R. C., Sohrabji, F., and Toran-Allerand, C. D. (1993). Neuronal colocalization of mRNAs for neurotrophins and their receptors in the developing central nervous system suggests a potential for sutocrine interactions. *Proceedings of the National Academy of Sciences of the USA*, **90**, 6439–43.

Mishina, M., Kurosaki, T., Tobimatus, T., Morimoto, Y., Noda, M., Yamamoto, T. *et al.* (1984). Expression of functional acetylcholine receptor from cloned cDNAs. *Nature*, **307**, 604–8.

Mishina, M., Tobimatsu, T., Imoto, K., Tanaka, K., Fujita, Y., Fukuda, K. *et al.* (1985). Location of functional regions of acetylcholine receptor α-subunit by site-directed mutagenesis. *Nature*, **313**, 364–9.

Mishina, M., Takai, T., Imoto, K., Noda, M., Takahashi, T., Numa, S. *et al.* (1986). Molecular distinction between fetal and adult forms of muscle acetylcholine receptor. *Nature*, **321**, 406–11.

Mishkin, M. (1978). Memory in monkeys severely impaired by combined but not by separate removal of amygdala and hippocampus. *Nature*, **273**, 297–8.

Mitchell, J. F. and Silver, A. (1963). The spontaneous release of acetylcholine from the denervated hemidiaphragm of the rat. *Journal of Physiology (London)*, **165**, 117–29.

Miyajima, A., Hara, T., and Kitamura, T. (1992a). Common subunits of cytokine receptor and the functional redundancy of cytokines. *Trends in Biochemical Sciences*, **17**, 378–82.

Miyajima, A., Kitamura, T., Harada, N., Yokota, T., and Arai, K. (1992b). Cytokine receptors and signal transduction. *Annual Review of Immunology*, **10**, 295–331.

Miyamoto, E., Kuo, J. F., and Greengard, P. (1969). Adnosine 3′.5′-monophosphate-dependent protein kinase from brain. *Science*, **165**, 63–5.

Miyata, Y., Kashihara, Y., Homma, S., and Kuno, M. (1986). Effects of nerve growth factor on the survival and synaptic function of Ia sensory neurons axotomized in neonatal rats. *Journal of Neuroscience*, **6**, 2012–8.

Montarolo, P. G., Goelet, P., Castellucci, V. F., Morgan, J., Kandel, E. R., and Schacher, S. (1986). A critical period for macromolecular synthesis in long-term heterosynaptic facilitation in *Aplysia*. *Science*, **234**, 1249–54.

Montarolo, P. G., Kandel, E. R., and Schacher, S. (1988). Long-term heterosynaptic inhibition in *Aplysia*. *Nature*, **333**, 171–4.

Montminy, M. R. and Bilezikjian, L. M. (1987). Binding

of a nuclear protein to the cyclic-AMP response element of the somatostatin gene. *Nature*, **328**, 175–8.

Montminy, M. R., Sevarino, K. A., Wagner, J. A., Mandel, G., and Goodman, R. H. (1986). Identification of a cyclic AMP- responsive element within the rat somatostatin gene. *Proceedings of the National Academy of Sciences of the USA*, **83**, 6682–6.

Monyer, H., Sprengel, R., Schoepfer, R., Herb, A., Higushi, M., Lomeli, H. *et al.* (1992). Heteromeric NMDA receptors: molecular and functional distinction of subtypes. *Science*, **256**, 1217–21.

Monyer, H., Burnashev, N., Laurie, D. J., Sakmann, B., and Seeburg, P. H. (1994). Developmental and regional expression in the rat brain and functional properties of four NMDA receptors. *Neuron*, **12**, 529–40.

Morel, N., Israel, M., and Manaranche, R. (1978). Determination of ACh concentration in *Torpedo* synaptosomes. *Journal of Neurochemistry*, **30**, 1553–7.

Morgan, J. I. and Curran, T. (1991). Stimulus-transcription coupling in the nervous ststem: involvement of the inducible proto-oncogenes *fos* and *jun*. *Annual Review of Neuroscience*, **14**, 421–51.

Mori, H., Masaki, H., Yamakura, T., and Mishina, M. (1992). Identification by mutagenesis of a Mg^{2+}-block site of the NMDA receptor channel. *Nature*, **358**, 673–5.

Mori, Y., Friedrich, T., Kim, M. S., Mikami, A., Nakai, J., Ruth, P. *et al.* (1991). Primary structure and functional expression from complementary DNA of a brain calcium channel. *Nature*, **350**, 398–402.

Morita, K. and Barrett, E. F. (1989). Calcium-dependent depolarizations originating in lizard motor nerve terminals. *Journal of Neuroscience* **9**, 3359–69.

Moriyoshi, K., Masu, M., Ishii, T., Shigemoto, R., Mizuno, N., and Nakanishi, S. (1991). Molecular cloning and characterization of the rat NMDA receptor. Nature, **354**, 31–7.

Morse, J. K., Wiegand, S. J., Anderson, K., You, Y., Cai, N., Carnahan, J. *et al.* (1993). Brain-derived neurotrophic factor (BDNF) prevents the degneration of medial septal cholinergic neurons following fimbria transection. *Journal of Neuroscience*, **13**, 4146–56

Moss, B. L., Schuetze, S. M., and Role, L. W. (1989). Functional properties and developmental regulation of nicotinic acetylcholine receptors on embryonic chicken sympathetic neurons. *Neuron*, **3**, 597–607.

Moss, S. J., Smart, T. G., Blackstone, C. D., and Huganir, R. L. (1992). Functional modulation of GAB_A receptors by cAMP- dependent protein phsophorylation. *Science*, **257**, 661–5.

Mu, X., Silos-Santiago, I., Carroll, S. L., and Snider, W. D. (1993). Neurotrophin receptor genes are expressed in distinct patterns in developing dorsal root ganglia. *Journal of Neuroscience*, **13**, 4029–41.

Mulderry, P. K. and Lindsay, R. M. (1990). Rat dorsal root ganglion neurons in culture express vasoactive intestinal polypeptide (VIP) independently of nerve growth factor. *Neuroscience Letters*, **108**, 314–20.

Mulle, C., Benoit, P., Pinset, C., Roa, M., and Changeux, J. P. (1988). Calcitonin gene-related peptide enhances the rate of desensitization of the nicotinic acetylcholine receptor in cultured mouse muscle cells. *Proceedings of the National Academy of Sciences of the USA*, **85**, 5728–32.

Muller, D., Joly, M., and Lynch, G. (1988). Contributions of quisqualate and NMDA receptors to the induction and expression of LTP. *Nature*, **242**, 1694–7.

Murachi, T. (1989). Intracellular regulatory system involving calpain and calpastatin. *Biochemistry International*, **18**, 263–94.

Murakami, F., Katsumaru, H., Saito, K., and Tsukahara, N. (1982). A quantitative study of synaptic organization in red nucleus neurons after lesion of the nucleus interpositus of the cat: an electron microscopic study involving intracellular injection of horseradish peroxidase. *Brain Research*, **242**, 41–53.

Murakami, F., Katsumaru, H., Maeda, J., and Tsukahara, N. (1984). Reorganization of corticorubral synapses following cross-innervation of flexor and extensor nerves of adult cat: a quantitative electron microscopic study. *Brain Research*, **306**, 299–306.

Murakami, F., Oda, Y., and Tsukahara, N. (1988). Synaptic plasticity in the red nucleus and learning. *Behavioral Brain Research*, **28**, 175–9.

Murase, K. and Randić, M. (1983). Electrophysiological properties of rat spinal dorsal horn neurones in vitro: calcium-dependent action potentials. *Journal of Physiology (London)*, **334**, 141–53.

Murphy, M., Reid, K., Hilton, D. J., and Bartlett, P. F. (1991). Generation of sensory neurons is stimulated by leukemia inhibitory factor. *Proceedings of the National Academy of Sciences of the USA*, **88**, 3498–501.

Nachmansohn, D. (1959). *Chemical and molecular basis of nerve activity*. Academic Press, New York.

Nakamura, T. and Gold, G. (1987). A cyclic nucleotide-gated conductance in olfactory receptor cilia. *Nature*, **325**, 442–4.

Nakamura, Y., Mizuno, N., Konishi, A., and Sato, M. (1974). Synaptic reorganization of the red nucleus after chronic deafferentation from cerebellorubral fibres; an electron microscopoic study in the cat. *Brain Research*, **82**, 298–301.

Nakanishi, S. (1992). Molecular diversity of glutamate receptors and implication for brain function. *Science* **258**, 597–603.

Nastuk, M., Lieth, E., Ma, J., Cardasis, C. A., Moynihan, E. B., McKechinie, B. A. *et al.* (1991). The putative agrin receptor binds ligand in a calcium-dependent manner and aggregates during agrin-induced acetylcholine receptor clustering. *Neuron*, **7**, 807–18.

Nathan, C. and Sporn, M. (1991). Cytokines in context. *Journal of Cell Biology*, **113**, 981–6.

Navone, F., Greengard, P., and De Camilli, P. (1984). Synapsin I in nerve terminals: Selective association with small synaptic vesicles. *Science*, **226**, 1209–11.

Nawa, H. and Patterson, P. H. (1990). Separation and partial characterization of neuropeptide-inducing factors in heart cell conditioned medium. *Neuron*, 4, 269–77.

Nawa, H., Yamamori, T., Le, T., and Patterson, P. H. (1991). The generation of neuronal diversity: analogies and homologies with hematopoiesis. *Cold Spring Harbor Symposia on Quantitative Biology*, **55**, 247–53.

Nawy, S. and Jahr, C. E. (1991). cGMP-gated conductance in retinal bipolar cells is suppressed by the photoreceptor transmitter. *Neuron*, 7, 677–83.

Nef, P., Oneyser, C., Alliod, C., Couturier, S., and Ballivet, M. (1988). Genes expressed in the brain define three distinct neuronal nicotinic acetylcholine receptors. *EMBO Journal*, 7, 595–601.

Neff, N. T., Prevette, D., Houenou, L. J., Lewis, M. E., Glicksman, M. A., Yin, Q. W. *et al* (1993). Insulin-like growth factors; putative muscle-derived trophic agents that promote motoneuron survival. *Journal of Neurobiology*, 24, 1578–88.

Neher, E. and Marty, A. (1982). Discrete changes of cell membrane capacitance observed under conditions of enhanced secretionin bovine adrenal chromaffin cells. *Proceedings of the National Academy of Sciences of the USA*, 79, 6712–6.

Neher, E. and Sakmann, B. (1976). Single-channel currents recorded from membrane of denervated frog muscle fibres. *Nature*, **260**, 779–802.

Neher, E. and Steinbach, J. H. (1978). Local anaesthetics transiently block currents through single acetylcholine receptor channels. *Journal of Physiology (London)*, 277, 153–76

Nestler, E. J. and Greengard, P. (1983). Protein phosphorylation in the brain. *Nature*, **305**, 583–8.

Neubig, R. R. and Cohen, J. B. (1980). Permeability control by cholinergic receptors in *Torpedo* postsynaptic membranes: agonist dose-response relations measured at second and millisecond times. *Biochemistry*, **19**, 2770–9.

New, H. V. and Mudge, A. W. (1986). Calcitonin gene-related peptide regulates muscle acetylcholine receptor synthesis. *Nature*, **323**, 809–11.

Nickolas, A., Lavidis, A., and Bennett, M. R. (1992). Probabilistic secretion of quanta from visulaized sympathetic nerve varicosities in mouse vas deferens. *Journal of Phsyiology (London)*, 454, 9–26.

Nicoll, R. A. (1988). The coupling of neurotransmitter receptors to ion channels in the brain. *Science*, 241, 545–51.

Nicoll, R. A., Kauer, J. A., and Malenka, R. C. (1988). The current excitement in long-term potentiation. *Neuron*, 1, 97–103.

Nishi, R. and Berg, D. K. (1979). Survival and development of ciliary ganglion neurones grown alone in cell culture. *Nature*, 277, 232–4.

Nishizuka, Y. (1986). Studies and perspective of protein kinase C. *Science*, **233**, 305–12.

Nitkin, R. M., Smith, M. A., Magill, C., Fallon, J. R., Yao, M. M., Wallace, B. G. *et al.* (1987). Identification of agrin, a synaptic organizing protein from *Torpedo* electric organ. *Journal of Cell Biology*, **105**, 2471–8.

Nja, A. and Purves, D. (1978). The effects of nerve growth factor and its antiserum on synapses in the superior cervical ganglion of the guinea-pig. *Journal of Physiology (London)*, 277, 53–75.

Noda, M., Takahashi, H., Tanabe, T., Toyosato, M., Furutani, Y., Hirose, T. *et al.* (1982). Primary structure of α-subunit precursor of *Torpedo californica* acetylcholine receptor deduced from cDNA sequence. *Nature*, **299**, 793–7.

Noda, M., Shimizu, S., Tanabe, T., Takai, T., Kayano, T., Ikeda, T. *et al.* (1984). Primary structure of *Electrophorus electricus* sodium channel deduced from cDNA sequence. *Nature*, **312**, 121–7.

Noda, M., Ikeda, T., Suzuki, H., Takeshima, H., Takahashi, T., Kuno, M. *et al.* (1986). Expression of functional sodium channels from cloned cDNA. *Nature*, **322**, 826–8.

Noguchi, K., Senba, E., Morita, Y., Sato, M., and Tohyama, M. (1989). Prepro-VIP and preprotachykinin mRNAs in the rat dorsal root ganglion cells following peripheral axotomy. *Molecular Brain Research*, 6, 327–30.

Nowak, L., Bregestovski, P., Ascher, P., Herbet, A., and Prochiantz, A. (1984). Magnisium gates glutamate-activated channels in mouse central neurones. *Nature*, 307, 462–5.

Nowycky, M. C., Fox, A. P., and Tsien, R. W. (1985). Three types of neuronal calcium channel with different calcium agonist sensitivity. *Nature*, **316**, 440–3.

Numa, S. (1987). Structure and function of ionic channels. *Chemica Scripta*, 27B, 5–19.

Numa, S. (1989). A molecular view of neurotransmitter receptors and ionic channels. *Harvey Lectures*, **83**, 121–65.

Numa, S. (1991). Neurotransmitter receptors and ionic channels: from structure to function. *Fidia Research Foundation Neuroscience Award Lectures*, **5**, 23–44.

Numa, S., Noda, M., Takahashi, H., Tanabe, T., Toyosato, M., Furutani, Y. *et al.* (1983). Molecular structure of the nicotinic acetylcholine receptor. *Cold Spring Harbor Symposia on Quantitative Biology*, **48**, 57–69.

Oblinger, M. M. and Lasek, R. J. (1988). Axotomy-induced alterations in the synthesis and transport of neurofilaments and microtubules in dorsal root ganglion cells. *Journal of Neuroscience*, **8**, 1747–58

O'Connor, T. and Wyttenbach, C. R. (1974). Cell death

in the embryonic chick spinal cord. *Journal of Cell Biology*, **60**, 448–59.

Ocorr, K. A., Walters, E. T., and Byrne, J. H. (1985). Associative conditioning analog selectively increases cAMP levels of tail sensory neurons in *Aplysia*. *Proceedings of the National Academy of Sciences of the USA*, **82**, 2548–52.

O'Dell, T. J., Hawkins, R. D., Kandel, E. R., and Arancio, O. (1991a). Tests of the roles of two diffusible substances in long-term potentiation: evidence for nitric oxide as a possible early retrograde messenger. *Proceedings of the National Academy of Sciences of the USA*, **88**, 11285–9.

O'Dell, T. J., Kandel, E. R., and Grant, S. G. N. (1991b). Long-term potentiation in the hippocampus is blocked by tyrosine kinase inhibitors. *Nature*, **353**, 558–60.

Ogura, A., Akita, K., and Kudo, Y. (1990). Non-NMDA receptor mediates cytoplasmic Ca^{2+} elevation in cultured hippocampal neurones. *Neuroscience Research*, **9**, 103–13.

O'Hara, P. J., Sheppard, P. O., Thogersen, H., Venezia, D., Haldeman, B. A., McGrane, V., Houamed, K. M., Thomsen, C., Gilbert, T., and Mulvihill, E. R. (1993). The ligand- binding domain in metabotropic glutamate receptors is related to bacterial periplasmic binding proteins. *Neuron*, **11**, 41–52.

Oiki, S., Danho, W., Madison, V., & Montal, M. (1988). M_2 delta, a candidate for the structure lining the ionic channel of the nicotinic cholinergic receptor. *Proceedings of the National Academy of Sciences of the USA*, **85**, 8703–7.

Okamoto, T., Murayama, Y., Hayashi, Y., Inagaki, M., Ogata, E., and Nishimoto, I. (1991). Identification of a G_s activator region of the β_2–adrenergic receptor that is autoregulated via protein kinase A-dependent phosphorylation. *Cell*, **67**, 723–30.

O'Lague, P. H., MacLeish, P. R., Nurse, C. A., Clasude, P., Furshpan, E. J., and Potter, D. D. (1975). Physiological and morphological studies on developing sympathetic neurons in dissociated cell culture. *Cold Spring Harbor Symposia on Quantitative Biology*, **40**, 399–407.

Olivera, B. M., Gray, W. R., Zeikus, R., McIntosh, J. M., Varga, J., Rivier, J. *et al.* (1985). Peptide neurotoxins from fish-hunting cone snails. *Science*, **230**, 1338–1343.

Olmstead, J. M. D. (1920). The nerve as a formative influence in the development of taste-buds. *Journal of Comparative Neurology*, **31**, 465–8.

Olsen, R. W. and Tobin, A. J. (1990). Molecular biology of $GABA_A$ receptors. *FASEB Journal*, **4**, 1469–80.

Oppenheim, R. W., Maderdrut, J. L., and Wells, D. J. (1982). Cell death of motoneurons in the chick embryo spinal cord. VI. Reduction of naturally occurring cell death in the thoracolumbar column of terni by nerve growth factor. *Journal of Comparative Neurology*, **210**, 174–89.

Oppenheim, R. W., Bursztajn, S., and Prevette, D. (1989). Cell death of motoneurons in the chick embryo spinal cord. XI. Acetylcholine receptors and synaptogenesis in skeletal muscle following the reduction of motorneurone death by neuromuscular blockade. *Development*, **107**, 331–41.

Oppenheim, R. W., Prevette, D., Tytell, M., and Homma, S. (1990). Naturally occurring and induced neuronal death in the chick embryo *in vivo* requires protein and RNA synthesis: evidence for the role of cell death genes. *Developmental Biology*, **138**, 104–13.

Oppenheim, R. W., Prevette, D., Qin-Wei, Y., Collins, F., and MacDonald, J. (1991). Control of embryonic motoneuron survival *in vivo* by ciliary neurotrophic factor. *Science*, **251**, 1616–8.

Oppenheim, R. W., Qin-Wei, Y., Prevette, D., and Yan, Q. (1992). Brain-derived neurotrophic factor rescues developing avian motoneurons from cell death. *Nature*, **360**, 755–7.

O'Shea, E. K., Klemm, J. D., Kim, P. S., and Alber, T. (1991). X-ray structure of the GCN4 leucine zipper, a to-stranded, parallel coiled coil. *Science*, **254**, 539–44.

Otani, S., Marshall, C. J., Tate, W. P., Goddard, G. V., and Abraham, W. C. (1989). Maintenance of long-term potentiation in rat dentate gyrus requires protein synthesis but not messenger RNA synthesis immediately post-tetanization. *Neuroscience*, **28**, 519–26.

Otsu, H., Yamamoto, A., Maeda, N., Mikoshiba, K., and Tashiro, Y. (1990). Immunogold localization of inositol 1,4,5-trisphosphate ($InsP_3$) receptor in mouse cerebellar Purkinje cells using three monoclonal antibodies. *Cell Structure and Function*, **15**, 163–73.

Ovchinnikov, Y. A. (1982). Rhodopsin and bacteriorhodopsin: structure-function relationships. *FEBS Letters*, **148**, 179–91.

Palade, G. E. (1954). Electron microscope observations of interneuronal and neuromuscular synapses. *Anatomical Record*, **118**, 335–6.

Palay, S. L. (1954). Electron microscope study of the cytoplasm of neurons. *Anatomical Record*, **118**, 336.

Palay, S. L. (1956). Synapse in the central nervous system. *Journal of Biophysics, Biochemistry and Cytology*, **2**, 193–202.

Palmer, E., Monaghan, D., and Cotman, C. (1989). Trans-ACPD, a selective agonist of the phosphoinositide-coupled excitatory amino acid receptor. *European Journal of Pharmacology*, **166**, 585–7.

Palmer, R. M. J., Ferrige, A. G., and Moncada, S. (1987). Nitric oxide release accounts for the biological activity of endothelium-derived relaxing factor. *Nature*, **327**, 524–6.

Patrick, J., Boulter, J., Deneris, E., Wada, K., Wada, E., Connolly, J. *et al.* (1989). Structure and function of neuronal nicotinic acetylcholine receptors deduced from cDNA clones. *Progress in Brain Research*, **79**, 27–33.

Patterson, P. H. (1978). Environmental determination of autonomic neurotransmitter functions. *Annual Review of Neuroscience*, 1, 1–17.

Patterson, P. H. (1990). Control of cell fate in a vertebrate neurogenic lineage. *Cell*, 62, 1035–8.

Patterson, P. H. and Nawa, H. (1993). Neuronal differentiation factors/cytokines and synaptic plasticity. *Neuron*, 10 (Supplement), 123–37.

Pavlov, I. P. (1906). The scientific investigation of the psychical faculties or processes in the higher animals. *Science*, 24, 613–9.

Pavlov, I. P. (1927). *Conditioned reflexes: An investigation of the physiological activity of the cerebral cortex*. Oxford University Press.

Payton, B. W., Bennett, M. V. L., and Pappas, G. D. (1969). Permeability and structure of junctional membranes at an elctrotonic synapse. *Science*, 166, 1641–43.

Peng, H. B., Baker, L. P., and Chen, Q. (1991). Induction of synaptic development in cultured muscle cells by basic fibroblast growth factor. *Neuron*, 6, 237–46.

Peng, Y. Y. and Frank, E. (1989a). Activation of $GABA_B$ receptors causes presynaptic inhibition at synapses between muscle spindle afferents and motoneurons in the spinal cord of bullfrogs. *Journal of Neuroscience*, 9, 1502–15.

Peng, Y. Y. and Frank, E. (1989b). Activation of $GABA_A$ receptors causes presynaptic and postsynaptic inhibition at synapses between muscle spindle afferents and motoneurons in the spinal cord of bullfrogs. *Journal of Neuroscience*, 9, 1516–22.

Penner, R. and Dreyer, F. (1986). Two different presynaptic calcium currents in mouse motor nerve terminals. *Pflügers Archiv*, 406, 190–7.

Peralta, E. G., Ashkenazi, A., Winslow, J. W., Smith, D. H., Ramachandran, J., and Capon, D. J. (1987). Distinct primary structures, ligand-binding properties and tissue-specific expression of four human muscarinic acetylcholine receptors. *EMBO Journal*, 6, 3923–9.

Perez-Reyes, E., Kim, H. S., Lacerda, A. E., Horne, W., Wei, X., Rampe, D. *et al.* (1989). Induction of calcium currents by the expression of the α_1-subunit of the dihydropyridine receptor from skeletal muscle. *Nature*, 340, 233–236.

Perney, T. M., Hirning, L. D., Leeman, S. E., and Millar, R. J. (1986). Multiple calcium channels mediate neurotransmitter release from peripheral neurones. *Proceedings of the National Academy of Sciences of the USA*, 83, 6656–9.

Peroukta, S. J. (1988). 5-Hydroxytryptamine receptor subtypes. *Annual Review of Neuroscience*, 11, 45–60.

Pestronk, A., Drachman, D. B., and Griffin, J. W. (1976). Effect of muscle disuse on acetylcholine receptors. *Nature*, 260, 352–3.

Petrenko, A. G., Perin, M. S., Daietov, B. A., Ushkaryov, Y. A., Geppert, M., and Südhof, T. C. (1991). Binding of synaptotagmin to the α-latrotoxin receptor implicates both in synaptic vesicle exocytosis. *Nature*, 353, 65–8.

Pilar, G. and Landmesser, L. (1972). Axotomy mimicked by localized colchicine application. *Science*, 117, 116–18.

Pilar, G. and Landmesser, L. (1976). Ultrastructural differences during embryonic cell death in normal and peripheral deprived ciliary ganglia. *Journal of Cell Biology*, 68, 339–56.

Pinney, D. F., Pearson-White, S. H., Konieczny, S. F., Latham, K. E., and Emerson, C. P. (1988). Myogenic lineage determination and differentiation: evidence for a regulatory gene pathway. *Cell*, 53, 781–93.

Pinsker, H. M., Hening, W. A., Carew, T. J., and Kandel, E. R. (1973). Long-term sensitization of a defensive withdrawal reflex in *Aplysia*. *Science*, 182, 1039–42.

Plummer, M. R., Logothetis, D. E., and Hess, P. (1989). Elementary properties and pharmacological sensitivities of calcium channels in mammalian peripheral neurons. *Neuron*, 2, 1453–63.

Pockett, S. and Slack, J. R. (1982). Source of the stimulus for nerve terminal sprouting in partially denervated muscle. *Neuroscience*, 7, 3173–6.

Pongs, O., Kecskemethy, N., Muller, R., Krah-Jentgens, I., Baumann, A., Kiltz, H. H. *et al.* (1988). Shaker encodes a family of putative potassium channel proteins in the nervous system of *Drosophila*. *EMBO Journal*, 7, 1087–96.

Popper, P., Abelson, L., and Micevych, P. E. (1992). Differential regulation of α-calcitonin gene-related peptide and preprocholesystokinin messenger RNA expression in α-motoneurons: effects of testosterone and inactivity induced factors. *Neuroscience*, 51, 87–96.

Potter, L. T. (1970). Synthesis, storage and release of $[^{14}C]$-acetylcholine in isolated rat diaphragm muscles. *Journal of Physiology (London)*, 206, 145–66.

Prestige, M. C. (1967). The control of cell number in the lumbar ventral horns during the development of *Xenopus laevis* tadpoles. *Journal of Embryology and Experimental Morphology*, 18, 359–87.

Pritchett, D. B., Sontheimer, H., Shivers, B. D., Ymer, S., Kettenmann, H., Schofield, P. R. *et al.* (1989). Importance of a novel GABA-A receptor subunit for benzodiazepine pharmacology. *Nature*, 338, 582–5.

Propst, J. W. and Ko, C. P. (1987). Correlations between active zone ultrastructure and synaptic function studied with freeze-fracture of physiologically identified neuromuscular junctions. *Journal of Neuroscience*, 7, 3654–64.

Pumplin, D. W., Reese, T. S., and Llinás, R. (1981). Are the presynaptic membrane particles calcium channels? *Proceedings of the National Academy of Sciences of the USA*, 78, 7210–3.

Purves, D. (1975). Functional and structural changes in mammalian sympathetic neurones following interruption of their axons. *Journal of Physiology (London)*, **252**, 429–63.

Purves, D. (1976). Functional and structural changes in mammalian sympathetic neurones following colchicine application to postganglionic nerves. *Journal of Physiology (London)*, **259**, 159–75.

Purves, D. (1988). *Body and brain. A trophic theory of neural connections*. Harvard University Press, Cambridge, Mass.

Purves, D. and Lichtman, J. W. (1985). *Principles of neural development*. Sinauer Associates Inc., Sunderland.

Purves, D. and Nja, A. (1976). Effect of nerve growth factor on synaptic depression following axotomy. *Nature*, **260**, 535–6.

Purves, D., Voyvodic, J. T., Magrassi, L., and Yawo, H. (1987). Nerve terminal remodeling visualized in living mice by repeated examination of the same neuron. *Science*, **238**, 1122–6.

Purves, D., Snider, W. D., and Voyvodic, J. T. (1988). Trophic regulation of nerve cell morphology and innervation in the autonomic nervous system. *Nature*, **336**, 123–8.

Radeke, M. J., Misko, T. P., Hsu, C., Herzenberg, L. A., and Shooter, E. M. (1987). Gene transfer and molecular cloning of the rat nerve growth factor receptor. *Nature*, **325**, 593–7.

Raftery, M. A., Hunkapiller, M. W., Strader, C. D., and Hood, L. E. (1980). Acetylcholine receptor: complex of homologous subunits. *Science*, **208**, 1454–7.

Rahamimoff, R., Meiri, H., Erulkar, S. D., and Barenholz, Y. (1978). Changes in transmitter release induced by ion-containing liposomes. *Proceedings of the National Academy of Sciences of the USA*, **75**, 5214–6.

Rahamimoff, R., DeRiemer, S. A., Sakmann, B., Stadler, H., and Yakir, N. (1988). Ion channels in synaptic vesicles from *Torpedo* electric organ. *Proceedings of the National Academy of Sciences of the USA*, **85**, 5310–4.

Raisman, G. (1977). Formation of synapses in the adult rat after injury: similarities and differences between a peripheral and a central nervous site. *Philosophical Transactions of the Royal Society B*, **278**, 349–59.

Raivich, G., Zimmermann, A., and Sutter, A. (1985). The spatial and temporal pattern of βNGF receptor expression in the developing chick embryo. *EMBO Journal*, **4**, 637–44.

Rane, S. G., Holz, G. G., and Dunlap, K. (1987). Dihydropyridine inhibition of neuronal calcium current and substance P release. *Pflügers Archiv*, **409**, 361–6.

Rang, H. P. (1975). Acetylcholine receptors. *Quarterly Review of Biophysics*, **7**, 283–399.

Rangel-Aldao, R. and Rosen, O. M. (1976). Mechanism of self-phosphorylation of adenosine 3′:5′-monophosphate-dependent protein kinase from bovine cardiac muscle. *Journal of Biological Chemistry*, **251**, 7526–9.

Ranscht, B. (1991). Cadherin cell adhesion molecules in vertebrate neural development. *Seminars in Neurosciences*, **3**, 285–96.

Ranson, S. W. (1914). Transplantation of the spinal ganglion with observations on the significance of the complex types of spinal gagnlion cells. *Journal of Comparative Neurology*, **24**, 547–58.

Ranvier, L. (1874). De quelques faits relatifs a l'histologie et à la physiologie des muscles stries. *Archives de physiologie normale et pathologique*, **6**, 1–15

Rao, M. S. and Landis, S. C. (1993). Cell interactions that determine sympathetic neuron transmitter phenotype and the neurokines that mediate them. *Journal of Neurobiology*, **24**, 215–32.

Rao, M. S., Tyrrell, S., Landis, S. C., and Patterson, P. H. (1992). Effects of ciliary neurotrophic factor (CNTF) and depolarization on neuropeptide expression in cultured sympathetic neurons. *Developmental Biology*, **150**, 281–93.

Rashevsky, N. (1938). *Mathematical biophysics*. University of Chicago Press.

Rassendren, F. A., Bloch-Gallego, E., Tanaka, H., and Henderson, C. E. (1992). Levels of mRNA coding for motoneuron growth-promoting factors are increased in denervated muscle. *Proceedings of the National Academy of Sciences of the USA*, **89**, 7194–8.

Rathjen, F. G. and Jessell, T. M. (1991). Glycoproteins that regulate the growth and guidance of vertebrate axons: domains and dynamics of the immunoglubulin/fibronectin type III subfamily. *Seminars in Neurosciences*, **3**, 297–307.

Rayport, S. G. and Schacher, S. (1986). Synaptic plasticity *in vitro*: cell culture of identified *Aplysia* neurons mediating short-term habituation and sensitization. *Journal of Neuroscience*, **6**, 759–63.

Redfern, P. A. (1970). Neuromuscular transmission in new-born rats. *Journal of Physiology (London)*, **209**, 701–9.

Redman, S. (1979). Junctional mechanisms at group Ia synapses. *Progress in Neurobiology* **12**, 33–83.

Redman, S. (1990). Quantal analysis of synaptic potentials in neurons of the central nervous system. *Physiological Reviews*, **70**, 165–98.

Redman, S. and Walmsley, B. (1983). Amplitude fluctuations in synaptic potentials evoked in cat spinal motoneurones at identified group Ia synapses. *Journal of Physiology (London)*, **343**, 135–45.

Regan, L. J., Sah, D. W. Y., and Bean, B. P. (1991). Ca^{2+} channels in rat central and peripheral neurons: high-threshold current resistant to dihydropyridine blockers and ω-conotoxin. *Neuron*, **6**, 269–80.

Reist, N. E., Werle, M. J., and McMahan, U. J. (1992). Agrin released by motor neurons induces the aggregation of acetylcholine receptors at neuromuscular junctions. *Neuron*, **8**, 865–8.

Rende, M., Muir, D., Ruoslahti, E., Hagg, T., Varon,

S., and Manthorpe, M. (1992). Immunolocalization of ciliary neurotrophic factor in adult rat sciatic nerve. *Glia*, 5, 25–32.

Revah, F., Bertrand, D., Galzi, J. L., Devillers-Thiery, A., Mulle, C., Hussy, N. *et al.* (1991). Mutations in the channel domain alter desensitization of a neuronal acetylcholine receptor. *Nature*, **353**, 846–9

Revel, J. P. and Karnovsky, M. J. (1967). Hexagonal array of subunits in intercellular junctions of the mouse heart and liver. *Journal of Cell Biology*, **33**, C7–C12.

Rieger, F., Grumet, M., and Edelman, G. M. (1985). N-CAM at the vertebrate neuromuscular junction. *Journal of Cell Biology*, **101**, 285–93.

Ritter, A. M., Lewin, G. R., Kremer, N. E., and Mendell, L. M. (1991). Requirement for nerve growth factor in the development of myelinated nociceptors *in vivo*. *Nature*, **350**, 500–2.

Roberts, W. M., Jacobs, R. A., and Hudspeth, A. J. (1990). Colocalization of ion channels involved in frequency selectivity and synaptic transmission at presynaptic zones of hair cells. *Journal of Neuroscience*, **10**, 3664–84.

Robitaille, R., Adler, E. M., and Charlton, M. P. (1990). Strategic location of calcium channels at transmitter release sites of frog neuromuscular synapses. *Neuron*, **5**, 773–9.

Rodbell, M. (1980). The role of hormone receptors and GTP-regulatory proteins in membrane transduction. *Nature*, **284**, 17–22.

Rodbell, M., Krans, H. M., Pohl, S. L., and Birnbaumer, L. (1971a). The glucagon-sensitive adenyl cyclase system in plasma membranes of rat liver. IV. Effects of guanyl nucleotides on binding of ^{125}I-glucagon. *Journal of Biological Chemistry*, **246**, 1872–6.

Rodbell, M., Birnbaumer, L., Pohl, S. L., and Krans, H. M. (1971b). The glucagon-sensitive adenyl cyclase system in plasma membranes of rat liver. V. An obligatory role of guanyl nucleotides in glucagon action. *Journal of Biological Chemistry*, **246**, 1877–82.

Rodriguez-Tebar, A., Jeffrey, P. L., Thoenen, H., and Barde, Y. A. (1989). The survival of chick retinal ganglion cells in response to brain-derived neurotrophic factor depends on their embryonic age. *Developmental Biology*, **136**, 296–303.

Rodriguez-Tebar, A., Dechant, G., and Barde, Y. A. (1990). Binding of brain-derived neurotrophic factor to the nerve growth factor receptor. *Neuron*, **4**, 487–92.

Rohrer, H. (1985). Nonneuronal cells from chick sympathetic and dorsal root sensory ganglia express catecholamine uptake and receptors for nerve growth factor during development. *Developmental Biology*, **111**, 95–107.

Romanes, G. J. (1946). Motor localization and the effects of nerve injury on the ventral horn cells of the spinal cord. *Journal of Anatomy*, **80**, 117–31.

Roper, S. and Ko, C. P. (1978). Synaptic remodelling in the partially denervated parasympathetic ganglion in the heart of the frog. In *Neuronal plasticity*, (ed. C. W. Cotman), pp. 1–25. Raven, New York.

Rosenfeld, M. G., Mermod, J. J., Amara, S. G., Swanson, L. W., Sanchenko, P. E., Rivier, J. *et al.* (1983). Production of a novel neuropeptide encoded by the calcitonin gene via tissue-specific RNA processing. *Nature*, **304**, 129–35.

Rosenthal, A., Goeddel, D. V., Nguyen, T., Lewis, M., Shih, A., Laramee, G. R. *et al.* (1990). Primary structure and biological activity of a novel human neurotrophic factor. *Neuron*, **4**, 767–73.

Rosenthal, J. (1969). Post-tetanic potentiation at the neuromuscular junction of the frog. *Journal of Physiology (London)*, **203**, 121–33.

Ross, E. M. (1989). Signal sorting and amplification through G protein-coupled receptors. *Neuron*, **3**, 141–52.

Rovainen, C. M. (1967). Physiological and anatomical studies on large neurons of central nervous system of the sea lamprey (*Petromyzon marinus*). II. Dorsal cells and giant interneurons. *Journal of Neurophysiology*, **30**, 1024–42.

Ruegg, M. A., Tsim, K. W. K., Horton, S. E., Kroger, S., Escher, G., Gensch, E. M. *et al.* (1992). The agrin gene codes for a family of basal lamina proteins that differ in function and distribution. *Neuron*, **8**, 691–9.

Ruit, K. G., Osborne, P. A., Schmidt, R. E., Johnson, E. M., and Snider, W. D. (1990). Nerve growth factor regulates sympathetic ganglion cell morphology and survival in the adult mouse. *Journal of Neuroscience*, **10**, 2412–9.

Ruit, K. G., Elliott, J. L., Osborne, P. A., Yan, Q., and Snider, W. D. (1992). Selective dependence of mammalian dorsal root ganglion neurons on nerve growth factor during embryonic development. *Neuron*, **8**, 573–87.

Rupp. F., Payan, D. G., Magill-Solc, C., Cowan, D. M., and Scheller, R. H. (1991). Structure and expression of a rat agrin. *Neuron*, **6**, 811–23.

Ruppersberg, J. P., Schroter, K. H., Sakmann, B., Stocker, M., Sewing, S., and Pongs, O. (1990). Heteromultimeric channels formed by rat brain potassium-channel proteins. *Nature*, **345**, 535–7.

Rutishauser, U. (1991). Pleiotropic biological effects of the neural cell adhesion molecule (NCAM). *Seminars in Neurosciences*, **3**, 265–70.

Saadat, S., Sendtner, M., and Rohrer, H. (1989). Ciliary neurotrophic factor induces cholinergic differentiation of rat sympathetic neurons in culture. *Journal of Cell Biology*, **108**, 1807–16.

Sakaguchi, M., Inaishi, Y., Kashihara, Y., and Kuno, M. (1991). Release of calcitonin gene-related peptide from nerve terminals in rat skeletal muscle. *Journal of Physiology (London)*, **434**, 257–70.

Sakaguchi, M., Inaishi, Y., Kashihara, Y., and Kuno, M. (1992). Degeneration of motor nerve fibers enhances the expression of calcitonin gene-related peptide in rat sensory neurons. *Neuroscience Letters*, **137**, 61–4.

Sakimura, K., Morita, T., Kushiya, E., and Mishina, M. (1992). Primary structure and expression of the γ2 subunit of the glutamate receptor channel selective for kainate. *Neuron*, **8**, 267–74.

Sakmann, B. (1992). Elementary steps in synaptic transmission revealed by currents through single ion channels. *Neuron*, **8**, 613–29.

Sakmann, B. and Brenner, H. R. (1978). Change in synaptic channel gating during neuromuscular development. *Nature*, **276**, 401–2.

Sakmann, B., Patlak, J., and Neher, E. (1980). Single acetylcholine-activated channels show burst-kinetics in presence of desensitizing concentrations of agonist. *Nature*, **286**, 71–3.

Sakmann, B., Methfessel, C., Mishina, M., Takahashi, T., Takai, T., Kurasaki, M. *et al.* (1985). Role of acetylcholine receptor subunits in gating of the channel. *Nature*, **318**, 538–43.

Sakurai, M. (1987). Synaptic modification of parallel fibre-Purkinje cell transmission in *in vitro* guinea-pig cerebellar slices. *Journal of Physiology* (*London*), **394**, 463–80.

Sakurai, M. (1990). Calcium is an intracellular mediator of the climbing fibre in induction of cerebellar long-term depression. *Proceedings of the National Academy of Sciences of the USA*, **87**, 3383–5.

Salmons, S. and Vrbova, G. (1969). The influence of activity on some contractile characteristics of mammalian fast and slow muscles. *Journal of Physiology* (*London*), **201**, 535–49.

Samuel, S. (1860). *Die trophischen Nerven. Ein Beitrag zur Physiologie und Pathologie*. Leipzig.

Sanes, J. R. and Covault, J. (1985). Axon guidance during reinnervation of skeletal muscle. *Trends in Neurosciences*, **8**, 523–8.

Sanes, J. R., Marshall, L. M., and McMahan, U. J. (1978). Reinnervation of muscle fiber basal lamina after removal of myofibers. *Journal of Cell Biology*, **78**, 176–98.

Sano, K., Enomoto, K., and Maeno, T. (1987). Effects of synthetic ω-conotoxin, a new type Ca^{2+}-antagonist, on frog and mouse neuromuscular transmission. *European Journal of Pharmacology*, 141, 235–41.

Sassone-Corsi, P., Sisson, J. C., and Verma, I. M. (1988). Transcriptional autoregualtion of the proto-oncogene *fos*. *Nature*, **334**, 314–9.

Sastry, B. R., Goh, J. W., and Auyeung, A. (1986). Associative induction of posttetanic and long-term potentiation in CA1 neurons of rat hippocampus. *Science*, **232**, 988–90.

Satoh, T., Ross, C. A., Villa, A., Supattapone, S., Pozzan, T., Snyder, S. H. *et al.* (1990). The inositon 1,4,5–trisphosphate receptor in cerebellar purkinje cells: quantitative immunogold labelling reveals concentration in an endoplasmic reticulum subcompartment. *Journal of Cell Biology*, **111**, 615–24.

Savion, N., Vlodavsky, I., and Gospodarowicz, D. (1981). Nuclear accumulation of epidermal growth factor in cultured bovine corneal enodthelial and granulosa cells. *Journal of Biological Chemistry*, **256**, 1149–54.

Schacher, S., Castellucci, V. F., and Kandel, E. R. (1988). cAMP evokes long-term facilitation in *Aplysia* sensory neurons that requires new protein synthesis. *Science*, 240, 1667–9.

Schäfer, E. A. (1900). *Text-book of physiology*. Young J. Pentland, London.

Schatzman, R. C., Raynor, R. L., and Kuo, J. F. (1983). N-(6-amino-hexyl)-5-chloro-1-naphtalensulfonamide (W-7), a calmodulin antagonist, also inhibits phospholipid-sensitive calcium-dependent protein kinase. *Biochimica et Biophysica Acta*, 755, 144–7.

Scheterson, L. C. and Bothwell, M. (1992). Novel roles for neurotrophins are suggested by BDNF and NT-3 mRNA expression in developing neurons. *Neuron*, 9, 449–63.

Schmalbruch, H. (1984). Motoneuron death after sciatic nerve section in newborn rats. *Journal of Comparative Neurology*, 224, 252–8.

Schmidt, J. and Raftery, M. A. (1973). Purification of acetylcholine receptors from *Torpedo californica* electroplax by affinity chromatography. *Biochemistry*, 12, 852–6.

Schmieden, V., Grenningloh, G., Schofield, P. R., and Betz, H. (1989). Functional expression in *Xenopus* oocytes of the strychnine binding 48 kd subunit of the glycine receptor. *EMBO Journal*, **8**, 695–700.

Schofield, P. R., Darlison, M. G., Fujita, N., Burt, D. R., Stephenson, F. A., Rodriguez, H. *et al.* (1987). Sequence and functional expression of the GABA-A receptor shows a ligand-gated receptor super-family. *Nature* **328**, 221–7.

Scholz, K. P. and Byrne, J. H. (1988). Intracellular injection of cAMP induces a long-term reduction of neuronal K^+ currents. *Science*, 240, 1661–6.

Schotzinger, R. J. and Landis, S. C. (1988). Cholinergic phenotype developed by noradrenergic sympathetic neurons after innervation of a novel cholinergic target *in vivo*. *Nature*, **335**, 637–9.

Schotzinger, R. J. and Landis, S. C. (1990). Acquisition of cholinergic and peptidergic properties by sympathetic innervation of rat sweat glands requires interaction with normal target. *Neuron*, **5**, 91–100.

Schuetze, S. M. and Role, L. W. (1987). Developmental regulation of nicotinic acetylcholine receptors. *Annual Review of Neuroscience*, 10, 403–57.

Schuman, E. M. and Madison, D. (1991). A requriement

for the intercellular messenger nitric oxide in long-term potenttiation. *Science*, **254**, 1503–6.

Schwab, M. E., Heumann, R., and Thoenen, H. (1982). Communication between target organs and nerve cells: retrograde axonal transport and site of action of nerve growth factor. *Cold Spring Harbor Symposia on Quantitative Biology*, **46**, 125–34.

Schwartz, J. H., Castellucci, V. F., and Kandel, E. R. (1971). Functioning of identified neurons and synapses in abodominal ganglion of *Aplysia* in absence of protein synthesis. *Journal of Neurophysiology*, **34**, 939–53.

Schwartz, J. H., Bernier, L., Castellucci, V. F., Palazzolo, M., Saitoh, T., Stapleton, A. *et al.* (1983). What molecular steps determine the time course of the memory for short-term sensitization in *Aplysia? Cold Spring Harbor Symposia on Quantitative Biology*, **48**, 811–9.

Schwartz, W. and Passow, H. (1983). Ca-activated K channels in erythrocytes and excitable cells. *Annual Review of Physiology*, **45**, 359–74.

Schwartzkroin, P. A. and Wester, K. (1975). Long-lasting facilitation of a synaptic potential following tetanization in the *in vitro* hippocampal slice. *Brain Research*, **89**, 107–19.

Scott, J., Selby, M., Urdea, M., Quiroga, M., Bell, G. I., and Rutter, W. J. (1983). Isolation and nucleotide sequence of a cDNA encoding the precursor of mouse nerve growth factor. *Nature*, **302**, 538–40.

Sealock, R., Wray, B. E., and Froehner, S. C. (1984). Ultrastructural localization of the Mr 43,000 protein and the acetylcholine receptor in *Torpedo* postsynaptic membranes using monoclonal antibodies. *Journal of Cell Biology*, **98**, 2239–44.

Seeburg, P. H. (1993). The molecular biology of mammalian glutamate receptor channels. *Trends in Neurosciences*, **16**, 359–65.

Seiler, M. and Schwab, M. E. (1984). Specific retrograde transport of nerve growth factor (NFG) from neocortex to nucleus basalis in the rat. *Brain Research*, **300**, 33–6.

Sejnowski, T. J., Chattarji, S., and Stanton, P. K. (1990). Homosynaptic long-term depression in hippocampus and neocortex. *Seminars in Neurosciences*, **2**, 353–63.

Sendtner, M., Kreutzberg, G. W., and Thoenen, H. (1990). Ciliary neurotrophic factor prevents the degeneration of motor neuorns after axotomy. *Nature*, **345**, 440–1.

Sendtner, M., Holtmann, B., Kolbeck, R., Thoenen, H., and Barde, Y. A. (1992a). Brain-derived neurotrophic factor prevents the death of motoneurons in newborn rats after nerve section. *Nature*, **360**, 757–9.

Sendtner, M., Stokli, K. A., and Thoenen, H. (1992b). Synthesis and localization of ciliary neurotrophic factor in the sciatic nerve of the adult rat after lesion and during regeneration. *Journal of Cell Biology*, **118**, 139–48.

Shapovalov, A. I. and Shiriaev, B. I. (1980). Dual modes of junctional transmission at synapses between single primary afferent fibres and motoneurones in the amphibian. *Journal of Physiology (London)*, **306**, 1–15.

Sheng, M. and Greenberg, M. E. (1990). The regulation and function of *c-fos* and other immediate early genes in the nervous system. *Neuron*, **4**, 477–85.

Sherrington, C. (1906). *The integrative action of the nervous system*. Yale University Press, New Haven.

Shibuki, K. and Okada, D. (1991). Endogenous nitric oxide release required for long-term synaptic depression in the cerebellum. *Nature*, **349**, 326–8.

Shigemoto, R., Abe, T., Nomura, S., Nakanishi, S., and Hirano, T. (1994). Antibodies inactivating $mGluR_1$ metabotropic glutamate receptor block long-term depression in cultured Purkinje cells. *Neuron*, **12**, 1245–55.

Shorey, M. L. (1909). The effect of the destruction of peripheral areas on the differentiatrion of the neuroblasts. *Journal of Experimental Zoology*, **7**, 25–64.

Shuster, M. J., Camardo, J. S., Siegelbaum, S. A., and Kandel, E. R. (1985). Cyclic AMP-dependent kinase closes the serotonin-sensitive K channels of *Aplysia* sensory neurones in cell-free membrane patches. *Nature*, **313**, 392–5.

Shyng, S. L. and Salpeter, M. M. (1990). Effect of reinnervation on the degradation rate of junctional acetylcholine receptors synthesized in denervated skeletal muscle. *Journal of Neuroscience*, **10**, 3905–15.

Siegelbaum, S. A., Camardo, J. S., and Kandel, E. R. (1982). Serotonin and cyclic AMP close single K channels in *Aplysia* sensory neurons. *Nature*, **299**, 413–8.

Siesjo, B. K. (1988). Historical overview: calcium, ischemia, and death of brain cells. *Annals of the New York Academy of Sciences*, **522**, 638–61.

Sigel, E. and Barnard, E. A. (1984). A gamma-aminobutyric acid/benzodiazepine receptor complex from bovine cerebral cortex. *Journal of Biological Chemistry*, **259**, 7219–23.

Sigel, E., Baur, R., Trube, G., Mohler, H., and Malherbe, P. (1990). The effect of subunit composition of rat brain $GABA_A$ receptors on channel function. *Neuron*, **5**, 703–11.

Silinsky, E. M. (1975). On the association between transmitter secretion and the release of adenine nucleotides from mammalian motor nerve terminals. *Journal of Phsyiology (London)*, **247**, 145–62.

Silinsky, E. M. (1980). Evidenced for specific adonsine receptors at cholinergic nerve endings. *British Journal of Pharmacology*, **71**, 191–4.

Silva, A. J., Stevens, C. F., Tonegawa, S., and Wang, Y. (1992). Deficient hippocampal long-term potentiation in α-calcium-calmodulin kinase II mutant mice. *Science*, **257**, 201–6.

Simon, M. I., Strathmann, M. P., and Gautam, N.

(1991). Diversity of G proteins in signal transduction. *Science*, 252, 802–8.

Singer, D., Biel, M., Lotan, I., Flockerzi, V., Hofmann, F., and Dascal, N. (1991). The roles of the subunits in the function of the calcium channel. *Science* 253, 1553–7.

Singer, S. J. (1975). Architecture and topography of biological membranes. In *Cell membranes; biochemistry, cell biology and pathology*, (ed. G. Weissmann and R. Claiborne), pp. 35–45. HP Publishing Co. Inc., New York.

Singer, S. J. and Nicolson, G. L. (1972). The fluid mosaic model of the structure of cell membranes. *Science*, 175, 720–731.

Sjöstrand, F. S. (1953). The ultrastructure of the retinal rod synapses of the guinea pig eye. *Journal of Applied Physics*, 24, 1422.

Sladeczek, F., Pin, J. P., Recasens, M., Bockaert, J., and Weiss, S. (1985). Glutamate stimulates inositol phosphate formation in striatal neurons. *Nature*, 317, 717–9.

Smart, T. G. (1992). A novel modulatory binding site for zinc on the $GABA_A$ receptor complex in cultured rat neurones. *Journal of Physiology (London)*, 447, 587–625.

Smeyne, R. J., Klein, R., Schnapp, A., Long, L. K., Bryant, S., Lewin, A. *et al.* (1994). Severe sensory and sympathetic neuropathies in mice carrying a disrupted *Trk*/NGF receptor gene. *Nature*, 368, 246–9.

Smith, D. O. (1991). Sources of adenosine released during neuromuscular transmission in the rat. *Journal of Physiology (London)*, 432, 343–54.

Smith, R. G., Vaca, K., McManaman, J., and Appel, S. H. (1986). Selective effects of skeletal muscle extract fractions on motoneuron development *in vitro*. *Journal of Neuroscience*, 6, 439–47.

Smith, S. and Augustine, G. J. (1988). Calcium ions, active zones and synaptic transmitter release. *Trends in Neurosciences*, 11, 458–65.

Smith, S. B., White, H. D., Siegel, J. B., and Krebs, E. G. (1981). Cyclic AMP-dependent protein kinase I: cyclic nucleotide binding, structural changes, and release of the catalytic subunits. *Proceedings of the National Academy of Sciences of the USA*, 78, 1591–5.

Snider, W. D. (1988). Nerve growth factor enhances dendritic arborization of sympathetic ganglion cells in developing mammals. *Journal of Neuroscience*, 8, 2628–34.

Snider, W. D. and Harris, G. L. (1979). A physiological correlate of disuse-induced sprouting at the neuromuscular junction. *Nature*, 281, 69–71.

Snider, W. D. and Thanedar, S. (1989). Target dependence of hypoglossal motoneurons during development and in maturity. *Journal of Comparative Neurology*, 279, 489–98.

Snutch, T. P. (1988). The use of *Xenopus* oocytes to probe synaptic communication. *Trends in Neurosciences*, 11, 250–6.

Snutch, T. P. and Reiner, P. B. (1992). Ca^{2+} channels: diversity of form and function. *Current Opinion in Neurobiology*, 2, 247–53.

Sommer, B. and Seeburg, P. H. (1992). Glutamate receptor channels: novel properties and new clones. *Trends in Pharmacological Sciences*, 13, 291–6.

Sommer, B., Kohler, M., Sprengel, R., and Seeburg, P. H.(1991). RNA editing in brain controls a determinant of ion flow in glutamate-gated channels. *Cell*, 67, 11–19.

Song, Y. and Huang, L. Y. M. (1990). Modulation of glycine receptor chloride channels by cAMP-dependent protein kinase in spinal trigeminal neurons. *Nature*, 348, 242–5.

Soppet, D., Escandon, E., Maragos, J., Middlemas, D. S., Reid, S. W., Blair, J. *et al.* (1991). The neurotrophic factors brain-derived neurotrophic factor and neurotrophin-3 are ligands for the *trk*B tyrosine kinase receptor. *Cell*, 65, 895–903.

Soreide, A. J. (1981). Varitations in the axon reaction in animals of different ages. A light microscopic study on the facial nucleus of the rat. *Acta Anatomica*, 110, 40–7.

Spencer, W. A., Thompson, R. F., and Neilson, D. R. (1966). Response decrement of the flexion reflex in the acute spinal cat and transient restoration by strong stimuli. *Journal of Neurophysiology*, 29, 221–39.

Spruce, A. E., Breckenridge, L. J., Lee, A. K., and Almers, W. (1990). Properties of the fusion pore that forms during exocytosis of a mast cell secretory vesicle. *Neuron*, 4, 643–54.

Squinto, S. P., Aldrich, T. H., Lindsay, R. M., Morrissey, D. M., Panayotatos, N., Bianco, S. *et al.* (1990). Identification of functional receptors for ciliary neurotrophic factor on neuronal cell lines and primary neurons. *Neuron*, 5, 757–66.

Squinto, S. P., Stitt, T. N., Aldrich, T. H., Davis, S., Bianco, S. M., Radziejewski, C. *et al.* (1991). *trk*B encodes a functional receptor for brain-derived neurotrophic factor and neurotrophin-3 but not nerve growth factor. *Cell*, 65, 885–93.

Squire, L. R. (1982). The neuropsychology of human memory. *Annual Review of Neuroscience*, 5, 241–73.

Stanley, E. F. (1989). Calcium currents in a vertebrate presynaptic neve terminal: the chick ciliary ganlion calyx. *Brain Research*, 505, 341–5.

Stanley, E. F. and Atrakchi, A. H. (1990). Calcium currents recorded from a vertebrate presynaptic terminal resistant to the dihydropyridine nifedipine. *Proceedings of the National Academy of Sciences of the USA*, 87, 9683–7.

Stanley, E. F. and Ehrenstein, G. (1985). A model for exocytosis based on the opening of calcium-activated potassium channels in vesicles. *Life Science*, 37, 1985–95.

Stanley, E. F. and Goping, G. (1991). Characterization of a calcium current in a vertebrate cholinergic presynaptic terminal. *Journal of Neuroscience*, 11, 985–93.

Stanton, P. K. and Sarvey, J. M. (1984). Blockade of long-term potentiation in rat hippocampal CA1 region by inhibitors of protein synthesis. *Journal of Neuroscience*, 4, 3080–8.

Stanton, P. K. and Sejnowski, T. J. (1989). Associative long-term depression in the hippocampus induced by hebbian covariance. *Nature*, **339**, 215–8.

Starling, E. H. (1912). *Principles of human physiology*. J. & A. Churchill, London.

Stehelin, D., Varmus, H. E., Bishop, J. M., and Vogt, P. K. (1976). DNA related to the transforming gene(s) of avian sarcoma viruses is present in normal avian DNA. *Nature*, **260**, 170–3.

Stemple, D. L. and Anderson, D. J. (1992). Isolation of a stem cell for neurons and glia from the mammalian neural crest. *Cell*, 71, 973–85.

Stent, G. S. (1973). A physiological mechanism for Hebb's postulate of learning. *Proceedings of the National of Academy of Sciences of the USA*, **70**, 997–1001.

Stern, P., Edwards, F. A., and Sakmann, B. (1992). Fast and slow components of unitary EPSCs on stellate cells elicited by focal stimulation in slices of rat visual cortex. *Journal of Physiology (London)*, **449**, 247–78.

Stevens, C. F. (1989). Strengthening the synapses. *Nature*, **338**, 460–1.

Stevens, L. M. and Landis, S. C. (1990). Target influences on transmitter choice by sympathetic neurons developing in the anterior chamber of the eye. *Developmental Biology*, **137**, 109–24.

Stockel, K., Schwab, M., and Thoenen, H. (1975). Comparison between the retrograde axonal transport of nerve growth factor and tetanus toxin in motor, sensory and adrenergic neurons. *Brain Research*, **99**, 1–16.

Stockli, K. A., Lottspeich, F., Sendtner, M., Masiakowski, P., Carroll, P., Gotz, R. *et al*. (1989). Molecular cloning, expression and regional distribution of rat ciliary neurotrophic factor. *Nature*, **342**, 920–3.

Stockli, K. A., Lillien, L. E., Naher-Noe, M., Breitfeld, G., Hughes, R. A., Thoenen, H. *et al*. (1991). Regional distribution, developmental changes and cellular localization of CNTF-mRNA abd protein in the rat brain. *Journal of Cell Biology*, **115**, 447–59.

Stoeckel, K., Guroff, G., Schwab, M., and Thoenen, H. (1976). The significance of retrograde axonal transport for the accumulation of systemically administered nerve growth factor (NGF) in the rat superior cervical ganglion. *Brain Research*, **109**, 271–84.

Strader, C. D., Dixon, R. A. F., Cheung, A. H., Candelore, M. R., Blake, A. D., and Sigal, I. S. (1987). Mutations that uncouple the β-adrenergic receptor from G and increase agonist affinity. *Journal of Biological Chemistry*, **262**, 16439–43.

Streit, W. J., Dumoulin, F. L., Raivich, G., and Kreutzberg, G. W. (1989). Calcitonin gene-related peptide increases in rat facial motoneurons after peripheral nerve transection. *Neuroscience Letters*, **101**, 143–8.

Strong, J. A., Fox, A. P., Tsien, R. W., and Kaczmarek, L. K. (1987). Stimulation of protein kinase C recruits covert calcium channels in *Aplysia* bag cell neurons. *Nature*, **325**, 714–7.

Stroud, R. M., McCarthy, M. P., and Shuster, M. (1990). Nicotinic acetylcholine receptor superfamily of ligand-gated ion channels. *Biochemistry*, **29**, 1109–23.

Stryer, L. (1986). Cyclic GMP cascade of vision. *Annual Review of Neuroscience*, **9**, 87–119.

Stuart, G. J. and Redman, S. J. (1992). The role of $GABA_A$ and $GABA_B$ receptors in presynaptic inhibition of Ia EPSPs in cat spinal motoneurons. *Journal of Physiology (London)*, **447**, 675–92.

Stümer, W., Conti, F., Suzuki, H., Wang, X., Noda, M., Yahagi, N. *et al*. (1989). Structural parts involved in activation and inactivation of the sodium channel. *Nature*, **339**, 597–603.

Südhof, T. C. and Jahn, R. (1991). Properties of synaptic vesicles involved in exocytosis and membrane recycling. *Neuron*, **6**, 665–77.

Südhof, T. C., Lottespeich, F., Greengard, P., Mehl, E., and Jahn, R. (1987). A synaptic vesicle protein with a novel cytoplasmic domain and four transmembrane regions. *Science*, **238**, 1142–4.

Sugita, S., Shen, K. Z., and North, R. A. (1992). 5-Hydroxytryptamine is a fast excitatory transmitter at 5-HT_3 receptors in rat amygdala. *Neuron*, **8**, 199–203.

Sugiyama, H., Ito, I., and Hirono, C. (1987). A new type of glutamate receptor linked to inositol phospholipid metabolism. *Nature*, **325**, 531–3.

Sumikawa, K., Parker, I., and Miledi, R. (1986). *Xenopus* oocytes as a tool for moleculoar cloning of the genes coding for neuromuscular receptors and voltage-operated channels. *Fortschritte der Zoologie*, **33**, 127–38.

Sutherland, E. W. (1972). Studies on the mechanism of hormone action. *Science*, **177**, 401–8.

Sutter, A., Riopelle, R. J., Harris-Warrick, R. M., and Shooter, E. M. (1979). Nerve growth factor receptors. Characterization of two distinct classes of binding sites on chick embryo sensory ganglia cells. *Journal of Biological Chemistry*, **254**, 5972–82.

Swandulla, D., Carbone, E., and Lux, H. D. (1991a). Do calcium channel classification account for neuronal calcium channel diversity? *Trends in Neurosciences*, **14**, 46–51.

Swandulla, D., Hans, M., Zipser, K., and Augustine, G. J. (1991b). Role of residual calcium in synaptic depression and posttetanic potentiation: fast and slow calcium signaling in nerve terminal. *Neuron*, **7**, 915–26.

Swope, S. L., Moss, S. J., Blackstone, C. D., and Huganir, R. L. (1992). Phosphorylation of ligand-gated ion channels: a possible mode of synaptic plasticity. *FASEB Journal*, **6**, 2514–23.

Tachabina, M. and Okada, T. (1991). Release of endogenous excitatory amino acids from on-type bipolar cells isolated from the goldfish retina. *Journal of Neuroscience*, 11, 2199–208.

Tachibana, M., Okada, T., Arimura, T., Kobayashi, K., and Piccolino, M. (1993). Dihydropyridine-sensitive calcium current mediates neurotransmitter release from bipolar cells of the goldfish retina. *Journal of Neuroscience*, 13, 2898–909.

Taga, T. and Kishimoto, T. (1992). Cytokine receptors and signal transduction. *FASEB Journal*, 6, 3387–96.

Taga, T., Hibi, M., Hirata, Y., Yamasaki, K., Yasukawa, K., Matsuda, T. *et al.* (1989). Interleukin-6 triggers the association of its receptor with a possible signal transducer, gp130. *Cell*, **58**, 573–81.

Takagi, S. F. (1989). *Human olfaction*. University of Tokyo Press.

Takahashi, K., Tsuchida, K., Tanabe, Y., Masu, M., and Nakanishi, S. (1993). Role of the large extracellular domain of metabotropic glutamate receptors in agonist selectivity determination. *Journal of Biological Chemistry*, **268**, 19341–5.

Takahashi, R., Yokoji, H., Misawa, H., Hayashi, M., Hu, J., and Deguchi, T. (1994). A null mutation in the human CNTF gene is not causally related to neurological diseases. *Nature* genetics, 7, 79–84.

Takahashi, T. (1992). The minimal inhibitory synaptic currents evoked in neonatal rat motoneurones. *Journal of Physiology (London)*, **450**, 593–611.

Takahashi, T. and Momiyama, A. (1991). Single-channel currents underlying glycinergic inhibitory postsynaptic responses in spinal neurones. *Neuron* 7, 1–20.

Takahashi, T. and Momiyama, A. (1993). Multiple types of calcium channel mediate central synaptic transmission. *Nature*, **366**, 156–8.

Takahashi, T., Momiyama, A., Hirai, K., Hishinuma, F., and Akagi, H. (1992). Functional correlation of fetal and adult forms of glycine receptors with developmental changes in inhibitory synaptic receptor channels. *Neuron*, **9**, 1155–61.

Takai, T., Noda, M., Mishina, M., Shimizu, S., Furutani, Y., Kayano, T. *et al.* (1985). Cloning, sequencing and expression of cDNA for a novel subunit of acetylcholine receptor from calf muscle. *Nature*, **315**, 761–4.

Takami, K., Kawai, Y., Uchida, S., Tohyama, M., Shiotani, Y., Yoshida, H. *et al.* (1985). Effect of calcitonin gene-related peptide on contraction of striated muscle in the mouse. *Neuroscience Letters*, **60**, 227–30.

Takami, K., Hashimoto, K., Uchida, S., Tohyama, M., and Yoshida, H. (1986). Effect of calcitonin gene-related peptide on the cyclic AMP level and contraction of isolated mouse diaphragm. *Japanese Journal of Pharmacology*, **42**, 345–50.

Takeichi, M. (1991). Cadherin cell adhesion receptors as a morphogenetic regulator. *Science* 251, 1451–1455.

Takeshima, H., Nishimura, S., Matsumoto, T., Ishida, H., Kangawa, K., Minamino, N. *et al.* (1989). Primary structure and expression from complementary DNA of skeletal muscle ryanodine receptor. *Nature*, **339**, 439–45.

Takeuchi, A. and Takeuchi, N. (1960). On the permeability of end-plate membrane during the action of transmitter. *Journal of Physiology (London)*, **154**, 52–67.

Takeya, T. and Hanafusa, H. (1983). Structure and sequence of the cellular gene homologous to the RSV *src* gene and the mechanism for generating the transforming virus. *Cell*, **32**, 881–90.

Tanabe, T., Takeshima, H., Mikami, A., Flockerzi, V., Takahashi, H., Kangawa, K. *et al.* (1987). Primary structure of the receptor for calcium channel blockers from skeletal muscle. *Nature*, **328**, 313–8.

Tanabe, T., Beam, K. G., Powell, J. A., and Numa, S. (1988). Restoration of excitation-contraction coupling and slow calcium current in dysgenic muscle by dihydropyridine receptor complementary DNA. *Nature*, **336**, 134–9.

Tanabe, Y., Masu, M., Ishii, T., Shigemoto, R., and Nakanishi, S. (1992). A family of metabotropic glutamate receptors. *Neuron*, **8**, 169–79.

Tang, J. and Landmesser, L. (1993). Reduction of intramuscular nerve branching and synaptogenesis is correlated with decreased motoneuron survival. *Journal of Neuroscience*, **13**, 3095–103.

Taniuchi, M., Clark, H. B., Schweitzer, J. B., and Johnson, E. M. (1988). Expression of nerve growth factor receptors by Schwann cells of axotomized peripheral nerves: ultrastructural location, suppression by axonal contact, and binding properties. *Journal of Neuroscience*, **8**, 664–81.

Tapscott, S. J., Davis, R. L., Thayer, M. J., Cheng, P. F., Weintraub, H., and Lassar, A. B. (1988). MyoD1: A nuclear phosphoprotein requiring a Myc homology region to convert fibroblasts to myoblasts. *Science*, 242, 405–11.

Tauc, L. (1982). Nonvesicular release of neurotransmitter. *Physiological Reviews*, **62**, 857–893.

Tauc, L., Hoffmann, A., Tsuji, S., Hinzen, D. H., and Faille, L. (1974). Transmission abolished on a cholinergic synapse after injection of acetylcholinesterase into the presynaptic neurone. *Nature*, **250**, 496–8.

Taylor, C. W. (1990). The role of G proteins in transmembrane signalling. *Biochemical Journal*, **272**, 1–13.

Taylor, J. M., Mitchell, W. M., and Cohen, S. (1972). Epidermal growth factor: Physical and chemical properties. *Journal of Biological Chemistry*, **247**, 5928–34.

Teyler, T. J. and DiScenna, P. (1987). Long-term potentiation. *Annual Review of Neuroscience*, **10**, 131–61.

Thesleff, S. (1955). The mode of neuromuscular block caused by acetylcholine, nicotine, decamethonium and succinylcholine. *Acta Physiologica Scandinavica* 34, 218–31.

Thoenen, H. (1991). The changing scene of neurotrophic

factors. *Trends in Neurosciences*, 14, 165–70.

Thomas, D. D. and Stryer, L. (1982). The transverse location of the retinal chromophore of rhodopsin in rod outer segment disc membranes. *Journal of Molecular Biology*, 154, 145–57.

Thomas, G. J. and Otis, L. S. (1958). Effects of rhiencephalic lesions on maze learning in rats. *Journal of Comparative Physiological Psychology*, 51, 161–6.

Thomas, K. R. and Capecchi, M. R. (1987). Site-directed mutagenesis by gene targeting in mouse embryo-derived stem cells. *Cell*, 51, 503–12.

Thomas, L. and Betz, H. (1990). Synaptophysin binds to physophilin, a putative synaptic plasma membrane protein. *Journal of Cell Biology*, 111, 2041–52.

Thomas, L., Hartung, K., Langosch, D., Rehm, H., Bamberg, E., Franke, W. W. *et al.* (1988). Identification of synaptophysin as a hexameric channel protein of the synaptic vesicle membrane. *Science*, 242, 1050–3.

Thomas, P., Supernant, A., and Almers, W. (1990). Cytosolic Ca^{2+}, exocytosis, and endocytosis in single melanotrophs of the rat pituitary. *Neuron*, 5, 723–33.

Thompson, R. F. (1967). *Foundations of physiological psychology*. Harper and Row, New York.

Thompson, R. F. and Spencer, W. A. (1966). Habituation: A model phenomenon for the study of neuronal substrates of behaviour. *Psychological Reviews*, 173, 16–43.

Titmus, M. J. and Faber, D. S. (1990). Axotomy-induced alterations in the electrophysiological characteristics of neurons. *Progress in Neurobiology*, 35, 1–51.

Torrey, T. W. (1934). The relation of taste buds to their nerve fibres. *Journal of Comparative Neurology*, 59, 203–20.

Torri-Tarelli, F., Passafaro, M., Clementi, F., and Sher, E. (1991). Presynaptic localization of α-conotoxin-sensitive calcium channels at the frog neuromuscular junction. *Brain Research*, 547, 331–4.

Toth, P. T., Bindokas, V. P., Bleakman, D., Colmers, W. F., and Miller, R. J. (1993). Mechanism of presynaptic inhibition by neuropeptide Y at sympathetic nerve terminals. *Nature*, 364, 635–9.

Tower, S. S. (1937). Function and structure in the chronically isolated lumbo-sacral spinal cord of the dog. *Journal of Comparative Neurology*, 67, 109–32.

Toyoshima, C. and Unwin, N. (1988). Ion channel of acetylcholine receptor reconstructed from images of postsynaptic membranes. *Nature* 336, 247–50.

Traynelis, S. F., Silver, R. A., and Cull-Candy, S. G. (1993). Estimated conductance of glutamate receptor channels activated during EPSCs at the cerebellar mossy fiber-granule cell synapse. *Neuron*, 11, 279–89.

Treisman, R. (1985). Transient accumulation of c-*fos* RNA following serum stimulation requires a conserved 5′ element and c-*fos* 3′ sequences. *Cell*, 42, 889–902.

Triller, A. and Korn, H. (1982). Transmission at a central inhibitory synapse. III. Ultrastructure of physiologically identified and stained terminals. *Journal of Neurophysiology*, 48, 708–36.

Trimble, W. S. and Scheller, R. H. (1988). Molecular biology of synaptic vesicle-associated proteins. *Trends in Neurosciences*, 11, 241–2.

Tsien, R. W., Lipscombe, D., Madison, D. V., Bley, K. R., and Fox, A. P. (1988). Multiple types of neuronal calcium channels and their selective modulation. *Trends in Neurosciences*, 11, 431–8.

Tsien, R. W., Ellinor, P. T., and Horne, W. A. (1991). Molecular diversity of voltage-dependent Ca^{2+} channels. *Trends in Pharmacological Sciences*, 12, 349–54.

Tsien, R. Y. (1988). Fluorescence measurement and photochemical manipulation of cytosolic free calcium. *Trends in Neurosciences*, 11, 419–24.

Tsim, K. W. K., Ruegg, M. A., Escher, G., Kroger, S., and McMahan, U. J. (1992). cDNA that encodes active agrin. *Neuron*, 8, 677–89.

Tsujimoto, T. and Kuno, M. (1988). Calcitonin gene-related peptide prevents disuse-induced sprouting of rat motor nerve terminals. *Journal of Neuroscience*, 8, 3951–7.

Tsujimoto, T., Umemiya, M., and Kuno, M. (1990). Terminal sprouting is not responsible for enhanced transmitter release at disused neuromuscular junctions of the rat. *Journal of Neuroscience*, 10, 2059–65.

Tsukahara, N. (1981). Synaptic plasticity in the mammalian central nervous system. *Annual Review of Neuroscience*, 4, 351–79.

Tsukahara, N. (1987). Cellular basis of classical conditioning mediated by the red nucleus in the cat. In *Synaptic function*, (ed. G. M. Edleman, W. E. Gall, and W. M. Cowan), pp. 447–89, John Wiley, New York.

Tsukahara, N., Hultborn, H., Murakami, F., and Fujito, Y. (1975). Electrophysiological study of new synapses and collateral sprouting in red nucleus neurons after partial denervation. *Journal of Neurophysiology*, 38, 1359–72.

Tsumoto, T. (1990). Long-term potentiation and depression in the cerebral neocortex. *Japanese Journal of Physiology*, 40, 573–93.

Tuttle, J. B., Suszkiw, J. B., and Ard, M. (1980). Long-term survival and development of dissociated parasympathetic neurons in culture. *Brain Research*, 183, 161–80.

Uchitel, O. D., Protti, D. A., Sanchez, V., Cherksey, B. D., Sugimori, M., and Llinás, R. (1992). P-type voltage-dependent calcium channel mediates presynaptic calcium influx and transmitter release in mammalian synapses. *Proceedings of the National Academy of Sciences of the USA*, 89, 3330–3.

Ullrich, A. and Schlessinger, J. (1990). Signal transduction by receptors with tyrosine kinase activity. *Cell*, 61, 203–12.

Ullrich, A., Gray, A., Berman, C., and Dull, T. (1983). Human β-nerve growth factor gene highly homologous to that of mouse. *Nature*, **303**, 821–5.

Umemiya, M., Araki, I., and Kuno, M. (1993). Electrophysiological properties of axotomized facial motoneurones that are destined to die in neonatal rats. *Journal of Physiology (London)*, **462**, 661–78.

Unwin, N. (1989). The structure of ion channels in membranes of excitable cells. *Neuron*, **3**, 665–76.

Unwin, N. (1993a). Neurotransmitter action: opening of ligand-gated ion channels. *Neuron*, **10** (Supplement), 31–41.

Unwin, N. (1993b). Nicotinic acetylcholine receptor at 9 Å resolution. *Journal of Molecular Biology*, **229**, 1101–24.

Unwin, N., Toyoshima, C., and Kubalek, E. (1988). Arrangement of the acetylcholine receptor subunits in the resting and desensitized states, determined by cryoelectron microscopy of crystallized *Torpedo* postsynaptic membranes. *Journal of Cell Biology*, **107**, 1123–38.

Usdin, T. B. and Fischbach, G. D. (1986). Purification and characterization of a polypeptide from chick brain that promotes the accumulation of acetylcholine receptors in chick myotubes. *Journal of Cell Biology*, **103**, 493–507.

Ushkaryov, Y. A., Petrenko, A. G., Geppert, M., and Südhof, T. C. (1992). Neurexins: synaptic cell surface proteins related to the α-latrotoxin receptor and laminin. *Science*, **257**, 50–6.

Valtorta, F., Jahn, R., Fesce, R., Greengard, P., and Cecarelli, B. (1988). Synaptophysin (p38) at the frog neuromuscular junction: its incorporation into the axolemma and recycling after intense quantal secretion. *Journal of Cell Biology*, **107**, 2717–27.

Vandaele, S., Fosset, M., Galizzi, J. P., and Lazdunski, M. (1987). Monoclonal antibodies that coimmunoprecipitate the 1,4–dihydropyridine and phenyalkylamine receptors and reveal the Ca^{2+} channel structure. *Biochemistry*, **26**, 5–9.

Van den Bosch, H. (1980). Intracellular phospholipase A. *Biochimica et Biophysica Acta*, **604**, 191–246.

Vanhoutte, P. M., Rubanyi, G. M., Miller, V. M., and Houston, D. S. (1986). Modulation of vascular smooth muscle contraction by the endothelium. *Annual Review of Physiology*, **48**, 307–20.

Verdoorn, T. A., Draguhn, A., Ymer, S., Seeburg, P. H., and Sakmann, B. (1990). Functional properties of recombinant rat $GABA_A$ receptors depend on subunit composition. *Neuron*, **4**, 919–28.

Verdoorn, T. A., Burnashev, V., Monyer, H., Seeburg, P. H., and Sakmann, B. (1991). Structural determinants of ion flow through recombinant glutamate receptor channels. *Science*, **252**, 1715–18.

Verhage, M., McMahon, H. T., Ghijsen, W. E. J. M., Boomsma, F., Scholten, G., Wiegant, V. M. *et al.* (1991). Differential release of amino acids, neuropeptides, and catecholamines from isolated nerve terminals. *Neuron*, **6**, 517–24.

Vicini, S. and Schuetze, S. M. (1985). Gating properties of acetylcholine receptors at developing rat endplates. *Journal of Neuroscience*, **5**, 2212–24.

Villar, M. J., Cortes, R., Theodorsson, E., Wiesenfeld-Hallin, Z., Schalling, M., Fahrenkrug, J. *et al.* (1989). Neuropeptide expression in rat dorsal root ganglion cells and spinal cord after peripheral nerve injury with special reference with galanin. *Neuroscience*, **33**, 587–604.

Villarroel, A., Herlitze, S., Koenen, M., and Sakmann, B. (1991). Location of a threonine residuein the α-subunit M_2 transmembrane segment that determines the ion flow through the acetylcholine receptor channel. *Proceedings of the Royal Society B*, **243**, 69–74.

Vincent, A., Lang, B., and Newsom-Davis, J. C. (1989). Autoimmunity to the voltage-gated calcium channel underlies the Lambert-Eaton myasthenic syndrome, a paraneoplastic disorder. *Trends in Neurosciences*, **12**, 496–502.

Von Gersdorff, H. and Mathews, G. (1994). Dynamics of synaptic vesicle fusion and membrane retrieval in synaptic terminals. *Nature*, **367**, 735–9.

Voyvodic, J. T. (1989). Peripheral target regulation of dendritic geometry in the rat superior cervical ganglion. *Journal of Neuroscience*, **9**, 1997–2010.

Vyskocil, F. and Illes, P. (1977). Non-quantal release of transmitter at mouse neuromuscular junction and its dependence on the activity of Na-K ATPase. *Pflügers Archiv*, **370**, 295–7.

Wagner, J. A., Carlson, S. S., and Kelly, R. B. (1978). Chemical and physical characterization of cholinergic synaptic vesicles. *Biochemistry*, **17**, 119–1206.

Wallace, B. G. (1988). Regulation of agrin-induced acetylcholine receptor aggregation by Ca^{++} and phorbol ester. *Journal of Cell Biology*, **107**, 267–278.

Wallace, B. G., Nitkin, R. M., Reist, N. E., Fallon, J. R., Moayeri, N. N., and McMahan, U. J. (1985). Aggregates of acetylcholinesterase induced by acetylcholine receptor-aggreating factor. *Nature*, **315**, 574–7.

Wallace, B. G., Zhican, Q., and Huganir, R. L. (1991). Agrin induces phosphorylation of the nicotinic acetylcholine receptor. *Neuron*, **6**, 869–78.

Walmsley, B. (1991). Central synaptic transmission: studies at the connection between primary afferent fibres and dorsal spinocerebellar tract (DSCT) neurones in Clarke's column of the spinal cord. *Progress in Neurobiology*, **36**, 391–423.

Walmsley, B., Edwards, F. R., and Tracey, D. J. (1988). Nonuniform release probabilities underlie quantal synaptic transmission at a mammalian excitatory central synapse. *Journal of Neurophysiology*, **60**, 889–908.

Walsh, F. S. and Doherty, P. (1991). Structure and function of the gene for neural cell adhesion molecule. *Seminars in Neurosciences*, 3, 271–84.

Wang, L. Y., Salter,M. W., and MacDonald, J. F. (1991). Regulation of kainate receptors by cAMP-dependent protein kinase and phosphatase. *Science*, 253, 1132–5.

Watanabe, M., Inoue, Y., Sakimura, K., and Mishina, M. (1992). Developmental changes in distribution of NMDA receptor channel subunit mRNAs. *Neuroreport*, 3, 1138–40.

Watanabe, M., Inoue, Y., Sakimura, K., and Mishina, M. (1993). Distinct distribution of five N-methyl-D-aspartate receptor channel subunit mRNAs in the forebrain. *Journal of Comparative Neurology*, 338, 377–90.

Watanabe, S., Nakayama, N., Yokota, T., Arai, K., and Miyajima, A. (1991). Colony-stimulating factors and cytokine receptor network. *Current Opinion in Biotechnology*, 2, 227–37.

Watanabe, S., Mui, A. L. F., Muto, A., Chen, J. X., Hayashida, K., Yokota, T. *et al.* (1993). Reconstituted human granulocyte-macrophage colony-stimulating factor receptor transduces growth-promoting signals in mouse NIH 3T3 cells: comparison with signalling in BA/F3 pro-B cells. *Molecular and Cellular Biology*, 13, 1440–8.

Watkins, J. C. and Evans, R. H. (1981). Excitatory amino acid transmitters. *Annual Review of Pharmacology*, 21, 165–204.

Watkins, J. C., Krogsgaard-Larsen, P., and Honore, T. (1990). Structure-activity relationships in the development of excitatory amino acid receptor agonists and competitive antagonists. *Trends in Pharmacological Sciences*, 11, 25–33.

Wayne, D. B. and Heaton, M. B. (1988). Retrograde transport of NGF by early chick embryo spinal cord motorneurones. *Developmental Biology*, 127, 220–3.

Webb, C. B. and Cope, T. C. (1992). Modulation of Ia EPSP amplitude: the effects of chronic synaptic inactivity. *Journal of Neuroscience*, 12, 338–44.

Weber, M. J. (1981). A diffusible factor responsible for the determination of cholinergic functions in cultured sympathetic neurones. *Journal of Biological Chemistry*, 256, 3447–53.

Weill, C. L., McNamee, M. G., and Karlin, A. (1974). Affinity-labelling of purified acetylcholine receptors from *Torpedo californica*. *Biochemical and Biophysical Research Communications*, 61, 997–1003.

Weinberg, R. A. (1983). A molecular basis of cancer. *Scientific American*, 249(5), 102–16.

Weinreich, D. (1971). Ionic mechanism of post-tetanic potentiation at the neuromuscular junction of the frog. *Journal of Physiology (London)*, 212, 431–46.

Welcher, A. A., Bitler, C. M., Radeke, M. J., and Shooter, E. M. (1991). Nerve growth factor binding domain of the nerve growth factor receptor. *Proceedings of the National Academy of Sciences of the USA*, 88, 159–63.

Werman, R. (1969). An electrophysiological approach to drug-receptor mechanisms. *Comparative Biochemical Physiology*, 30, 997–1017.

Weskamp, G. and Reichardt, L. F. (1991). Evidence that biological activity of NGF is mediated through a novel subclass of high affinity receptors. *Neuron*, 6, 649–63.

Westbrook, G. L. and Mayer, M. L. (1987). Micromolar concentrations of Zn^{2+} antagonize NMDA and GABA responses of hippocampal neurones. *Nature*, 328, 640–3.

Wheeler, D. B., Randall, A., and Tsien, R. W. (1994). Roles of N-type and Q-type Ca^{2+} channels in supporting hippocampal synaptic transmission. *Science*, 264, 197–9.

Whitehouse, P. J., Price, D. L., Struble, R. G., Clark, A. W., Coyle, J. T., and Delon, M. R. (1982). Alzheimer's disease and senile dementia: loss of neurons in the basal forebrain. *Science*, 215, 1237–9.

Whittaker, V. P., Essman, W. B., and Dowe, G. H. C. (1972). The isolation of pure cholinergic synaptic vesicles from the electric organs of elasmobranch fish of the family. *Biochemical Journal*, 128, 833–46.

Wiedenmann, B., and Franke, W. W. (1985). Identification and localization of synaptophysin, an integral membrane glycoprotein of Mr 38,000 characteristic of presynaptic vesicles. *Cell*, 41, 1017–28.

Wiesel, T. N. (1982). Postnatal development of the visual cortex and the influence of environment. *Nature*, 299, 583–91.

Wiesel, T. N. and Hubel, D. H. (1963). Single-cell responses in striate cortex of kittens deprived of vision in one eye. *Journal of Neurophysiology*, 26, 1003–17.

Wiesel, T. N. and Hubel, D. H. (1965). Comparison of the effects of unilateral and bilateral eye closure on cortical unit responses in kittens. *Journal of Neurophysiology*, 28, 1029–40.

Wigstrom, H. and Gustafsson, B. (1984). A possible correlate of the postsynaptic condition for long-lasting potentiation in the guinea pig hippocampus *in vitro*. *Neuroscience Letters*, 44, 327–32.

Wiklund, L., Toggenburger, G., and Cuenod, M. (1982). Aspartate: possible neurotransmitter in cerebellar climbing fibres. *Science*, 216, 78–80.

Williams, L. R., Varon, S., Peterson, G. M., Wictorin, K., Fischer, W., Bjorklund, A. *et al.* (1986). Continuous infusion of nervee growth factor prevents basal forebrain neuronal death after fimbria-fornix transection. *Proceedings of the National Academy of Sciences of the USA*, 83, 9231–5.

Williams, M. E., Brust, P. F., Feldman, D. H., Patthi, S., Simerson, S., Maroufi, A. *et al.* (1992). Structure and functional expression of an α-conotoxin-sensitive human N-type calcium channel. *Science*, 257, 389–95.

Wisden, W., Errington, M. L., Williams, S., Dunnett, S. B., Waters, C., Hitchcock, D. *et al.* (1990). Differential expression of immediate early genes in the hippocampus and spinal cord. *Neuron*, 4, 603–14.

Witzemann, V., Brenner, H. R., and Sakmann, B. (1991). Neural factors regulate AChR subunit mRMAs at rat neuromuscular synapses. *Journal of Cell Biology*, **114**, 125–41.

Wojtowicz, J. M. and Atwood, H. L. (1986). Long-term facilitation alters transmitter releasing properties at the crayfish neuromuscular junction. *Journal of Neurophysiology*, **55**, 484–98.

Wojtowicz, J. M. and Atwood, H. L. (1988). Presynaptic long-term facilitation at the crayfish neuromuscular junction: Voltage-dependent and ion-dependent phases. *Journal of Neuroscience*, **8**, 4667–74.

Wojtowicz, J. M., Marin, L., and Atwood, H. L. (1989). Synaptic restructuring during long-term facilitation at the crayfish neuromuscular junction. *Canadian Journal of Physiology and Pharmacology*, **67**, 167–71.

Wong, J. and Oblinger, M. M. (1990). A comparison of peripheral and central axotomy effects on neurofilaments and tubulin gene expression in rat dorsal root ganglion neurons. *Journal of Neuroscience*, **10**, 2215–22.

Wong, J. and Oblinger, M. M. (1991). NGF rescues substance P expression but not neurofilament or tubulin gene expression in axotomized sensory neurons. *Journal of Neuroscience*, **11**, 543–52.

Wood, S. J., Pritchard, J., and Sofroniew, M. V. (1990). Re-expression of nerve growth factor receptor after axonal injury recapitulates a develpmental event in motor neurons: Differential regulation when regeneration is allowed or prevented. *European Journal of Neuroscience*, **2**, 650–7.

Wyllie, A. H. (1980). Glucocorticoid-induced thymocyte apoptosis is associated with endogenous endonuclease activation. Nature, **284**, 555–6.

Wyllie, A. H., Kerr, J. F. R., and Currie, A. R. (1980). Cell death: the significance of apoptosis. *International Review of Cytology*, **68**, 251–306.

Xie, X. Y., and Barrett, J. N. (1991). Membrane resealing in cultured rat septal neurons after neurite transection: evidence for enhancement by Ca^{2+}-triggered protease activity and cytoskeletal disassembly. *Journal of Neuroscience*, **11**, 3257–67.

Yamamori, T. (1991). Localization of cholinergic differentiation factor/leukemia inhibitory factor mRNA in the rat brain and peripheral tissues. *Proceedings of the National Academy of Sciences of the USA*, **88**, 7298–302.

Yamamori, T. (1992). Molecular mechanisms for generation of neural diversity and specificity: roles of polypeptide factors in development of postmitotic neurons. *Neuroscience Research*, **12**, 545–82.

Yamamori, T. and Sarai, A. (1992). Coevolution of cytokine receptor families in the immune and nervous systems. *Neuroscience Research*, **15**, 151–61.

Yamamori, T., Fukada, K., Aebersold, R., Korsching, S., Fann, M. J., and Patterson, P. H. (1989). The cholinergic neuronal differentiation factor from heart cells is identical to leukemia inhibitory factor. *Science*, **246**, 1412–6.

Yamamoto, C., Higashima, M., Sawada, S., and Kamiya, H. (1991). Quantal components of the synaptic potential induced in hippocampal neurons by activation of granule cells, and the effect of 2-amino-4-phosphonobutyric acid. *Hippocampus*, **1**, 93–106.

Yamamoto, K. K., Gonzalez, G. A., Biggs, W. H., and Montminy, M. R. (1988). Phosphorylation-induced binding and transcriptional efficacy of nuclear factor CREB. *Nature*, **334**, 494–8.

Yamasaki, K., Taga, T., Hirata, Y., Yawata, H., Kawanishi, Y., Seed, B. *et al.* (1988). Cloning and expression of the human interleukin-6 (BSF-2/IFN 2) receptor. *Science*, **241**, 825–8.

Yan, H. and Chao, M. V. (1991). Disruption of cysteine-rich repeats of the p75 nerve growth factor receptor leads to loss of ligand binding. *Journal of Biological Chemistry* **266**, 12099–104.

Yan, Q., Snider, W. D., Pinzone, J. J., and Johnson, E. M. (1988). Retrograde transport of nerve growth factor (NGF) in motoneurons of developing rats: Assessment of potential neurotrophic effects. *Neuron*, **1**, 335–43.

Yan, Q., Elliott, J., and Snider, W. D. (1992). Brain-derived neurotrophic factor rescues spinal motor neurons from axotomy-induced cell death. *Nature*, **360**, 753–5.

Yanker, B. A. and Shooter, E. M. (1979). Nerve growth factor in the nucleus: interaction with receptors on the nuclear membrane. *Proceedings of the National Academy of Sciences of the USA*, **76**, 1269–73.

Yarden, Y. and Ullrich, A. (1988). Growth factor receptor tyrosine kinases. *Annual Review of Biochemistry*, **57**, 443–78.

Yau, K. W. and Baylor, D. A. (1989). Cyclic GMP-activated conductance of retinal photoreceptor cells. *Annual Review of Neuroscience*, **12**, 289–327.

Yawo, H. (1987). Changes in the dendritic geometry of mouse superior cervical ganglion cells following post-ganglionic axotomy. *Journal of Neuroscience*, **7**, 3703–11.

Yawo, H. and Chuma, N. (1993). Preferential inhibition of ω-conotoxin-sensitive presynaptic Ca^{2+} channels by adenosine autoreceptors. *Nature*, **365**, 256–258.

Yawo, H. and Kuno, M. (1983). How a nerve fibre repairs its cut end: involvement of phospholipase A_2. *Science*, **222**, 1351–3.

Yawo, H. and Momiyama, A. (1993). Re-evalutation of calcium currents in pre- and postsynaptic neurones of the chick ciliary ganglion. *Journal of Physiology (London)*, **460**, 153–72.

Yellen, G. (1982). Single Ca-activated nonselective cation channels in neuroblastoma. *Nature*, **296**, 357–9.

Yellen, G., Jurman, M. E., Abramson, T., and MacKinnon, R. (1991). Mutations affecting internal TEA blockade identify the probable pore-forming region of a K^+ channel. *Science*, **251**, 939–42.

Yip, H. K. and Johnson, E. M. (1984). Developing dorsal

root ganglion neurons require trophic support from their central processes: evidence for a role of retrogradely transported nerve growth factor from the central nervous system to the periphery. *Proceedings of the National Academy of Sciences of the USA*, **81**, 6245–9.

Yool, A. J. and Schwarz, T. L. (1991). Alteration of ionic selectivity of a K^+ channel by mutation of the H5 region. *Nature* **349**, 700–4.

Yoon, K. W. and Rothman, S. W. (1991). The modulation of rat hippocampal synaptic conductances by baclofen and γ-aminobutyric acid. *Journal of Physiology (London)*, 442, 377–90.

Yoshikami, D., Bagaboldo, Z., and Olivera, B. M. (1989). The inhibitory effects of omega-conotoxin on Ca channels and synapses. *Annals of the New York Academy of Sciences*, **560**, 230–48.

Young, J. D. E., Cohn, Z. A., and Gilula, N. B. (1987). Functional assembly of gap junction conductance in lipid bilayers: Demonstration that the major 27 kd protein forms the junctional channel. *Cell*, **48**, 733–43.

Young, J. D. E. (1989). Killing of target cells by lymphocytes: A mechanistic view. *Physiological Reviews*, **69**, 250–314.

Young, J. Z. (1951). Growth and plasticity in the nervous system. *Proceedings of the Royal Society B*, **139**, 18–37.

Yovell, Y., Kandel, E. R., Dudai, Y., and Abrams, T. W. (1987). Biochemical correlates of short-term sensitization in *Aplysia*: temporal analysis of adenylate cyclase stimulation in a perfused-membrane preparation. *Proceedings of the National Academy of Sciences of the USA*, **84**, 9285–9.

Zakeri, Z. F., Quaglino, D, Latham, T., and Lockshin, R. A. (1993). Delayed internucleosomal DNA fragmentation in programmed cell death. *FASEB Journal*, 7, 470–8.

Zalutskey, R. A. and Nicoll, R. A. (1990). Comparison of two forms of long-term potentiation in single hippocampal neurons. *Science*, **248**, 1619–24.

Zampighi, G. A., Hall, J., and Kreman, M. (1985). Purified lens junctional protein forms channels in planar lipid films. *Proceedings of the National Academy of Sciences of the USA*, **82**, 8468–72.

Zengel, J. E. and Magleby, K. L. (1982). Augmentation and facilitation of transmitter release: a quantitative description at the frog neuromuscular junction. *Journal of General Physiology*, **80**, 583–611.

Zhang, N., Walberg, F., Laake, J. H., Meldrum, B. S., and Ottersen, O. P. (1990). Aspartate-like and glutamate-like immunoreactivities in the inferior olive and climbing fibre system: a light microscopic and semiquantitative electron microscopic study in rat and baboon (*Papio anubis*). *Neuroscience*, **38**, 61–80.

Zheng, F. and Gallagher, J. P. (1992). Metabotropic glutamate receptors are required for the induction of long-term potentiation. *Neuron*, **9**, 163–72.

Zhu, P. C., Thureson-Klein, A., and Klein, R. L. (1986). Exocytosis from large dense cored vesicles outside the active synaptic zones of terminals within the trigeminal subnucleus caudalis: a possible mechanism for neuropeptide release. *Neuroscience*, **19**, 43–54.

Zhuo, M., Small, S. A., Kandel, E. R., and Hawkins, R. D. (1993). Nitric oxide and carbon monoxide produce activity-dependent long-term synaptic enhancement in hippocampus. *Science*, **260**, 1946–50.

Zhuo, M., Hu, Y., Schultz, C., Kandel, E. R., and Howkins, R. D. (1994). Role of guanylyl cyclase and cGMP-dependent protein kinase in long-term potentiation. *Nature*, **368**, 635–39.

Zimmerberg, J., Curran, M., Cohen, F. S., and Brodwick, M. (1987). Simultaneous electrical and optical measurements show that membrane fusion precedes secretory granule swelling during exocytosis of beige mouse mast cells. *Proceedings of the National Academy of Sciences of the USA*, **84**, 1585–9.

Zimmerman, A. and Sutter, A. (1983). β-Nerve growth factor receptors on glial cells. Cell-cell interaction between neurones and Schwann cells in cultures of chick sensory ganglia. *EMBO Journal*, 2, 879–85.

Zingsheim, H. P., Neugebauer, D. C., Barrantes, F. J., and Franck, E. J. (1980). Structural details of membrane-bound acetylcholine receptor from *Torpedo marmorata*. *Proceedings of the National Academy of Sciences of the USA*, 77, 952–6.

Zucker, R. S. (1989). Short-term synaptic plasticity. *Annual Review of Neuroscience*, 12, 13–31.

Zucker, R. S. and Lara-Estrella, L. O. (1983). Post-tetanic decay of evoked and spontaneous transmitter release and a residual-calcium model of synaptic facilitation at crayfish neuromuscular junction. *Journal of General Physiology*, **81**: 355–72.

INDEX